T0135326

Communications
in Computer and Information Science
1848

Rationale

The CCIS series is devoted to the publication of proceedings of computer science conferences. Its aim is to efficiently disseminate original research results in informatics in printed and electronic form. While the focus is on publication of peer-reviewed full papers presenting mature work, inclusion of reviewed short papers reporting on work in progress is welcome, too. Besides globally relevant meetings with internationally representative program committees guaranteeing a strict peer-reviewing and paper selection process, conferences run by societies or of high regional or national relevance are also considered for publication.

Topics

The topical scope of CCIS spans the entire spectrum of informatics ranging from foundational topics in the theory of computing to information and communications science and technology and a broad variety of interdisciplinary application fields.

Information for Volume Editors and Authors

Publication in CCIS is free of charge. No royalties are paid, however, we offer registered conference participants temporary free access to the online version of the conference proceedings on SpringerLink (http://link.springer.com) by means of an http referrer from the conference website and/or a number of complimentary printed copies, as specified in the official acceptance email of the event.

CCIS proceedings can be published in time for distribution at conferences or as post-proceedings, and delivered in the form of printed books and/or electronically as USBs and/or e-content licenses for accessing proceedings at SpringerLink. Furthermore, CCIS proceedings are included in the CCIS electronic book series hosted in the SpringerLink digital library at http://link.springer.com/bookseries/7899. Conferences publishing in CCIS are allowed to use Online Conference Service (OCS) for managing the whole proceedings lifecycle (from submission and reviewing to preparing for publication) free of charge.

Publication process

The language of publication is exclusively English. Authors publishing in CCIS have to sign the Springer CCIS copyright transfer form, however, they are free to use their material published in CCIS for substantially changed, more elaborate subsequent publications elsewhere. For the preparation of the camera-ready papers/files, authors have to strictly adhere to the Springer CCIS Authors' Instructions and are strongly encouraged to use the CCIS LaTeX style files or templates.

Abstracting/Indexing

CCIS is abstracted/indexed in DBLP, Google Scholar, EI-Compendex, Mathematical Reviews, SCImago, Scopus. CCIS volumes are also submitted for the inclusion in ISI Proceedings.

How to start

To start the evaluation of your proposal for inclusion in the CCIS series, please send an e-mail to ccis@springer.com.

Mayank Singh · Vipin Tyagi · P.K. Gupta ·
Jan Flusser · Tuncer Ören
Editors

Advances in Computing and Data Sciences

7th International Conference, ICACDS 2023
Kolkata, India, April 27–28, 2023
Revised Selected Papers

 Springer

Editors
Mayank Singh
Consilio Research Lab
Tallinn, Estonia

Vipin Tyagi
Jaypee University of Engineering
and Technology
Guna, India

P.K. Gupta
Jaypee University of Information Technology
Waknaghat, India

Jan Flusser
Institute of Information Theory
and Automation
Prague, Czech Republic

Tuncer Ören
University of Ottawa
Ottawa, ON, Canada

ISSN 1865-0929   ISSN 1865-0937 (electronic)
Communications in Computer and Information Science
ISBN 978-3-031-37939-0   ISBN 978-3-031-37940-6 (eBook)
https://doi.org/10.1007/978-3-031-37940-6

This Springer imprint is published by the registered company Springer Nature Switzerland AG
The registered company address is: Gewerbestrasse 11, 6330 Cham, Switzerland

Preface

It is with great pleasure and enthusiasm that we present the proceedings of the 7th International Conference on Advances in Computing and Data Sciences. This conference has established itself as a premier platform for researchers, academicians, and professionals from around the world to exchange ideas, discuss advancements, and explore the frontiers of computing and data sciences. The rapid progress in computing technologies and the exponential growth of data in various domains have revolutionized the way we live, work, and interact with the world. The 7th International Conference on Advances in Computing and Data Sciences aimed to bring together experts and practitioners from diverse backgrounds to delve into the latest research findings, breakthroughs, and challenges in these fields.

This conference received an overwhelming response with contributions from researchers across the globe. The submissions underwent a rigorous peer-review process, ensuring the selection of high-quality papers that cover a wide range of topics related to computing and data sciences. The conference program encompassed a rich tapestry of keynote speeches and technical sessions, providing a comprehensive platform for knowledge dissemination and networking.

The theme of this year's conference was "Exploring Innovations in Computing and Data Sciences." The papers featured in this proceedings volume presented at the 7th International Conference on Advances in Computing and Data Sciences (ICACDS 2023), held during April 27–28, 2023, at Brainware University, Kolkata, West Bengal, India, showcase innovative research in areas such as artificial intelligence, machine learning, big data analytics, cloud computing, computer vision, natural language processing, and many more.

The Program Committee of ICACDS 2023 is extremely grateful to the authors who showed an overwhelming response to the call for papers, with over 464 papers submitted in the tracks of Advanced Computing and Data Sciences. All submitted papers went through a double-blind peer-review process, and finally 47 papers were accepted for presentation in conference. We would like to express our sincere gratitude to the conference organizing committee, program chairs, session chairs, and the technical program committee for their dedication and tireless efforts in ensuring the success of this conference. We extend our appreciation to all the authors who submitted their valuable work and the reviewers for their insightful feedback and constructive suggestions. We also extend our thanks to our keynote speakers, whose expertise and vision have enlightened us and set the tone for stimulating discussions. Their valuable insights have undoubtedly inspired researchers and professionals to embark on new journeys of exploration and innovation.

The Organizing Committee of ICACDS 2023 is indebted to Sri. Phalguni Mookhopadhayay, Chancellor, Brainware University, India for the confidence that he gave to us during organization of this international conference, and all faculty members and staff of Brainware University, India, for their support in organizing the conference and for making it a grand success.

We would also like to thank Debadatta Pal and Suprativ Saha, Brainware University, India; Hemant Gupta, Carleton University, Canada; Nishant Gupta, Sharda University, India; Mahesh Kumar, Divya Jain, Kunj Bihari Meena, Neelesh Jain, and Nilesh Patel, JUET Guna, India; Vibhash Yadav, REC Banda, India; Lavanya Sharma, Amity University, Noida, India; Poonam Tanwar and Rashmi Agarwal, MRIIS, India; Rohit Kapoor, SK Info Techies, Noida, India; and Tarun Pathak, Consilio Intelligence Research Lab, India, for their support.

Special thanks to the staff of Springer Nature for support and guidance throughout the conference.

Our sincere thanks to Consilio Intelligence Research Lab, India; the GISR Foundation, India; SK Info Techies, India; Adimaginz Marketing Services, India; and Print Canvas, India, for sponsoring the event.

We hope that the proceedings of the 7th International Conference on Advances in Computing and Data Sciences will serve as a valuable resource for researchers, academicians, and practitioners in their pursuit of knowledge and innovation. May this volume inspire new collaborations, spark novel ideas, and contribute to the advancement of computing and data sciences for the betterment of society.

May 2023

Mayank Singh
Vipin Tyagi
P.K. Gupta
Jan Flusser
Tuncer Ören

Organization

Steering Committee

Buddha Chandrashekhar,	All India Council for Technical Education, India
Alexandre Carlos Brandão Ramos	UNIFEI, Brazil
Mohit Singh	Georgia Institute of Technology, USA
H.M. Pandey	Edge Hill University, UK
M.N. Hooda	BVICAM, India
S.K. Singh	IIT BHU, India
Jyotsna Kumar Mandal	University of Kalyani, India
Ram Bilas Pachori	IIT Indore, India

Chief Patron

Phalguni Mookhopadhayay	Brainware University, India

Patrons

Sankar Gangopadhyay	Brainware University, India

Honorary Chairs

Suman Majumder	Brainware University, India
Mahua Pal	Brainware University, India

General Chairs

Jan Flusser	Institute of Information Theory and Automation, Czech Republic
Mayank Singh	Consilio Research Lab, Estonia

Advisory Board Chairs

Vipin Tyagi	JUET-Guna, India
P.K. Gupta	JUIT-Solan, India
Shailendra Mishra	Majmaah University, Kingdom of Saudi Arabia

Technical Program Committee Chairs

Tuncer Ören	University of Ottawa, Canada
Viranjay M. Srivastava	University of Birmingham, UK
Ling Tok Wang	National University of Singapore, Singapore
Ulrich Klauck	Aalen University, Germany
Anup Girdhar	Sedulity Group, India
Arun Sharma	Indira Gandhi Delhi Technical University for Women, India
Mahesh Kumar	JUET, India

Conference Chairs

Sameer Kumar Jasra	University of Malta, Malta
Debdutta Pal	Brainware University, India

Conveners

Suprativ Saha	Brainware University, India
Hemant Gupta	Carleton University, Canada

Co-conveners

Arun Agarwal	Delhi University, India
Ritesh Prasad	Brainware University, India
Gaurav Agarwal	ABES Engineering College, India

Organizing Chairs

Shashi Kant Dargar   KARE, India
Sitikantha Chattopadhyay   Brainware University, India

Organizing Co-chairs

Abhishek Dixit   Tallinn University of Technology, Estonia
Nileshkumar Patel   JUET, India
Vibhash Yadav   Rajkiya Engineering College Banda, India
Neelesh Kumar Jain   JUET, India
Nishant Gupta   Sharda University, India

Organizing Secretaries

Tarun Pathak   Consilio Intelligence Research Lab, India
Rohit Kapoor   SKIT, India

Program Committee

A.K. Nayak   Computer Society of India, India
A.J. Nor'aini   Universiti Teknologi MARA, Malaysia
Aaradhana Deshmukh   Aalborg University, Denmark
Abdel Badeeh Salem   Ain Shams University, Egypt
Abdelhalim Zekry   Ain Shams University, Egypt
Abdul Jalil Manshad Khalaf   University of Kufa, Iraq
Abhhishek Verma   Indian Institute of Information Technology and
                                   Management, Gwalior, India
Abhinav Vishnu   Pacific Northwest National Laboratory, USA
Abhishek Gangwar   Centre for Development of Advanced Computing,
                                   India
Aditi Gangopadhyay   IIT Roorkee, India
Adrian Munguia   AI Mexico, Mexico
Amit K. Awasthi   Gautam Buddha University, India
Antonina Dattolo   University of Udine, Italy
Arshin Rezazadeh   University of Western Ontario, Canada
Arun Chandrasekaran   National Institute of Technology Karnataka, India
Arun Kumar Yadav   National Institute of Technology Hamirpur, India
Asma H. Sbeih   Palestine Ahliya University (PAU), Palestine

Brahim Lejdel	University of El Oued, Algeria
Chandrabhan Sharma	University of the West Indies, Trinidad and Tobago
Ching-Min Lee	I-Shou University, Taiwan
Deepanwita Das	National Institute of Technology Durgapur, India
Devpriya Soni	Jaypee Institute of Information Technology, India
Donghyun Kim	Georgia State University, USA
Eloi Pereira	University of California, Berkeley, USA
Felix J. Garcia Clemente	Universidad de Murcia, Spain
Gangadhar Reddy Ramireddy	Rajarajeswari College of Engineering, India
Hadi Erfani	Islamic Azad University, Iran
Harpreet Singh	Alberta Emergency Management Agency, Canada
Hussain Saleem	University of Karachi, Pakistan
Jai Gopal Pandey	CSIR-Central Electronics Engineering Research Institute, India
Joshua Booth	University of Alabama in Huntsville, USA
Khattab Ali	University of Anbar, Iraq
Lokesh Jain	Delhi Technological University, India
Manuel Filipe Santos	University of Minho, Portugal
Mario José Diván	National University of La Pampa, Argentina
Megat Farez Azril Zuhairi	Universiti Kuala Lumpur, Malaysia
Mitsunori Makino	Chuo University, Japan
Moulay Akhloufi	Université de Moncton, Canada
Naveen Aggarwal	Panjab University, India
Nawaz Mohamudally	University of Technology, Mauritius
Nileshkumar R. Patel	Jaypee University of Engineering and Technology, India
Nirmalya Kar	National Institute of Technology Agartala, India
Nitish Kumar Ojha	Indian Institute of Technology Allahabad, India
Paolo Crippa	Università Politecnica delle Marche, Italy
Parameshachari B.D.	GSSS Institute of Engineering and Technology for Women, India
Patrick Perrot	Gendarmerie Nationale, France
Prathamesh Churi	NMIMS Mukesh Patel School of Technology Management and Engineering, India
Pritee Khanna	Indian Institute of Information Technology, Design and Manufacturing Jabalpur, India
Purnendu Shekhar Pandey	G.L. Bajaj Institute of Management & Research, India
Quoc-Tuan Vien	Middlesex University London, UK
Rubina Parveen	Canadian All Care College, Canada
Saber Abd-Allah	Beni-Suef University, Egypt

Sahadeo Padhye	Motilal Nehru National Institute of Technology, India
Sarhan M. Musa	Prairie View A&M University, USA
Shamimul Qamar	King Khalid University, Saudi Arabia
Shashi Poddar	University at Buffalo, USA
Shefali Singhal	Madhuben & Bhanubhai Patel Institute of Technology, India
Siddeeq Ameen	University of Mosul, Iraq
Sotiris Kotsiantis	University of Patras, Greece
Subhasish Mazumdar	New Mexico Tech, USA
Sudhanshu Gonge	Symbiosis International University, India
Tomasz Rak	Rzeszow University of Technology, Poland
Vigneshwar Manoharan	Bharath Corporate, India
Xiangguo Li	Henan University of Technology, China
Youssef Ouassit	Hassan II University, Morocco

Sponsor

Consilio Intelligence Research Lab, India

Co-sponsors

GISR Foundation, India
Print Canvas, India
SK Info Techies, India
Adimaginz Marketing Services, India

Contents

Text Based Traffic Signboard Detection Using YOLO v7 Architecture

Ananya Negi [iD], Yash Kesarwani [iD], and P. Saranya[(✉)] [iD]

Department of Computing Technologies, SRM Institute of Science and Technology,
Kattankulathur, Tamil Nadu 603203, India
saranyap@srmist.edu.in

Abstract. Recent developments in computer vision and deep learning technology have increased the prevalence of advanced driver assistance systems (ADAS). ADAS technologies aim to reduce traffic accidents and make driving safer. The proposed work is an additional ADAS feature or can help the driver navigate better through roads while focusing more on the roads. The system uses a small camera mounted at the front of the car, and images from that are then fed into the YOLOv7 model, which can run on jetson nano or other such computing hardware. In the proposed model, the results we have achieved have an overall accuracy of 86% with the system and speed at which it can perform efficiently, ranging from object detection to reading the data on the sign boards.

Keywords: Assistance systems · portable · Optical · recognition

1 Introduction

Most auto accidents result from human error, preventable with advanced driver assistance systems (ADAS). ADAS helps to prevent fatalities and injuries by reducing the number of car accidents and their severity. It has become increasingly crucial for decision-makers and planners to learn how to gather traffic sign information efficiently and effectively as autonomous driving technology gains popularity [11]. A user driving at speed or in a traffic-congested area may miss the text-based traffic signboards, which can neglect important information. These terminal boards are crucial as they contain destination navigation or essential information related to the road, which may be crucial for the driver's safety. The proposed model helps identify the text written on the road signboards and displays this information right on a heads-up display or near the instrument cluster screen, making it easy to read the information at a faster accuracy and speed whilst the driver can focus on the road ahead. This can save much time for the user and reduce the risk of road accidents. All autonomous machines/vehicles are designed to travel on the road. As roads become more congested with other vehicles, it becomes difficult for drivers and self-driving cars to see all the traffic signs along the way. It does not ignore the possibility of missing a vital sign that could lead to a fatal accident. To help users address this issue, camera-based traffic sign recognition systems will be introduced as part of ADAS. Automated driving should recognise all road markers important for driving safety, such as traffic signs, which typically come in a wide variety but are few due to the existing poor coverage of roadside cooperating devices [1, 6].

M. Singh et al. (Eds.): ICACDS 2023, CCIS 1848, pp. 1–11, 2023.
https://doi.org/10.1007/978-3-031-37940-6_1

The proposed model uses YOLO v7 for object localization inside an image. When comparing all recognized real-time item detectors with 30 FPS or more on GPU V100, YOLOv7 has an accuracy of 56.8% AP, outperforming them in both speed and accuracy within the range of five to one hundred and sixty FPS. While YOLO operates in a single fully connected network, algorithms like Faster RCNN determine prospective regions of interest using a Region Proposal Network and then detect each region independently. All predictions are performed using the layer. Therefore, using a region proposal network method result in multiple iterations on the same image, while YOLO requires only one iteration. The second part of character recognition uses a CNN-based OCR model. Using optical pictures of characters as input, optical character recognition (OCR) generates the corresponding characters as output. It can be used for various things, including robotics, traffic monitoring, and digitizing printed material. Convolutional neural networks (CNN), a well-liked deep neural network design, can be used to implement OCR. After character segmentation, we use OCR to detect the individual characters and concatenate the output into a single string [7].

There are two main hurdles while dealing with the detection use case. The first is the detection of text-based sign boards, and the second is reading the text written on the signboards; as there are many approaches for solving these tasks, our considerations for choosing the most optimal method were that it had to be fast enough to be used in cars cruising on highway speeds and still have accuracy for reading the text on the boards. The project uses two models. The first model is used to extract the text from the road signboards and the second model uses the OCR system to detect the text. This extracted data from the OCR model is then concatenated to make a coherent sentence, i.e., information. A manual customized dataset has been created and used in the model. Increasing the accuracy of the model depends on the collective dataset. Challenges faced during the implementation of the model include increasing the model's efficiency by expanding the dataset. The potential challenge includes adverse weather conditions and fog.

The proposed paper introduces the improved object detection algorithm YOLO v7 and how it has been implemented in the proposed model for faster and more accurate object detection. It has been used in the model to detect traffic text-based signboards. These text-based signboards include essential information for user navigation and safety purposes. After the character segmentation has been performed, the second part of the model includes an OCR system that detects every letter and concatenates it into a single output. The CNN working in the model has been displayed. A thorough comparison with the previous models using similar versions of the algorithm and research papers has been shown in the proposed paper.

2 Related Work

In the computer learning algorithm, the photos are categorized using Support Vector Machine (SVM) based on their attributes. Gábor Baláz [1] proved heuristics in his decision-making model, locating TCDs with a 3D coordinate frame fixed to the ego vehicle and detecting TCDs with a decent probability and error rate. Precisions for distance-based recognition increased from 78.9% to 88.9%. Since there were only a few samples, the number should be significantly raised. To separate pertinent data, this system uses image processing techniques. While moving, the system's accuracy declines. The research paper mentioned above contains some flaws. Light and the surroundings have a negative impact on the system. Sometimes real-time streaming video images have high or low contrast; in these situations, the algorithm cannot recognize the traffic sign. The voice signal in the text-to-speech converter API occasionally has a slight delay. The software prototype used to implement the RTD approach described in this article adequately identified the different road types for European (continental) and UK roads.

In a published study, Narendra Sahu [2] looked at whether the OCR technique would work best for a mobile platform while scanning prescription drug labels. Caution is advised when comparing the power usage and energy efficiency of the proposed TSR accelerator to the prior developments. The project suggested a low-cost real-time TSR accelerator for embedded systems built on CNN. Both the software and the hardware for this design were optimized. It has been established that Standard Deep Learning, or Tesseract in this case, is the OCR technique that offers the best efficiency of time, resources, and accuracy. StanDL revealed that it had extremely acceptable resource costs comparable to approaches that did not use deep learning, was the most accurate method and processed the most photos in less than one second [10].

Road signs are necessary for ensuring efficient and secure circulation. Refraining from posting and incorrect signage interpretation are two critical contributors to traffic accidents. A model system that uses a convolutional neural network was proposed by Sampada P S [3]. The location of the traffic signs is also maintained in a database so that the driver can be alerted in advance of the following traffic sign. The system aims to protect drivers, passengers, and pedestrians from harm. The German Traffic Sign Benchmarks Dataset, which includes 51,900 photos of traffic signs in 43 categories, was employed in the model. However, this model has no real-time detection of the traffic sign board. It is necessary to upload images to the website. Because it is a static representation of a website, accuracy is good. Several CNN models were investigated, and the one that produced the best results on the GTSRB dataset was used [12].

Due to a lack of automatic annotation systems, most city establishments in developing cities are digitally unlabelled. As a result, location-based applications like Google Maps, Uber, and others are rarely used in such places. By combining two specialized pre-training methods and a runtime hyperparameter value selection approach, Sadrul Islam Toaha and Sakib Bin Asad [4] suggested a Faster R-CNN-based localization. They are using the SVSO (Street View Signboard Objects) signboard dataset that was created. Significantly better than cutting-edge techniques like Faster R-CNN, YOLOv4, and YOLOv5. The essential task for error-free information retrieval from such city streets or locations where trajectory services like uber, google maps, etc., remain neglected is accurate signboard detection in natural scene photos. Because sign boards differ significantly from country to country, extracting information from signboard images is still challenging. The variations include noise, brightness, recorded angle, and background type—information extraction problems from photos of billboards with low image quality and numerous colour fluctuations. The Street View Signboard Objects (SVSO) dataset, which now contains 5000 signboard photos of businesses, is the largest signboard dataset that is publicly accessible and is used in this study. Each picture has a 32-bit bit depth and is in PNG le format.

To help VIB people cross the road safely, the research addressed the issue of traffic sign recognition. Traffic sign detection and recognition are accomplished using a unique shape-specific feature extraction method and Random Gradient Succession with Momentum (RGSM). M. Sudha [5] suggested a model in which traffic sign recognition and detection are accomplished by utilising innovative shape-specific feature extraction techniques and unique Random Gradient Succession with Momentum (RGSM) techniques. The trained output labels will then be classified by the CNN classifier, which will subsequently translate the traffic sign into the audio signal during the training and testing phases. Considering the state of Indian roads, it is crucial to design an automatic guidance system for visually impaired persons and identify traffic signs. For those who are blind, it is pretty challenging because the traffic sign could be overlooked due to its prominent placement. It utilises the TSRD dataset. The TSRD has 58 sign classes and 6164 images of traffic signs. The proposed technique can be improved regarding characteristics like accuracy, precision, specificity, F-score, Jaccard coefficient, Kappa, and Dice coefficient.

Research published and models proposed by the pre-existing papers and previous works have limited time-speed efficiency and various shortcomings. These are slower, and some require data input in a static website which may need to be more efficient for moving accurate time object detection. The weather, fog and improper surroundings also adhere to the text or object detection accuracy. Lightning and Image contrast become another significant issue in accuracy. The proposed model overcomes and improves considerably on many of these parameters. The model works on YOLO v7 for object detection, which has been recently introduced in the community and performs better than any pre-existing CNN model. With the help of a faster and more accurate object detection algorithm, the model can capture and provide text detection on the traffic signboards significantly faster, which has real-time importance since the capturing and coherence of the written data has to be done as accurately and quickly possible through a moving vehicle [13].

3 Implementation

In the proposed model, a devised ADAS system has been implemented to prevent road accidents and improve the signaling and direction location of the users driving on the road. The main aim of the model is to detect text-based traffic signboards on the roads with crucial information, which not only helps for proper navigation but also may be crucial to prevent any future accidents or mishaps. i.e., road limit, construction work, roundabout texts etc. The previous models used YOLO detection algorithms for detecting signboards and text detection. The proposed model consists of two parts. The first section uses the YOLO v7 model, which is the latest object detection algorithm and uses it for the board and character detection. Individual characters are detected and fed into another model, part two of the system. The provided text is characterized and concatenated into a single string to make a coherent sentence.

In Fig. 1 shows a clear demonstration of the system architecture combining two models. The primary step is data collection. The proposed model has an input of self-collected data images from live road photos and video captures. About 150 images have been collected and annotated in this manner. These images are self-annotated with the help of Roboflow. An expansion tool was used to expand the data further to increase the dataset size for increased accuracy in board detection. The road sign dataset is fed into the modules at first. The parameters and pre-training model weights are set for the training model. The dataset then goes through data preprocessing and further data partitioning. The YOLO v7 algorithm helps traffic signboard detection and identifying text boards on the roadside. The preprocessed detected data is then fed into a CNN-based OCR system where individual characters are read and recognized and later concatenated into a single string as an output. Three key steps are present in every OCR system: Preprocessing, feature extraction, and classification and recognition are the first three steps. The image is improved and made cleaner by using picture binarization, noise reduction, skew detection and correction, text segmentation, and other preprocessing techniques. The process of measuring certain information from the data most pertinent to the specific classification task is known as feature extraction. The fundamental goal of classification is to convert the segmented text in a document image into an equivalent text representation.

The data is split into train and test folders after being tagged. The user can customize the split ratio. However, the most popular split is (80–20). Training will not start if an image is corrupted in any way. Training will not start if any label file is corrupt since Yolov7 will disregard those pictures and label files. The validation set, a set of data that is not part of the training set, is used to evaluate how well our model is performing while it is being trained. The outcomes of this validation procedure can be used to change the hyperparameters and settings of the model. It functions similarly to a critic telling us where our training is going. At the conclusion of each epoch, the model is tested on the validation set after being tested on the training set. To prevent overfitting, which occurs when our model excels at identifying samples in the training set but struggles to generalise to new data and make precise classifications, the dataset was divided into a validation set.

Large-scale and small-scale traffic sign detection accuracy depends on the model. Here YOLO v7, one of the fastest and most accurate models, captures the image with utmost accuracy. Annotation of the dataset has to be done manually. YOLO v7 is a one-stage algorithm. The one-stage detection methods are SSD, YOLO, etc. The enhancements, as mentioned earlier, maximize detection accuracy while maintaining inference speed, resulting in an accuracy good enough to compete with Faster R-CNN algorithms while requiring less inference time. The cutting-edge Yolo algorithm is ideal for detecting traffic signs because of its real-time speed and followed accuracy. Object detection models usually consider the depth, breadth, and resolution of the network that was used to train the network. In YOLOv7, the concatenation of layers while scaling the network's depth and width simultaneously happens. Studies on ablations demonstrate that this method maintains the ideal model design while scaling for various sizes.

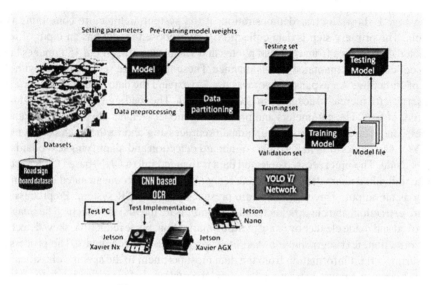

Fig. 1. Flowchart of the proposed model

3.1 YOLO v7 Architecture

The following paper uses the latest version of the object detection model, i.e., YOLO v7. YOLO v7 is a SOTA (State of the Art) object identification algorithm. It is much faster than previously documented object detection techniques, so that it can be used in real-time detection applications. Examples include AI memory, automatic license plate readers, and more. Among the various object detection models that are efficient for specific use cases, its latest version, his YOLOv7, stands out. According to the researchers, it had the best accuracy among known real-time object detectors with 56.8% AP and outperformed all other known object detectors in terms of speed and accuracy. Scaled-YOLOv4, Scaled-YOLOv5, DETR, Deformable DETR, DINO-5scale-R50, and Vit-Adapter-B were all slower and less accurate than YOLOv7, although SWINL Cascade-Mask R-CNNR did very well. This performed better than transform-based detectors like CNN.

In Fig. 2, a demonstrated version of the YOLO v7 architecture has been shown which is an enhanced version of the previous YOLO models. In the proposed model YOLO v7 has been used for the real time fast object detection, i.e., capturing text-based traffic road signboards. The most recent YOLO algorithm outperforms all earlier object detection algorithms and YOLO iterations in terms of speed and precision. The YOLO v7 algorithm in the proposed paper is a flexible research framework written in different low-level languages. The SxS region in each of the N grids created by the YOLO algorithm is the same size. Each of these N grids is responsible for finding and locating the object it contains. The object label, the likelihood that the object will be in the cell, and the B bounding box coordinates about the object's cell coordinates are thus predicted by these grids.

In comparison to other neural networks, YOLOv7 performs substantially better on small datasets without any pre-learned weights and with equipment that costs much less.

The YOLOv7 E-ELAN architecture improves model learning by merging cardinality to achieve the capacity to continuously boost the learning ability of the network without altering the initial gradient route. In YOLOv7, compound model scaling for a concatenation-based model is presented. By employing the compound scaling method, the optimum structure of the model can be maintained while preserving its original design characteristics. The structural components of a YOLO are a head, a neck, and a backbone. In YOLOv7, compound model scaling for a concatenation-based model is presented. By employing the compound scaling method, the optimum structure of the model can be maintained while preserving its original design characteristics. The structural components of a YOLO are a head, a neck, and a backbone. Due to the fact that YOLOv7 was motivated by Deep Supervision, a technique often employed in training deep neural networks, it is not limited to just one head. The auxiliary head serves to support middle-layer training while the lead head oversees creating the final product.

In the model YOLOv7 is exclusively used for the board detection and capture. It provides both speed and accuracy for the real time board detection and capture. The images captured are then manually annotated to focus on the text-based region of the board. This way YOLOv7 provides speed and accuracy while driving in a user vehicle or an autonomous car. YOLOv7 improves object identification technology by inferring more quickly and precisely than its predecessors (i.e., YOLOv5).

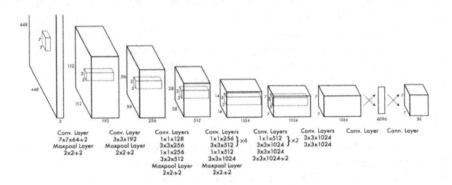

Fig. 2. Architecture of YOLO v7 model

4   Results and Discussions

The proposed model detected the signboards and text written on them with almost an accuracy of 86%. It gave more accuracy and confidence, as seen in the previous papers and other modules. The model performed well when used on a real-world dataset collected manually. The project can be improved in confidence by using more data images and training it for future enhancement. The proposed project can be extended to actual world applications by using it in IoT projects and rear cameras for driving vehicles.

4.1 Dataset

About 150 images from across the web were gathered to create the novel dataset specifically for the model. Roboflow was used to label the data and the expansion tool to increase the dataset to 300 images and divide them into test and validation sets with a ratio of 178 test images, ten validation images, and 10 test images. After fine-tuning the accuracy, it was trained on YOLO v7 models. The dataset available for the text-based traffic road signboards is very few on any online platforms. Here for the model, a self-devised, manually annotated dataset has been used to feed into the detection algorithm [9].

4.2 Performance Evaluation Matrix

When seeking to address classification problems, confusion matrices are a frequently employed assessment. The confusion matrix for the suggested model is shown in Fig. 3. It can be used to address problems with both binary classification and multiclass classification. Counts between the expected and observed values are displayed in confusion matrices. The result "TN" stands for True Negative and represents the quantity of cases that were accurately identified as being negatively classed. Similar to this, the abbreviation "TP" stands for True Positive and represents the number of accurately detected positive cases. The abbreviation "FP" stands for "real positive instances mistakenly classified as negative," and "FN" stands for "real positive examples mistakenly classified as positive." One of the most often employed metrics in categorization is accuracy. Alternative metrics based on the confusion matrix are also pertinent for evaluating performance because accuracy may be deceiving when applied to unbalanced datasets.

4.3 F1 Score

In Fig. 4, a graph of the F1 score vs. confidence for machine learning evaluation is shown. It evaluates the precision of a model. It incorporates a model's recall and precision ratings. The accuracy statistic shows how often a model predicts accurately over the entire dataset. The most popular evaluation statistic used to assess an algorithm's performance is model correctness. Usually, accuracy was the primary parameter utilized to compare machine learning models. Accuracy, on the other hand, merely measures the proportion of times a model is correctly predicted across the entire dataset, which is still valid if the dataset is class-balanced. An alternative machine learning evaluation statistic called the F1 score evaluates a model's predictive ability by focusing on its performance inside each class rather than its overall performance, as is done by accuracy. A typical F1 score ranging from 0.5- 0.8 is considered OK. An F1 score of 0.8–0.9 is considered good and F1 score > 0.9 is considered very good. A model's precision and recall scores are combined into this one metric. Our model shows an F1 of 0.8 which comes under the good region.

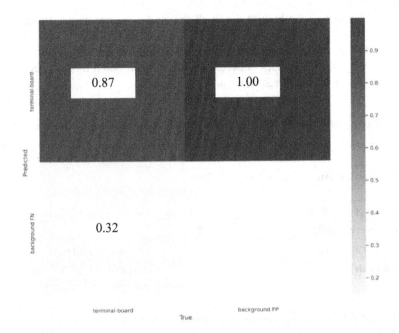

Fig. 3. Confusion matrix

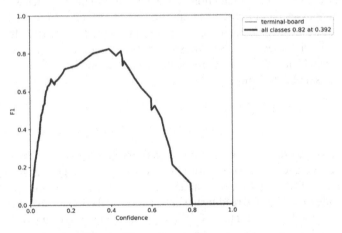

Fig. 4. F1 score vs confidence matrix

4.4 Comparison Between Proposed Model and Previous Models

In Table 1. a comparison between the previous research paper models and our proposed model has been made. On closer analysis based on parameters such as accuracy and speed, the proposed model performs better regarding text-based signboard detection. This is primarily because of the advanced technology of YOLO v7 detection, whose accuracy and speed outperform all the previous models. The accuracy of our model stands at 86% which is the highest yet in all the models alongside its efficiency and speed. To evaluate object identification methods like Fast R-CNN, YOLO, Mask R-CNN, etc., Mean Average Precision (mAP) is utilized. The average accuracy (AP) values are calculated using recall values between 0 and 1. The proposed model's accuracy as measured by Mean Average Precision (mAP)@.5 is 0.77.

Table 1. Comparison of performance analytics of the proposed model with previous models

Author	Accuracy	mAP (Mean average precision)
Daria Snegireva and Anastasiia Perkova, (*2021*)	72%	-
Gábor Balázs et al. (*2022*)	78.9%	-
Sampada P S et al. (*2022*)	98.5%	-
Xinyue Ren et al. (*2022*)	-	39.8% increased mAP
Proposed model	86%	mAP@.5:.95:100% 0.698

As indicated by the above comparative metrics, our model's accuracy is better than those of the previous models. However, accuracy of a bad model can also be high, since accuracy is a metric that measures the number of predictions that are correct as opposed to the total number of predictions made. It is a useful metric when we have equal distribution of classes on our classification. It doesn't work efficiently for imbalanced classes. To tackle this issue, we have compared our model on mAP (Mean average precision), which is based on sub metrics of confusion metrics, recall, precision and Intersection over Union (IoU), The mAP value of our v7 model is higher than that of previous models and the v5 model.

5   Conclusion

The proposed model detected the signboards and the text written on them with high accuracy and efficiency. The results we obtained using the suggested model have an overall accuracy of 86% and the system and speed at which it can operate effectively across object identification and reading the information on the sign boards. The mAP@.5 for the proposed model stands at 0.77. For future enhancements, the output of the model, i.e., concatenated string, can be used through an embedded IOT system to alert the user or voice output the processed information. This technology can be well integrated with car systems both externally and internally with the help of a camera device for text board capture and identification purposes. YOLO v7 helps in faster object detection, which is crucial for text-based detection through a moving vehicle since the time frame for detection and object analysis is very small. This model in the ADAS systems can even be configured with Arduino and embedded in manufactured vehicles. It could be a revolutionary driving safety system with proper actual-world implementation.

References

1. Gábor Balázs, G., Gyulai, C.: "Road type detection based on traffic sign and data", Article ID 6766455. Journal of Advanced Transportation, Hindawi Sds (2022)
2. Sahu, N., Sonkusare, M.: A study on optical character recognition technique. Int. J. Comput. Sci. Info. Technol. Cont. Eng. (2017)
3. Sampada, P.S., Shakeela, A., Singh, S., Supriya, J., Kavya, M.: Traffic sign board recognition and voice alert system using convolution neural network. Int. J. Eng. Res. Technol. (IJERT) RTCSIT **10**(12) (2022)
4. Toaha, S.I., et al.: Automatic signboard detection and localization in densely populated developing cities. Signal Processing: Image Communication **109** (2022)
5. Sudha, M., Galdis pushparathi, D.: Traffic sign detection and recognition using RGSM and a novel feature extraction method. Peer-to-Peer Netw. Appl. **14**, 2026–2037 (2021)
6. Ciuntu, V., Ferdowsi, H.: Real-time traffic sign detection and classification using machine learning and optical character recognition. IEEE Int. Conf. Electro Info. Technol. (EIT) **2020**, 480–486 (2020). https://doi.org/10.1109/EIT48999.2020.9208309
7. Snegireva, D., Perkova, A.: Traffic sign recognition application using Yolov5 architecture.2021 International Russian Automation Conference (RusAutoCon), Sochi, Russian Federation, pp. 1002–1007 (2021). https://doi.org/10.1109/RusAutoCon52004.2021.9537355
8. Ren, X., Zhang, W., Wu, M., Li, C., Wang, X.: Meta-YOLO: meta-learning for few-shot traffic sign detection via decoupling dependencies. Appl. Sci. **12**, 5543 (2022). https://doi.org/10.3390/app12115543
9. Srivastava, A.: Fast detection of multiple objects in traffic scenes with a common detection framework. Int. J. Res. Appl. Sci. Eng. Technol. **9**, 642–648 (2021). https://doi.org/10.22214/ijraset.2021.37386
10. Zhu, Y., Liao, M., Yang, M., Liu, W.: Cascaded segmentation-detection networks for text-based traffic sign detection. IEEE Trans. Intell. Transp. Syst. **19**(1), 209–219 (2018). https://doi.org/10.1109/TITS.2017.2768827. Jan.
11. Hu, J., Wang, Z., Chang, M., Xie, L., Xu, W., Chen, N.: PSG-"Yolov5: a paradigm for traffic sign detectionand recognition algorithm based on deep learning." Symmetry **14**, 2262 (2022). https://doi.org/10.3390/sym14112262
12. Belongie, S., Malik, J., Puzicha, J.: Shape matching and object recognition using shape contexts. California Univ San Diego La Jolla Dept of Computer Science and Engineering (2002)
13. Maldonado-Bascón, S., Lafuente-Arroyo, S., Gil-Jimenez, P., Gómez-Moreno, H., Lómez-Ferreras, F.: Road-sign detection and recognition based on support vector machines. IEEE Trans. Intell. Transp. Syst. **8**, 264–278 (2007)

Energy Preserving ABE-Based Data Security Scheme for Fog Computing

Sandeep Kumar[✉] [iD] and Ritu Garg [iD]

Department of Computer Engineering, NIT Kurukshetra, Kurukshetra, India
{sandeep_32113220,ritu.59}@nitkkr.ac.in

Abstract. A cutting-edge technology called fog computing makes it possible to offer services and applications for IoT/end devices at the network's edge quickly and effectively. As the fog computing paradigm is driven from the cloud computing paradigm, the security and privacy concerns have been also passed down to fog. But fog and cloud differ in terms of architecture and capabilities, thus the available cloud solutions are irrelevant to the fog. Moreover, fog is resource constrained in nature as well as it has limited energy as it runs over batteries. Therefore there is a need to develop an energy preserving lightweight security scheme to secure the data as well as to provide a fine grained access control mechanism. Thus, in this article we have developed a cryptosystem based on Chosen-Ciphertext Attribute-based encryption (ABE) with dynamic key change in order to secure the data and to prohibit unauthorized access in an energy efficient manner. We have evaluated and analyzed the sturdiness and the security of our proposed cryptosystem considering the energy consumption and mitigating attacks as primary aspects. Additionally, our cryptosystem is effective and energy efficient compared to other techniques.

Keywords: Fog Computing · Data Security · Energy Consumption · Access Control Policy

1 Introduction

Fog computing was inaugurated as an add-on to the cloud by Cisco in 2014, in order to preserve the latency sensitive data and to tackle the tremendous incoming traffic from the IoT devices as well as the other edge devices [1, 2]. Moreover, the IoT devices' volatile data that is continuously collected is kept on the fog level while the significant data is offloaded to the cloud to do training/analytics [3]. As the fog handles the continuous data flow and requests from the IoT devices and end users, it is quite impossible to identify whether a certain request/data has been received from a compromised node/adversary or not. Therefore, fog computing has been subjected to many security concerns such as data security, unauthorized access, privacy, intrusion detection, authentication etc. Data is considered as fuel for this era of internet, as most of the data processing for IoT is done over the fog, thus it is critical to protect the data and its integrity over the fog. Additionally, most of the data breach attacks are performed by gaining the unauthorized access. Thus, among the aforementioned security concerns, data security and unauthorized access are

M. Singh et al. (Eds.): ICACDS 2023, CCIS 1848, pp. 12–22, 2023.
https://doi.org/10.1007/978-3-031-37940-6_2

critical issues that need to be addressed on a severe note. There are plethora of techniques for data security and access control mechanisms available for cloud computing but they are unlikely applicable to fog as both differ in terms of architecture and capabilities (like energy, computation power, storage etc.). Moreover, fog is resource constrained in nature and it also doesn't have uninterrupted energy supply as most of them get energy from rechargeable sources like solar batteries etc. Thus, there is a need to develop an energy preserving lightweight security scheme to secure the data as well as to provide a fine grained access control mechanism. Attribute-based Encryption (ABE) sounds suitable for securing data over the fog as compared to other schemes like AES, DES etc. because ABE is less computation intensive and will consume less energy as well as it also provides access control policy. Therefore, in this paper we have proposed a lightweight cryptosystem inspired by Chosen-Ciphertext Attribute-based encryption (CP-ABE) and to attain the better utilization of energy we have also introduced the concept of dynamic key change with it, in dynamic key change the key sizes will going to vary with the different energy levels of the fog nodes i.e., we have used CP-ABE with dynamic key change in order to secure the data and to prohibit unauthorized access in an energy efficient manner.

The contribution of this research article is given below:

- Proposed a lightweight cryptosystem inspired by CP-ABE,
- Provided dynamic key change for better energy utilization,
- Developed a fine-grained access control policy in order to protect data from the adversary, trying to gain access.

The remainder of this article is given as follows; we have analyzed the different data security and access control methods in Sect. 2. Section 3 contains the preliminaries for our proposed cryptosystem. Further, the proposed lightweight cryptosystem ABE is discussed in the Sect. 4. In the Section 5 we have analyzed and evaluated our proposed security model. Finally, we have concluded our work in the Sect. 6.

2 Related Work

This section consists of several techniques outlined in literature to tackle data security concerns. In Ref. [1], authors have presented attribute-based encryption (ABE) along with outsourced decryption while using chosen-ciphertext instead of using traditional ABE schemes, the chosen-ciphertext security is widely regarded as a cryptosystem's greatest level of security, and it can withstand undiscovered sophisticated attacks on critical data in fog computing [1]. But the authors have not considered energy consumption as an important parameter as well as access control policy was not addressed. In research work [4], the authors have proposed a combination of two cryptography algorithms i.e. ECC for generating keys and for encryption they have used the Blowfish algorithm. The presented model in [4] was quite secure but it was computation intensive as it was using the hybrid implementation so it lacks efficiency in terms of usage of the energy. In Ref. [5] the authors have presented a solution using De-duplication of the encrypted data, deduplication provides immunity against the data theft but it does not show any security towards brute force attack as its lags proper access control mechanism. In the paper [6], the authors have developed an encryption method which uses the Paillier technique but the issue of energy usage was not appropriately addressed by the authors. Apart from encryption based solutions some authors have also provided blockchain and machine learning based solutions like in [7], the authors have offered a data protection technique in which blockchain technology has been used along with the software-defined networking(SDN). The model developed in [7] was comparatively firm from encryption based schemes but it was only suitable for fog nodes with higher configuration as blockchain and SDN are resource and computation hectic. In Ref. [8], the authors came up with a neutrosophic ranking technique known as Preference Ranking Organization Method for Enrichment Evaluation which provides security and validation via a real case study [8]. The detailed overview of the existing work is summarized in the Table 1. Additionally, Table 1 contains the observed findings, key objective in terms of energy and access control policy as well as the resilience against the attacks.

In the above discussed methods none of the authors have considered energy as a key parameter in their proposed scheme. As fog has limited energy source as well they also failed to provide an access control mechanism to prohibit unauthorized access. Thus in this paper, we have developed a CP-ABE based lightweight technique with dynamic key change in order to conserve energy and to also provide a fine grained access control mechanism for more details see Sect. 4.

3   Preliminaries

3.1   Attribute Based Encryption

Sahai and Waters [9] developed Attribute-based encryption (ABE), which enhances identity-based encryption by linking private keys and ciphertexts to set of descriptive qualities. When the two pieces of information sufficiently overlap, decryption is then feasible. Further, its two variants i.e., Key policy ABE and CP-ABE are presented [10]. In key-policy ABE schemes (KP-ABE), ciphertexts are annotated with attribute sets, and access structures that define which ciphertexts the user will be allowed to decode are linked to private keys. In CP-ABE, attribute sets are assigned to private keys and senders can provide an access policy that receivers' attribute sets must adhere to decrypt the message [10]. CP-ABE is considered to be more expressive and flexible than KP-ABE in terms of specifying access policies. Additionally, CP-ABE allows for the decryption of ciphertexts without the need for a central authority, while KP-ABE requires a central authority to distribute keys. Moreover, CP-ABE is a pairing based encryption scheme therefore, we have utilized the bilinear map and the IP addresses of the IoT/ end devices have been used as the attribute associated with the private key in order to get a fine access control mechanism.

3.2   Bilinear Map/Pairing

Let G_1, G_2 are the groups of prime order p. If k is a bilinear group, i.e. $G_1 \times G_2$, then it must follow these properties:

- Bilinearity: $\forall$ a, b $\in G_1$, k $(a^u, b^v) = $ k $(a, b)^{uv}$, where u,v $\in Z_p$.

Table 1. Comparative analyses of existing work and our proposed work.

S.No	Ref	Findings	Objective		Resilience against attacks
			Energy	Access Control	
1	Zuo et al. [1]	CP-ABE with outsourced decryption	-	✔	Brute Force attack
2	Chaudhary et al. [4]	Combination of two cryptography algorithms i.e. ECC for generating keys and for encryption they have used the Blowfish algorithm	-	-	Brute Force and Man in the Middle attack
3	Koo et al. [5]	De-duplication of the encrypted data	-	-	Data Theft attack
4	Khalid et al. [6]	An encryption method which uses the Paillier technique	-	-	Man in the Middle attack
6	Muthanna et al. [7]	Developed a data protection technique in which blockchain technology has been used along with the software-defined networking (SDN)	-	✔	Data Theft and Brute force attack
7	Abdel et al. [8]	A neutrosophic ranking technique known as Preference Ranking Organization Method	-	-	Man in the Middle and Data Theft attack
8	Our Work	Chosen-Ciphertext Attribute-based encryption (CP-ABE) with dynamic key change	✔	✔	Man in the Middle and Brute Force attacks

- Non-Degeneracy: means if g_1 is a generator of G_1 as well as g_2 is a generator of G_2 then, k (g_1, g_2) is a generator of G_1 x G_2.
- Efficiency: i.e. k (g_1, g_2) must be an efficiently calculable function [11].

In this paper, we have used a bilinear map in the configuration setup phase, which outputs the global parameters as well as the public master key.

3.3 Access Control Structures

Definition 1. Let {$P1, P2,... Pn$} be a set of parties. A collection $A \subseteq 2^{\{P1,P2,...,Pn\}}$ is said to be monotone if for any two sets B and C, if B is an element of A and B is a subset of C, then C must also be an element of A. An access structure is a collection of non-empty subsets of {$P1, P2,... Pn$}, denoted by A, where $A \subseteq 2^{\{P1,P2,...,Pn\}} \setminus \{\emptyset\}$. The authorized sets in A are the subsets that belong to A, while the unauthorized sets are the subsets that do not belong to A. A monotone access structure is a collection of authorized sets that also satisfies the monotonicity property.

Basically, it indicates whether the provided attributes belong to the respective user/IoT or not. It is also responsible for attaining fine-grained access control [12].

4 Proposed Security Model

In this section, we have discussed the proposed architecture and security model for protecting data over fog level.

4.1 Proposed Architecture

We have proposed the three-layer-architecture in which IoT devices and other end users are present in the first layer, which transmits the data to the second layer, which consists of fog nodes and fog broker. The third layer of our presented architecture is cloud level, as shown in Fig. 1. In the proposed architecture, the fog broker is responsible for monitoring and controlling the fog nodes. Moreover, it is also accountable for transmitting the data to the cloud in the encoded format if the data is subscribed to the cloud.

4.2 Proposed Security Model

We have utilized the concept of CP-ABE with dynamic key generation to conserve the fog nodes' energy. The reason behind choosing the CP-ABE is, it is a lightweight encryption algorithm which also provides fine-grained access control policy as the attributes of the IoT devices/end user are associated with the access structure. Moreover, as fog nodes are resource constrained devices, CP-ABE is sound and feasible encryption technique. And to utilize it in energy efficient way we have introduced the concept of dynamic key change. Our proposed security model consists of four phases, i.e., Pairing Setup, Key Generation, Encryption and Decryption.

The Pairing Setup and Key Generation process took place over the fog broker, according to the energy level of the fog nodes, as the fog broker regularly monitors the energy levels of the fog nodes. Further, Encryption and Decryption occur on the fog nodes.

4.2.1 Pairing Setup

On receiving the input argument π of the prime field, the broker initializes the pairing setup, i.e. selects the bilinear map (bp), a function (f) which maps the user's/IoT device's

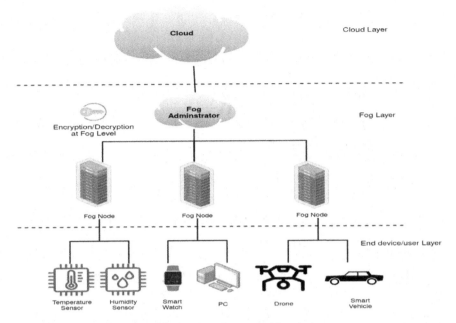

Fig. 1. Three-Layer-Architecture

identity to the bp and τ which maps the user's/IoT device's attributes to the attribute to the attribute authority. We have used IP address as the user's identity.

This phase returns the global parameters (gp) and master security key (msk).

```
Algorithm 1:    Pairing Setup

begin
     gp = (bp, p, f, τ, A_id, I_u )    //p = prime number
     for I_u ∈A_id   //A_id= authorities of user's attributes
        msk_id= (a_id, b_id)         //I_u = user's identity, i.e.
                                     IP address
     end for
end.
```

4.2.2 Key Generation

The size of the keys generated varies based on the energy level of the fog nodes. We have considered three energy levels, i.e. below 30% (i.e. key size is 128 bytes), less than 60% (i.e. key size is 256 bytes) and 85% or above (i.e. key size is 384 bytes). The key generation algorithm takes 'msk' and a random prime number 'n' as input and produces the public key (pk) and the private key (sk).

```
Algorithm 2:      Key Generation

begin
     for i ∈ S_{i,aid}              //S_i= set of user's identity
         k_{i,id}= bp^{aid}.f(i)^{bid}
         k_i= bp^n                  //n = random prime number
         if energy_lev <= 0.3*battery_cap
             pk_{id,i} = (bp(n)^{aid}, n^{bid})
             sk_{id,i} = (k_{i,id}, k_i)
         else if energy_lev <= 0.6*battery_cap
             pk_{id,i} = (bp(n)^{aid}, n^{bid})
             sk_{id,i} = (k_{i,id}, k_i)
         else
             pk_{id,i} = (bp(n)^{aid}, n^{bid})
             sk_{id,i} = (k_{i,id}, k_i)
     end for
     Initialize access structure (As) ∀ device/user.
end.
```

4.2.3 Encryption

The encryption took place over the fog nodes; the encryption algorithm takes the public key (pk), access structure (As) and a random prime number p as input and outputs the cipher text.

```
Algorithm 3:      Encryption

begin
     for i ∈ S_{user}
         C_1 = sc_{pk}(msg)     //sc  =  symmetric   cryptographic
                                 scheme, msg = data
         C_2 = p.bp(g_i, g_{i+1})^Π    //p = random prime number
         Cipher_text = (As_i, {C_1, C_2}//As = access structure
     end for
end.
```

4.2.4 Decryption

Private key is required in the decryption algorithm to retrieve the original data. But prior to it, if the user's identity does not match the respective access structure, then it flashes the unauthorized operation message, else the user can decrypt the data.

```
Algorithm 4:       Decryption

begin
    for i ∈ S_ct
        if I_{i,u} ∈ As_i
            if ct == C_1                     //ct = cipher text,
                m`_1= (C_{1,i}, sk_i)
            else
                m`_2= (C_{2,i}, msk_i)
                M`= (m`_1, m`_2)             //M`= retrieved data.
        else
            print "unauthorized operation"
    end for
end.
```

The observed time complexity of the proposed CP-ABE cryptosystem with dynamic key change is O(N). The container's number of items is indicated by the letter "N." The complexity can be O(1), if random access method is used to crack the cryptological container. Moreover, the complexity varies according to the key size.

5   Performance Evaluation

In this section, we have evaluated and analyzed the sturdiness and the security of our proposed cryptosystem. In order to evaluate our developed cryptosystem, we have considered the energy consumption and mitigating attacks as primary aspects. The security and resistance of our scheme is simulated and tested over the Ubuntu 20.04 x64 - 4 GB RAM laptop along with Raspberry Pi4 as fog node whose configuration is as follows:

- Processor - Quad core Cortex-A72 (64-bit SoC).
- RAM - 2 GB.
- Clock speed - 1.5 GHz.
- Minimum energy requirement - 5 V/3 A.

5.1   Energy Consumption

As we have designed our cryptosystem with dynamic key change i.e., the size of the keys varies with respect to the energy levels of the fog nodes. We have considered three energy levels with three different key sizes such that if the energy of fog node is below 30% i.e., key of 128 bytes is used or if it is less than 60% then we have utilized a key of 256 bytes else if it is 85% or above the key of 384 bytes will be considered. Further, computing resources (i.e., CPU, storage etc.) will be consumed accordingly. Moreover, computation overhead will also vary with the key size. Thus, the computation overhead is directly proportional to the energy consumption.

To calculate the total energy consumption of our cryptosystem we have considered all modules like chipset, CPU, battery, RAM, sensors/actuators. Energy consumption is calculated using $E = I*V*C$ where, E is the node's energy in watt-hour, I is the current in ampere, V is the apparatus's voltage and C is the number of cycles [13, 14]. Additionally, the Pairing Setup and Key Generation process took place over the fog broker which also assists in energy preservation of fog nodes. Moreover, compared to other approaches as shown in Fig. 2, the output of our developed cryptosystem performs better in terms of energy preservation.

5.2   Mitigating Attacks

Here we have demonstrated how our proposed cryptosystem is resistant to man-in-the-middle attack and brute force attack.

Definition 2 Man-in-the-Middle (MITM). The adversary hijacks communication among two parties and reads, modifies, or inserts new information into the transmission without the knowledge of either side [15].

Our proposed cryptosystem mitigates MITM attacks by allowing for the creation of ciphertexts that can only be decrypted by specific sets of users with specific attributes i.e., IP address of IoT devices/ end users. Moreover, our proposed technique provides fine grained access control so if somehow it enters the network it is impossible to decrypt or modify the data. Thus, our model successfully mitigated the MITM.

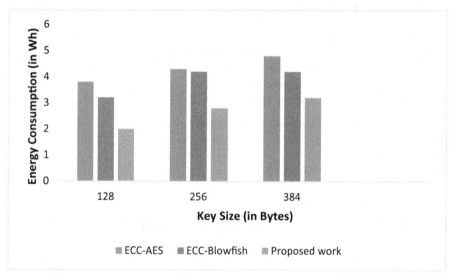

Fig. 2. Comparative Analysis of Existing Work with Our Proposed Work

Definition 3 Brute Force Attack Are the attacks that involves trying many different inputs or combinations of inputs in an attempt to guess the correct one. Brute force attacks can be used to obtain unauthorized access to computer systems, network [16].

As the proposed cryptosystem allows for fine-grained access control by encrypting data with a set of attributes, and only allowing access to users who possess a set of attributes that match the attributes linked to the ciphertext. Moreover, it is based on the use of pairing-based cryptography, which is secure against classical brute force attacks. Therefore, our model is resistant to brute force attacks.

6 Conclusion

Fog computing paradigm was inaugurated as an add-on feature to the cloud to handle massive network traffic produced by the IoT sensors/actuators as well to preserve the latency sensitive data. Similar to the cloud, fog also faces security concerns such as data security, unauthorized access, privacy, intrusion detection, authentication etc., but among these concerns data security and unauthorized access are critical issues. Thus there is a need to develop a lightweight cryptosystem keeping the constrained nature of fog in mind. Therefore, to address the issues of data security and unauthorized access we have developed a cryptosystem based on Chosen-Ciphertext Attribute-based encryption (ABE) with dynamic key change(i.e., different key sizes will be used on the basis of energy levels of the fog) in order to secure the data and to prohibit unauthorized access in an energy efficient manner. To evaluate and analyze the sturdiness and security of our model we have considered energy and attacks mitigation as the prime aspects. Additionally, our cryptosystem is effective and energy efficient compared to other techniques.

References

1. Zuo, C., Shao, J., Wei, G., Xie, M., Ji, M.: CCA-secure ABE with outsourced decryption for fog computing. Futur. Gener. Comput. Syst. **78**, 730–738 (2018)
2. Kumar, N., Misra, S., Rodrigues, J.J., Obaidat, M.S.: Coalition games for spatio-temporal big data in internet of vehicles environment: a comparative analysis. IEEE Internet Things J. **2**(4), 310–320 (2015)
3. Iorga, M., Feldman, L., Barton, R., Martin, M.J., Goren, N. S., Mahmoudi, C.: Fog computing conceptual model (2018)
4. Chaudhary, R., Kumar, N., Zeadally, S.: Network service chain-ing in fog and cloud computing for the 5G environment: data management and security challenges. IEEE Commun. Mag. **55**(11), 114–122 (2017)
5. Koo, D., Hur, J.: Privacy-preserving deduplication of encrypted data with dynamic ownership management in fog computing. Futur. Gener. Comput. Syst. **78**, 739–752 (2018)
6. Khalid, T., Khan, A.N., Ali, M., Adeel, A., Shuja, J.: A fog-based security framework for intelligent traffic light control sys-tem. Multimedia Tools and Applications **78**(17), 24595–24615 (2019)
7. Muthanna, A., et al.: Secure and reliable IoT networks using fog computing with software-defined networking and blockchain. J. Sens. Actuator Netw. **8**(1), 15 (2019)

8. Abdel-Basset, M., Manogaran, G., Mohamed, M.: A neutro-sophic theory-based security approach for fog and mobile-edge computing. Comput. Netw. **157**, 122–132 (2019)

9. Dan, B., Matthew, F.: Identity-based encryption from the weil pairing. SIAM J. Comput. **32**(3), 586–615 (2003)

10. Sahai, A., Waters, B.: Fuzzy identity-based encryption. In: Annual international conference on the theory and applications of cryptographic techniques, pp. 457–473. Springer, Berlin, Heidelberg (2005, May)

11. Boneh, D., Franklin, M.: Identity-based encryption from the Weil pairing. In: Annual international cryptology conference, pp. 213–229. Springer, Berlin, Heidelberg (2001, August)

12. Beimel, A.: Secure schemes for secret sharing and key distribution (1996)

13. Jindal, P., Singh, B.: Optimization of the security-performance tradeoff in RC4 encryption algorithm. Wireless Pers. Commun. **92**(3), 1221–1250 (2017)

14. Deb, P.K., Mukherjee, A., Misra, S.: CEaaS: Constrained Encryption as a Service in Fog-Enabled IoT. IEEE Internet Things J. **9**(20), 19803–19810 (2022)

15. Yuvaraj, N., Raja, R.A., Karthikeyan, T., Praghash, K.: Improved authentication in secured multicast wireless sensor network (MWSN) using opposition frog leaping algorithm to resist man-in-middle attack. Wireless Pers. Commun. **123**(2), 1715–1731 (2022)

16. Verma, R., Dhanda, N., Nagar, V.: Enhancing Security with In-Depth Analysis of Brute-Force Attack on Secure Hashing Algorithms. In: Proceedings of Trends in Electronics and Health Informatics: TEHI 2021, pp. 513–522. Springer Nature Singapore, Singapore (2022)

An Approach for Effective Object Detection

Rohit Bisht[1] and Nileshkumar Patel[2(✉)]

[1] Ultimate Kronos Group, Noida, India
rohitbisht@outlook.com
[2] Jaypee University of Engineering and Technology, Guna, India

Abstract. The ability to detect and recognize objects in images or videos is a vital aspect of computer vision, with applications in fields like surveillance, autonomous driving, robotics, and medical imaging. Object detection aims to locate and identify objects of interest within an image or video, and with the emergence of deep learning techniques, the accuracy and efficiency of this process have greatly improved. Modern approaches such as Faster R-CNN, YOLO, and SSD have demonstrated remarkable results in real-time object detection and recognition. However, despite these advancements, challenges such as occlusion, varying scales, cluttered backgrounds, and changes in viewpoint still pose significant obstacles. Further research and development of more sophisticated algorithms are necessary to address these challenges and accurately detect and recognize objects under diverse conditions. This paper seeks to improve the current object detection techniques' performance by identifying their shortcomings and proposing solutions to enhance targeted performance metrics.

Keywords: Object Detection · Object recognition · CNN

1 Introduction

1.1 Object Recognition

Object Recognition in its literal sense means identification of objects present in a targeted scene. The provided view can be a digital image or a real time visual interaction with the surrounding environment. Living beings with their gift of sight can easily derive meaningful information from their visual interactions, with just one look at the target image humans can detect if an object of interest present in it or not. Some birds of prey e.g. Eagles can recognize their game from miles away. The effective mechanism for Object detection and identification possessed by living beings is difficult to replicate artificially and still remains a much-coveted benchmark for computer vision.

In case of machines, object recognition generally refers to a group of computer vision tasks that include recognizing things in digital pictures. 'Object recognition' is sometimes referred to as "object detection" however object recognition takes a broader approach towards truly identifying an object along with its class and is not limited to simply differentiating objects from background. Object Recognition involves figuring out objects in images using customized logic with a goal to enable machines to recognize images just like us humans.

The first step is to classify the input image by putting a label on it. This labelling can be done on the basis of any metric i.e., probability, loss, accuracy, etc. depending upon the target input image. For example, an image of a dog can be classified as a class label 'dog' or an image of wolf can be classified as a class label 'wolf' with some probability. This is called Image Classification.

Next step involves figuring out the locality of our object of interest in the input image This involves an algorithm processing the input image and finally drawing a rectangular boundary around the area of interest in the image. In other words, location of object in the output with the help of a bounding box (position, height, and width).

1.2 Object Detection

Object detection refers to the task of identifying and locating objects within an image or video frame, and drawing a bounding box around each object. Object detection typically involves a multi-stage process. The first stage is to use a feature extraction algorithm to identify key visual features within an image or video frame, such as edges, corners, and color contrasts. These features are then fed into a machine learning algorithm, such as a convolutional neural network, which analyzes the features extracted from the image and produces a set of bounding boxes around any objects that it detects. It also assigns a probability score to each box, indicating how confident it is that the box contains an object of interest. The boxes with the highest probability scores are typically selected as the final output of the object detection algorithm.

Object detection is a challenging problem due to the variability in object appearance, size, and orientation, as well as the presence of occlusions and clutter in the scene. This paper will attempt to increase performance of current Object Detection techniques (performance of targeted metrics) after identifying the shortcomings in current techniques.

2   Related Work

Convolutional neural networks have revolutionized the field of computer vision, enabling accurate and efficient image classification, object detection, and other visual recognition tasks. At their core, CNNs are inspired by the visual cortex of animals, and consist of multiple layers of filters that extract increasingly complex features from an input image. The architecture of CNN allows it to optimize numerous related tasks at the same time (for example, rapid R-CNN combines classification and bounding box regression into a multitask learning method) [1].

Object detection systems based on deep learning have advanced rapidly in recent years. The algorithms can be broken down into two categories. One-stage detectors like YOLO and two-stage detectors like Faster R-CNN are available (Faster Region Based Convolutional Neural Networks). Though YOLO and its variants aren't as accurate as two-stage detectors, they outpace their counterparts in terms of speed by a wide amount. When confronted with standard-sized objects, YOLO performs admirably, but it fails to detect little objects. The accuracy drops dramatically when dealing with things with faces that appear to have large scale dynamic features [2].

2.1 R-CNN Model Family

The RCNN family of object detection models have revolutionized the field of computer vision, enabling accurate and efficient object detection in images and videos. RCNN stands for Region-based Convolutional Neural Network, and the family includes several variations, including Fast R-CNN, Faster R-CNN, and Mask R-CNN. The R-CNN was first described in the 2014 publication "Rich feature hierarchies for accurate object detection and semantic segmentation" by Ross Girshick et al. from UC Berkeley. At the heart of the RCNN family is the idea of using a region proposal algorithm to identify potential object locations within an image or video frame, and then using a convolutional neural network (CNN) to classify and refine these regions. This two-stage approach allows RCNN models to achieve high accuracy while maintaining computational efficiency.

In 2015, study titled "Fast R-CNN" by Ross Girshick, then at Microsoft Research, presented an update to overcome R-CNN's speed difficulties, citing the effectiveness of the algorithm Fast R-CNN was the first model in the family to introduce the region of interest (RoI) pooling layer, which enables the use of a single CNN to classify multiple regions within an image, rather than using a separate CNN for each region. This approach dramatically reduces computational complexity and improves accuracy, making Fast R-CNN a popular choice for object detection.

Shaoqing Ren, et al. at Microsoft Research improved the model architecture for both training and detection speed in the 2016 study "Faster R-CNN: Towards Real-Time Object Detection with Region Proposal Networks." Faster R-CNN introduced a region proposal network (RPN) that generates region proposals directly from the CNN feature map, rather than using a separate algorithm. This further improves computational efficiency and allows for real-time object detection in video streams.

2.2 YOLO – Real Time Object Detection

Joseph Redmon and colleagues first proposed the YOLO model in year 2015 in their publication, "You Only Look Once: Unified, Real-Time Object Detection" offered a new approach to object detection with its ability to make predictions directly from the full image, without the need for region proposal algorithms or post-processing steps. YOLO considers object detection to be a regression issue with spatially separated bounding boxes and associated class probabilities. It uses a single convolutional neural network (CNN) to simultaneously predict bounding boxes and class probabilities for objects in an input image.

YOLO is able to achieve high accuracy and real-time performance, making it a popular choice for applications such as self-driving cars, surveillance, and robotics. To achieve this, YOLO divides the input image into a grid of cells and predicts bounding boxes and class probabilities for each cell. Each bounding box includes a confidence score, which reflects the model's confidence in its prediction, as well as a set of coordinates that define the box's position and size. The class probabilities reflect the likelihood that an object of a given class is present within the box.

YOLO v2 and YOLO v3

In the publication of 2016, "YOLO9000: Better, Faster, Stronger" [3] by Joseph Redmon and Ali Farhadi revised the model in an effort to increase the model performance.

Though this variant of the model is popularly referred to as YOLO v2, but reportedly this model when was trained on two object identification datasets in parallel was capable of predicting 9,000 object classes, and so this model also came to be referred as 'YOLO9000'.

One of the major improvements in YOLOv2 is the use of anchor boxes or priors, which allow the model to predict bounding boxes of different shapes and sizes. This enables YOLOv2 to better detect objects of varying sizes and aspect ratios, improving its overall accuracy. The priors or bounding box predictions are adjusted iteratively to allow changes which have small influence on the predictions, resulting in a more stable model. Instead of explicitly forecasting position and size using boundary boxes, offsets for moving and reshaping the pre-defined anchor boxes relative to a grid cell are calculated by a logistic function (Fig. 1).

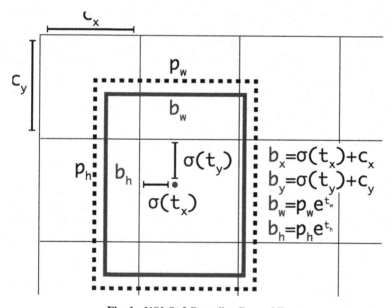

Fig. 1. YOLOv2 Bounding Boxes [4]

Another key innovation in YOLOv2 is the use of batch normalization, which accelerates the training process and improves the model's ability to generalize to new data. YOLOv2 also uses a new network architecture called Darknet-19, which is smaller and faster than the previous architecture used in YOLOv1.

In the study "YOLOv3: An Incremental Improvement", Joseph Redmon and Ali Farhadi presented more enhancements to the model. While YOLOv2 and YOLOv3 share many similarities, YOLOv3 represents a significant improvement over its predecessor in terms of accuracy and robustness, thanks to its deeper architecture, skip connections, and spatial pyramid pooling feature.

Scale and Ratio Aware Model

The difficulty of recognizing small object like pedestrians from and image when the picture ratio of different photographs varies and scaling the images introduces noise was addressed using the scale and ratio aware model of YOLO. When aspect ratios are uneven, the multiscaling method increases noise, resulting in a problem to detect small pedestrian ratios in photos, today most deep learning algorithms use multiscaling [5].

Ratio-aware algorithms are provided to dynamically modify the input layer length and width hyperparameters of YOLOv3, addressing the issue of significant aspect ratio discrepancies. Ratio-aware techniques allow for dynamic aspect ratio adjustments to match those required by current images and combines information from all resolutions. Information from images of different resolutions (low- and high-resolution data) is fused using multiresolution fusion. In RA- YOLO multiresolution fusion is employed to overcome the problem of misdetection of astonishingly small walkers in photos since the original and local images produce low- and high-resolution pedestrian identification information.

YOLO v5 Model

YOLO v5 like its predecessors though is a single shot detector comprises of 3 components, a Cross Stage Partial Network (CSPNet) Backbone is used to extract key features from an input image, a PANet neck to extract feature pyramids which aids in the identification of the same objects of varying scales and working on unknown data. A model head identical to that of the preceding YOLO V3 and V4 is employed for final detection process of objects. After applying anchor boxes to features, it generates final output vectors with class probabilities, objectness scores, and bounding boxes.

In YOLO v5, the middle/hidden layers use the Leaky ReLU activation function, whereas the final detection layer uses the sigmoid activation function.

Summary:

- CNN is a multi-pipeline technique with numerous benefits that is employed in a variety of research domains, including image classification, image retrieval, face recognition, pedestrian detection, and video analysis.
- CNN's architecture allows it to optimize numerous related tasks at once (for example, rapid R-CNN integrates classification and bounding box regression into a multitask learning method) [1].
- Because the entire detection pipeline is a single network in the case of YOLO, it can be optimized end-to-end based on detection performance.

3 Proposed Approach

3.1 Problem Statement

Object identification is an important part of many deep learning models, and it has undergone a lot of radical changes in recent years.

- The main problem has been the fact that many applications require real-time object detection.
- Modeling object detection with deep learning presents two issues.
- Because the size and number of objects can vary, the network must be able to handle this variation.
- The number of possible bounding box combinations is enormous, therefore these networks are computationally intensive. Compiling the output at real-time speed has become a challenge.

Solution Overview
There is no such thing as the fastest and most accurate model. We have to make a choice between speed and precision. Some models with more accuracy have lesser speed, and vice versa (Table 1).

Table 1. Model combination experiments on VOC 2007 [6]

Method	mAP	Combined	Gain
Fast R-CNN	71.8	-	-
Fast R-CNN (2007 Data)	66.9	72.4	0.6
Fast R-CNN (VGG-M)	59.2	72.4	0.6
Fast R-CNN (CaffeNet)	57.1	72.1	0.3
YOLO	63.4	75	3.2

Next Steps

- Run tests to determine a quantitative metric for assessing performance.
- Conduct independent testing to determine which YOLO algorithm produces the best results.
- Use the same data set to compare alternative methodologies and uncover gaps.

Dataset Identified: public blood cell detection dataset.

This is a blood cell photo dataset that was first made public by cosmicad and akshaylambda.

Actions Taken

- Conducted tests to determine a quantitative metric for assessing performance.
- Conducted independent experiments to determine which YOLO algorithm produced the best results.
- Identified gaps by comparing different ways utilizing the same data set.

YOLOv4 vs YOLOv5 PERFORMANCE

There are 364 photos in all, divided into three categories: white blood cells, red blood cells, and platelets. There are 4888 labels divided into three classes. Seventy percent (255 photos) of the 364 images were chosen for training, twenty percent (73 images) for validation, and ten percent (36 images) for testing.

Preprocessing and data augmentation were applied to the Training set photos in order to improve item detection accuracy.

Preprocessing:

Resize: Stretch to 416x416.

Augmentations:

Training example outputs: 3
    Flip: Horizontal, Vertical.
    90° Rotate: Clockwise, Counter-Clockwise, Upside Down.
    Crop: 0% Minimum Zoom, 15% Maximum Zoom.
    Hue: Between -25° and + 25°
    Saturation: Between -25% and + 25%
    Brightness: Between -15% and + 15%
    Exposure: Between -20% and + 20%

The above process culminates in a final Training Set of 765 photos (416x416). Google Colab Infrastructure was chosen to host the test environment in order to evaluate performance metrics between YOLO4 and YOLO5. Google Research's Colaboratory, or "Colab" for short, is a product. Google Colab is an online available paid infrastructure service which one can use for developing and testing the machine learning models.

We get additional session time and more powerful GPUs with the subscription version of the above service. This is what I used to do this test.

Details of Test Environment

GPU Type: Tesla P100-PCIE-16GB.

ARCH Value: -gencode arch = compute_60,code = sm_60Memory: 17.070817 GB.

Dataset Details

Training Set: 765 images (416x416).

Validation Set: 73 images.
Testing Set: 36 images.

Execution Results

The Class wise Average Precision for predicting classes of images is displayed in Table 2 below-

Table 2. Class wise Average Precision

Class	YOLOv4	YOLOv5
Platelets	83.64	88.2
RBC	77.87	87.8
WBC	97.62	97.1

Performance metrics comparison between YOLO4 and YOLO5 is displayed in Table 3 below-

Table 3. YOLO4 vs YOLOv5 Performance comparison

Parameter	YOLOv4	YOLOv5
mAP@0.50	86.376667	91.0333333
Precision	0.59	0.822
Recall	0.85	0.928
Prediction Time	20.844 ms	9ms

Inference from YOLO Algorithm Comparison

When using the BCCD Dataset on the same machine architecture, YOLOv5 outperforms YOLOv4.

- mAP@0.50 for YOLOv4 is 4.6566663 percent lower than for YOLOv5.
- YOLOv5 also exceeded YOLOv4 in terms of Precision and Recall, with a 23.2 percent and 7.8 percent difference, respectively.
- When compared to YOLOv4, which had a forecast time of 20.844ms, YOLOv5 had a prediction time of 9ms.
- The average FPS for YOLOv5 was 111, while the average FPS for YOLOv4 was 48.
- The YOLO4 model took roughly 7–8 h to train (max batch size = 6000), whereas the YOLOv5 model took about 13 min (for 100 epochs)
- YOLO v5 is a compact version. A weights file for YOLO v5 is 27 gigabytes in size. The YOLO v4 (Darknet architecture) weights file is 171 MB. YOLO v5 is almost 90% smaller than YOLO v4.

- Now that we've shown that YOLOv5 is, in fact, the quicker model, we'll try to fine-tune it even more by changing the model's settings to get better results.

Most of the research done in the area to improve performance of YOLO involves enriching the training data by either collecting it from multiple diverse sources and by using data augmentation methods. Not much has been tweaked in the model or hyper parameters that is involved in training of model.

The idea here is to attempt to observer increase in performance of YOLO by tweaking i.e. making small changes to default model parameters or hyperparameters.

Tweaking Hyperparameters to Improve Model Performance

- Conducting additional research to better grasp the YOLOv5 algorithm's hypo parameters
- Fine-tune algorithm parameters and confirm any noticeable performance gains.

Results and Observations

Multiple executions were carried out targeting various hyper parameters, eventually we settled on targeting two of them, warmup_epochs and batch size. Hyper parameters that deal with data augmentation were left untouched as it is already well established that enrichment of training data using these methods or increasing diversity of training data will lead to better results.

Warmup Epochs: This usually entails using a low learning rate for a specific amount of training steps (warmup steps). we then employ the "normal" learning rate or learning rate scheduler after the warmup steps. In other words, it is a phase in which we start with a learning rate that is substantially lower than the "initial" learning rate and steadily increase it over a few iterations or epochs until it matches the "initial" learning rate.

Results before updating parameters
(Table 4)

Table 4. YOLOv5 Execution Default Results

Class	Precision	Recall	mAP@0.5	mAP@.5:.95
All	0.852	0.925	0.901	0.59
Platelets	0.792	0.961	0.849	0.438
RBC	0.798	0.815	0.887	0.612
WBC	0.964	1	0.968	0.721

Warmup Epochs = 3, Batch size = 16, and Epochs = 100

Results after updating parameters
(Table 5)

Table 5. YOLOv5 Execution Results After Updating Parameters

Class	Precision	Recall	mAP@0.5	mAP@.5:.95
All	0.856	0.904	0.917	0.614
Platelets	0.802	0.906	0.887	0.477
RBC	0.801	0.807	0.888	0.619
WBC	0.967	1	0.975	0.747
Warmup Epochs = 5, Batch size = 15, and Epochs = 100				

Difference
(Table 6)

Table 6. Difference in Results

Class	Precision	Recall	mAP@0.5	mAP@.5:.95
All	0.004	-0.021	**0.016**	**0.024**
Platelets	0.01	-0.055	**0.038**	**0.039**
RBC	0.003	-0.008	**0.001**	**0.007**
WBC	0.003	0	**0.007**	**0.026**

General understanding and Information regarding models and how to conduct tests with YOLOv4 and YOLOv5 has been accumulated from [7–34].

Conclusion

Improvement in mAP0.5:0.95 is observed for the YOLO algorithm after making minor alteration to the default values of warmup_epochs and batch size. Though the improvement is not that high, nevertheless it is still an improvement when using the same data set.

All tests were carried out using Tesla P100 GPU on Google Collab platform. As stated earlier to not much independent research work have been done and published regarding performance improvement of YOLO by tweaking algorithm's original hyperparameter. This study is a proof of concept that this line of thought to improve algorithm's performance is something that calls for a deeper study.

Future Work

The results obtained after a performing a very primitive test after small tweaks in algorithm's hyper parameters look very promising. However, though we observed an improvement in the mAP of the YOLO algorithm, there was also a slight increase in detection time was when multiple tests were executed with different values of the two targeted parameters.

A deep dive into the concept of hyperparameter finetuning can still yield significant improvement in performance of algorithm. Detailed comments on the merits and demerits of the approach will only be possible once we explore this line of thought to its fullest extent.

References

1. Zhao, Z.-Q., et al.: Object detection with deep learning: a review. IEEE Transactions on Neural Networks and Learning Systems **30**(11), 3212–3232 (November 2019)
2. Sharma, V.: Face Mask Detection using YOLOv5 for COVID-19 (2020)
3. Joseph Redmon, A.F.: YOLO9000: Better, Faster, Stronger (2016). [Online]. Available: https://pjreddie.com/media/files/papers/YOLO9000.pdf. Accessed 30 May 2022
4. Brownlee, J.: A Gentle Introduction to Object Recognition With Deep Learning. [Online]. Accessed 31 March 2022
5. Lin, W.-Y., Hsu, W.-Y.: Ratio-and-Scale-Aware YOLO for Pedestrian Detection. IEEE Transactions on Image Processing **30**, 934–947 (November 2020)
6. Redmon, J., et al.: You Only Look Once: Unified, Real-Time Object Detection. In: IEEE Conference on Computer Vision and Pattern Recognition (2016)
7. "Test Automation," [Online]. Available: https://www.ibm.com/topics/automation. Accessed 31 March 2022
8. [Online]. Available: https://github.com/AlexeyAB/darknet/issues/5920#issuecomment-642 812152. Accessed 31 March 2020
9. "Negative opinion of Deep Learning developers about Yolov5-Ultralytics," [Online]. Available: https://www.reddit.com/r/MachineLearning/comments/h0ddia/news_yolov5_is_here_s tateoftheart_object/. Accessed 31 March 2020
10. "YOLO5 is Here," [Online]. Available: https://blog.roboflow.com/yolov5-is-here. Accessed 31 March 2022
11. [Online]. Available: https://www.taiwannews.com.tw/en/news/3957400. Accessed 31 March 2022
12. [Online]. Available: https://github.com/amzn/distance-assistant. Accessed 31 March 2022
13. [Online]. Available: https://github.com/BMW-InnovationLab. Accessed 31 March 2022
14. [Online]. Available: https://www.forbes.com/sites/janakirammsv/2020/01/19/apple-acq uires-xnorai-to-bolster-ai-at-the-edge/#20a12e943975. Accessed 31 March 2022
15. [Online]. Available: https://pjreddie.com/darknet/yolo/. Accessed 31 March 2022
16. Redmon, J., et al.: YOLOv3: An Incremental Improvement. University of Washington, [Online]. Available: https://pjreddie.com/media/files/papers/YOLOv3.pdf. Accessed 31 March 2022
17. Redmon, J., et al.: YOLOv4: Optimal Speed and Accuracy of Object Detection. University of Washington, [Online]. Available: https://arxiv.org/pdf/2004.10934.pdf. Accessed 31 March 2022
18. "Scaled-YOLOv4," [Online]. Available: https://arxiv.org/abs/2011.08036. Accessed 31 March 2022
19. Jin, Z., et al.: DWCA-YOLOv5: An Improve Single Shot Detector for Safety. Journal of Sensors (2021)
20. Ali, S.M.: Comparative Analysis of YOLOv3, YOLOv4 and YOLOv5 for Sign Language Detection, vol. 7, no. 4 (2021)
21. Li, S., et al.: YOLO-FIRI: Improved YOLOv5 for Infrared. IEEE Access **9**, 141861–141875 (2021)

22. Jie, X., et al.: Improved YOLOv5 Network Method for Remote Sensing Image Based Ground Objects Recognition (2022)
23. Tan, S., et al.: Improved YOLOv5 Network Model and Application in Safety Helmet Detection. In: IEEE International Conference on Intelligence and Safety for Robotics (ISR) (2021)
24. Abhinu, C.G., et al.: Multiple Object Tracking using Deep Learning with YOLO V5. Int. J. Eng. Res. Technol. (IJERT) **9**(Special Issue) (2021)
25. Kim, J.-H., et al.: Object Detection and Classification Based on YOLO-V5 with Improved Maritime Dataset. Journal of Marine Science and Engineering **10** (2022)
26. Jeong, D.: Road Damage Detection Using YOLO with Smartphone Images. In: IEEE International Conference on Big Data (Big Data) (2022)
27. Zhou, F., et al.: Safety Helmet Detection Based on YOLOv5. In: IEEE International Conference on Power Electronics, Computer Applications (ICPECA)
28. Beera, S., et al.: The Yolo V5 Based Smart Cellphone Detector. Natural Volatiles and& Essential Oils **8**(6), 3437–3455 (2021)
29. He, K., Girshick, R., Sun, J., Ren, S.: Faster R-CNN: Towards Real-Time Object Detection with Region Proposal Networks. Computer Vision and Pattern Recognition (4 June 2015)
30. Girshick, R., Lin, T.-Y., et al.: Feature Pyramid Networks for Object Detection. Facebook AI Research (FAIR) (19 April 2017)
31. Jacob Solawetz, J.N.S.S.: How to Train YOLOv4 on a Custom Dataset. Roboflow, 21 May 2020. [Online]. Available: https://blog.roboflow.com/training-yolov4-on-a-custom-dataset/. Accessed 30 May 2022
32. Jacob Solawetz, J.N.: How to Train YOLOv5 On a Custom Dataset. roboflow, 10 June 2020. [Online]. Available: https://blog.roboflow.com/how-to-train-yolov5-on-a-custom-dataset/. Accessed 30 May 2022
33. Roboflow: YOLOv4-Darknet-Roboflow. [Online]. Available: https://colab.research.google.com/drive/1mzL6WyY9BRx4xX476eQdhKDnd_eixBlG. Accessed 30 May 2022
34. Roboflow: Roboflow-Train-YOLOv5. Roboflow, [Online]. Available: https://colab.research.google.com/drive/1gDZ2xcTOgR39tGGs-EZ6i3RTs16wmzZQ. Accessed 30 May 2022

An Algorithm for Solving Two Variable Linear Diophantine Equations

Mayank Deora$^{(\boxtimes)}$

Indian Statistical Institute Kolkata, Kolkata, India
mbdeora_r@isical.ac.in

Abstract. In cryptography algorithms like RSA, Elliptic Curve Cryptography, calculation of modulo multiplicative inverse of an integer is required. Calculation of modulo multiplicative inverse problem is a specialized form of solving Diophantine equation. Existing Extended Euclid's Algorithm solves the Diophantine equation using top down and bottom up method. It firstly, calculates the intermediate values using top down approach then calculates the coefficient values using bottom up approach. The proposed algorithm calculates the intermediate values and coefficient values using only top-down approach. Hence in some cases proposed algorithm takes less computation time to compute the coefficient values of Diophantine equation. The proposed algorithm has been analysed rigorously to obtain a general expression for its efficiency in average case. The results obtained after implementation of the proposed algorithm verify that in some cases proposed algorithm takes lesser time than the existing algorithm.

Keywords: Greatest Common Divisor · Modulo Multiplicative Inverse · Diophantine Equations · Extended Euclid's Algorithm

1 Introduction

Diophantine equation [5] in two variables, have the following general form,

$$ax + by = c, \tag{1}$$

where x and y are integer variables, a, b and c are integer constants. If there exists atleast one solution of a diophantine equation, then it has infinitely many solutions. All the solutions of Eq. 1 can be found with formula, $X = x + kb/GCD(a, b)$ and $Y = y - ka/GCD(a, b)$, where X and Y are the solutions and $k \in \mathbb{Z}$. $GCD(a, b)$ denotes the greatest common divisor of a and b. Note that,

Integer solutions are possible if and only if either $c = 0$ or if c is a multiple of $GCD(a, b)$ due to Bezout's Lemma [6, p. 10].

In diophantine equation $ax - by = c$, if $c = 1$ i.e. if a and b are co-prime integers, then x in solution (x, y) is the modulo multiplicative inverse of a with

M. Singh et al. (Eds.): ICACDS 2023, CCIS 1848, pp. 35–48, 2023.
https://doi.org/10.1007/978-3-031-37940-6_4

respect to b. The rest of this section discusses detailed applications of solving diophantine equation in cryptography as well as in other areas.

In cryptography algorithms, encryption algorithms transform or encrypt plain text to cipher text, which is converted back to the plain text using the decryption algorithm. Encryption and Decryption algorithms use a key to encrypt or decrypt text. Asymmetric key cryptography algorithms use different keys for encryption and decryption called the public key and private key. RSA [8] and Elliptic Curve Cryptography [9] are well-known examples of asymmetric key cryptography algorithms.

In RSA algorithm [8] and Elliptic Curve Cryptography [9], for public key generation, modulo multiplicative inverse of an integer is computed. In [4], solution of Diophantine equations have been applied in tuning the extent of Look Ahead in the process of Hiding Decision Tree Rules. The Extended Euclid's algorithm was adapted for computing the multiplicative inverse of a binary polynomial over $GF(2^m)$ by Berlekamp in [1].

The rest of this paper is organized as follows. Section 2 gives a detailed discussion of Extended Euclid's Algorithm and mentions about other methods for solving the two variable diophantine equations. Section 3 describes the proposed algorithm along with its correctness proof and time complexity. Section 4 and 5 discuss implementation and results of the proposed algorithm respectively. Section 6 concludes the paper along with giving future directions of research from this problem.

2   Related Work

Extended euclid's algorithm [3] is a recursive algorithm. In EXTENDED EUCLID(a, b), greatest common divisor, $GCD(a, b)$ of a and b is computed using euclid's gcd algorithm. In each recursion, two integers, (a, b) are passed as input and that particular recursion outputs x and y, such that $ax + by = GCD(a, b)$. Hence it can be considered as an extension of euclid's gcd algorithm. Extended Euclid's Algorithm is as follows.

Algorithm 1 EXTENDED EUCLID(a, b)

    Input: a, b $(a > b)$ **Output:** x, y such that $ax + by = \text{GCD}(a, b)$
    a, b, d, x, y are integers.
1: **if** $b == 0$ **then**
2:    **return** $(a, 1, 0)$
3: **else**
4:    $(d', x', y') = $ EXTENDED EUCLID$(b, a \bmod b)$
5:    $d = d'$
6:    $x = y'$
7:    $y = x' - \lfloor a/by' \rfloor$
8:    **return** (d, x, y)

Note that, the above algorithm makes same number of recursive calls as the euclid's gcd algorithm for computing $GCD(a, b)$. Appendix of this paper contains an execution of extended euclid's algorithm on an example equation.

For solving Diophantine equation, $ax + by = c_1$, firstly $ax' + by' = GCD(a, b)$ is solved for x' and y', using extended euclid's algorithm. Then, $ax + by = c_1$ is solved as, $x = x'.c_1/GCD(a, b)$ and $y = y'.c_1/GCD(a, b)$.

In [7], an algorithm for modular division $\frac{y}{x}$ mod m has been proposed, which is equivalent to computing x^{-1} mod m, on putting $y = 1$. This algorithm takes $2 \log x - 1$ time in worst case.

3 Proposed Work

An idea of the proposed algorithm is given here followed by Sect. 3.1, which presents the proposed algorithm. Consider the problem of solving diophantine equation, (Eq. 1).

$$\frac{c - ax}{b} = y \tag{2}$$

Suppose that k_1 be an integer such that

$$a = k_1 b + (a \bmod b) \tag{3}$$

(k_1 is a positive integer). Dividing Eq. (3) by b, the resultant equation is:

$$\frac{a}{b} = k_1 + \frac{(a \bmod b)}{b} \tag{4}$$

Since $a \bmod b < b$, so $a \bmod b$ is not divisible by b. Value of $\frac{a}{b}$ is substituted in Eq. 2:

$$\frac{c}{b} - \left(k_1 + \frac{a \bmod b}{b}\right) x = y \tag{5}$$

i.e.

$$\frac{1}{b}(c - (a \bmod b)x) - k_1 x = y \tag{6}$$

Suppose there is an integer, x' such that

$$c - (a \bmod b)x = x'b \tag{7}$$

Thus Eq. (6) can be rewritten as follows:

$$x' - k_1 x = y$$

This gives us the integer solution, x and y for the original diophantine equation Eq. (1). Equation (7), can be rewritten as follows:

$$(a \bmod b)x + bx' = c \tag{8}$$

In Eq. 8, x and x' are the only variables. Computing their values gives us value of y also. So the problem becomes recursive with parameters of next recursion as $(b, a \bmod b, c)$. The previous parameters were (a, b, c).

3.1 Algorithm

In the proposed algorithm, Algorithm 2, inputs are three integers a, b and c. In line 1 of Algorithm 2, $f(a, b, c)$ is assigned to y. Then $x = (c - by)/a$ gives the solution for $ax + by = c$. Line 4 and line 7 are the base cases of the recursion. If $b(\neq 0)$ is not a divisor of $c - a$, then line 10 makes a recursive function call to $f(b, a \bmod b, c)$. Note that g represents the greatest common divisor of a and b (a, b are from Eq. 1).

Algorithm 2 Algorithm to solve 2-variable Linear Diophantine equation

---

    Input: a, b, c ($a > b, b \neq 0$)
    Output: x, y such that $ax + by = c$
    a, b, c, x, y are integers.
1: $y \leftarrow f(a, b, c)$ ▷ General term for diophantine equation $Y = y + ka/\mathrm{GCD}(a, b)$,
    where k is any integer
2: $x \leftarrow (c - by)/a$ ▷ General term for diophantine equation $X = x + kb/\mathrm{GCD}(a, b)$,
    where k is any integer
3: **procedure** $f(a, b, c)$
4:     **if** b==0 **then**
5:         PRINT ("c is not a multiple of g")
6:         **exit**
7:     **else if** $(c - a) \bmod b == 0$ **then**
8:         **return** $(c - a)/b$
9:     **else**
10:         **return** $(c - f(b, a \bmod b, c) \times a)/b$

The diophantine equation, $1759x + 550y = 5$ is solved using proposed algorithm. Table 1 depicts the computation of x and y.

Table 1. $1759x + 550y = 5$, Solution as $x = -5$ and $y = 16$ using proposed algorithm

a	b	y
550	109	-5
1759	550	16

The y values in table are return values, of the function $f(a, b, c)$. In $f(1759, 550, 5)$, no base condition is satisfied, so it subsequently calls $f(550, 109, 5)$. In $f(550, 109, 5)$ one of the two base conditions is satisfied, because $5 - 550 = -545$ is divisible by 109 and -5 is returned (y value in the first row of Table 1). Thus $f(1759, 550, 5)$ returns $(5 - (-5) \times 1759)/550 = 16$ (y value in the second row of Table 1). An alternative illustration of the solution is as follows:

$$y = \frac{5 - \left(\frac{5-550}{109}\right) \times 1759}{550}$$

3.2   Correctness Proof

Theorem 1. *Algorithm 2 computes integers x and y such that $ax + by = c$, for inputs a, b and c, if and only if c is a multiple of $GCD(a, b)$.*

This theorem has been established through Lemma 1 and Lemma 2, in the remainder of this section.

Lemma 1. *If $f(a, b, c)$ returns "c is not a multiple of g" then c is not a multiple of g.*

Proof. In function call $f(a', b', c)$, if $b' = 0$, then it returns "c is not a multiple of g". Inputs to the proposed algorithm are a, b and c, where $b \neq 0$. If $b' = 0$, in $f(a', b', c)$ then it is not the initial recursive call.

If b' is 0 in $f(a', b', c)$, and if $f(a'', b'', c)$ is previous recursive call, then $b' = a'' \bmod b'' = 0$ i.e. $a'' = kb''$ for some integer k. This implies that $g = GCD(a'', b'') = b''$.

Assume that c is a multiple of g, then in the previous call to f, $(c - a'')$ would be a multiple of b'', so $f(a'', b'', c)$ would return $(c - a'')/b''$, but it returns $(c - f(a', b', c) \times a'')/b''$. So this is a contradiction, therefore c is not a multiple of g.

The inputs to Algorithm 2, a and b are strictly decreasing with each recursive call. If Algorithm 2 does not return from line 8, then b will become zero in the last recursive call because b is an integer and it is strictly decreasing. Hence, if Algorithm 2 does not return from line 8, then it must exit from line 6.

This proves that the Algorithm 2 must terminate. Thus, if c is a multiple of g, then Algorithm 2 must return some value of x, y, otherwise c is not a multiple of g (Lemma 1). The following lemma proves that the values x, y returned by Algorithm 2 are the solution to the diophantine equation.

Lemma 2. *If c is a multiple of g, then algorithm returns integers x and y, such that $ax + by = c$.*

Proof. Assume that the last recursive call to f in Algorithm 2 returns some value, say y_1 from line 8. y_1 must be an integer due to the condition in line 7. Now we prove using mathematical induction that in line 1, value assigned to y is an integer.

Assume that for some input a'', b'', c, $f(a'', b'', c)$ calls $f(a', b', c)$, and $f(a', b', c)$ returns an integer. $f(a', b', c)$ is same as $f(b'', a'' \bmod b'', c)$. $f(a'', b'', c)$ returns the following value:

$$(c - f(b'', a'' \bmod b'', c) \times a'')/b''$$

$$= \frac{c - \frac{(c - k_1 b'')}{a'' \bmod b''} \times a''}{b''}$$

Due to induction hypothesis, it is clear that k_1 is an integer. Without loss of generality, we can assume that $a'' = kb'' + a'' \bmod b''$, where $k \geq 0$ is an integer.

$$= \frac{c - \frac{(c-k_1 b'')}{a'' \bmod b''} \times (kb'' + a'' \bmod b'')}{b''}$$

$$= \frac{c - a'' \bmod b'' \times \frac{(c-k_1 b'')}{a'' \bmod b''} - kb'' \times \frac{(c-k_1 b'')}{a'' \bmod b''}}{b''}$$

We know that $f(a', b', c) = \frac{(c-k_1 b'')}{a'' \bmod b''}$ is an integer. Let $x' = \frac{(c-k_1 b'')}{a'' \bmod b''}$, then it simplifies the above expression as follows:

$$= \frac{c - a'' \bmod b'' \times \frac{(c-k_1 b'')}{a'' \bmod b''}}{b''} - kx'$$

without replacing the remaining occurrence of $\frac{(c-k_1 b'')}{a'' \bmod b''}$, with x', we can simplify the above expression further to the following:

$$= k_1 - kx'$$

i.e.

$$f(a'', b'', c) = k_1 - kx'$$

Thus, $f(a'', b'', c)$ is an integer and the x and y values returned by algorithm are also integers. This completes the proof.

3.3   Time Complexity

We analyze proposed algorithm relative to extended euclid's algorithm [3]. Extended euclid's algorithm makes $O(\log_\phi b)$ recursive calls (ϕ is an irrational number known as golden ratio [3]) and the time complexity of extended euclid's algorithm is directly proportional to number of recursive calls. Like extended euclid's algorithm, in the proposed algorithm also, the parameters in first recursive call are a, b, then the parameters in next recursive call are $b, a \bmod b$. Base case of recursion in worst case of proposed algorithm and the only base case of extended euclid's algorithm are same, ($b = 0$). So, maximum number of recursive calls required in proposed algorithm is same as number of recursive calls required in extended euclid's algorithm. Since the starting parameters of both algorithms is same and they execute further with same parameters, we can compare the number of recursive calls in both algorithms. Both algorithms differ in their stopping conditions.

Inside each recursive call in extended euclid's algorithm (Algorithm 1), one comparison takes place at line 1 ($b == 0$). Two division operations occur (at line 4 and line 7). One subtraction operation occurs at line 7. Let time required for comparison, division and subtraction is c_1, c_2 and c_3 respectively. If the number of recursive calls, required is k then $T_{EE} = O(k(c_1 + 2c_2 + c_3))$ is the time complexity of Algorithm 1. Here c_1, c_2 and c_3 are constants representing fixed time durations.

Inside each recursive call in proposed algorithm (Algorithm 2), in worst case, two comparisons take place (at lines 4 and 7). in worst case, three division operations take place (one at line 7, two divisions at line 10). In worst case, two subtraction operations occur (at line 7 and line 10). So if the number of recursive calls is $k - A$, then $T_P = O((k - A)(2c_1 + 3c_2 + c_3))$ is the time complexity of Algorithm 2. k is the maximum number of recursive calls in proposed algorithm and $0 \leq A \leq k - 1$. The time complexities of the two algorithms are compared as follows to get value of A for which proposed algorithm will perform better i.e.

$$T_P < T_{EE}$$

$$\implies (k - A)(2c_1 + 3c_2 + c_3) < k(c_1 + 2c_2 + c_3)$$

$$\implies k\frac{c_1 + c_2}{2c_1 + 3c_2 + c_3} < A \tag{9}$$

Since time requirements of comparison and subtraction are same $c_1 = c_3$. If n is the number of bits in the operands, then $c_1 = c_3 = O(n)$. Time requirement for division operation is $c_2 = O(n^2)$. Hence from inequality 9

$$k\frac{n + n^2}{3n + 3n^2} < A$$

$$\frac{k}{3} < A \tag{10}$$

So the difference (A) between the number of calls in Extended Euclid's algorithm(k) and that in proposed algorithm must be greater than one third of k, for proposed algorithm to be more efficient.

Let a' and b' be the values of a and b in the last recursive call of proposed algorithm, Algorithm 2. Let Algorithm 2 is making T recursive calls. Algorithm 2 stops making recursive calls in one of the following 2 cases-

1. $(c - a') \bmod b' = 0 \implies c - a' = kb'$: If $(c < b')$, then EEA(a', b') will call EEA(b', c). If $(c < b')$, Extended Euclid's Algorithm (EEA) will make $T + O(\log c)$ recursive calls therefore $T = O(\log b) - O(\log c)$. So, according to inequality 10, if $\log c$ is greater than $\frac{\log b}{3}$, then the proposed algorithm takes lesser time as compared to time taken by Extended Euclid's algorithm. If $c = 1$ (applicable in finding multiplicative inverse), then $T = O(\log b)$. If $(c > b')$ and suppose that $c = k_1 b' + k_2$, where $k_2 < b$, then from here EEA(b', k_2) will be called making $O(\log k_2)$ recursive calls to reach to solution. Thus $T = O(\log b) - O(\log k_2)$. According to inequality 10, if $\log k_2$ is greater than $\frac{\log b}{3}$, then the proposed algorithm may take lesser time than time taken by Extended Euclid's algorithm. If $(c = b')$, then EEA$(b', 0)$ will be called, thus $T = O(\log b) - 1 = O(\log b)$.

2. $b' = 0$: T is equal to the number of recursive calls made by EEA. This is because EEA$(a', 0)$ will return a'.

Thus, proposed algorithm is more efficient only when c is a multiple of $GCD(a, b)$ i.e. in base case 2. When a and b are coprime to each other or when both of them are prime, then any value of c would be multiple of $GCD(a, b)$.

3.4 Average Case Analysis

$f(1759, 550, 5)$ takes 2 recursive calls. There are more values of c for which f takes 2 recursive calls. Those values are in the general form $5 + 109k_1$ for any integer k_1. Likewise equations with $109 + 550k_2 = c$ take one recursive call. This is because, $1759 \mod 550 = 109, 550 \mod 109 = 5, 109 \mod 5 = 4, 5 \mod 4 = 1$. So if $c = 5 + 109k_1$, then in the second recursive call $(c - a) = 5 + 109k_1 - 550$, which is divisible by 109.

Average case analysis is based on above observation. If $b = O(\phi^k)$ then $a \mod b = O(\phi^{k-1})$ and so on [3]. Assume that N is any positive integer. We analyse the algorithm for all the values of c between 1 to N. For $O\left(N/\phi^k\right)$ values there will be only 1 recursive call. This is because that in the proposed algorithm, the first recursive call checks whether $(c - a) \mod b = 0$ i.e. $(c - a) \mod O(\phi^k) = 0$. Let

$$c = a \mod b + k_2 b, \tag{11}$$

then putting this value of c and $x = 1$ in Eq. 5 gives:

$$k_2 - k_1 = y, \tag{12}$$

producing the solution in only one recursive call. Number of integers between 1 to N, which are in form of Eq. 11 is $O\left(\frac{N}{\phi^k}\right)$ because

$$a \mod O(\phi^k) + k_2 O(\phi^k) = N \implies k_2 = \frac{N - a \mod O(\phi^k)}{O(\phi^k)} \approx O\left(\frac{N}{\phi^k}\right)$$

This is equal to the number of multiples of $O(\phi^k)$ between 1 to N. For solving Eq. 1, let

$$S_i = \{c | 1 \leq c \leq N, \text{ proposed algorithm takes i recursive calls}\}$$

An integer $c_1 \in S_i$ if and only if $c_1 \notin S_{i-1}$ because if equation is solvable in $i - 1$ recursive calls then algorithm will not enter into i^{th} recursion. $|S_1| = O\left(\frac{N}{\phi^k}\right)$. Let

$$A_i = \{c | c = O(\phi^{k-i}) + k_i O(\phi^{k-(i-1)})\}$$

here k_i is any constant integer.

$$|S_1| = |A_1|, |S_2| = |A_2| - |A_1 \cap A_2| \implies S_2 = A_2 \cap \overline{A_1}$$

The number of integers which are in both A_2 and A_1 are $O\left(\frac{N}{\phi^{k-1}\phi^k}\right)$ i.e. the number of integers between 1 to N, which are multiples of both $O(\phi^k)$ and

$O(\phi^{k-1})$. To get this we have divided N by the maximum possible LCM of $O(\phi^k)$ and $O(\phi^{k-1})$.

$|S_2| = O\left(\frac{N}{\phi^{k-1}}\right) - O\left(\frac{N}{\phi^{k-1}\phi^k}\right)$. Similarly

$$|S_3| = O\left(\frac{N}{\phi^{k-2}}\right) - O\left(\frac{N}{\phi^{k-2}\phi^k}\right) - O\left(\frac{N}{\phi^{k-2}\phi^{k-1}}\right) + O\left(\frac{N}{\phi^{k-2}\phi^k\phi^{k-1}}\right)$$

S_3 set represents the integers which are multiple of only $O\left(\phi^{k-2}\right)$ and not a multiple of $O\left(\phi^{k-1}\right)$ or/and $O\left(\phi^k\right)$. Here principle of inclusion-exclusion from set theory [2] has been applied. Number of multiples which were subtracted twice are later added once again. Generalizing thus

$$|S_i| = |A_i| + \sum_{\phi \neq m \subseteq \{1,....,i-1\}} (-1)^{|m|} |A_i \bigcap_{j \in m} A_j| \tag{13}$$

Equation 13 has been obtained by applying inclusion-exclusion principle from set theory. In the summation term m is a subset of $1, ..., i-1$. For each subset m, there is a summation corresponding to a set cardinality of $|m|+1$ (set intersection with A_i also). In Eq. 13, in summation, intersection is between A_i and A_j for all j in selected m. $|S_i|$ can be rewritten as follows:

$$|S_i| = O\left(\frac{N}{\phi^{k-(i-1)}}\right) + \sum_{\phi \neq m \subseteq \{0,....,i-2\}} (-1)^{|m|} O\left(\frac{N}{\phi^{k-(i-1)}\phi^{\sum_{j \in m} k-j}}\right) \tag{14}$$

We can write the average case complexity of proposed algorithm as follows:

$$T_{avg} = \frac{1|S_1|}{N} + \frac{2|S_2|}{N} + ... + \frac{k|S_k|}{N}$$

In $|S_i|$, the later terms except the first one i.e. $O\left(\frac{N}{\phi^{k-(i-1)}}\right)$, can be neglected because the first term is asymptotically larger than other terms. Values of $|S_i|$ are substituted and neglecting terms other than the first one, gives:

$$T_{avg} \leq \sum_{i=1}^{k} \frac{i}{O\left(\phi^{k-(i-1)}\right)}$$

i.e.

$$T_{avg} \leq \theta(k) \implies T_{avg} = O(\log_\phi b) = O(\log b)$$

4   Evaluation

For implementation of proposed algorithm python programming language has been used. For comparison with proposed algorithm, Extended Euclid's Algorithm has also been implemented in python. To show that the number of recursive calls in proposed algorithm doesn't depend only on a and b, but it depends on c also, a different implementation has been used.

For comparison, 99 equations are formed using fixed prime value of $a = 5915587277$, and b is a random value varying between 1000 and 100000. b varies between 1000 and 100000 with step size of 1000, i.e. initially value of b is 1000, then 1000 is added for getting new value of b, this process is repeated untill b becomes greater than 100000. c is also taken as a prime number 4127. Then both algorithms are run to count the number of recursive calls. a is taken as prime value greater than value of b, to make sure that the equation is always solvable. c is assigned 4127, just for experimental purpose, it is not any special value. In another implementation, values of a and b are fixed with values 43207 and 17922 respectively. c takes on values from 1 to 17922, both inclusive.

5   Results

Implementation of comparison of the two algorithms resulted in the graph, Fig. 1, depicting number of recursive calls in both algorithms for the same equations. On horizontal axis there is equation number (equation no. is representing values of b) and on vertical axis there are number of recursive calls taken for solving the equation. Cross denotes number of recursive calls in extended euclid's algorithm and dot denotes number of recursive calls in proposed algorithm. In graph it is observed that the number of recursive calls is increasing with the value of b. This increment is in both the algorithms, but there is an envelope of difference between the plot for these two algorithms. This difference gets significantly larger sometimes i.e. the number of recursive calls in proposed algorithm is significantly smaller than that in extended euclid's algorithm. Sometimes the difference becomes smaller, but always the number of recursive calls in proposed algorithm is lesser than that in extended euclid's algorithm. The number of recursive calls in proposed algorithm may vary between 1 and the highest number of recursive calls i.e. one less than the number of recursive calls in Extended Euclid's algorithm. This variation of number of recursive calls has been discussed in average case analysis Section, Sect. 3.4.

Result of implementation for $a = 43207$, $b = 17922$ and $c \in [1, 17922]$, (closed interval from 1 to 17922) is depicted in Fig. 2. In this graph, the horizontal axis represents number of recursive calls and vertical axis represents values of c. Figure 2 shows that for solving diophantine equation with different values of c, with fixed a and b , the proposed algorithm takes different number of recursive calls. According to Fig. 2, Extended Euclid's algorithm will solve any of the diophantine equations in 11 recursive calls but proposed algorithm will make lesser number of recursive calls for some values of c. Extended Euclid's algorithm solves $ax + by = GCD(a, b)$; then to solve $ax' + by' = c'GCD(a, b)$, x and y are multiplied by c'. Thus Extended Euclid's algorithm always takes the same number of recursive calls for equation $ax + by = c$, for fixed a and b but for any c. In proposed algorithm, the number of recursive calls depend not only on a and b, but on c also.

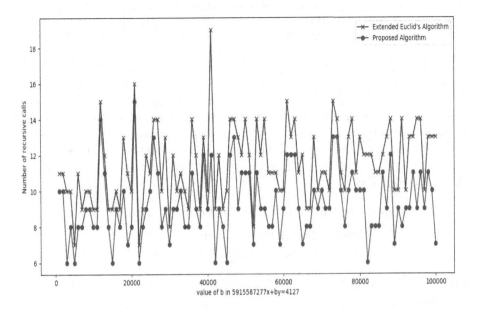

Fig. 1. Result of comparison of Extended Euclid's Algorithm and Proposed Algorithm. Cross denotes number of recursive calls in Extended Euclid's Algorithm and dot denotes number of recursive calls in Proposed algorithm.

6   Conclusion

For solving 2-variable linear diophantine equation, Extended Euclid's algorithm is one of the existing algorithms. Extended Euclid's algorithm is prominently used in public key cryptography algorithms. Section 3.4 shows that there exist remarkable amount of inputs with a, b as fixed and c varying, on which the proposed algorithm takes lesser number of recursive calls than that in extended euclid's algorithm. The number of arithmetic operations in each recursive call is more than that in extended euclid's algorithm, but proposed algorithm is more efficient when inequality 10 is satisfied.

For finding modulo mulitplicative inverse, the proposed algorithm makes only one or two recursive calls lesser than Extended Euclid's Algorithm as shown in Sect. 3.3. For reducing the number recursive calls, we can follow a different strategy as follows. Randomly choose two integers x_1 and y_1 followed by computing $ax_1 - by_1$. Let $ax_1 - by_1 = c_1$. Then use Algorithm 2 to solve $ax - by = c_1 + 1$. Then compute $x - x_1$ and $y - y_1$ for the modulo multiplicative inverse of a with respect to b. By applying this reduction, Algorithm 2 takes n number of recursive calls, where $1 \leq n \leq k$, where k is the number of recursive calls taken by extended euclid's algorithm. Section 3.4 discusses this behavior of the algorithm. So, by taking advantage of this behavior, we can use the proposed algorithm in public key cryptography algorithm for efficiently computing public key.

Worst case asymptotic time complexity of the proposed algorithm is same as that of the existing algorithms. Average case analysis shows that if the extended euclid's algorithm solves the equation in k recursive calls, then the proposed algorithm may solve that equation within 1 to k recursive calls. For average case time complexity, more rigorous analysis may be possible to come to a better efficiency result.

Implementation results show that the proposed algorithm does not make more number of recursive calls than that in extended euclid's algorithm. A different implementation result shows that the number of recursive calls and thus time complexity not only depends upon a and b but it depends on c also.

A   Appendix

A.1   Example Used for Extended Euclid's Algorithm

Consider the following diophantine equation:

$$1759x + 550y = 1 \tag{15}$$

Equation (15) has been solved in Table 2 and Table 3 using Extended Euclid's algorithm.

Table 2. Top down procedure to compute intermediate values of a and b for corresponding recursion

a	b
1759	550
550	109
109	5
5	4
4	1
1	0

top-down

Table 3. Bottom-up procedure to compute coefficient values x and y for respective values of a and b, such that $ax + by = d$

$\lfloor a/b \rfloor$	x	y	d
3	-111	355	1
5	22	-111	1
21	-1	22	1
1	1	-1	1
4	0	1	1
–	1	0	1

bottom-up

A.2   Evaluation

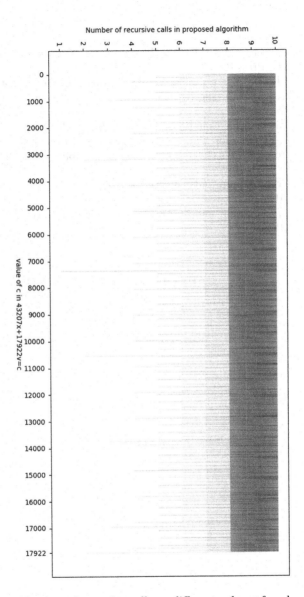

Fig. 2. Different Number of recursive calls on different values of c where c varies from 1 to 17922, with fixed values of a and b

References

1. Berlekamp, E.: Algebraic Coding Theory. World Scientific, Singapore (1968)
2. Comtet, L.: Advanced Combinatorics: The Art of Finite and Infinite Expansions, pp. 176–178. Springer, Heidelberg (2012)
3. Cormen, T.H., Leiserson, C.E., Rivest, R.L., Stein, C.: Introduction to Algorithms, pp. 935–938. MIT Press (2009)
4. Feretzakis, G., Kalles, D., Verykios, V.S.: On using linear diophantine equations to tune the extent of look ahead while hiding decision tree rules. arXiv preprint arXiv:1710.07214 (2017)
5. Mordell, L.J.: Diophantine Equations. Academic Press, Cambridge (1969)
6. Seroul, R.: Programming for Mathematicians. Springer, Heidelberg (2000)
7. Shantz, S.C.: From Euclid's GCD to montgomery multiplication to the great divide (2001)
8. Stallings, W.: Cryptography and Network Security, 7/E, pp. 294–297. Pearson Education India (2017)
9. Stallings, W.: Cryptography and Network Security, 7/E, pp. 321–333. Pearson Education India (2017)

Deep Watcher: A Surveillance System Using Deep Learning for the COVID-19 Pandemic

Rohil Kulshreshtha[1] and J. Jayapradha[2]([⊠])

[1] Department of Computer Science and Engineering, SRM Institute of Science and Technology, Kattankulathur, Tamil Nadu 603203, India
[2] Department of Computing Technologies, SRM Institute of Science and Technology, Kattankulathur, Tamil Nadu 603203, India
jayapraj@srmist.edu.in

Abstract. Several solutions have been proposed to combat the COVID-19 pandemic. In the absence or limited availability of medical resources, World Health Organization has recommended several safety measures. These measures were proposed to control the infection rate and keep current medical resources from depleting. Non-pharmaceutical intervention strategies such as wearing a mask and maintaining social distance are still being employed to combat the COVID-19 sickness. To contribute to this idea of human safety, our work aims to develop a model for detecting non-mask faces quickly and people who are not maintaining social distance in public. The proposed model uses computer vision and artificial intelligence to detect masks and distance between people. Also, a proposition has been made to increase localization performance during detection using the bounding box transformation. The combination of face mask detection and the social distance detection paradigm suggested in this paper is ideal for video surveillance equipment.

Keywords: COVID-19 · Social Distance Detection · Face Mask Detection · Artificial Intelligence

1 Introduction

Distinguished in Wuhan, China, on 12th December 2019, SARS-CoV-2 is a virus that causes a sickness known as COVID-19. The World Health Organization (WHO) recommended using hygiene, social distancing (SD), and face masks to prevent or halt the spread of the COVID-19 virus [1–4]. The face mask detection (FMD) technique makes wearing a face mask mandatory for everyone. FMD is a technique for determining whether someone is wearing a mask. Face recognition studies started in 2001, with the development of handmade attributes and the application of standard machine learning techniques to train verified and capable detection and identification classifiers. But these methods have some significant drawbacks. These so-called drawbacks include feature complexity and low accuracy rates. Face detection approaches based on convolutional neural networks (CNN) have gained much traction recently to get a better performance output [5–7].

M. Singh et al. (Eds.): ICACDS 2023, CCIS 1848, pp. 49–59, 2023.
https://doi.org/10.1007/978-3-031-37940-6_5

Although many academics have worked hard to develop fast methods for face recognition, "detection of the face under the mask" and "detection of the mask over the face" are not interchangeable. According to the existing literature, only a small percentage of research focuses on detecting masks on the face [8–10]. Our study aims to create a technique that can accurately detect face masks in public places to prevent further COVID virus transmission and contribute to public health. But one of the limitations is that detecting face masks in public is difficult because the available sample data is very small, making it difficult to train the model. Therefore, transfer learning is used in such a scenario to transfer learned models for face detection that will help detect masks. Social distance (physical distance) measures cover a wide range of activities aiming to slow or halt the virus's spread by preserving physical space between people. Several study articles have been published to examine whether the "FMD" rule has been broken [11–15]. It significantly reduced the risk of disease transmission and helped to contain the outbreak. The YOLOv3 model has been used to identify people. A deep-sort approach has been used to enclose detected people in bounding boxes and track them. Drones have also been used to detect SD using the YOLOv3 algorithm. Object detectors have also been used to compare and detect people's photographs. Employees who break the SD guideline can receive real-time voice alerts.

As shown in the above cases, artificial intelligence (AI) applications played a critical role in preventing the widespread COVID-19 outbreak. A lot of study has been done to keep track of SD in public settings. Because a bird's eye view provides a better overall picture and can be used to calculate individual distance, this study aims to use deep learning to measure SD. The following sections comprise the organisation of the work discussed in the article. The pertinent research and conclusions are discussed in Sect. 2. The work's contribution is covered in Sect. 3, and the proposed model is covered in Sect. 4. Section 5 presents the algorithm of the proposed work. Section 6 assesses the proposed technique's experimental result analysis. Finally, Sect. 7 contains the conclusion of the work with some enhancements.

2   Related Works

2.1 Literature Survey

The researcher reviewed various studies about the " SD" guideline used in the fight against COVID-19 [16] to assist in the containment of the epidemic or reduce the danger of transmission. The YOLOv3 model was used in crowded areas to track the 2-m distance between individuals for automated detection of human pairs. However, the images taken for detection proved that eclipsed individuals were difficult to see. [17].

The study employed two datasets [18, 19], with individuals wearing and without masks, and employed CNNs using the Alex-Net architecture in various stages, including face identification, data pre-processing, and picture categorization. This research might be used in several settings, including shopping complexes, schools, grocery stores, and other public places, to detect those not following the guidelines and make it simpler for security officials to keep track of everyone who enters. According to the study, mask recognition might be employed in surveillance cameras to hunt for individuals who aren't wearing masks.

The study proposed a model that used OpenCV DNN to detect masks over faces. The suggested approach may be used with security cameras to identify people not wearing face masks, preventing COVID-19 transmission. The memory score predicted the capacity to find all positive samples, whereas the f1 score indicated the test's accuracy [20–22].

The system includes a pre-trained model [23, 24]. Faces in photos and video streams are detected, the area of interest is extracted and the facemask analyzer is used to classify faces or images in the streaming content as wearing a mask, wrongly wearing a mask, or not wearing a mask. The input is then categorized as having a mask, having an incorrectly worn mask, or not having a mask. The Facemasknet model was developed, identifying face-masked and non-face-masked photos with an accuracy of 98.6%.

AI-based approaches are used in object detection algorithms. YOLO is a probability-based technique that operates on items in grids or bounding boxes. The paper compares some of the many object identification algorithms in-depth, highlighting some of the key features of each [25].

Researchers' most essential study task is object identification, which has numerous applications. All applications, from surveillance to agriculture, require excellent object detection to achieve their goals [26]. The most significant task in image processing and computer vision is object detection. Because deep learning is the primary reason for the increased evolution of object detection, the study briefly discusses and highlights the importance of deep learning for object detection [27].

A new object recognition network was presented [28]. After features were amplified by two consideration maps and fed into the layer division component, two tasks were completed. As a result, two task attributes were used to encode the proposed techniques effectively. The attention network for object localization creates an attention map for object recognition and localization in the initial stage. Finally, the attention map for localization is created using the sigmoid activation function. The model is proposed for the suggested technique and YOLOv3 [29]. Some photos containing ground truth were taken, and a comparative graph was generated. Based on this graph, the suggested research surpasses YOLOv3 in contexts of border-box accuracy. When it comes to object identification, precision is crucial to detecting an object [30].

2.2 Inference from Literature Survey

From the above study, the conclusion was made that many object detection algorithms exist and each has its strengths and weaknesses. In addition to object detection algorithms, some mathematical calculations also need to be performed to calculate the distance between the detected objects. There are real-time-based applications of object detection as well but sticking to the other way at first was better because that is easier to implement and then once the framework is ready, a shift can be made to a real-time-based application. Despite the models' efficiency, it is feasible that the system may forecast incorrectly when analyzing real-world data because the datasets are skewed. Challenges faced by object detection algorithms can be universal or specific to the method being used.

3 Contribution

1. A model is suggested to determine whether someone is hiding their face or keeping their distance from others. To meet the demands of both kinds of detection services better, a model called deep watcher was proposed.
2. Deep Watcher creates an all-in-one surveillance system that takes care of face masks and SD.
3. Visual data representation techniques have also been incorporated and AI-assisted algorithms are used for detection purposes to enhance the system's speed, precision, and efficiency.

4 Proposed Work

Deep Watcher works on a technology that combines face mask and SD detection services into a single piece of technology and adds certain visual analyzing features to study the results. The modules' workings and the proposed model's components are discussed below.

4.1 Input Module

The video is fed as input to the DeepWatcher model. The task of the input module is to take this video and pass it on to the following modules to perform an analysis on it in a frame-by-frame manner. But before that, the colab notebook needs to be provided with access to the drive folder where the video has been stored. Once access is provided, the link can be used to retrieve the location of the video file to access. The input module's primary function is to grab the file from the drive and pass it on to the next modules for processing after it is assigned to a variable so that it can be easily analyzed. Figure 1 depict the input video.

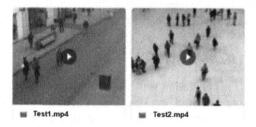

Fig. 1. Input Video

4.2 Distance Detection Module

The video file is passed on to the distance detection module. Here, the input module passes one frame from the video on which the analysis and detection need to be made. Then it calculates the distance between the people and predicts whether they are following norms of ethical distancing or not. The people following the guidelines are highlighted with a green box and the ones not following the guidelines are highlighted with a red box around them. Once the predictions are made on a frame, calculations are made on the number of violations happening in the given frame and, thus, the distance that the people are maintaining falls under violations, as shown in Fig. 2, and store this data accordingly for future use.

4.3 Face-Mask Detection Module

In order to determine if the persons in the video are wearing face masks in accordance with COVID requirements, it uses data from a frame of the previous module and works on the frames of the video feed. It checks for the face of the people and makes a red bounding box over their faces if they don't have their masks on correctly. Otherwise, if the people are wearing the mask, then it will show a green box and for the case where the model cannot predict whether the person detected by the model is wearing a mask or not, it will show no boxes.

Fig. 2. Detection and Visual Modules

4.4 Visual Detail Module

The violation data stored in the previous modules, i) the total count of the people, ii) the number of violations, and iii) the count of people that can't be judged based on their positioning and other reasons, will be taken into consideration in this module and will be displayed at the bottom of every frame. Also, if the distance measure falls under the violation category, it will be shown under various severity measures categorized by yellow, orange, and red-colored lines.

4.5 Output Module

Once all the processes are completed for the distance detection, facemask detection, visual detail, the output module will combine all the frames assessed by all the modules into a single video file. Once all the frames have been incorporated into a single video file, the output module's final task is to upload this video with all the analysis details. Figure 3 depicts the output module video.

Fig. 3. Output Video

Fig. 4. Object Detection Services in action

4.6 Centroid of the Bounding Box

The junction of two diagonals is where the rectangle's centroid is located. A width (y/2) and a height (x/2) are where diagonals intersect. As a bounding box technique will be used, all the detected objects will have a rectangular outline. After obtaining the rectangular outline, the centroid of the rectangle will be used as a fixed point from where the distance will be calculated between two objects or two people using the distance formula. Further, this distance will be highlighted using the visual details module, as shown in Fig. 4.

4.7 Distance Formula

The length of the line segment connecting those two points, for example, is the distance between them. The distance between two locations is computed using the Pythagoras theorem on a two-dimensional plane and two points on a three-dimensional plane. Using the centroids, the distance formula will be applied to them to find the distance between two people to verify whether people are following SD. Figure 5 depicts the architecture of the proposed model. The proposed algorithm is explained below:

5   DeepWatcher Algorithm

Step 1: Initialize variables and constant values for processing
   confidence_threshold = 0.3, minimum_distance = 50

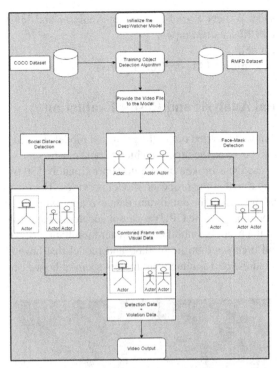

Fig. 5. Architecture Diagram

Step 2: extract frames from the video and change the processor to GPU

    n_frames = vs.get(cv2.CAP_PROP_FRAME_COUNT)

    Edit -> Notebook Settings -> Hardware Accelerator -> GPU

Step 3: Process the frames

(a) Apply SD detection services

        centroid_rectangle = [(length / 2), (breadth / 2)] + offset from origin

    (i) Calculate the distance between the centroids of the boxes

           distance_people = square root (power ((a1 - a2), 2) + power ((b1 - b2), 2))

    (ii) Check whether the distance between people is within the safe range or not

        distance < minimum_distance

    (iii) Change the color of the bounding box accordingly

(b) Apply face-mask detection services on the frame

        idxs = opencv.deepneuralnetworks.Boxes()

        Change the color of the bounding box accordingly

(c) Save all the collected data

Step 4: Display the gathered info as an informative giveaway at the bottom of the frame

    cv2.putText (FR,"SDA and Mask Monitoring wrt. COV-19″)

Step 5: Combine the frames into one single video file

Writer = cv2.VideoWriter ('Social-Distancing-Analyser-and-Mask-Monitoring-AI-system-wrt-Covid-19/Results/test.mp4')

    writer. Write(frame).

Step 6: Upload the combined file on Google Drive

6 Experimental Analysis and Result Evaluation

The experiments that were carried out in the proposed model showed promising results. A 21-s long video was used as the input in which people can be seen walking on a street. The video that was used was broken down into approximately 530 frames according to the 30 fps processing rate that the proposed model provides. The time it took to process the whole video and display all the details and data was approximately 52 min. It adds up to about 10 frames per minute. The GPU processor accelerates and enhances the process by 2 times, making the process smoother and faster simultaneously. Overall the results were promising and it shows that the method and model that have been proposed are working perfectly and the performance parameters seem to be well-tuned as well.

Fig. 6. Snapshot of the Input Video

Fig. 7. Snapshot of the Output Video

Figure 6 shows a snapshot from the input video. It is a generic video of people strolling along the pavement that was most likely recorded by a CCTV camera. This video has been fed to the Deep Watcher model to perform analysis. Once processed, this video will give the desired output shown in Fig. 7. In the output video, one can see how both the detection techniques have performed their analysis and provided the output video feed. In the output video, boxes can be seen around the people. Those boxes represent the people that the object detection algorithm has detected. Once these people have been detected, the distance between them is calculated, and accordingly, the color of the boxes is changed. Green boxes represent people who adhere to SD guidelines, while red boxes represent the opposite. The colored lines also represent the safe distance between people. A similar way of representing green-colored boxes over people's faces shows that people are wearing masks, and the red-colored boxes suggest otherwise. It can also be seen that the data is being displayed below the video feed on a label that was collected while the distance detection and FMD modules were processing the frame.

The COCO dataset has been used to train the model for SD detection. This dataset has been used because it has pre-built classes that differentiate many real-world objects and human beings. Then, the Masked Face Detection (RMFD) dataset was used for mask detection. It has over 5000 images of masked and unmasked people. These are the two datasets that have been used in the project and all the calculations and results are based on the data present in these datasets. These particular datasets were chosen because they have a large number of photos and item classifications, which have an impact on the proposed model's overall effectiveness and the output in accordance.

The accuracy achieved by the Deep Watcher model is almost 90% based on the dataset used to train the model. In addition, the prediction also depends on the quality of the video that has been taken as the input video to perform all the analysis and calculations. It can be seen that the objects in the video are very small, which leads to an even smaller-sized face that a camera can capture. In such a situation, it becomes hard for object detection and predictive algorithms to function correctly and give precise and accurate results.

7   Conclusion and Future Enhancements

The proposed model combined the two stand-alone technologies of FMD and SD detection in an efficient method. The system uses bounding boxes to display all of the people who have been detected. It also gives the specifics of all the different types of breaches occurring at any given time in a specific frame, such as people who are not wearing masks or those who are not adhering to the SD norms. The system takes a long time to process the video frame-by-frame, and it would be more efficient if the detection could be done in real-time. However, one of the primary issues with real-time surveillance is that it requires more powerful processors to run since it requires very high-level computational operations to be completed quickly. The proposed system does not currently allow for real-time surveillance. It is because it requires more computing power to control the process. As a result, one of the future additions that can significantly improve the efficiency of our system is to make it work in real time. Rather than analyzing the video all at once and then providing the results later, this will actively provide details on all violations as they occur. Thermal imaging is another improvement that can be made to the existing system. This can be used to monitor people's temperatures. Thermal imaging will be integrated into the monitoring system, eliminating the need to verify each individual's temperature and minimizing physical labor. Aside from that, some tweaks can be made to the surveillance system to forecast or recreate the face of a person hiding behind a mask. This will aid us in locating those who are breaking the rules and ensuring that they are dealt with appropriately.

References

1. Social distancing, surveillance, and stronger health systems as keys to controlling COVID-19 Pandemic, PAHO Director say - PAHO/WHO I Pan American Health Organization.
2. Garcia Godoy, L.R., et al.: Facial protection for healthcare workers during pandemics: a scoping review. BMJ Global Health **5**(5), e002553 (2020)
3. Eikenberry, S.E., et al.: To mask or not to mask: Modeling the potential for face mask use by the general public to curtail the COVID-19 pandemic. Infect Dis Model. **5**, 293–308 (2020)
4. Wearing surgical masks in public could help slow the COVID-19 pandemic advance: Masks may limit the spread of diseases including influenza, rhinoviruses.
5. Zhang, D., Hu, J., Li, F., Ding, X., Kumar Sangaiah, A., Sheng, V.S.: Small object detection via precise region-based fully convolutional networks. Computers Materials and Contiua **69**(2), 1503–1517 (2021)
6. Li, H., Liu, S.-M., Yu, X.-H., Tang, S.-L., Tang, C.-K.: Coronavirus disease 2019 (COVID-19): current status and future perspectives. International Journal of Antimicrob. Agents **55**(5), 105951 (2020)
7. Jin, Y.-H., et al.: A rapid advice guideline for the diagnosis and treatment of 2019 novel coronavirus (2019-nCoV) infected pneumonia (standard version). Miltary Medical Research **7**(1), 4 (2020)
8. Xu, X., et al.: Evolution of the novel coronavirus from the ongoing Wuhan outbreak and modeling of its spike protein for risk of human transmission. Science China Life Sciences **63**(3), 457–460 (2020). https://doi.org/10.1007/s11427-020-1637-5
9. Wang, D., et al.: Clinical characteristics of 138 hospitalized patients with 2019 novel Coronavirus-infected pneumonia in Wuhan, China. JAMA **323**(11), 1061–1069 (2020)
10. Holshue, M.L., et al.: First case of 2019 novel Coronavirus in the United States. N. Engl. J. Med. **382**(10), 929–936 (2020)
11. Huang, J., et al.: Speed/accuracy trade-offs for modern convolutional object detectors. In: 2017 IEEE Conference on Computer Vision and Pattern Recognition (CVPR) (2017)
12. Ahmed, A., et al.: Characterization of infection-induced SARS-CoV-2 seroprevalence amongst children and adolescents in North Carolina. Epidemiology and Infection, 1–25 (2023)
13. Dong, Y., et al.: Epidemiology of COVID-19 among children in China. Pediatrics **145**(6), e20200702 (2020)
14. Mahmoudi, S., et al.: Epidemiology, virology, clinical features, diagnosis, and treatment of SARS-CoV-2 infection. Journal of Experimental Clinical Meical. **38**(4), 649–668 (2021)
15. Lu, C.-W., Liu, X.-F., Jia, Z.-F.: 2019-nCoV transmission through the ocular surface must not be ignored. Lancet **395**(10224), e39 (2020)
16. Karaman, O., Alhudhaif, A., Polat, K.: Development of smart camera systems based on artificial intelligence network for social distance detection to fight against COVID-19. Appl. Soft Comput. **110**(107610), 107610 (2021)
17. Sethi, S., Kathuria, M., Kaushik, T.: Face mask detection using deep learning: An approach to reduce risk of Coronavirus spread. Journal of Biomed. Informatics **120**(103848), 103848 (2021)
18. Nowrin, A., Afroz, S., Rahman, M.S., Mahmud, I., Cho, Y.-Z.: Comprehensive review on facemask detection techniques in the context of covid-19. IEEE Access **9**, 106839–106864 (2021)

19. Gupta, S., Sreenivasu, S.V.N., Chouhan, K., Shrivastava, A., Sahu, B., Manohar Potdar, R.: Novel Face Mask Detection Technique using Machine Learning to control COVID'19 pandemic. Mater Today **80**, 3714–3718 (2023)
20. Nagrath, P., Jain, R., Madan, A., Arora, R., Kataria, P., Hemanth, J.: SSDMNV2: A real time DNN-based face mask detection system using single shot multibox detector and MobileNetV2. Sustainable Cities Society **66**(102692), 102692 (2021)
21. Vinh, T.Q., Anh, N.T.N.: Real-time face mask detector using YOLOv3 algorithm and Haar cascade classifier. In: 2020 International Conference on Advanced Computing and Applications (ACOMP) (2020)
22. Maharani, D.A., Machbub, C., Rusmin, P.H., Yulianti, L.: Improving the capability of real-time face masked recognition using cosine distance. In: 2020 6th International Conference on Interactive Digital Media (ICIDM) (2020)
23. Inamdar, M., Mehendale, N.: Real-time face mask identification using facemasknet deep learning network. SSRN Electronics Journal (2020)
24. Vrij, A., Hartwig, M.: Deception and lie detection in the courtroom: the effect of defendants wearing medical face masks. J. Appl. Res. Mem. Cogn. **10**(3), 392–399 (2021)
25. Malhotra, P., Garg, E.: Object detection techniques: A comparison. In: 2020 7th International Conference on Smart Structures and Systems (ICSSS) (2020)
26. Ramachandran, A., Sangaiah, A.K.: A review on object detection in unmanned aerial vehicle surveillance. Int. J. Cogni. Comp. Eng. **2**, 215–228 (2021)
27. Arulprakash, E., Aruldoss, M.: A study on generic object detection with emphasis on future research directions. J. King Saud Uni. Comp. Info. Sci. **34**(9), 7347–7365 (2022)
28. Kim, J.U., Man Ro, Y.: Attentive layer separation for object classification and object localization in object detection. In: 2019 IEEE International Conference on Image Processing (ICIP) (2019)
29. Blue, S.T., Brindha, M.: Edge detection based boundary box construction algorithm for improving the precision of object detection in YOLOv3. In: 2019 10th International Conference on Computing, Communication and Networking Technologies (ICCCNT) (2019)
30. World Health Organization et al.: Coronavirus disease 2019 (covid-19): situation report

Multiple Linear Regression Based Analysis of Weather Data for Precipitation and Visibility Prediction

Gurwinder Singh[(✉)] and Harun

Department of AIT-CSE, Chandigarh University, Chandigarh, Punjab, India
{gurwinder.e11253,harun.e11421}@cumail.in

Abstract. When considering the operational rendition and safety of a road, factors such as traffic condition, vehicle characteristics, and driver behaviour are just as important as the weather condition. In particular, the visibility and, by extension, the frequency with which accidents occur on a certain route or the probability of getting involved in accidents are affected by weather conditions like fog and rain. In this paper, a multiple linear regression based analysis of weather data is performed for predicting the precipitation and visibility so that different stockholder can be facilitated. This analysis method looks at the relationship between several independent variables (such as temperature, humidity, cloud-cover) and a dependent variable (precipitation or visibility). To evaluate the accuracy of the regression models, several evaluation metrics, including the mean square error (MSE), root mean square error (RMSE), and the mean absolute error (MAE), as well as the R-squared value have been employed on a large dataset containing 10 years of weather data, with 4018 samples. The obtained results proves that the multiple linear regression models can provide more accurate predictions of future weather conditions, benefiting a wide range of industries and individuals who depend on it for their operations and decision making.

Keywords: Multiple Linear regression · Weather forecasting · Prediction · Regression learner · Precipitation

1 Introduction

The weather can significantly impact the road surface, visibility, and overall driving conditions, which in turn can affect the likelihood of accidents and the severity of their outcomes. Therefore, it is crucial to take into account the weather forecast and its potential effects when assessing the safety of a road. Weather phenomena like fog and rain can decrease visibility, leading to an escalation in the occurrence and seriousness of accidents on a road, thereby heightening the probability of being involved in a collision.

The impact of climate on traffic accidents has been the subject of a number of studies. According to research by Pisano et al. [15], fog contributes to about

ⓒ The Author(s), under exclusive license to Springer Nature Switzerland AG 2023
M. Singh et al. (Eds.): ICACDS 2023, CCIS 1848, pp. 60–71, 2023.
https://doi.org/10.1007/978-3-031-37940-6_6

2% of all traffic accidents, while about 75% of weather-related collisions take place on wet roads. In their study on weather-related fatal crashes, Ashley et al. [3] found that the vast majority of such incidents happened on highways in the early morning hours during the coldest times of the year. Numerous other studies have been conducted on the effects of fog or smoke on vehicle collisions like Trick et al. [18], Hassan and AbdelAty [8], Mueller et al. [14], Ahmed et al. [2] and Theofilatos and Yannis [17]. These studies nave addressed issues such as reduced visibility due to fog, impaired vision which leads to reduced the information processing and difficulty in taking appropriate action while driving, the condition of the road (wet, icy, slushy, snowy, or muddy) which reduce the efficiency of braking and increase the risk of collisions etc.

Under foggy conditions, factors including delayed visibility-related advisory, rapid visibility reduction, driver conduct and braking efficacy are dependent on tyre and road surface conditions, all contribute to an increase in the incidence of crashes and the likelihood of being involved in a collision. Through the timely provision of visibility-related warnings or advisories, these issues could be prevented and improved the roadway safety. Putting in place weather monitoring devices equipped with cutting-edge technology at regular intervals to record the state of the atmosphere (including visibility) or precisely estimate visibility takes time and money, which limits the timeliness of information on visibility. Drivers could be informed with the actual or predicted visibility condition before setting out on a journey, or at any point during a trip to affect the driver's behaviour.

Since practitioners cannot consistently install technology/weather monitoring sensors, as it is critical to develop and assess the effectiveness of visibility prediction models to determine their suitability for the same. Rainfall in the preceding hours explains the rapid changes in visibility over the short intervals, while the other two variables assist explain the temporal and geographical aspects. Unfortunately, a robust model capable of accurate prediction could not be developed due to a lack of prior research into the effects of these explanatory variables and other meteorological data on visibility.

Among the most effective methods for making predictions are given by [6,11,12,19] and reviewed by [7,13] as well. And use of Regression Learner feature of Matlab by [4,5,9,10,16] has substantial benefit for different prediction applications.

1.1   Motivation

Multiple linear regression is a statistical technique that allows for the analysis of the relationship between a dependent variable and a group of independent variables, which are also known as predictors or explanatory variables. In the context of meteorology, this technique is commonly used to investigate how a set of independent meteorological factors may be related to an observable outcome such as precipitation or visibility. Through multiple linear regression, we can make predictions about how changes in the independent variables may impact the dependent variable. Using the values of the independent variables, multiple linear regression can provide predictions for the future values of the target variable. It can also shed light on the relative relevance of various weather factors in explaining the variance in the target variable.

1.2 Objective

The primary objective of this paper is to compare decision tree and linear regression models for prediction of precipitation and evaluate its predictability on a short scale. The obtained empirical results and conclusions can be utilised by practitioners to anticipate visibility as well as rainfall at the road network, disseminate information via tools such as variable or dynamic message signs located on roadways, as well as travel advisory software etc.

Accurate weather forecasting has a significant impact on various sectors of life, including transportation, energy, and agriculture. It helps improve safety, efficiency, and decision-making by providing advance warning of adverse weather conditions, optimizing energy generation and usage, and anticipating changes in food supply and prices. Accurate weather forecasting can benefit airlines, wind and solar power generators, farmers, and other stakeholders by providing useful information for planning and decision-making.

2 Data Understanding

In the analysis of weather data for precipitation and visibility prediction, data understanding is crucial. This is because the accuracy and reliability of the predictions are directly dependent on the quality and relevance of the data used for analysis. Data understanding is essential as it helps ensure the accuracy and reliability of the predictions and helps to identify any biases or limitations in the data that may affect the results.

For this purpose, data of 10 years period 2012–2022 with a total of 4018 samples representing six characteristics (including cloud cover, humidity, pressure, temperature, visibility, and precipitation) is collected from source [1]. Then, the collected data goes through some basic steps, such as data cleaning, integrating and exploring the data, removing irrelevant or inconsistent data points, filling in missing data points, fixing mistakes, and transforming the data so it used for analysis.

3 Data Visualization

Data visualization plays a crucial role in analysing weather data, including precipitation and visibility prediction. It helps to identify patterns, trends, and outliers, communicate complex information, improve decision-making, and support model. To gain a deeper understanding of weather data and make better-informed decisions, we have produced some important summary charts and graphs as explained in following subsections.

3.1   Statistical Summary

A statistical summary of data provides a condensed overview of the important characteristics of a dataset. It includes various measures of central tendency (mean, median, mode) and variability (standard deviation, range, interquartile range), as well as other measures, such as skewness and kurtosis, which describe the shape and distribution of the data. Particularly, a five-number summary as mentioned in Fig. 1 is very helpful.

	cloudcover	humidity	pressure	tempC	visibility	precipMM
mean	15.953211	44.303883	1009.863116	30.967397	9.757093	3.093853
std	18.418416	18.934211	6.721878	7.449118	0.615727	8.731276
min	0.000000	5.000000	993.000000	11.000000	5.000000	0.000000
25%	2.000000	29.000000	1004.000000	25.000000	10.000000	0.000000
50%	9.000000	43.000000	1010.500000	32.000000	10.000000	0.000000
75%	25.000000	59.000000	1016.000000	36.000000	10.000000	1.900000
max	98.000000	96.000000	1026.000000	50.000000	10.000000	172.200000

Fig. 1. Statistical summary of parameters

3.2   Correlation Chart

This correlation chart Fig. 2a, shows the pairwise correlations between different weather variables, specifically cloud cover, humidity, pressure, temperature, visibility, and precipitation. The values in the table represent the correlation coefficient, which quantifies how closely two variables are linearly related, including the direction and strength of that relationship. Here are some observations from the correlation chart: Cloud cover and humidity are positively correlated (0.55). This suggests that as cloud cover increases, so does the relative humidity in the air. Pressure and humidity have a weak negative correlation (-0.04). Temperature and pressure have a moderately strong negative correlation (-0.82). This suggests that as temperature increases, atmospheric pressure tends to decrease. Visibility and cloud cover have a strong negative correlation (-0.50). This suggests that as cloud cover increases, visibility decreases. Precipitation and temperature have a weak negative correlation (-0.01) and so on with other values.

	cloudcover	humidity	pressure	tempC	visibility	precipMM
cloudcover	1					
humidity	0.547188	1				
pressure	-0.23094	-0.03845	1			
tempC	-0.07876	-0.3478	-0.81789	1		
visibility	-0.50035	-0.52801	0.253427	0.01575	1	
precipMM	0.44524	0.46938	-0.23207	-0.00873	-0.81637	1

(a) Correlation matrix

	cloudcover	humidity	pressure	tempC	visibility	precipMM
cloudcover	339.1536					
humidity	190.7778	358.4151				
pressure	-28.5846	-4.89245	45.1724			
tempC	-10.8027	-49.0429	-40.9432	55.47554		
visibility	-5.67289	-6.15416	1.048631	0.07222	0.379025	
precipMM	71.58406	77.57852	-13.617	-0.56785	-4.38778	76.2162

(b) Covariance matrix

Fig. 2. Correlation and Covariance Charts

Figure 2b displays the covariances between every possible pair of variables and is symmetrical. The variance-covariance matrix is a symmetric matrix that shows the variances of each variable along the diagonal and the covariances between pairs of variables in the off-diagonal elements. Cloud cover has the largest variance (339.23), indicating that it varies the most across time or space. Humidity also has a relatively high variance (358.49), suggesting that it also varies significantly. Pressure has a relatively low variance (45.17), indicating that it is relatively stable over time or space. Temperature has a moderate variance (55.47), indicating that it is not as variable as cloud cover or humidity, but still varies significantly.

The Pairplot (Fig. 3) displays the correlations between pair-wise variables in a dataset. This makes for an attractive visualisation and aids comprehension by condensing numerous data points into one simple graphic. In the pairwise comparison method, each option is matched up against each other option, or "head-to-head". The decision-maker often pair-ranks the options, deciding which of two options is preferable or whether they are equally preferable.

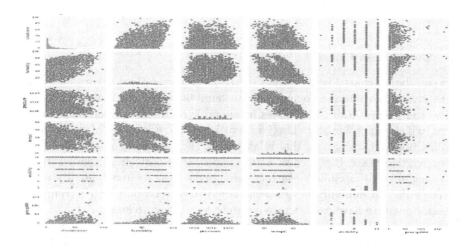

Fig. 3. pair-plot of each pair of parameters

3.3   Distribution Charts

Distribution plots are important for analyzing weather data because they help identify data skewness, assess normality, compare distributions, and highlight extreme events. They provide a visual representation of the characteristics of weather data, which is useful for making informed decisions based on the information provided by the data (Figs. 4 and 5).

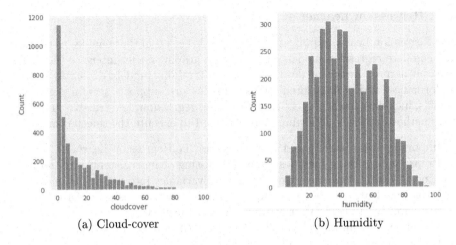

(a) Cloud-cover   (b) Humidity

Fig. 4. Distribution charts of Cloud cover and Humidity

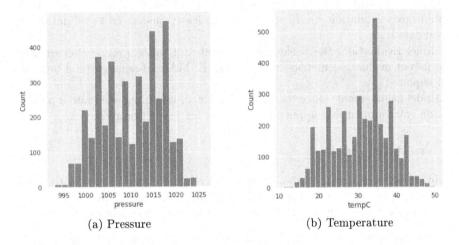

(a) Pressure   (b) Temperature

Fig. 5. Distribution charts of Pressure and Temperature

4   Methodology

As part of Methodology, we have used the Matlab Regression Learner feature for the analysis of weather data for precipitation and visibility prediction. The methodology involves importing weather data, selecting the appropriate regression model, dividing the data into training and testing sets, training the model using the training set, and then assessing its performance on the validation set.

4.1   Regression Learner

The Regression Learner application in MATLAB is a tool that enables us to train and compare regression models. It provides a graphical user interface (GUI) for selecting and preprocessing data, selecting and training models, evaluating model performance, and making predictions. This feature supports several regression models, including linear regression, polynomial regression, and tree-based regression methods etc. The following steps are used to execute the selected models:

- Importing Data: the dataset is imported in the Regression Learner.
- Preprocessing Data: includes removing missing or inconsistent data, normalizing the data, and converting categorical variables into numerical ones.
- Exploring Data: the data is explored and visualized using different plots, histograms, and correlation plots as depicted in Sect. 3.
- Feature Selection: the relevant features from the dataset are selected that can help in improving the model's performance.
- Training Model: the regression models and decision tree models are trained.
- Model Validation: the trained model using different techniques such as k-fold cross-validation, train-test split, and leave-one-out cross-validation are validated.
- Model Evaluation: the performance of different models are evaluated using different evaluation metrics such as RMSE, MAE, R-squared, and adjusted R-squared.
- Model Deployment: Once the model is trained, it is integrated into a production environment and applied to new data for prediction purposes.

4.2   Evaluation Metrics

In Regression Learner in Matlab, tree and regression models are evaluated using key metrics such as Mean squared error (MSE), Mean absolute error (MAE), Root mean squared error (RMSE), and Coefficient of determination (R-squared).

- Mean Squared Error (MSE) evaluates the average squared difference between the predicted and actual values in a model, allowing for an assessment of the model's predictive ability. A lower MSE value indicates better performance by the model in predicting the actual values.
- Mean Absolute Error (MAE) is a metric that computes the average absolute difference between the predicted and actual values in a model. It measures the degree of deviation between the predicted and actual values. It is less sensitive to outliers than MSE.
- Root Mean Squared Error (RMSE) is a metric frequently used in regression models that calculates the square root of the average squared difference between the predicted and actual values. It is similar to MSE but is expressed in the same units as the target variable.
- The Coefficient of determination (R-squared) is a crucial metric in regression analysis, representing the percentage of variance in the dependent variable

that can be explained by the independent variables. R-squared values range from 0 to 1, with 0 indicating no explanatory power and 1 representing a perfect explanation of the variability.

In summary, these evaluation metrics are essential in assessing the performance of tree and regression models in Regression Learner in Matlab. They help in selecting the best model, tuning the hyperparameters, and improving the model's accuracy. These are tabulated in Table 1, where x and y are vectors with D dimensions, and x_i represents the value of the ith sample of the x variable.

Table 1. Error Criteria

Criteria	Description	Formula
MSE	Mean Absolute Error	$\sum_{i=1}^{D} |x_i - y_i|$
MAE	Mean Squared Error	$\sum_{i=1}^{D} (x_i - y_i)^2$
RMSE	Mean Squared Error	$\sqrt{\sum_{i=1}^{D} (x_i - y_i)^2}$
R^2	Coefficient of Determination	$1 - \frac{sum squared regression(SSR)}{total sum of squares(SST)}$

5 Results and Analysis

The MATLAB Regression Learner feature is important because it provides an easy-to-use interface for creating and evaluating multiple linear regression models. The feature also includes a range of advanced options, such as the ability to specify regularization methods, cross-validation folds, and evaluation metrics. As per proposed approach, the Tree Models and various variants of Linear Regression Models are selected. The performance of the MAE, MSE, and RMSE criteria is compared using the 5-fold cross-validation method, and the number of delay steps is modified from 3 to 10 during model training. The obtained results are summarized in the Table 2 and Table 3.

Table 2. Performance measures of Tree Models

	Fine	Medium	Coarse
RMSE	5.3207	4.9568	5.0247
R-squared	0.63	**0.68**	0.67
MSE	28.31	24.57	25.248
MAE	1.6343	1.5173	**1.5107**

The Table 2 shows the performance measures of three different tree models, namely Fine Tree, Medium Tree, and Coarse Tree. The performance measures are presented in terms of four different criteria: RMSE, R-squared, MSE, and

MAE. In the table, the Medium Tree model has the best performance according to the R-squared and MAE criteria, with values of 0.68 and 1.5173, respectively. The Fine Tree model has the highest RMSE, which indicates that it has the highest deviation between predicted and actual values. The Coarse Tree model has the best performance according to the MSE criterion, with a value of 25.248. Overall, the table allows us to compare the performance of the different tree models based on multiple criteria.

Table 3. Performance measures of Multiple-Linear-Regression Models

	Linear	Interactions	Robust	Stepwise
RMSE	5.0329	4.7344	7.9715	**4.7254**
R-squared	0.67	0.71	0.17	0.71
MSE	25.33	22.414	63.544	**22.33**
MAE	1.5816	1.5409	4.7254	1.5393

The Table 3, shows the performance measures of different linear regression models. The models compared are Linear Regression, Interactions Linear Regression, Robust Linear Regression, and Stepwise Linear Regression. The best performance measures are highlighted in bold font. From the table, the Stepwise Linear Regression model performs the best in terms of RMSE, MSE, and MAE, while the Interactions Linear Regression model has the highest R-squared value. The Robust Linear Regression model performs the worst in all measures. Also the table indicates that the Interactions Linear model has the lowest RMSE and the highest R-squared, indicating that it is the most accurate model. The Stepwise Linear model also has a low RMSE and a high R-squared, making it a good alternative to the Interactions Linear model. The Robust Linear model has the highest RMSE and the lowest R-squared, indicating that it is the least accurate model.

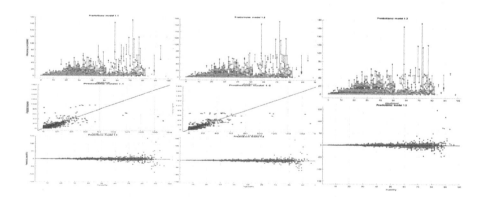

Fig. 6. Plot of results of Tree Models

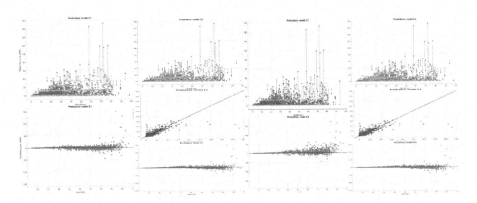

Fig. 7. Plot of results of Regression Models

A prediction chart, in Fig. 6, also known as an "Actual vs. Predicted Plot", is a graphical representation of the difference between the actual and predicted values for a dependent variable in a tree model. The chart plots the actual values on the y-axis and the predicted values on the x-axis. The points in the chart are falling close to a 45-degree line, indicating a good fit between the actual and predicted values. Similarly points are observed in the Fig. 7 as well.

Further, a residual plot, bottom plots in Fig. 6 & Fig. 7, is a scatter plot that displays the residuals, or the differences between the actual and predicted values, on the vertical axis and the predicted values, or the fitted values, on the horizontal axis. The purpose of a residual plot is to check for any patterns or trends in the residuals, which are examined with respect to non-constant variance. As visible, the residuals are randomly scattered around zero, this suggests that the regression model is a good fit for the data and that the residuals are independent and normally distributed.

The acquired results demonstrate that multiple linear regression is able to produce more accurate forecasts of future weather conditions, which is advantageous to a variety of companies and individuals that rely on it for operations and decision-making.

6   Conclusion and Future Scope

In this work, the effectiveness of using linear regression with multiple weather data-variables has been evaluated for predicting precipitation or visibility. The model makes use of Matlab's Regression Learners feature, which contains a weighted moving average filter and an exponential moving average filter based on linear and Gaussian kernels, to enhance the precision of its predictions. The approach investigates the potential of a 10-year dataset with six variables to enhance the forecasting capability. We use a 5-fold cross-validation technique to compare the performance of MAE, MSE, and RMSE criteria when training the model, and variation in the amount of delay stages from 3 to 10. This approach generates a graph to show the estimated versus the actual regression coefficients for the specified minimum error model. The study also supplies the correctness of the model and the coefficient of Determination of correlation (i.e., R-squared), which is a measurement of the strength of the relationship between input and output.

Additionally, future studies could investigate the effectiveness of incorporating other machine learning techniques, such as neural networks, in predicting weather patterns. Further research could also focus on expanding the dataset used in this paper to include more variables to enhance the forecasting capabilities of the models. Finally, the future scope could be in the development of practical applications of the findings of this approach in various sectors, such as transportation, energy, and agriculture, to improve safety, efficiency, and preparedness for weather-related events.

References

1. India meteorological department ministry of earth sciences government of India (2023). https://mausam.imd.gov.in/responsive/rainfall-statistics.php
2. Ahmed, M.M., Abdel-Aty, M., Lee, J., Yu, R.: Real-time assessment of fog-related crashes using airport weather data: a feasibility analysis. Accid. Anal. Prev. **72**, 309–317 (2014)
3. Ashley, W.S., Strader, S., Dziubla, D.C., Haberlie, A.: Driving blind: weather-related vision hazards and fatal motor vehicle crashes. Bull. Am. Meteor. Soc. **96**(5), 755–778 (2015)
4. Dhakal, S., Gautam, Y., Bhattarai, A.: Evaluation of temperature-based empirical models and machine learning techniques to estimate daily global solar radiation at Biratnagar airport, Nepal. Adv. Meteorol. **2020**, 1–11 (2020)
5. Ekici, S., Unal, F., Ozleyen, U.: Comparison of different regression models to estimate fault location on hybrid power systems. IET Gener. Transm. Distrib. **13**(20), 4756–4765 (2019)
6. Fang, T., Lahdelma, R.: Evaluation of a multiple linear regression model and SARIMA model in forecasting heat demand for district heating system. Appl. Energy **179**, 544–552 (2016)
7. Gad, I., Hosahalli, D.: A comparative study of prediction and classification models on NCDC weather data. Int. J. Comput. Appl. **44**(5), 414–425 (2022)
8. Hassan, H.M., Abdel-Aty, M.A.: Predicting reduced visibility related crashes on freeways using real-time traffic flow data. J. Saf. Res. **45**, 29–36 (2013)
9. Hrehova, S., Husár, J.: Selected application tools for creating models in the matlab environment. In: Perakovic, D., Knapcikova, L. (eds.) FABULOUS 2022. LNICST, vol. 445, pp. 181–192. Springer, Cham (2022). https://doi.org/10.1007/978-3-031-15101-9_13
10. Karamdel, S., Liang, X., Faried, S.O., Shabbir, M.N.S.K.: A regression model-based short-term PV power generation forecasting. In: 2022 IEEE Electrical Power and Energy Conference (EPEC), pp. 261–266. IEEE (2022)
11. Kavitha, S., Varuna, S., Ramya, R.: A comparative analysis on linear regression and support vector regression. In: 2016 Online International Conference on Green Engineering and Technologies (IC-GET), pp. 1–5. IEEE (2016)

12. Mahabub, A., Habib, A.-Z.S.B., Mondal, M.R.H., Bharati, S., Podder, P.: Effectiveness of ensemble machine learning algorithms in weather forecasting of Bangladesh. In: Abraham, A., Sasaki, H., Rios, R., Gandhi, N., Singh, U., Ma, K. (eds.) IBICA 2020. AISC, vol. 1372, pp. 267–277. Springer, Cham (2021). https://doi.org/10.1007/978-3-030-73603-3_25

13. Maulud, D., Abdulazeez, A.M.: A review on linear regression comprehensive in machine learning. J. Appl. Sci. Technol. Trends **1**(4), 140–147 (2020)

14. Mueller, A.S., Trick, L.M.: Driving in fog: the effects of driving experience and visibility on speed compensation and hazard avoidance. Accid. Anal. Prev. **48**, 472–479 (2012)

15. Pisano, P.A., Goodwin, L.C., Rossetti, M.A.: US highway crashes in adverse road weather conditions. In: 24th Conference on International Interactive Information and Processing Systems for Meteorology, Oceanography and Hydrology, New Orleans, LA (2008)

16. Pizzulli, V., Telesca, V., Covatariu, G.: Analysis of correlation between climate change and human health based on a machine learning approach. In: Healthcare 2021, vol. 9, p. 86 (2021)

17. Theofilatos, A., Yannis, G.: A review of the effect of traffic and weather characteristics on road safety. Accid. Anal. Prev. **72**, 244–256 (2014)

18. Trick, L.M., Toxopeus, R., Wilson, D.: The effects of visibility conditions, traffic density, and navigational challenge on speed compensation and driving performance in older adults. Accid. Anal. Prev. **42**(6), 1661–1671 (2010)

19. Vlachogianni, A., Kassomenos, P., Karppinen, A., Karakitsios, S., Kukkonen, J.: Evaluation of a multiple regression model for the forecasting of the concentrations of NOx and PM10 in Athens and Helsinki. Sci. Total Environ. **409**(8), 1559–1571 (2011)

Optimal and Event Driven Adaptive Fault Diagnosis for Arbitrary Network

Pradnya Chaudhari$^{(\boxtimes)}$ ⓘ, Anjusha Joshi ⓘ, Supriya Kelkar ⓘ, Anupama Joshi ⓘ, and Soniya Durgude ⓘ

MKSSS's Cummins College of Engineering for Women, Department of Computer Engineering, Pune, India

{Pradnya.Chaudhari,Anjusha.Joshi,Supriya.Kelkar,Anupama.Joshi, Soniya.Durgude}@cumminscollege.in

Abstract. Distributed computing system consists of numerous nodes that run as a single system. However, failures in nodes are unavoidable, which results in a node being marked as faulty. The system's performance is impacted by these failures. So, fault diagnosis is a crucial part of the distributed computing system. This paper proposes a new algorithm called Optimal and Event Driven Adaptive Fault Diagnosis in Distributed System (OED-AFD) to identify faulty nodes in the system. This algorithm discovers the dynamic network along with detecting faulty nodes. The algorithm ensures that every node knows the status of all the nodes in the system at the end of every diagnostic cycle. The proposed algorithm is initiated either periodically or when an event, such as a new node entry or a repaired node re-entry is detected by the existing nodes of the system. The laboratory observations indicate that the proposed algorithm discovers and diagnoses any arbitrary distributed system using a minimal number of messages as compared to algorithms and methods proposed earlier by the authors.

Keywords: Fault detection · Fault diagnosis · Arbitrary network · Adaptive algorithm · Distributed system · Distributed network · Computer Network

1 Introduction

Distributed computing system composes of numerous individual nodes located at different locations, performing different tasks and connected through a network. The network can be partially or fully connected. The system may experience numerous faults while performing various tasks. These faults may occur at various layers, such as the physical layer, the data-link layer, the application layer, or the network management level [1]. These faults impact the system's performance if they are not detected in time.

The literature presents several algorithms for fault diagnosis. These algorithms are mainly adaptive in nature [1–23]. The fault diagnosis algorithm is adaptive when it adjusts its testing scheme according to the current fault conditions as and when found in the distributed system [1]. This implies that the algorithm does not adhere to a set

M. Singh et al. (Eds.): ICACDS 2023, CCIS 1848, pp. 72–84, 2023.
https://doi.org/10.1007/978-3-031-37940-6_7

testing protocol but rather adjusts to any new fault scenario and successfully detects faulty nodes in the system.

The fault diagnosis algorithms are either periodic [1–21] or event-driven [22, 23] in nature. The periodic algorithms execute their fault diagnostic cycle at regular intervals. Any faults occurring between the two diagnostic cycles are detected in the next subsequent cycle. However, the event-driven algorithms are responsive to the events occurring in the system. They trigger the fault diagnostic cycle as a response to the event, thus updating the status of the node that caused the event.

This paper introduces a new algorithm, Optimal and Event Driven Adaptive Fault Diagnosis for Arbitrary Networks (OED-AFD) to address the issues such as optimal diagnostic message exchange between the nodes, dependency of the system on one or few nodes during diagnosis and adaptation of the algorithm to any arbitrary network model. The proposed algorithm executes on all the nodes simultaneously. At the end of every diagnostic cycle, each node has knowledge about the network and knows about every other node's fault status. In addition to periodically diagnosing the faults, the algorithm gets triggered whenever an event such as the re-entry of a repaired node or the entry of a new node occurs.

2 Related Work

The literature has various algorithms and techniques that are proposed for fault diagnosis in distributed systems. These can be divided into various categories based on methods used for fault diagnosis such as comparison based [1–7], accuracy technique [8], graph-based [10–12], clustering technique [13, 14], active probing method [16–18], event-triggered [22, 23] and others [9, 15, 19–21].

AFDCAN detects faulty nodes from an automotive network based on Controller Area Network (CAN). The number of communication messages and test rounds required for a complete diagnostic cycle is well-defined and bounded. It allows the re-entry of repaired faulty nodes and the entry of new nodes during the diagnosis cycle [1]. The L-AFD works by electing a leader and sub-leaders, based on the node's connectivity percentage. The diagnostic information is sent to the elected leader from all the nodes. The diagnostic cycle ends with the last leader having diagnostic information of the complete network and broadcasting it in the network [2]. The algorithm in [3] identifies a coordinator pair depending on the reachability of nodes. This coordinator pair diagnose faulty nodes based on the acknowledgments received. The algorithm works only on a static network, but it ensures that the efficiency is not affected by a single point of failure [3].

DP-AFD works in two stages, namely, leader election and fault diagnosis. Each fault-free node becomes a leader once in every diagnostic cycle and the information about the faulty nodes and network is passed from one leader to the next leader [4]. A Nearly Optimal Algorithm is a comparison-based model for arbitrary topology. The algorithm uses MM* model. This results in an increase in communication messages. Also, the system is vulnerable as the diagnostic information is collected by a central observer [5].

Every node in the ADSD algorithm uses comparison-based method for diagnosis.

This accounts to large number of tests, thereby increasing the network traffic [6]. Hi-ADSD is based on [6] and runs on a completely connected system and groups nodes in large logical clusters using the technique of clustering [7].

In the fault diagnosis based on accuracy technique, initiators are selected based on their reachability in a round-robin manner. The last initiator broadcasts the list of faulty nodes so that the complete network has information about the faulty nodes [8]. DisCaRia, assists system administrators in fault resolution along with fault diagnosis. The system uses peer-to-peer technology to explore distributed environments for fault knowledge resources [9].

An optimal algorithm diagnoses faulty nodes by creating a spanning tree from the non-faulty nodes. It is less reliable as the packet size grows when the diagnostic information advances up the tree and there is only one central observer gathering all of the diagnostic information [10]. The DNR algorithm determines the reachability of every node from every other node in the partitionable arbitrary and dynamic networks. More bandwidth is consumed because of the dead messages that are present in the network and there is latency due to message validation [11]. A graph-based fault diagnosis technique has a multi-relational graph representing the availability of testing records of the systems. The vertices of the graph are records of non-repeated failed availability tests. Similar vertices are grouped using a clustering method to get the fault cause [12].

A Novel Fault Diagnosis Algorithm is a two-phase online diagnostic algorithm that is based on tracking of cluster heartbeat that results from the normal workload execution [13]. [14] is a CTCN model-based algorithm, which localizes and diagnose faults using the temporal information of the alarm messages. However, this model is applicable only in cases with finite delays and is less accurate in cases with time-varying delays [14].

The health condition of the network is inferred by sending data packets called probes in the active probing method [16]. The probabilistic min search selection method is used to locate the defects in the probabilistic probe selection technique. The method updates the state of the node for iterations as per the states of identified nodes [17]. Traffic Dynamics-Aware Probe selection method selects probes and probe stations synchronously for fault diagnosis. However, this method relies on the premise that the association between probes and network nodes are fully deterministic [18].

[19] proposes a fault detection strategy based on observing internal network failure symptoms in peripheral elements of the system [19]. A fault localization algorithm uses passive measures based on Boolean particle swarm optimization for fault isolation [20]. FIFPDPMC algorithm uses a probabilistic approach to diagnose the intermittent faults in a randomly generated network [21]. In [22], each agent detects its fault along with faults that occurred in neighbors, and the reference model output is tracked by all agents.

After studying the literature, some of the observations made by the authors are, i) single point of failure, ii) dependency on a single node or pair of nodes, iii) congestion in the network caused by a large number of network communications, iv) inability to adapt to arbitrary network, v) traffic overhead.

Therefore, the authors propose OED-AFD algorithm to address the above-mentioned observations or gaps. OED-AFD can adapt to any arbitrary network, and it is adaptive in nature. It is event-triggered and has a t-diagnosability of (n-2), where the total number of nodes is shown by n. OED-AFD use less number of messages for fault detection, thus, not contributing to the congestion of the network.

3 Proposed Optimal and Event Driven Adaptive Fault Diagnosis (OED-AFD) Algorithm

3.1 Fault Model

The OED-AFD employs the following fault model. The fault model, as illustrated below, compares the self-test results, exchanged between nodes using data frames, to determine if the node is fault-free or faulty.

$$f(r, B) = 1, 0, -1 \tag{1}$$

where, the self-test result of node n_i is denoted by r and B is the result of self-test performed by node n_i kept in the buffer of node n_j. 1 indicates that n_i is a faulty node, 0 indicates that n_i is a fault-free node and -1 denotes that node n_i could not send its self-test results to node n_j within n_j's waiting time. If node n_i doesn't respond or node n_j doesn't receive an acknowledgment from n_i, then node n_i is considered faulty at the end of the diagnostic cycle. In the algorithm, busy status (-1) has lower priority than faulty (1) and fault-free (0) statuses.

3.2 Assumptions

- Every node is only aware of its neighbors and does not have information about the network beyond its neighbors.
- A node is considered fault-free if the self-test results are correct and the data-frame is received by its neighbors.
- Faults can either be temporary or permanent.
- The re-entry of the repaired node or entry of a new node is considered as an event and triggers the next diagnostic cycle.
- Node failures and communication link failures are not differentiated.

3.3 OED-AFD Algorithm

Network discovery is an important part of fault diagnosis for an arbitrary network. The proposed OED-AFD discovers the network based on the resource discovery algorithm, namely, Name Dropper [24]. However, OED-AFD uses a unique fault detection method, while discovering the network simultaneously.

In OED-AFD, the diagnostic cycle is executed either periodically or due to events such as re-entry of the repaired node or new node entry. Hence, this algorithm is known as "Event-Driven". During the ongoing diagnostic cycle, the algorithm detects a new node entry. However, the re-entry of the repaired node is not handled during the ongoing diagnostic cycle of the algorithm. But re-entry of the repaired node triggers an event that generates a fresh diagnostic cycle. When the diagnostic cycle is triggered periodically, every node executes OED-AFD in parallel. A new node or repaired node starts the diagnostic cycle, whenever the diagnostic cycle is triggered by an event.

Every node performs a self-test at the beginning of the algorithm. A collection of arithmetic, logical, and memory operations constitute a self-test, which is used to evaluate the accuracy of the CPU [2]. At the end of the diagnostic cycle, every node has information about every other node in the network along with its fault or fault-free status. The OED-AFD algorithm is adaptive in nature since it is able to detect fault nodes in any arbitrary network.

The data structures that need to be initialized as pre-requisite are *Nodes_List* (Fig. 1) and *Neighbors_List* (Fig. 2). *Neighbors_List*, as the name suggests, is the list of host-names of the neighbors of a node, say n_i. *Nodes_List* is like a dictionary with key-value pairs. Here, the key is the hostname of the discovered node n_j by the node n_i and the value is the status of the discovered node n_j. *Nodes_list* is considered complete by n_i, when it has the statuses of all the nodes in the network.

req_frame and *ack_frame* can be generalized as data-frame. Figure 3 shows the generic layout of the data frame. Here, for *req_frame* the *REQ/ACK bit* is 1 and for *ack_frame*, it is 0. The source address contains the hostname of the sender node, and the destination address is the hostname of the receiver node. The result of the source node's self-test is referred to as the self-test result in the data-frame and is shown in pseu-docode as the Result. *Nodes_List* is the *Nodes_List* of the sender node. *Complete_Bit* and *Event_Bit* are boolean values and are initialized to 0 (false) in the beginning. *Complete_Bit* is set to 1 (true) when the *Nodes_List* of a node is completed. *Event_Bit* is set to 1 (true), when a node detects an event or when it receives a data-frame from one of its neighbors with *Event_Bit* set to 1.

The pseudo-code for the fault diagnosis algorithm when a diagnostic cycle is trig-gered periodically is depicted in Fig. 4. In Fig. 4, the neighboring nodes are indicated by n_i and n_j. Figure 5 shows the pseudo-code for the second part of the algorithm, i.e., when the diagnostic cycle is triggered by an event.

Nodes' IP Address/Hostname	Status (Faulty / Fault-free / Busy)
....	....
Node's IP Address/Hostname n(i)	Status n(i)

Fig. 1. Nodes_List

Nodes' IP Address/ Hostname
....
Node's IP Address/Hostname n(i)

Fig. 2. Neighbors_List

REQ/ACK Bit	Source Address	Destination Address	Self_Test Result	Nodes_List	Complete_Bit	Event_Bit

Fig. 3. Data frame

```
Require: Initialized data structures

Perform self-test
while Complete_Bit(ni) ≠ 1 do
    if send req_frame(ni, nj) then
        if nj is free then
            if Result(ni) = Result(nj) then
                Nodes_List(nj)(ni) ⇐ 0
                Merge(Nodes_List(ni), Nodes_List(nj))
                Send ack_frame(nj, ni)
            else
                Nodes_List(nj)(ni) ⇐ 1
            end if
        else if nj is busy/non-responsive then
            Nodes_List(nj)(ni) ⇐ −1
            if nj is fault-free & nj becomes free then
                Send ack_frame(nj, ni)
            end if
        end if
    end if
    if receive ack_frame(nj, ni) then
        if ni is busy then
            send ack_frame(ni, nj), when nj becomes free
        else if ni is free then
            if Result(nj) = Result(ni) then
                Nodes_List(ni)(nj) ⇐ 0
                Merge(Nodes_List(nj), Nodes_List(ni))
            else
                Nodes_List(ni)(nj) ⇐ 1
            end if
        end if
    end if
    Rotate neighbors_list in cyclic manner
    if Nodes_List is complete then
        Complete_List ⇐ 1
    end if
end while
```

Fig. 4. OED-AFD Part 1 – Periodic Execution (Runs on every node simultaneously)

3.4 Fault Scenarios

Scenario 1: System with Some Faulty Nodes –
Figure 6 shows a generic arbitrary network that utilizes the proposed OED-AFD algorithm for discovering the network and fault diagnosis. This system diagram is similar to the diagram used by these authors as in [4]. In the proposed work, authors have used nine nodes instead of ten nodes with different connections between the nodes as compared to the system diagram used in [4]. Here, N3 is computationally faulty, and N9 is a non-responsive node. One test round in OED-AFD consists of sending and receiving of *req_frame* and *ack_frame* between the pair of nodes formed based on *Neighbors_List*.

The diagnostic cycle starts with all the nodes performing self-test and initializing the data structures. Each node sends *req_frame* to the first neighbor from its *Neighbors_List*. On receiving the *req_frame*, the neighbor compares its self-test result with the self-test

Ensure: *Complete_bit* of all nodes existing in network set to 1.
Require: n_k is new node or repaired node or node communicating
event occurrence with n_i as neighbor.

```
if receive req_frame(n_k → n_i) then
    if n_k ∉ Neighbors_List(n_i) then    // New Node detected
        Neighbors_List(n_i) ⇐ Neighbors_List(n_i) ∪ n_k
    else if (Nodes_List(n_i).value(n_k) == 1) or
        (Complete_Bit(n_i) = 1 & Event_Bit(n_k) = 1) then
        // Re-entry of Repaired Node or Event occurred detected
    end if
    if New Node or Re-entry of Repaired Node or Event detected then
        Event_Bit(n_i) ⇐ 1
        Complete_Bit(n_i) ⇐ 0
        Nodes_List(n_i)(all nodes) ⇐ NULL
    end if
    if Result(n_k) = Result(n_i) then
        Nodes_List(n_i).update(n_k ⇐ 0)
        Merge(Nodes_List(n_k), Nodes_List(n_i))
        Send ack_frame(n_i, n_k)
    else
        Nodes_List(n_i).update(n_k ⇐ 1)
    end if
end if

// Event propagation from node n_i to other nodes in the network
if Complete_Bit(n_i) ≠ 1 then
    if send req_frame(n_i → n_j) then
        // Processed similarly to Part 1
    end if
end if
// Continue sending and receiving data_frames till Complete_Bit = 1
```

Fig. 5. OED-AFD Part 2 – Event Triggered Execution

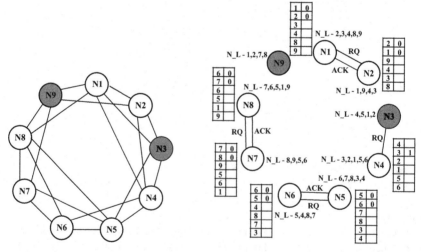

Fig. 6. Generic Arbitrary Network   **Fig. 7.** Case 1: Test Round 1

result received in *req_frame*. If the results are the same, status of the sender is updated
as fault-free (0) by the receiver node and *Node_List* in *req_frame* is merged with the

Node_List of the receiver node. Then, *ack_frame* is sent back from receiver to sender of the *req_frame*. This interaction between sender and receiver is termed as a test round. These steps are carried out between nodes N1 & N2, N5 & N6, and N7 & N8 in the first test round as they are each other's first neighbor as per the Neighbors_List.

N3 being computationally faulty, it is marked as faulty (1) by N4 and it does not send *ack_frame* to N3. N9 is not detected in the first round, as it is not the first neighbor of any of the nodes in the network. As N9 is non-responsive, it does not send *req_frame* to its neighbors or reply with *ack_frame* to requests received from its neighbors. The *Neighbors_List* of each node undergoes a cyclic rotation at the end of each test round, Refer to Fig. 7 for the first test round.

The second test round follows steps similar to the first test round, starting with all the nodes sending *req_frame* to the first node in *Nodes_List*. Nodes N4, N5, N6, and N8 wait for *ack_frame* from N2, N7, N4, and N6 respectively for a specific waiting period. As N2, N7, N4, and N6 are busy, their status is set to busy (-1) in *Nodes_List* of N4, N5, N6, and N8 respectively. N2 gets information about N3 being faulty from N4. N9 is marked as busy (-1) by N2 and N7, as these nodes don't receive *ack_frame* from N9. In the next test round, as soon as the busy nodes get free, they send *ack_frame* to those nodes from whom they had received *req_frame*. Refer to Fig. 8 for the second test round.

These test rounds continue till every node has its *Complete_Bit* set to 1. After *Complete_Bit* is set to 1, status of N9 is changed from busy (-1) to faulty (1). This completes one diagnostic cycle (Refer to Fig. 9). On completion of each diagnostic cycle, every node in the network has information about every other node with status.

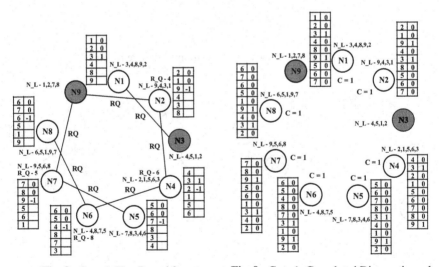

Fig. 8. Case 1: Test Round 2   **Fig. 9.** Case 1: Completed Diagnostic cycle

Scenario 2: Entry of New Node or Re-entry of Repaired Node–
Consider N10 as a new node added to the generic arbitrary network with the neighbor as N3 (Refer to Fig. 10). Here, it is assumed that all the nodes in the network is fault-free.

All the existing nodes in the network have *Complete_Bit* set to 1, as they have completed a diagnostic cycle.

When N10 enters the network, it starts a new diagnostic cycle, by sending the *req_frame* to its neighbor N3. As *Neighbors_List* of N3 does not contain N10, N3 detects N10 as a new node and adds N10 to its *Neighbors_List*. As the event of a new node entry is detected, N3 toggles its *Event_Bit* and *Complete_Bit*. Also, the statuses of all the nodes in the *Nodes_List* of N3 are re-initialized. The first test round is conducted between N3 and N10 (Refer to Fig. 11).

For the second test round, as N3 has an incomplete *Nodes_List*, it sends *req_frame* to its first neighbor N2 from its *Neighbors_List* with Event_Bit set to 1. When N2 receives *req_frame*, its data structures are modified in a similar manner as N3 (Refer to Fig. 12). In the next test round, N2 and N3 both will send *req_frame* to their next neighbors in *Neighbors_List*.

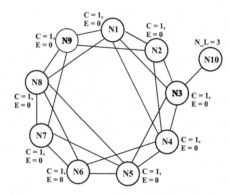

Fig. 10. Case 2: New Node Entry/Repaired Node Re-entry

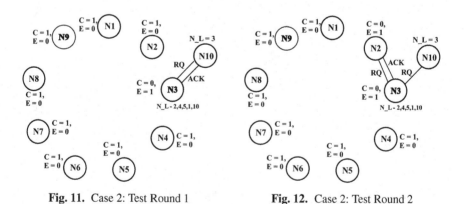

Fig. 11. Case 2: Test Round 1   **Fig. 12.** Case 2: Test Round 2

The diagnostic cycle continues, with all the nodes changing values of *Complete_Bit* and *Event_Bit*, testing their neighbors, and merging the *Nodes_List* of the sender, if the sender is fault-free (as per OED-AFD algorithm Fig. 6). When the *Nodes_List* of a node is completed, its *Complete_Bit* is set to 1 and its *Event_Bit* is re-set to 0. Thus, at the end of the diagnostic cycle, every node, including N10, has information and the status of all the nodes in the network.

4   Results and Discussion

The proposed OED-AFD algorithm is implemented on two arbitrary networks consisting of 9 and 18 computer systems. The hostname in the algorithm is interpreted as an IP address for the implementation purpose. The IP address of node N1 is 172.16.20.101, of N2 is 172.16.20.102, and so on.

Figure 13 shows four cases representing the number of data frames sent by fault-free nodes for different fault conditions with an experimental arbitrary network of 18 computer systems. It can be observed that the t-diagnosability of the OED-AFD is 16 (n-2, where n denotes the number of nodes in the network). There are two network topologies used for conducting experiment, one for case 1 and case 3 and the other for case 2 and case 4. In case 1 and case 2 faulty nodes are assumed to be non-responsive, whereas in case 3 and case 4 they are assumed to be computationally faulty.

As seen in Fig. 13 (cases 1 and 2), when faulty nodes are non-responsive, with the rise in number of faulty nodes, the data frames sent by nodes also increase. This pattern persists until the number of faulty and fault-free nodes is become equal. After this, there is a reduction in the number of data frames. The reason for the increase in data frames initially is due to fault-free nodes repeatedly sending *req_frame* to the non-responsive nodes. These non-responsive nodes are tagged as busy (-1) during the diagnostic cycle. When there are more faulty nodes then fault-free nodes, the non-responsive faulty nodes do not send *req_frame*, reducing the total number of data frames sent. When faulty nodes are considered computationally faulty (case 3 & case 4 of Fig. 13), it is observed that with an increase in the number of faulty nodes, the number of data frames sent decreases. This behavior of OED-AFD in case 3 & case 4 is different than that of its behavior in case 1 & case 2 as shown in Fig. 13. This behavior is due to the following reason. The computationally faulty nodes are marked as faulty (1) after a test round. When a node marks its neighbor as faulty, it does not send *req_frame* to that neighbor, thus reducing the data frames.

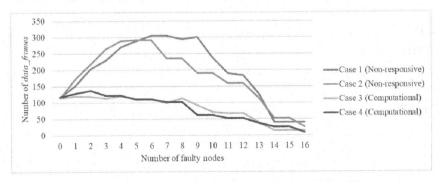

Fig. 13. Number of data frames used in different fault conditions

Figure 14 shows the time taken to complete one diagnostic cycle and the number of test rounds required for a complete diagnostic cycle for different fault conditions is shown in Fig. 15. It can be seen that the time taken and the number of test rounds to a complete diagnostic cycle increases with the rise in the number of faulty nodes. This trend continues till the faulty nodes are less than the faulty-free nodes. After that, the time taken and the number of test rounds start decreasing. The same reason as before serves as the basis for this behavior.

The number of test rounds, data frames used, and time taken for a diagnostic cycle depends on the degree of the nodes, order of neighboring nodes in the *Neighbors_List* together with the number of faulty nodes and total number of nodes in the system. As the total number of nodes and faulty nodes increase or decrease in the system, the number of data frames used, test rounds, and time taken also change accordingly.

This algorithm is compared with earlier proposed algorithms [2–4, 8] of the authors (Table 1). The number of communication messages required to discover and diagnose the arbitrary network is much less as compared to other algorithms [2–4, 8], when algorithms are executed for the system diagram as shown in Fig. 6.

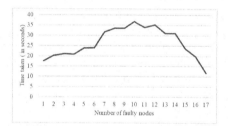

Fig. 14. Time taken for one diagnostic cycle v/s Number of faulty nodes

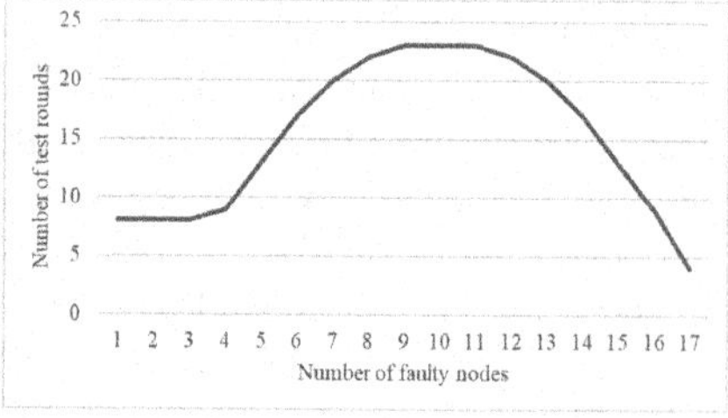

Fig. 15. Number of test rounds required v/s Number of faulty nodes

Table 1. Comparison of proposed OED-AFD with other algorithms

	Required network type	Execution type	Average number of messages	Broadcasting required
L-AFD [2]	Static	Periodic	77	Yes
C-AFD [3]	Static	Periodic	236	Yes
DP-AFD [4]	Dynamic	Periodic	65	Yes
FFD-A [8]	Static	Periodic	70	Yes
OED-AFD	Dynamic	Periodic & Event Triggered	41	No

5 Conclusions

The proposed OED-AFD correctly diagnoses every faulty node in any arbitrary network. The algorithm is executed in parallel on all the nodes of the network.

OED-AFD does not require broadcasting of the results at the end of the diagnostic cycle as in the case of [2–4]. This is because in the case of OED-AFD, at the end of diagnostic cycle, each node has information about every other node in the network. Thus, a huge number of messages required for broadcasting are not needed in the case of OED-AFD, thereby saving the network bandwidth and reducing the network traffic.

OED-AFD uses a definite number of messages and a definite number test rounds of to discover and diagnose any arbitrary distributed system. It is found that the proposed algorithm performs better with respect to communication messages required to discover and diagnose the arbitrary network.

The time taken by a complete diagnostic cycle of OED-AFD is also definite. OED-AFD is optimal in terms of the communication messages required and the time required for diagnosis.

OED-AFD supports the new nodes entry during the diagnostic cycle. OED-AFD is executed periodically. However, a new diagnostic cycle of OED-AFD is also triggered when an event such as a re-paired node re-entry or a new node entry occurs in between the two diagnostic cycles.

It is required to have at least two connected fault-free nodes for the execution of OED-AFD. Also, the algorithm does not detect the partitioning of the network. Therefore, the algorithm gets executed independently at every partition. The future scope of the OED-AFD may be to overcome the above-mentioned limitations.

References

1. Kelkar, S., Kamal, R.: Adaptive fault diagnosis algorithm for controller area network. IEEE Trans. On Ind. Electron. Soc. **61**(10), 5527–5537 (2014)
2. Manghwani, J., Taware, R., Kelkar, S., Chinde, P., Alwani, S.: Leader based adaptive fault diagnosis algorithm for distributed systems. In: 2017 International Conference on Information, Communication, Instrumentation and Control (ICICIC), pp. 1–6. Indore (2017)
3. Kelkar, S., Yeole, D.G., Sinkar, M.B., Jagtap, P.B., Zagade, D.S.: Coordinator-based adaptive fault diagnosis algorithm for distributed computing systems. In: 2017 International Conference on Advances in Computing, Communications and Informatics (ICACCI), pp. 745–751. Udupi (2017)
4. Sarna, L., Shenolikar, S., Kulkarni, P., Deshpande, V, Kelkar, S.: Distributed periodic approach for adaptive fault diagnosis in distributed systems ArXiv abs/1812.07782 (2018)
5. Ziwich, R., Duarte, E.P.: A nearly optimal comparison based diagnosis algorithm for systems of arbitrary topology. IEEE Trans. Parallel Distrib. Syst. **27**(11), 3131–3143 (2016)
6. Bianchini, R.P., Buskens, R.: An adaptive distributed system-level diagnosis algorithm and its implementation. In: Proceedings FTCS-21, pp. 222–229 (1991)
7. Duarte, E.P., Nanya, T.: A hierarchical adaptive distributed system-level diagnosis algorithm. IEEE Trans. Comput. **47**(1), 3445 (1998)
8. Kulkarni, P., Deshpande, V., Sarna, L., Shenolikar, S., Kelkar, S.: Fault diagnosis in distributed systems using accuracy technique. ArXiv abs/1812.07771 (2018)

9. Tran, H.M., Schonwalder, J.: DisCaRia – distributed case-based reasoning system for fault management. IEEE Trans. Netw. Serv. Manage. **12**(4), 540–553 (2015)
10. Bagchi, A., Hakimi, S.L.: An optimal algorithm for distributed system-level diagnosis. In: Proceedings 21st IEEE International Symposium on Fault-Tolerant Computing, Montreal, Canada (1991)
11. Duarte, E.P., Weber, A., Fonseca, K.V.O.: Distributed diagnosis of dynamic events in partitionable arbitrary topology networks. IEEE Trans. Parallel Distrib. Syst. **23**(8), 1415–1425 (2012)
12. Zhao, L., Liu, Z., Liu, W., He, H., Wang, Y.: G-FDDS: a graph-based fault diagnosis in distributed systems. In: 2nd IEEE International Conference on Computational Intelligence and Applications, pp. 559–567 (2017)
13. Punyotoya, S., Khilar, P.: A novel fault diagnosis algorithm for k connected distributed clusters. In: International Conference on Industrial Electronics, Control and Robotics December, pp. 101–105 (2010)
14. Cui, Y., Shi, J., Wang, Z.: Fault propagation reasoning and diagnosis for computer networks using cyclic temporal constraint network model. IEEE Trans. Syst. Man Cybern. Syst. **47**(8), 1–14 (2017)
15. Pourmoghadam, M.R., Sedaghat, Y., Ghodsollahee, I.: Improving the fault tolerance and efficiency of CAN communication networks based on bus redundancy. In: 2020 10th International Conference on Computer and Knowledge Engineering (ICCKE), pp. 549–554 (2020)
16. Lu, L., Zhengguo, X., Wang, W., Sun, Y.: A new fault detection method for computer networks. Reliab. Eng. Syst. Saf. **114**, 45–51 (2013)
17. Qi, X., Li, J., Wang, Z., Liu, L.: Probabilistic probe selection algorithm for fault diagnosis in communication networks. Comput. Netw. **198**, 108365 (2021)
18. Tayal, A., Sharma, N., Hubballi, N., et al.: Traffic dynamics-aware probe selection for fault detection in networks. J. Netw. Syst. Manage. **28**(4), 1055–1084 (2020)
19. Vargas-Arcila, A.M., Corrales, J.C., Sanchis, A., Gallón, Á.R.: Peripheral diagnosis for propagated network faults. J. Netw. Syst. Manage. **29**(2), 1–23 (2021)
20. Gontara, S., Boufaied, A., Korbaa, O.: Fault localization algorithm in computer networks based on the boolean particle swarm optimization. In: 2019 IEEE International Conference on Systems, Man and Cybernetics (SMC), pp. 4347–4352 (2019)
21. Song, J., Lin, L., Huang, Y., Hsieh, S.-Y.: Intermittent fault diagnosis of split-star networks and its applications. IEEE Trans. Parallel Distrib. Syst. **34**(4), 1253–1264 (2023). https://doi.org/10.1109/TPDS.2023.3242089
22. Hajshirmohamadi, S., Sheikholeslam, F., Davoodi, M., Meskin, N.: Event-triggered simultaneous fault detection and consensus control for linear multi-agent systems. In: 2016 Second International Conference on Event-based Control, Communication, and Signal Processing (EBCCSP), pp. 1–7 (2016)
23. Zhao, Z., Wang, Z., Zou, L., Wang, Y., Guo, J.: Event-triggered fault estimation for networked systems with redundant channels. In: 2020 39th Chinese Control Conference (CCC), pp. 4492–4497 (2020)
24. Harchol-Balter, M., Leighton, T., Lewin, D.: Resource discovery in distributed networks. In: Proceedings of the Eighteenth Annual ACM Symposium on Principles of Distributed Computing – PODC (1999)

Analysis of Routing in IOT

Garima Shrivastava[1] and Mahesh Kumar[2(✉)] ⓘ

[1] Infosys Bangaluru, Bangalore, India
[2] Computer Science and Engineering, Jaypee University of Engineering and Technology, A.B
Road, Guna 473226, India
mahesh.chahar@gmail.com

Abstract. Internet has grown exponentially in recent years. Now-a-days, everything that performs some function is supposed to be connected to the internet. Most
of the devices in everyday life are not only small in size but also lack processing
capabilities. This becomes a bottleneck in bringing these devices online. Internet
of things often referred to as IOT is a concept that involves tools, technologies and
methodologies to connect these devices to the internet. In order to achieve this
objective, many standards and protocols are devised so far to ensure proper functionality of IOT networks. IOT networks consist of usually a very large number of
devices with minimal computation and storage capabilities. Hence, IOT networks
should be designed in such a way to support a high degree of scalability. Another
key characteristic of IOT networks is wireless connectivity. Many technologies
like Bluetooth, Wi-Fi, Zigbee, etc. are widely used to establish connectivity among
these devices. The fact that IOT consist of so many devices lead to generation of a
huge amount of data. Routing this data with limited capacities becomes one of the
core issues of IOT. One of the most widely used routing protocol for IOT is RPL.
To gain a better understanding of the applicability and operationality of routing
protocol like RPL, this paper suggests analysis through simulation using Cooja
simulator over Contiki operating system. The simulation result helps in improving decision making process regarding the choice of right version of the routing
protocol according to given network environment.

1 Introduction

Internet of things often abbreviated as IOT, has become a major area of consideration
for many researchers in networking domain. "Internet of things" is a term used to
describe a world that consists of a huge number of devices with internet connectivity.
The basic reason the IOT environments are different from the conventional networking
environments is due to the constrained nature of the devices [1]. These devices are
usually sensors and actuators that lack storage and computation resources with memory
and RAM ranging up to few kilo bytes. Such devices are called reduced functional
devices (RFDs). An example of such devices can be temperature sensors deployed in a
room etc. On the other hand, the conventional mainstream networks usually consist of
fully functional devices (FFDs). These FFDs doesn't lack any computation or storage
resources [1]. Example of these devices are smart phones, personal desktops etc. Hence

the traditional protocols and techniques cannot be deployed in IOT environments to achieve full functionality of such systems. This gives rise to a whole new field of research and development in this area.

2   IOT Network Stack

Given the wide differences between the traditional networks consisting of fully functional devices and IOT constrained networks, it is logical enough to devise a separate network stack for such environment as the conventional ones would not be able to fully justify the structure and operationality of IOT systems.

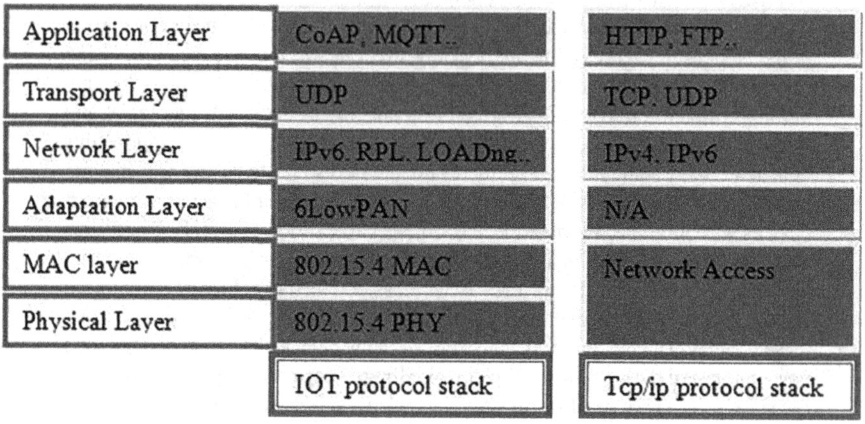

Application Layer	CoAP, MQTT..	HTTP, FTP..
Transport Layer	UDP	TCP, UDP
Network Layer	IPv6, RPL, LOADng..	IPv4, IPv6
Adaptation Layer	6LowPAN	N/A
MAC layer	802.15.4 MAC	Network Access
Physical Layer	802.15.4 PHY	
	IOT protocol stack	Tcp/ip protocol stack

Fig. 1. IOT vs. Tcp/ip protocol stack

Various protocols have been devised for IOT giving rise to a whole new IOT protocol stack [2].

Figure 1 shows the basic difference between the protocol stacks of IOT and conventional TCP/IP.

Application layer: IOT applications use light weight protocols for data transfer like CoAp,MQTT etc. CoAP stands for constrained application protocol. CoAP is very similar to HTTP in functionality however incurs way less overhead comparatively [2]. CoAP allows data transfer in reliable and non-reliable modes. MQTT stands for message queuing transport telemetry protocol. This protocol works in client-broker mode. The clients can subscribe to a topic by requesting the broker. The broker in turn manages subscription to various topics and their validity.

Transport layer: Transport layer is responsible for establishing process to processcommunication. IOT networks being constrained in nature, highly relies on UDP protocol for connection establishment due to light weight nature of the protocol.

Network layer: The core objective of network layer is to route packets over the network.Traditional Tcp/ip networks uses routing protocols like RIP, OSPF etc. These routing protocols demand very high processing and memory resources. Therefore it is logical enough to devise energy aware routing protocols for IOT systems with an objective to save power consumption that can operate on low memory, low power without compromising with the functionality of IOT systems. Certain routing protocols for IOT networks are RPL, LOADng etc.

Adaptation layer: IOT systems run over IPv6 because of the large address space available underthis protocol. For IOT networks, an adaptation layer is defined by IEEE 802.15.4 group [3]. This group proposed 6LowPAN as the adaptation standard. 6LowPAN stands for IPv6 over low power personal area network. This standard provides rules and specifications to support ipv6 addresses over low power and lossy networks (LLNs). The need for this adaptation layer arises due to the very large size of IPv6's maximum transmission unit (MTU) which is 1280 octets. However, the MTU size suggested by 802.15.4 is 127 octets [3]. Another area of consideration under this standard is to achieve compatibility between the existing IPv4 protocols which is still the most widely used protocol over the mainstream internet. Adaptation layers incorporate IPv6 packets by performing basic operations like header compression, fragmentation and link layer forwarding. It also supports mesh under routing which means that the fragmented packets can reach the destination for reassembly via different routes.

802.15.4 MAC: MAC stands for medium access control. As the name suggests, 802.15.4 MAClayer provides link level specifications for channel access and channel management for low power and lossy networks. Channel access is associated with sensing the channel periodically to detect and transmission or possible collision [4]. This layer allows a node to associate or disassociate from a network. It does so by generating beacon frames to enable network discovery. Another special feature provided by this layer for RFDs is duty cycling. Duty cycle is defined as the fraction of time the radio is turned on. Duty cycling enables a device to go in sleep mode when the channel is idle for too long, hence saving power.

802.15.4 PHY: 802.15.4 PHY standard provides specification for the physical layer ofconstrained LLNs. These networks are often established using wireless connectivity. This standard provides the suggested characteristics of these wireless channels. Two PHY options are provided by this standard with frequency band as the basic difference. 2.4 GHz band with 250 kb/s transmission rate has worldwide availability [5]. The 868/915 MHz PHY specifies operation in the 868 MHz band in Europe with 20 kb/s data rate and 915 MHz ISM band in the United States with 40 kb/s data rate. 802.15.4 PHY standard also enables devices to have small 8 bit or extended 64 bit addresses.

3 Routing in IOT

One of the key characteristic of IOT systems is scalability. IOT networks are highly scalable due to very large number of interconnected devices. Discovering paths and forwarding packets to their intended destination keeping energy awareness in consideration, is a critical and crucial task in IOT environments [6]. This goal cannot be achieved by using routing protocols designed for fully functional devices. Another reason that the same can't be used in constrained networks is that all protocols must comply with 802.15.4 standard so that desired compatibility between protocols can be attained without compromising the full functionality of IOT systems. This is to ensure that the network remains operational and fully functional [7]. For this purpose, new routing protocols are designed specifically for low power and lossy networks. These routing protocols not only comply with 802.15.4 standard but also inhabit characteristics like energy awareness, small and less complex code footprint along with enough flexibility to accommodate future improvements.

3.1 The RPL Protocol

RPL stands for routing over low power and lossy [6]. This protocol is one of the most widely used routing protocols in IOT networks defined by IETF ROLL group. RPL is distance vector, source routing protocol. Distance vector routing protocol identifies optimal paths to a destination by keeping track of the distance to that destination. The distance usually signifies hop count but can assume any other costs as well like ETX or link energy etc. [8]. Source routing usually means that the nodes can partially or completely state the route to the destination.

RPL functions by building a DODAG (Destination oriented directed acyclic graph). DODAG consist of a sink node and many sender nodes. The traffic can flow in both directions [9]. The sink node usually collects all the data from the sender nodes. RPL builds a DODAG with the help of an objective function. This objective function calculates a predefined metric/metrics over the links and chooses the best optimum path that minimizes the value of objective function [10]. The metrics used can be hop count or ETX etc. One can also use combination of such metrics. RFC 6551 suggests creating a Routing Metric/Constraint Type for Routing Metric/Constraint object types that ranges from 0 to 255 where value 0 is undefined [11].

Value	Meaning
1	Node state and attribute
2	Node energy
3	Hop count
4	Link throughput
5	Link latency
6	Link quality level
7	Link ETX
8	Link color

Two of the most used metrics are hop count and ETX. Hop count is defined as the minimum number of hops to reach a certain destination [12]. ETX is defined as the minimum number of transmission a node needs to make in order for the successful delivery of a packet.

RPL carry all this information with the help of three kinds of control packets:

DIO: DODAG information object (DIO) is transmitted by nodes in order to join a DODAG and obtaining necessary information regarding DODAG.

DIS: DODAG information solicitation (DIS) is sent by nodes in a DODAG to solicit information out from a neighbor.

DAO: DODAG advertisement object (DAO) is used for upward communication in an object.

This includes information about topology changes.

4 Analysis of RPL

Keeping in consideration the widening popularity of RPL protocol, it is wise enough to analyze it to obtain full functionalities of the protocol. RPL protocol in itself can perform under various objective functions. In order to achieve a better understanding of the appropriate scenarios for these different objective functions, simulation proves to be a good practice to follow. Simulation results will help in deciding which objective function is more suitable for given network conditions.

4.1 Experimental Setup and Simulation

RPL protocol is analyzed using cooja simulator on Contiki operating system. Contiki is an open source operating system specifically designed for IOT environments. The simulation will be carried out over two different objective functions viz. MRHOF and OF0.

Cooja/Contiki: Cooja simulator provides a real time environment to perform analysis of routing protocol. It also provides emulated hardware of sensor devices like sky motes, z1 motes etc. Energy aware radio propagation modes are also available with this simulator.

Key features:

- Memory allocation
- Full IP networking
- Power awareness
- Cooja Network Simulator
- Emulated hardware platforms

MRHOF: Minimum rank hysteresis objective function (MRHOF) is an energy aware objective function that computes the optimal path with minimum cost by considering ETX as metric. MRHOF does so by preventing excessive churn in the network. It.

OF0: Objective function zero (OF0) is an objective function which is more static in nature as it considers hop count as it's metric. It finds and compares the rank of the nodes and chooses the best preferred parent based on this rank (see Table 1).

DIO minimum interval: DIO packets are sent periodically for the establishment of DODAG. DIO minimum interval decides the lowest possible time lapse between the transmissions of consecutive DIO packets. DIO minimum interval is managed by trickle timer algorithm in cooja/Contiki which is set to approximately 4 s.

UDGM-Distance loss: Unit disk graph model-distance loss is a radio propagation mode supported by Contiki. In this mode the reception range is shown in the form of a green circular disk around the node. Distance loss means that the reception ratio of a node decreases with the increase in distance as is shown as a grey circular disk around a node (see Fig. 2).

Table 1. Experimental Parameters

Operating system	Contiki
Simulator	Cooja
Objective functions	MRHOF,OF0
DIO minimum interval	~ 4 s
Number of nodes	10
Mote type	Sky mote
Radio propagation mode	UDGM-Distance loss
Simulation time	1 h 13 min

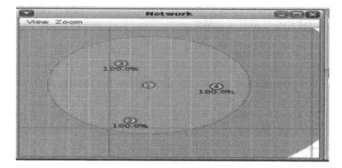

Fig. 2. UDGM-Distance loss

4.2 Simulation Results

First the RPL network is simulated using MRHOF objective function. Following results are obtained:

A. DODAG MRHOF

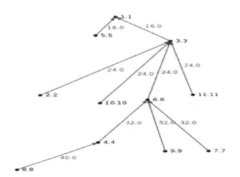

Fig. 3. DODAG MRHOF

The above DODAG shows the DODAG created by MRHOF objective function. Much of the traffic goes through nodes 3.3 and 6.6 making them more heavily loaded than other nodes. This indicates towards a possibility of these nodes dying out faster than any other node (see Fig. 4).

B. Historical power consumption MRHOF

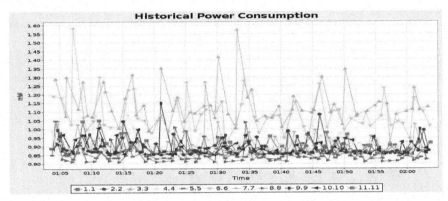

Fig. 4. Historical Power Consumption MRHOF

The above graph shows comparatively higher power consumption by nodes 3.3 and 6.6. It is clear from Fig. 3 that these two nodes are heavily loaded than other nodes. This indicates that power consumption of a node increases with increase in its child nodes (see Fig. 5).

C. Beacon Interval

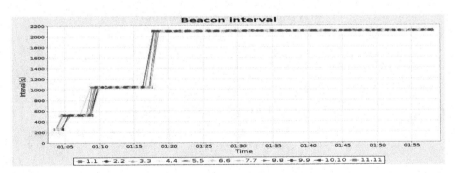

Fig. 5. Beacon interval

Beacon interval is defined as the time lapse between two consecutive beacon frames. Beacon frames are necessary for a node to associate or disassociate from a network. The above graph shows rapidly increasing beacon interval. This is so because initially, beacon frames must be sent within a short interval in order to keep network convergence time small. But as the network is setup, keeping beacon interval large can save excessive network traffic (see Fig. 6).

D. DODAG OF0

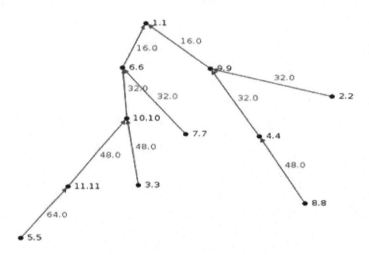

Fig. 6. DODAG OF0

As is clear from the above graph, topology has changed with the change in objective function from MRHOF to OF0. Topology from OF0 shows a comparatively even load distribution than in MRHOF.

E. Historical power consumption OF0

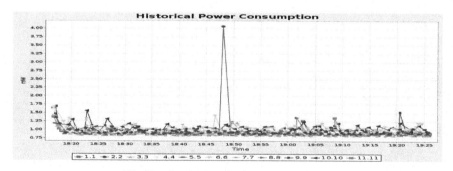

Fig. 7. Historical Power Consumption OF0

Considerable difference can be observed between the power consumption between MRHOF and OF0 objective functions. In Fig. 7, OF0 shows comparatively very less power consumption. One possible reason can be static nature of hop count metric used in OF0.

4.3   Comparison Between MRHOF and OF0

i. Routing Metric comparison

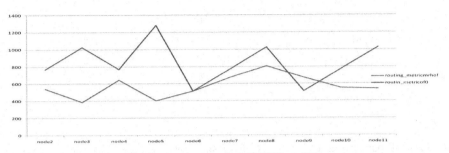

Fig. 8. Routing metric: MRHOF vs. OF0

As can be observed from the above graph, there lies a significant difference between the routing metric values of both objective functions. The curve obtained from OF0 is significantly higher than the MRHOF. It can also be noted that the curve attains a very sharp high peak at leaf nodes when OF0 is used. This indicates that the larger the network grows in size, OF0 returns significantly increasing values of routing metric which is undesirable (see Fig. 8).

ii. Transmit duty cycle comparison

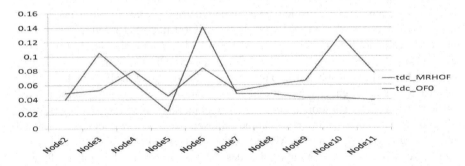

Transmit duty cycle attains sharp peaks at nodes 3 and 6 in case of MRHOF. The reason being, these two nodes are exposed to comparatively large amount of load. However in case of OF0 both the amount of load as well as distance from sink might be the possible reason behind such pattern of transmit duty cycle (see Table 2).

iii. Comparison summary

The comparison summary shows considerable difference in the values of routing metrics and average transmit duty cycle of both the functions. From the above data, it can be stated that not only a carefully planned topology can help in even load distribution in

Table 2. Comparison Summary

S.No.	Parameters	MRHOF	OF0
1	Average received packets	61.900	70.300
2	Average lost packets	0.2	0.1
3	Average transmit duty cycle	0.059	0.069
4	Average listen cycle	0.687	0.689
5	Average beacon interval	30 min. 09 s	29 min. 37 s
6	Average routing metric	571.711	844.800

the network but also efficient values of transmit duty cycles which relates to power consumption directly leading to increased lifetime of the sensor nodes. It is also found that there exists a tradeoff between routing metric and power consumption. For smaller network OF0 is preferable and for larger networks, MRHOF is found to be more cost effective.

5 Conclusion

Simulation results helps in exploiting full applicability and functionality of a routing protocol. It was observed that network density impacts the lifetime of a node. A carefully planned topology would help in achieving even load distribution which in turn would increase the lifetime of a node. It was also observed that duty cycle plays a very crucial role in routing. Unsynchronized duty cycle can become the possible cause of packet loss. Another factor that is considerable is routing metric. The main goal of an objective function is to minimize the routing metric. In case of OF0, the value of routing metric was found to be comparatively very large than MRHOF which is not desirable. However, OF0 results in significantly low power consumption than MRHOF. The basic reason MRHOF incurs high power overheads is due to the energy aware nature of the ETX metric. As the link characteristics might change with time, this results in varying values of the routing metric which would incur more processing and computation overheads. However, when the topology is very large, trade-off can be achieved between power consumption and routing metric. In such scenarios, MTHOF proves to be a better choice while when the topology consists of a small number of nodes, OF0 is observed to be more preferable. This is so because OF0 relies upon hop count for cost calculation. Therefore, as the topology increase in size, leaf nodes would incur more cost in terms of routing metric. Hence, in case of shorter topologies, it is wiser to use OF0 as the routing metric values will be low along with less power consumption.

Another more efficient approach to achieve full functionality of the RPL protocol would be to use a threshold value of routing metric below which OF0 can be used as the objective function and above that protocol would switch to MRHOF. This will result in achieving better trade-off between power consumption and routing metric. The threshold value can be identified with the help of available research data in this area and further experimentation.

References

1. Sethi, P., Sarangi, S.R.: Internet of Things: architectures, protocols, and applications. J Electr. Comput. Eng. **2017**, 1–25 (2017)
2. Atzori, L., Iera, A., Morabito, G.: The Internet of Things: a survey. Comput. Netw. **54**(15), 2787–2805 (2010)
3. Kushalnagar, N., Montenegro, G.: 6LoWPAN: Overview, Assumptions, Problem Statement and Goals (2005). https://tools.ietf.org/html/draft-kushalnagar-lowpan-goals-assumptions-00
4. https://standards.ieee.org/findstds/standard/802.15.4-2015.html IEEE 802.15.4
5. Warakar, P., Chavan, N.: Wireless integrated system using IEEE 802.15.4. Int. J. Eng. Res. Appl. **2**, 395–399 (2012)
6. Lotte Steenbrink Ausarbeitung Fakultät Technik Routing in the Internet of Things. Hamberg University of Applied sciences http://www.inet.haw-hamberg.de/teaching/ss-2014/master-projekt/aw1_Iotte_steenbrink.pdf (2014)
7. Xin, H.-M., Yang, K.: Routing protocol analysis for internet of things. In: Information Science and Control Engineering 2nd International Conference (2015)
8. Parasuram, A., Culler, D., Katz, R.: An analysis of the rpl routing standard for low power and lossy networks. Electr. Eng. Comput. Sci (2016). http://www2.eecs.berkley.edu/Pubs/TechRepts/2016/EECS-2016-106.pdf
9. Vucini´c, M., Tourancheau, B., Duda, A.: Performance comparison of the RPL and LOADng routing protocols in a home automation scenario. Wirel. Commun. Networking Conf. https://arxiv.org/abs/1401.0097. (2014)
10. Dhumane, A., Prasad, R., Prasad, J.: Routing issues in Internet of Things: a survey. In: IMECS, vol. 1 (2016)
11. Winter, T., et al.: RPL: Routing protocol for low-power and lossy networks. IETF (2012)
12. Zaatouri, I., et al.: Study of routing metrics for low power and lossy network. In: Smart, Monitored and Controlled Cities International Conference (2017)

Autism Children Behavioural Identification from Facial Regions Through Thermal Image Interpretations

Kandukuri Muniraja Manjunath[✉] [ID] and Vijayaraghavan Veeramani[ID]

Department of Electronics and Communication Engineering, Vignan's Foundation for Science, Technology and Research (Deemed to Be University), Vadlamudi, Guntur, Andhra Pradesh 522213, India
kmanju286@gmail.com

Abstract. When Autism children are treated with drugs, there are a lot of problems facing the physician. To prescribe appropriate medicine, usually at clinics, drug dosage levels are anticipated based upon the behavioural activity of the children with the help of direct health care staff, but till today, behaviour assessment is observed at clinics for just a few minutes to assess the behavioural activity of the children, with feedback collected from the parents and school teachers by word-of-mouth consideration only drug has to be prescribing. This is not enough to prescribe appropriate medicine. The physician's need for continuous monitoring is a vital problem. To solve this problem by fixing thermal video cameras wherever the children are. The continuous monitoring needs to identify the aggressive behaviour of the particular children. The thermal images cropped from that video that are going to be used to analyse the children's behaviour in facial emulations are eye, forehead, mouth, nose, and cheek region pixels. Those images are affected by scaled and unscaled noise. The proposed efficient filtering techniques are discrete wavelet transformation (DWT) and stationary wavelet transformation (SWT). To remove the scaled and unscaled noises for an accurate assessment of facial emotions. The proposed DWT, SWT filtering properties for assessing accurate mood, emotion, and repetitive behaviours of autistic children have more than 94% accuracy compared to specified parameters Aberrant Behaviour Checklist (ABC).

Keywords: Autism spectrum disorder · thermal camera · threshold values · de-noising

1 Introduction

An image is obtained from a camera or thermal imaging equipment that cannot be used directly. Those images are corrupted by random variations called noise. The noise-affected image may be defined by several kinds of sprinkle noise, brackish noise, and impulse noise, which cause poor contrast in an image. Filtering techniques include smoothing, sharpening, and edge enhancement. In preprocessing, suppress either variation frequencies of a thermal image to enhance either edge detection or correction of

© The Author(s), under exclusive license to Springer Nature Switzerland AG 2023
M. Singh et al. (Eds.): ICACDS 2023, CCIS 1848, pp. 96–108, 2023.
https://doi.org/10.1007/978-3-031-37940-6_9

the image, emphasising certain intensity levels to remove unwanted components of the image [1]. Thermal images are used in several applications, such as military, industrial, and medical. The use of thermal images in medicine to detect several diseases based on skin, figure, and body temperature can reproduce the presence of hypertension in underlying tissues. A wide range of disorders are susceptible to being recognised via thermal imaging, including those where blood flow is altered because of a brain abnormality that causes autism developmental disorder (ASD) in children or where skin temperature may show the presence of inflammation in deeper tissues.

A thermal imaging system is also called thermography, which is simply converting infrared (IR) radiation (heat) into visible objects in the form of a digital image of a scene and measuring the infrared radiation caused by thermally reflected pixels emitted by the radiation. The thermal images captured by a thermal camera, which consists of a sensor that absorbs IR radiation [2], are called thermal images. Thermal radiation consists of the long wavelengths of light just away from the visual perception of EM radiation, which carries power, but infrared absorption increases its temperature, causing it to emit IR light called blackbody emission, and its temperature is able to sense the IR radiation, which allows creating a thermal image. Wherever the children's moments are there, we are fixing a thermal imaging camera at highly secured places when the children exhibit particularly aggressive behaviour. The children's images only crop out from that video, and we are going to assess the continuous monitoring of the children's behaviour analysis to assess facial emotions identifies the thermally reflected pixel values through thermal image and filtering techniques for DWT and SWT [3].

2   Contributions

2.1   Discrete Wavelet Transformation (DWT)

The DWT exhibits oscillatory behaviour for a short period with a zero integral value. The possibility of time sifting, scaling, and orthogonality of translation applies to images. However, the Fourier transform uses single functions, dilations, and translation for the generation of orthogonal basis.

2.2   Stationary Wavelet Transformation (SWT)

The SWT is used to create thermal images and has levels of high-pass and low-pass filters. The SWT is not decimated and is similar in length to the two original sequences. The SWT filters, however, are altered at each level by padding them with zeros.

2.3   Hard and Soft Thresholding

Methods of Hard and Soft Threshold, which are divided further into The Bal.sparsity-norm(sqrt) parameter, which defines the multiplicative threshold rescaling option accessible for 2-D stationary wavelet, is fixed from threshold to penalise high, medium, and low. For examining any type of thresholding operation, starting from the coefficients of thresholding from minimum to maximum at each level as depicted in Fig. 1. After denoising steps, thresholds are based on a subset of the coefficients in the stationary wavelet transformation (SWT).

Denoising of an image with a defined noise structure that is either unscaled or scaled, with the horizontal, vertical, and diagonal coefficients taken into consideration as in Eq. (1)

$$p = d + n \tag{1}$$

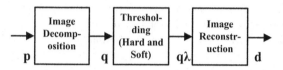

Fig. 1. Blocks of de-noising following steps

Signals having zero mean noise in both scaled and unscaled noise, constant spectral power density (SPD) variance, and discrete domain noise removal are described in [4]. Wavelet coefficients to denoise and residue threshold adhering to regulations, such as for denoising, assuming independent coefficients for the hard and soft thresholds Let (•) represent the forward and (−1) the inverse wavelet coefficient transformations, respectively. Let D (•) represent the soft denoising threshold operator, which signifies the symbol. Denoising p [n] with wavelet coefficients is intended to recover d [n]. The steps that the wavelet transform takes are listed in Eq. (2).

$$Q = \omega(P) \tag{2}$$

Wavelet transforms Threshold condition

$$d = \omega^{-1}(q\lambda) \tag{3}$$

where the parameter d stands for wavelet coefficients and set value for the threshold. This approach, known as "the signal recovery from threshold condition," restores the thermal image through a noisy equation as

$$\lambda = \sigma\sqrt{2\log(N)} \tag{4}$$

where σ denotes mean absolute deviation (MAD) and N is noisy signal, σ can be expressed as

$$\sigma = (1/0.6745)\, median\,(adj) \tag{5}$$

To eliminate scaled and unscaled noises from facial movements, de-noising methods like SWT and DWT wavelet transformations need certain threshold levels and coefficients [5].

3   Literature Survey

Table 1 lists the benefits and drawbacks of the suggested method in comparison to alternative approaches. A summary of the research on thermal image filtering methods that solve various problems.

Table 1. Existing methods and their merits and demerits

Year/Author	Ref.	Noise type	Existing methods	Benefits and Drawbacks
[2017]/(Liu et al.)	[8]	A multiplicative hum	Wavelet and bilateral filters	Better than older adaptive techniques for thermal picture denoising's inadequate evaluation metrics
[2020]/(Li et al.)	[6]	Ambient noise	Gabor filtration	Failing to respond to images of different colours
[2016]/(Shin and Huang.)	[9]	Sporadic noise	Filtering for spatial forecasting	Reduce the clutter and noise-making items
[2020]/(Wang et al.)	[10]	Squeaky noise	Directed filter	Enhances tracking, object detection, and stability after leftover infrared pictures have been filtered
[2016]/(Mao et al.)	[11]	Unreliable noise	Space and chromatic filtering	Tuning is the fundamental problem with image pixel-to-pixel fluctuation
[2018]/(Zing et al.)	[12]	Noise stripes	Sharp filter, spatial filtering, and a two-step filter	Multiple images of noise cannot be eliminated by infrared image conservation and better improvement
Proposed method		Scaled and unscaled noises and also remove artefacts in facial skin emotions	SWT DWT and sub-wavelet coefficients	Repetitive behaviour analysis of thermal imaging for facial emotions identifies gaps and aids in the prescription of the proper dosage levels

4 Problem Statement

Acquiring images or videos from a thermal camera for analysis using pre-processing techniques by applying DWT and SWT wavelet transformations to those thermal images using efficient filtering techniques to remove scaled and unscaled noises, as shown

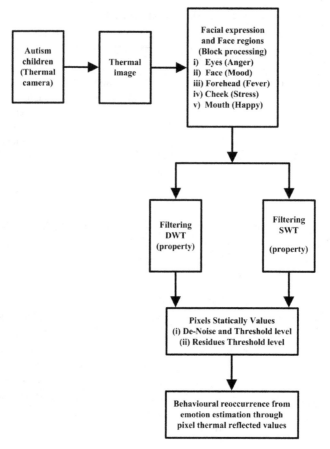

Fig. 2. Pre-processing of thermal image following steps

in Fig. 2. In pre-processing, it identifies the scaled and unscaled facial emotions and the behaviour of each region. The autism child's emotions different emotions through thermal images.

By applying filtering techniques to the facial emotions of the thermal image, scaled and unscaled noises are removed. The following contributions such as,

- Images of autistic children are collected from a thermal camera and then sent into a denoising process that incorporates the SWT and DWT algorithms for denoising to remove scaled and unscaled noises, respectively.
- Then, several face emotions and thermally reflected pixel values from a thermal camera were examined, including cheek region pixels (92.0) for sadness, mouth region pixels (95.5) for joy, eye region pixels (97.0) for frustration, and face region pixels (97.0) for expression. Pixels (98.3) for hyperactive are extracted and used to detect children's emotions using pixel thermal reflected values, and
- Identified emotions are input into SVM classification for behaviour analysis. Figure 2 shows the identification of different facial emotions from thermal image de-noising

and threshold levels by applying filtering techniques for DWT and SWT to remove the scaled and unscaled noises and identify hard and soft threshold levels for facial emotions with ambient temperature. De-noise images of different wavelet types and threshold [6] levels, distribution of histograms, and cumulative histogram wavelet coefficients at various sub-bands such as haar, sym, db, and coif are shown in Figs. 3 and 4. Thermally reflected image pixel parameters, such as pixels in the cheek area

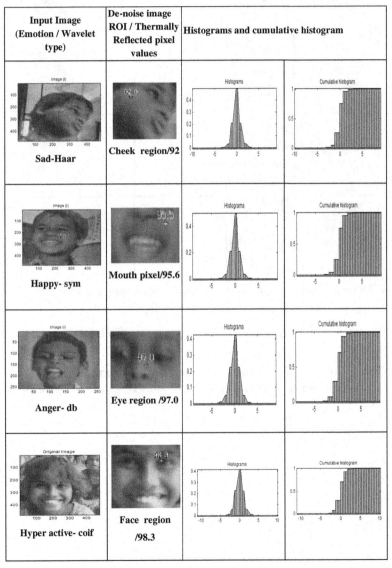

Fig. 3. Performance measures for stationary wavelet transformations with various wavelet coefficients.

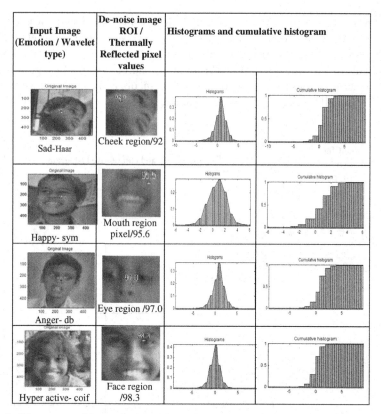

Fig. 4. Discrete wavelet transforms performance metrics for different wavelet coefficients

for depression, mouth region pixels for joy, eye section pixels for upset, and face area pixels for restlessness.

Subjective Comparisons between DWT and SWT Performance metrics of DWT Table 2 shows de-noise and threshold levels compares De-Noise and Threshold levels of Table 3.

Table 2. Threshold level performance metrics for DWT

Child ID	DWT Wavelet 2D	De-Noise Threshold Levels		Standard Threshold Levels	
		Point 1	Point 2	Mean	Deviation
CH-1	haar-2D	3.911	5.062	−0.002135	1.128
	db	3.405	5.079	−0.002235	1.128
	sym	3.405	5.079	−0.001865	1.4
	bior	3.405	5.079	−0.002235	1.188
	rbio	3.405	5.079	−0.002435	1.788
	dmey	3.988	5.826	0.8995	1.145

Table 3. Threshold Levels Performance Metrics for SWT

Child ID	Stationary Wavelet Transform Denoising 2D	De-Noise Threshold Levels		Standard Threshold Levels	
		Point 1	Point 2	Mean	Deviation
CH-1	haar-2D	4.042	4.476	0.7601	1.38
	db	4.042	4.476	0.7601	1.38
	sym	4.782	4.576	0.7756	1.124
	coif	4.882	4.876	0.7469	1.127
	bior	4.782	4.776	0.7601	1.138
	rbio	4.782	4.476	0.7701	1.38
	dmey	5.098	4.915	0.7912	0.8236

5   Experimental Results and Discussion

In this section, the results of behavioural analysis based on facial emotions using SVM that has been enhanced by denoising techniques like SWT and DWT are presented. PSNR and SSIM were used to assess the quality of denoised thermal images.

$$PSNR = 10log_{10}\left(\frac{255^2}{Mean\ Square\ Error}\right)(dB) \tag{6}$$

$$SSIM(a, b) = f(l(a, b), c(a, b), s(a, b)) \tag{7}$$

where a is a reference image and b is a denoised image; $l(a, b)$ is a luminance comparison; $c(a, b)$ is a contrast comparison; $s(a, b)$ is a structure comparison.

The proposed methods, as shown in Table 4, can effectively denoise an image with both scaled and unscaled noise. We tested our proposed results in denoising and compared them to existing methods such as BM3D and WNNM. Figure 5 shows the graphical

Table 4. PSNR and SSIM for scaled and unscaled noise

Methods	Scaled Noise		Unscaled Nosie	
	PSNR	SSIM	PSNR	SSIM
BM3D	20.57	0.355	30.02	0.766
WNNM	18.73	0.288	29.98	0.760
SWT (ours)	24.23	0.523	34.12	0.824
DWT (ours)	26.82	0.721	37.34	0.888

representation of PSNR and SSIM analysis for different denoising methods. From Fig. 5, we can see that our proposed methods, such as SWT and DWT, achieved more satisfactory results in terms of PSNR and SSIM than existing methods such as BM3D and WNNM.

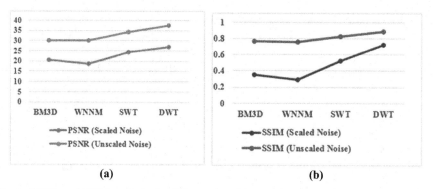

(a) (b)

Fig. 5. PSNR and SSIM analysis for different denoising methods (a) PSNR analysis (b) SSIM analysis

SVM is a machine learning algorithm based on statistical learning theory. It is a method of supervised classification centered on structural risk minimization. The SVM algorithm is capable of processing nonlinear, high-dimensional, unbalanced small sample data [7] with good generalization. Images of autistic children are captured in our work and then fed into a denoising step that includes the SWT and DWT algorithms for denoising scaled and unscaled noise, respectively. Then different face ROI emotions are extracted, like the eyes, mouth, nose, forehead, and cheek, that are used to find the emotions of children by pixel thermal reflected values, and identified emotions are fed into SVM classification for behaviour analysis. For performance measurement of behaviour analysis based on SVM, the following metrics are used:

$$Accuracy = \frac{TP + TN}{TP + TN + FP + FN} \tag{8}$$

$$Precision = \frac{(TP)}{(TP + FP)} \tag{9}$$

$$Sensitivity = \frac{(TP)}{(TP + FN)} \tag{10}$$

$$Specificity = \frac{(TN)}{(TN + FP)} \tag{11}$$

$$F_\varepsilon = (1 + \varepsilon^2)\frac{Recall \times Precision}{\varepsilon^2.Precision + Recall} \tag{12}$$

The above performance metrics and their analyzed results after denoising by SWT and DWT with SVM are shown in Tables 5 and 6. After SWT and DWT denoising, different facial emotions of thermal imaging emotions are identified through the thermal reflectance values, and then SVM-based behavior analysis is conducted, whose performance measures such as accuracy 90.24% for SWT with SVM and 92.1% for DWT with SVM) are shown in Tables 5, 6. When comparing the two methods, DWT presented the highest accuracy.

Table 5. Performance metrics measure for behaviour analysis after SWT denoising using SVM

Performance Metrics	Results (%)
Accuracy	90.24
Precision	89.65
Sensitivity	89.07
Specificity	94.52
F-score	89.36

Table 6. Performance metrics measure for behaviour analysis after DWT denoising using SVM

Performance Metrics	Results (%)
Accuracy	92.10
Precision	92.91
Sensitivity	91.24
Specificity	96.72
F-score	91.67

Behavioural analysis using SVM with SWT and DWT is graphically shown in Fig. 6. Table 7 and Fig. 7 show the comparison of the outcomes before and after denoising, such as SWT-based denoising and DWT-based denoising, for behavioural analysis accuracy using SVM.

Table 7 displayed an accuracy of 86.92% for behavior analysis based on SVM, which is done before denoising thermal imaging. And 90.24% and 92.10% accuracy for SVM-based behavioral analysis after applying denoising methods such as SWT and DWT.

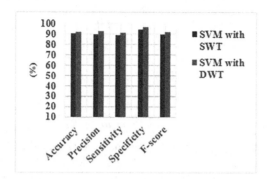

Fig. 6. Behavioural analysis using SVM with SWT and DWT

Table 7. A comparison of the outcomes before and after denoising for behavioural analysis accuracy using SVM

Method	SVM – Accuracy (%)
Behavior analysis Accuracy (before Denoise)	86.92
Behavior analysis Accuracy (after Denoise using SWT)	90.24
Behavior analysis Accuracy (after Denoise using DWT)	92.10

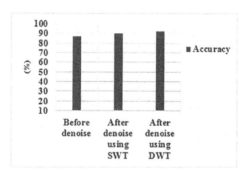

Fig. 7. Applying SVM, compare behavioural analysis prior to and after denoising.

Finally, Table 7 demonstrated the behavior analysis after denoising, which resulted in better performance.

6   Conclusion

In this research study, wavelet filtering techniques are applied for the pre-processing of thermal images for autistic children to remove scaled and unscaled noises affected by thermal images. In each image cropped from thermal video frames, the facial emotions of autistic children are identified through repetitive behaviour analysis of thermal images affected by internal and external noises. Those images are obtained by applying filters that are able to remove scaled and unscaled noise. In this paper, we propose filtering techniques such as discrete and stationary wavelet transformations (DWT and SWT) under different wavelet coefficients such as haar, sym, db, and coif. Wavelet coefficients identify pixels from various facial emotions based on their mood assessment through the aggressiveness of Autistic child behaviour is analysed through the artistic behaviour check list (ABC) according to WHO standards for measuring emotional and behavioural problems. Measurement of emotional and behavioural analysis is obtained through continuous monitoring of a child in a social environment by fixing a camera and capturing images. The behaviour analysis of an autistic child is assessed through emotions. The proposed filtering wavelets for DWT and SWT properties for assessing mood, emotion, and repetitive behaviours of autistic children have more than 92.10% accuracy compared to ABC and also differentiate based on skin temperature. Performs classification for the eye pixel region and other classifications of performance metrics as shown in Table 2 and Table 3 by applying different wavelet transformations to measure the performance metrics of patient 1. With feature scope, we have to apply denoise convolutional neural networks (DnCNN) to achieve better accuracy for facial region analysis.

References

1. Olbrycht, R., Kałuża, M.: Optical gas imaging with uncooled thermal imaging camera – impact of warm filters and elevated background temperature. IEEE Trans. Ind. Electron. **67**(11), 9824–9832 (2019). https://doi.org/10.1109/TIE.2019.2956412

2. Cao, Y., Yang, M.Y., Tisse, C.: Effective strip noise removal for low-textured infrared images based on 1D guided filtering. IEEE Trans. Circuits Syst. Video Technol. **26**(12), 2176–2188. IEEE (2015). https://doi.org/10.1109/TCSVT.2015.2493443

3. Fuentes, D., Yung, J., Hazle, J.D., Weinberg, J.S., Stafford, R.J.: Kalman filtered MR temperature imaging for laser induced thermal therapies. IEEE Trans. Med. Imaging **31**(4), 984–994 (2011). https://doi.org/10.1109/TMI.2011.2181185

4. Funk, C.C., Theiler, J., Roberts, D.A., Borel, C.C.: Clustering to improve matched filter detection of weak gas plumes in hyperspectral thermal imagery. IEEE Trans. Geosci. Remote Sensing **39**(7), 1410–1420 (2001). https://doi.org/10.1109/36.934073

5. Norouzzadeh, Y., Rashidi, M.: Image de-noising in wavelet domain using a new thresholding function. In: Proc. IEEE International Conference on Information Science and Technology, pp. 721–724, IEEE, 2011. https://doi.org/10.1109/ICIST.2011.5765347

6. Li, H.A., et al.: Medical image coloring based on gabor filtering for internet of medical things. IEEE Access **8**, 104016–104025 (2020). https://doi.org/10.1109/ACCESS.2020.2999454

7. Ding, S., Hua, X.: Recursive least squares projection twin support vector machines for nonlinear classification. Neurocomputing **130**, 3–9 (2014). https://doi.org/10.1016/j.neucom.2013.02.046

8. Liu, Y., Wang, Z., Si, L., Zhang, L., Tan, C., Xu, J.: Anon-referenceimagedenoising method for infrared thermal image based on enhanced dual-tree complex wavelet optimized by fruitfly algorithm and bilateral filter. Appl. Sci. **7**(11), 1190 (2017). https://doi.org/10.3390/app7111190

9. Shin, J., Huang, L.: Spatial prediction filtering of acoustic clutter and random noise in medical ultrasound imaging. IEEE Trans. Med. Image **36**, 396–406 (2016). https://doi.org/10.1109/TMI.2016.2610758

10. Wang, E., Jiang, P., Li, X., Cao, H.: Infrared stripe correction algorithm based on wavelet decomposition and total variation-guided filtering. J. Euro. Opti. Soc.-Rapid Pub. **16,** 1–12 (2020). https://doi.org/10.1186/s41476-019-0123-2

11. Mao, H., et al.: MEMS-Based Tunable-Fabry—perot filters for adaptive multispectral thermal imaging. J. Microelectromech. Sys. **25,** 227–235. https://doi.org/10.1109/JMEMS.2015.2509058

12. Zeng, Q., Qin, H., Yan, X., Yang, S., Yang, T.: Single infrared image-based stripe nonuniformity correction via a two-stage filteringmethod. Sensors **18,** 1–19 (2018). https://doi.org/10.3390/s18124299

Synthesis of Elementary Cellular Automata for Targeted Cache Applications

Sutapa Sarkar[1]([✉])[iD] and Mousumi Saha[2][iD]

[1] The Neotia University, Sarisha, West Bengal, India
sutapa321@gmail.com
[2] National Institute of Technology Durgapur, Durgapur, West Bengal, India
mousumi.saha@cse.nitdgp.ac.in

Abstract. Stephen Wolfram's Cellular Automata (CA) is a powerful universal computing tool. It can provide design solution for various cache applications and also other discrete physical systems with the exploitation of single length cycle attractor(s). Analysis and synthesis are needed to employ a CA in various application domain. That requires various characterization tools for one-dimensional 3-neighbourhood null boundary and periodic boundary CA like De Bruijn graph, State transition diagram, Reachability tree etc. This research paper targets to focus on analysis and synthesis of single single-length cycle attractor CA (SACA), two single-length cycle attractor CA (TACA) and multiple single-length cycle attractor CA (MACA). The novelty of the paper is to provide formal analysis of the properties of SACA, TACA and MACA into a objective-specific manner, using state transition diagram (STD). A subset of ECA rules are provided here for the design space exploration of on-chip cache applications.

Keywords: Elementary cellular automata · State transition diagram · Single single-length cycle attractor cellular automata · Two single-length cycle attractor cellular automata · Multiple single-length cycle attractor cellular automata

1 Introduction

Stephen Wolfram's proposed Cellular Automata or Elementary Cellular Automata (ECA) is simple in construction though capable to model complex discrete physical [1,2] and biological systems [3]. Analysis and synthesis which are reversible processes, are prerequisites to employ this 1-D, 3-neighbourhood CA in various application domain. For a one-dimensional CA, various characterization tools are usually employed to get the statistics of reachable states, non-reachable states, successor and predecessor states and their relationship to obtain self-loops. For a one-dimensional CA, different characterization tools available such as De Bruijn graph, Pair graph, Subset graph, Cyclic graph, State transition

M. Singh et al. (Eds.): ICACDS 2023, CCIS 1848, pp. 109–121, 2023.
https://doi.org/10.1007/978-3-031-37940-6_10

diagram, Reachability tree etc. The algebraic properties of CA can be reflected through the derived state transition diagrams (STDs) [1], which in turn gives the idea of reversible and irreversible properties of CA [5]. The complete state space evolution representation through STDs, can explore the attractor states. Attractors of unit cycle length only, attracts the interest of a large number of researchers [2,5].

CA can be synthesized with state transition diagrams to evolve with an identified subset of single length cycle attractors(s) applied in various applications for pattern classifiers [2], fault tolerant designs [7,13], digital medical image processing applications [3], memory testing [8], associative memory designing etc. The characterization of single length cycle (fixed-point) attractors, recognizes SACA, TACA and MACA [13]. SACA, TACA and MACA can give design solutions of run time, real life problems through a cost-effective VLSI implementation [2,12].

Towards CA based VLSI circuits and systems design, properties of 1-D null boundary and periodic boundary CA are explored [15–17]. The formal analysis/synthesis of the properties of SACA, TACA and MACA puts them into the target designs [13]. A subset of ECA rules are provided for the design space exploration. Here in this paper, we have derived the properties of SACA, MACA and TACA in the context of active and passive Rule mean terms (RMTs). The research paper is comprised of five relevant sections. In Sect. 2, preliminaries of ECA, it's classification and various applications are given. In Sect. 3, characterization tools like De Bruijn graph, state transition diagram and reachability tree, are introduced. Analysis/synthesis of CA properties are discussed in Sect. 4 and the document is concluded with Sect. 5.

2 Elementary Cellular Automata

A CA is a spatially extended decentralized and natural computing model which comprises of a number of cells. Each cell is capable to exhibit global dynamics of CA by local interaction depending on transition function (f_i). The next state (NS) (S_i^{t+1}) of each CA cell (i^{th} cell), at time step ($t+1$), is effected by the PSs of just left neighbor (S_{i-1}^t), self (S_i^t) and just right neighbor (S_{i+1}^t) in ECA (Fig. 1). The NS (transition function) is (Eq. 1):

$$S_i^{t+1} = f_i(S_{i-1}^t, S_i^t, S_{i+1}^t) \tag{1}$$

According to boundary conditions, CA can be of different types like fixed boundary, null boundary, reflexive boundary, adiabatic boundary, periodic and intermediate boundary [10]. In fixed boundary condition, extreme cells at left (C_n) and right (C_1) are tied up to a logic state '0' or '1'. For a null boundary CA (NBCA), the logic states are tied to '0' for both the extreme cells. In a periodic boundary CA (PBCA), the CA takes a circular structure where left most cell terminals are connected to rightmost cell terminals. The vector representation of null and periodic boundary CA are given below:

$$0 \ (S_{n-1}^t, S_{n-2}^t, S_{n-3}^t \ S_{n-4}^t \ \cdots \ S_{i+3}^t S_{i+2}^t \ S_{i+1}^t, \ S_i^t, \ S_{i-1}^t S_{i-2}^t \ \cdots \ S_3^t \ S_2^t, \ S_1^t) \ 0$$

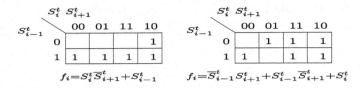

Fig. 1. Transition function representation for CA rules (a) 222 (b) 244

$$S_1^t \ (S_n^t, \ S_{n-1}^t, \ S_{n-2}^t \ S_{n-3}^t \ \cdots \ S_{i+3}^t S_{i+2}^t \ S_{i+1}^t, \ S_i^t, \ S_{i-1}^t S_{i-2}^t \ \cdots \ S_3^t \ S_2^t, \ S_1^t) \ S_n^t$$

From the perspective of switching theory, the combinations of PSs of $(i-1)^{th}$, i^{th} and $(i+1)^{th}$ cells are called Rule mean terms (RMTs). RMTs are therefore, three variable minterms of switching function. The row-1 of Table 1 gives the possible RMTs of ECA. If the NS function (f_i) of the ith cell is represented in the form of a truth table, the decimal equivalent of its output is convention- ally termed as *Rule*. There are 2^3 neighbourhood configurations for a 2-state (binary), 3-neighborhood CA, where 2^{2^3} unique mappings from each of these neighbourhood configurations are possible. Each mapping denoting a CA rule. Few such CA rules 60, 102, 135, 150, 184, 192, 207, 232, 254 and 255 are given in the Table 1. The NSs of RMT 4 (column 100 of Table 1) is '0' for rule 102, 135, 192, 207 and 232 whereas '1' for 60, 150, 184, 254 and 255.

Table 1. Few example ECA rules

PS	111	110	101	100	011	010	001	000	Rule
RMT	(7)	(6)	(5)	(4)	(3)	(2)	(1)	(0)	
NS	0	0	1	1	1	1	0	0	60
	0	1	1	0	0	1	1	0	102
	1	0	0	0	0	1	1	1	135
	1	0	0	1	0	1	1	0	150
	1	0	1	1	1	0	0	0	184
	1	1	0	0	0	0	0	0	192
	1	1	0	0	1	1	1	1	207
	1	1	1	0	1	0	0	0	232
	1	1	1	1	1	1	1	0	254
	1	1	1	1	1	1	1	1	255

2.1   Classification of Wolfram's CA

The characterization of CA behavior puts insight into the power of modelling of a CA. ECA has 256 (2^{2^3}) possible CA rules as it is binary in nature with

Table 2. Wolfram's classification of ECA rule

class I
0, 8, 32, 40, 128, 136, 160, 168
class II
1, 2, 3, 4, 5, 6, 7, 9, 10, 11, 12, 13, 14, 15, 19, 23, 24, 25, 26, 27, 28, 29, 33, 34, 35, 36,
37, 38, 41, 42, 43, 44, 46, 50, 51, 56, 57, 58, 62, 72, 73, 74, 76, 77, 78, 94, 104, 108, 130, 132
134, 138, 140, 142, 152, 154, 156, 162, 164, 170, 172, 178, 184, 200, 204, 232
class III
18, 22, 30, 45, 60, 90, 105, 122, 126, 146, 150
class IV
54, 106, 110

alphabet set $\{0, 1\}$. CA is classified into four different classes by Wolfram as per their dynamics: (1) Class 1: this class of CA cells evolve to a homogeneous state; (ii) Class 2: showing periodic structures; (iii) Class 3: CAs show chaotic or pseudo-random behavior, and (iv) Class 4: Complex patterns within localized structures distinguishes global structure suitable for modern computation. This class of CA is considered to have universal computational capacity. Wolfram's classification was based on complexity analysis (Table 4).

Recently, the complexity of the discrete dynamical systems are quntified using information-theoretic tools. Transfer entropy (TE) and Schreiber's measure are the two valuable parameters of measure to information exchange among interacting parts. In another work, the CA rules are classified quantitatively under three different classes. Classification is based on the amount of information content or processing capacity. Depending on initial conditions and Schreiber's transfer entropy (TE), ECA rules are classified as given in [4]. The most populated class I_1 contains a very less amount of information (TE $\simeq 0$) for simple and complex outputs. For class I_1, TE is high for complex inputs and low for simple inputs. Class I_3 is almost unaffected by inputs and always shows high value.

2.2 Application of ECA

ECA configurations has become an attractive model for addressing problems of physical sciences and biological sciences. CA can be used to develop parallel multiplier, prime number sieves, parallel computational machine etc. In biological systems, CA can be used to detect tumor, image processing, abnormal heart pumping behavior, cancer detection, modelling immune systems etc. It also finds applications in genetic algorithm. Further, a subset of ECA are chosen to apply in different domain targeted for VLSI implementation (refer Table 3). The application domains are Pattern Recognition, image processing, Data Compression, Cryptosystem, Error Correcting Codes, Authentication, VLSI Circuit Testing, Cellular Mobile Network, Sensor Network etc. CAs are found applications in various homogeneous or heterogeneous CMPs [7] [13]. CA can also be used in different cache applications for protocol processor design, data migration, fault tolerance etc. [8,18]. Das et al. proposed a data migration protocol which is implemented through CA. Design for a cache-coherence controller based on CA was suggested by Dalui et al. The Protocol processor (PP) proposed for MSI, MOSI or MOESI protocols works on the state (modified, shared and invalid) identification of the cache blocks. According to Dalui, this PP concept can be extended to include stuck-at fault-tolerant designs at the granularity of cache line. Stuck-at fault tolerant circuits and fault diagnosis are designed through NBCA and PBCA using uniform/nonuniform CA rules. Therefore, the vast field of applications of CA motivates us to characterize CA rules.

Table 3. Comparative analysis between ECA based cache applications

Scheme	Contribution	Special features
NBCA-based fault tolerant design (TACA/SACA rules used 254 and 255) [14]	Applicable in multi-bit stuck-at faults with error correction pointers to reduce overhead	Physical implementation in embedded systems is quite complex as requires in-built libraries to create pointers dynamically
NBCA-based fault diagnosis scheme, (TACA/SACA rules used 254 and 255, 192 and 207) [7]	It tolerates multi-bit stuck-at faults with dynamic partitioning for reduced space overhead than others	Innovative architectural solution, but limited number of tolerable faults and not applicable for complex faults
PBCA-based data migration scheme [9]	It reduces cache block access latency	Unique hybrid PBCA based approach (TACA used 232 and 184)
PBCA-based wear leveling scheme (TACA with 232, 226 rule) [12,17]	Enhances cell lifetime of emerging nonvolatile cache	Complementary to stuck-at fault tolerant scheme to improve two-fold cell durability
PBGA-based wear leveling scheme (TACA rules 232, 226) [16]	Spatial locality aware (SWear) enhances cell lifetime of hybrid cache	Complementary to fault tolerant scheme to improve cell durability with both read/write policy

This research paper focuses on analysis/synthesis of single length cycle attractor CA for objective-specific manner. Characterization tools of 1-D CA for identification of CA attractors are discussed in the next section.

3   Characterization Tools for Cellular Automata

CA operation is analogous with finite state machine [13]. Modelling of finite state automata is represented by the switching mechanism, where automata states are on and off. Transition between the states may or may not occur by the application of external triggers. From the above mentioned charateristics of a CA, the comprehensible fact is that, CA comes up with the varieties of dynamics, rules, shapes, boundary conditions etc. CA models can have the capacity of massive parallelism and universal computation, if it is reversible CA (RCA) [11]. RCA follows the fundamental laws of physics. Therefore, miniaturization and hardware implementation is more energy saving (avoiding $KTln2$ J of one bit erasure energy where K is Boltzman's constant and T is temperature in Kelvin) than the irreversible CA implementation. Irreversible CA also comes up with it's own advantages those can be used in various cache applications [8,12].

This research paper is dealt with finite CA. So, the lattice dimension ($Z \subseteq Z^D$) is also finite. The subset (Z) forms an additive subgroup under modulo addition 'n' for $Z_n = \{0, 1, \dots, n-1\}$. A CA is considered to be linear, if it's transition function is a linear map. A linear CA follows the properties: (i) Additivity, and (ii) Homogeneity. For, a CA to be reversible, the global (G) and local (F) transition function must be inverse function ($G \circ F = F \circ G = I$ (identity function)) to each other. This section analyzes the characterization of CA cell rules that helps to identify the reversible/irreversible CA structures.

Theorem 1. *A CA is a reversible CA if it contains only cyclic states in its state transition diagram otherwise it is irreversible CA.*

Theorem 2. *For finite configurations, 'G' is injective if and only if it is surjective but vice versa is not true.*

This theorem is well known as Garden-of-Eden theorem. Therefore, injectivity, bijectivity and reversibility are identical [11].

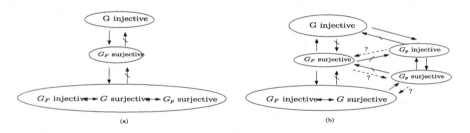

Fig. 2. Relation between injectivity, surjectivity properties of (a) 1-D CA (b) 2-D CA

Theorem 3. *1-D CA with reversibility property can be used as a universal computing tool [11].*

There exists algorithms to test for a CA to be injective and surjective. K, Sutner introduced Bruijn graph for algorithmic decision of injectivity and surjectivity. The relation between injectivity and surjectivity properties of G, G_p and G_F are pictorially represented in Fig. 2 [11]. To apply a CA in any application domain, characterization of CA is essential. A list of charaterization tools are introduced by the researchers like Bruijn graph, Pair graph, Subset graph, Cyclic graph, State Transition diagram, Reachability Tree, characterization diagram, Next State RMT Transition Diagram [18]. In this research article, State transition diagram (STD) and reachability tree (RT) have been used for analysis of CA state-space evolution (Fig. 6).

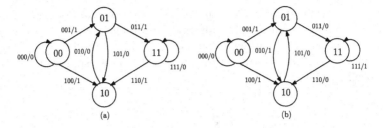

Fig. 3. Bruijn graph (a) For rule 90 (b) rule 150

3.1   State Transition Diagram

Space time evolution of CA makes it possible to characterize it's behavior. The sequence of states generated (state transitions) during its evolution with time, directs the CA behavior. The flow diagram of generating STDs is given in Fig. 4. An STD may contain cyclic and non-cyclic states (a state is called cyclic if it lies in a cycle) of a CA and based on this, the CA can be categorized as reversible or irreversible CA.

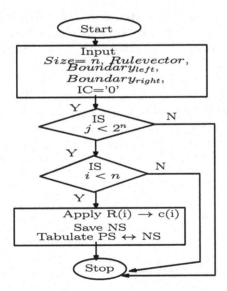

Fig. 4. Flow diagram of State transition generation

The state space evolution of a irreversible CA is given in Fig. 5. It has the single-length $10 \rightarrow 10$ and multi-length $8 \rightarrow 11 \rightarrow 8$ cycles. Such cycles are the *attractors* of the CA. An attractor forms a basin containing the states that lead to the attractor. In Fig. 5, 10-basin consists of two states 7 and 10. The maximum distance traversed to reach an attractor from any other state is defined as the depth of the CA. The CA of Fig. 5 has the depth 7 which associates the time delay in a CA based operation. The alternative naming of STDs are flow graph, basins of attraction or network of attraction etc. To draw a STD of an n-cell CA for both NBCA and PBCA, it requires exhaustive 2^n state-space exploration i.e. the complexity is $O(2^n)$.

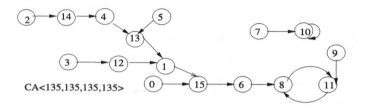

Fig. 5. State transition diagram of an irreversible CA

3.2   Reachability Tree

Reachability tree is defined to characterize the CA states. It is a binary tree and represents the reachable, non-reachable as well as cyclic or acyclic states of a CA. Each node of the tree is constructed with the possible RMT(s) of a rule at this level. The left edge of a node is considered as the 0-edge (states with NS 0) and the right edge is the 1-edge (states with NS 1). Root node is at level 0 and the leaf nodes are at level n for an n-cell CA. The nodes of level i are constructed following the selected RMTs of R_{i+1} for the next state computation of cell (i+1). The number of leaf nodes are the count of reachable states of the CA. An RMT sequence (n bit binary string) represents an attractor state. Structure of a reachability tree is shown in Fig. 6.

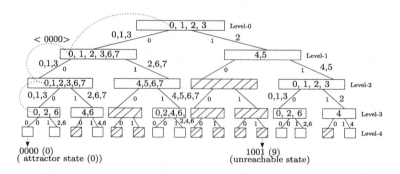

Fig. 6. Reachability Tree for 4-cell uniform NBCA

4   Analysis and Synthesis of Cellular Automata

In this section, CA properties are explored to characterize fixed-point attractors of ECA. The proposed theory is based on the properties of active and passive RMTs to identify SACA, TACA and MACA. The chosen characterization tool is STD.

Theorem 4. *The formation of single length cycle attractor is established by the same present state and the next state numeric values.*

As per Theorem 4, there must be at least one single/multiple RMT(s) that do not change it's state to form single length cycle attractor(s). So, it requires passive RMT within the set of RMTs $\{0, 1, 4, 5\}$ must produce 0 value at it's next state. Alternately, there must be at least one RMT within the set of RMTs $\{2, 3, 6, 7\}$ is to be passive with the next state value 1. For example, for a given passive RMT $S_{i-1}^t S_i^t S_{i+1}^t$ for S_i^t is 0 or 1, S_{i-1}^{t+1} is also 0 or 1 accordingly for any considered rule R_i. Therefore, it is implied that such an RMT is responsible to change the present state to next state cell transition value $(S_i^t \rightarrow S_i^{t+1})$ as $0 \rightarrow 0$ or $1 \rightarrow 1$ respectively. For rule 254 (Table 1, the RMTs 0, 1, 2, 3, 4, 5, 6 are 1 and RMT 7 is 0. For rule 255 (Table 1), RMTs 0, 1, 2, 3, 4, 5, 6, 7 are 1. So, the CA rules 254 and 255 can be configured to form of single length cycle attractors as per property 1.

Property 1. A rule Ri can contribute to the formation of single length cycle attractor(s) if at least next states of one of the RMTs 0, 1, 4 or 5 is 0, or the next states of one of the RMTs 2, 3, 6 or 7 is 1 [12].

STD of a irreversible CA is given in the Fig. 7. From, the STDs of 4-cell uniform or non-uniform CA rule vectors $<254, 254, 254, 254>$, $<254, 254, 255, 254>$ and $<254, 254, 254, 255>$, it is confirmed that, it is forming single length cycle attractors $15 \rightarrow 15$ and $0 \rightarrow 0$, $15 \rightarrow 15$, $15 \rightarrow 15$ respectively. Therefore it is essential for a CA rule (Ri) to obey Property 1 to form fixed point attractor(s).

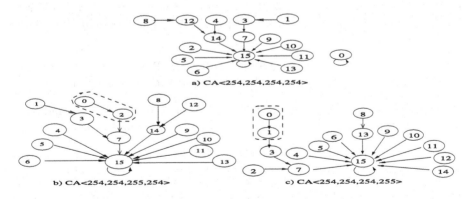

Fig. 7. STD of 4-cell NBCA with various hybridization

4.1   Analysis of Single Single Length Cycle Attractor CA

STD can be used as a analysis/synthesis tool for uniform/nonuniform NBCA or PBCA. State transition representation gives the details of nonreachable (with no predecessor state(s)), reachable (with predecessor state(s)) and reachable and cyclic or acyclic states. The presence of multi-length cycles (cycle length > 1) must be detected/eliminated to identify SACA, TACA and MACA. STDs can report the single length cycle attractors, hence subset of SACA rules can be identified from ECA rules. State space exploration of ECA reveals a subset of 41-tuple SACA rules (Table 4). In Fig. 7(b) and (c), STDs of NBCA with nonuniform rule vectors $<254, 254, 255, 254>$ and $<254, 254, 254, 255>$ are shown. Both the CAs forming a SACA (having single single length cycle attractor) with nonzero attractor $15 \rightarrow 15$. However, zero/nonzero attractor(s) in STDs are identified for objective-specific application like memory testing [8], cache coherence controller design [6], fault tolerance designs [12] and others. For an example, in stuck-at fault tolerant design, $0 \rightarrow 0$ attractor is used to detect nonfaulty memory blocks, with uniform SACA rules 254 and 192. Whereas faulty status is detected through non-zero attractor (e.g $15 \rightarrow 15$) [12].

Table 4. Uniform rules forming SACA construction

Group of CA	SACA rules
2	34, 48
3	2, 16, 32, 42, 56, 98, 112, 162, 176
4	0, 10, 15, 24, 40, 66, 80, 85, 96, 130, 144, 160, 170, 184, 226, 240, 255
5	8, 64, 128, 138, 143, 152, 168, 194, 208, 213, 224
6	136, 192

Property 2. The uniform CA rule 192 conforms SACA with active RMT set $\{7, 6\}$ and passive RMT set $\{5, 4, 3, 2, 1, 0\}$ in it's 3-neighborhood null-boundary construction.

Property 3. Uniform rule 34 forms SACA with active RMT 4 and passive RMT set $\{7, 6, 5, 4, 3, 2, 1, 0\}$ in it's 3-neighborhood null-boundary construction.

Property 4. Uniform rule 226 forms SACA with active RMT set $\{7, 6\}$ and passive RMT set $\{5, 4, 3, 2, 1, 0\}$ in it's 3-neighborhood null-boundary construction.

Property 5. Uniform rule 152 forms SACA with active RMTs 6,4,2 and passive RMT set $\{7, 5, 3, 1, 0\}$ in it's 3-neighborhood null-boundary construction.

The reported functional analysis on SACA initiates the conceptualization of CA properties of fixed-point attractor CA. Those CAs are synthesized to construct scalable architectures of CMPs.

4.2   Analysis of TACA and MACA

However synthesis of SACA rules are extended to define TACA rules to apply in different CMPs applications like fault diagnosis, CA based memory testing with March algorithm, data migration etc. Fault diagnosis is also achieved through PBCA based TACA rules having non-zero or all zeros attractor. Uniform/ non-uniform TACA based data migration [9] and wear leveling [12] techniques are proposed for on-chip cache applications of CMPs. Density classification task is performed with TACA for various other applications of image processing, networking etc. [2].

Property 6. Uniform rule 220 forms TACA with active RMT 4 and passive RMTs 7, 6, 5, 4, 3, 2, 1, 0 in it's 3-neighborhood null-boundary construction.

Property 7. A null boundary CA with uniform rule 38 forms a TACA with passive RMT set $\{4, 2, 1, 0\}$ and active RMT set $\{7, 6, 5, 3\}$ (Fig. 8a).

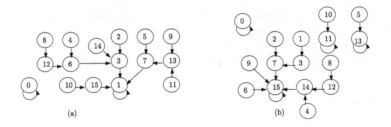

(a)   (b)

Fig. 8. STD of uniform NBCA (a) TACA (b) MACA

MACA rules are already applied for fault diagnosis, pattern classifier, networking etc. All the designs are based on multiple nonzero and all zero attractors. A few NBCA MACA rules 212, 222, 116 and 244 are introduced and analysed in Table 5. The attractors are detected through STDs to set design options.

Property 8. A Uniform 3-neighborhood NBCA with CA rule 244 forms an MACA for RMTs 5, 4, 3 being active and RMTs 7, 6, 2, 1, 0 being passive.

Property 9. A Uniform 3-neighborhood NBCA with CA rule 13 forms an MACA for RMTs 7, 6, 1 being active and RMTs 5, 4, 3, 2, 1, 0 being passive.

Property 10. A Uniform 3-neighborhood NBCA with CA rule 69 forms an MACA for RMTs 7, 3, 0 being active and RMTs 6, 5, 4, 2, 1 being passive.

Though STDs can successfully establish SACA, TACA and MACA construction along with it's attractor states but it is difficult to analyse CA with STDs for higher number of n-cell cellular structure.

Table 5. Comparative analysis between various CA rules with variable cell size

Configuration	Rule	Number of Attractors					
		3-cell	4-cell	5-cell	6-cell	7-cell	8-cell
NBCA	212	3	3	4	4	5	5
Uniform	244	2	2	2	2	2	2
	116	2	2	2	2	2	2
	222	3	4	6	8	14	22

5   Conclusion

This research article targets to signify the characterization of single-length cycle attractor(s) of NBCA. Wolfram's CA rules are distinguished as SACA, TACA and MACA. State transition diagram is used to formulate the properties of single-length cycle attractor CA with the context of active and passive RMTs. It is also established that though various characterization tools are available to be dealt with 1-D 3-neighborhood CAs, but STDs are more explanatory when the CA is required to be synthesized for scalable VLSI design architectures of realtime application domain, in a objective-specific manner. CA-based machine designing through analysing STDs are future scope of this current research work.

References

1. Martin, O., Odlyzko, A.M., Wolfram, S.: Algebraic properties of cellular automata. Commun. Math. Phys. **93**, 219–258 (1984)
2. Das, S., Mukherjee, S., Naskar, M., Sikdar, B.K.: Characterization of single cycle CA and its application in pattern classification. Electron. Notes Theor. Comput. Sci. **52**, 181–203 (2009)
3. Singh, H., Kumar, Y.: Cellular automata based model for e-healthcare data analysis. Int. J. Inf. Syst. Model. Des. **10**(3), 1–18. https://doi.org/10.4018/ijismd
4. Borriello, E., Walker, S.I.: An information-based classification of elementary cellular automata. Complexity. https://doi.org/10.1155/2017/1280351
5. Choudhury, P.P., et al.: Classification of cellular automata rules based on their properties. Computer Science, Mathematics, arXiv (2009)
6. Dalui, M., Sikdar, B.K.: A cellular automata based self-correcting protocol processor for scalable CMPs. Microelectron. J. **62**, 108–119 (2017)
7. Sarkar, S., Saha, S., Sikdar, B.K.: Multi-bit fault tolerant design for resistive memories through dynamic partitioning. In: 2017 IEEE East-West Design & Test Symposium (EWDTS), Novi Sad, pp. 1–6 (2017)
8. Saha, M., Das, B., Sikdar, B.K.: Periodic boundary cellular automata based test structure for memory. In: IEEE East-West Design & Test Symposium (EWDTS), pp. 1–6 (2017). https://doi.org/10.1109/EWDTS.2017.8110050
9. Das, B., Dalui, M., Kamilya, S., Das, S., Sikdar, B.K.: Synthesis of Periodic Boundary CA for Efficient Data Migration in Chip-Multiprocessors. IEEE (2013)

10. Chaudhuri, P.P., Chowdhury, D., Nandi, S.R., Chatterjee, S.: Additive Cellular Automata - Theory and Applications, (1). IEEE Computer Society Press, USA (1997)
11. Kari, J.: Reversible cellular automata. In: De Felice, C., Restivo, A. (eds.) DLT 2005. LNCS, vol. 3572, pp. 57–68. Springer, Heidelberg (2005). https://doi.org/10.1007/11505877_5
12. Sarkar, S., Ghosh, M., Sikdar, B.K., Saha, M.: Periodic boundary cellular automata based wear leveling for resistive memory. IAENG Int. J. Comput. Sci. **47**(2), 310–321 (2020)
13. Sarkar, S., Sikdar, B.K., Saha, M.: Cellular Automata-Based Multi-bit Stuck-at-Fault Diagnosis for Resistive Memory (2021). http://www.jzus.zju.edu.cn/iparticle.php?doi=10.1631/FITEE.2100255
14. Sarkar, S.: Multi-bit stuck-at fault recovery system with error correction pointer. In: Proceedings of the third International Conference on Communication and Electronics Systems. IEEE (2018)
15. Saha, M., Sarkar, S., Sikdar, B.K.: Cellular automata based fault tolerant resistive memory design. In: ISED (2016)
16. Sarkar, S., Sikdar, B.K., Saha, M.: An efficient wear leveling for hybrid cache using periodic boundary cellular automata. Solid State Technol. **63**(2s), 10144–10160 (2020)
17. Sarkar, S.: Cellular automata based wear leveling in resistive memory. In: 2019 IARIA in the Proceeding of Fourteenth International Conference on Systems, Valencia, Spain (2019)
18. Dalui, M., Chakraborty, B., Das, N., Sikdar, B.K.: NSRT diagram for identification of SACA and TACA rules in null-boundary. https://doi.org/10.1142/S0129183122500711

Optimal Perfect Phylogeny Using ILP and Continuous Approximations

B. E. Pranav Kumaar⬤, A. Aadharsh Aadhithya⬤, S. Sachin Kumar(✉),
Harishchander Anandaram, and K. P. Soman

Center for Computational Engineering and Networking (CEN), Amrita Vishwa
Vidyapeetham, Coimbatore, India
s_sachinkumar@cb.amrita.edu

Abstract. The concept of perfect phylogeny is observed to be non-universal, however, in some studies, it is shown that it provides great insights from a biological standpoint. Most natural phylogenies are imperfect but this is debated by the presence of noise in data acting as a disguise, the solution to uncover the underlying perfect phylogeny in the simplest approach is known as Minimum Character Removal (MCR) problem. Another variant of the solution achieved by a change in perspective and computational methodology is known as Maximal Character Compatibility (MCC) problem. Both MCR and MCC problems are solved optimally with reasonable efficiency by using Integer Linear Programming (ILP) technique. A central part of MCC solution involves solving the Max-clique problem of the generated representation allowing numerous clique-solving algorithms to generate various solutions. The comparison between the solution's optimality and the run-time of these methods substantiates the goal of the different algorithms being applied. The above methods are formulated and implemented using modern programming languages like Python and Matlab. This exploration introduces the problem of perfect phylogeny, its ILP formulations, and its solution along with comparisons between different methods in consideration.

Keywords: Integer Linear Programming · Perfect Phylogeny ·
Non-Negative Matrix Factorization · Minimum Character Removal ·
Maximal Character Compatibility

1 Introduction

The usage of databases and ontologies fundamentally reshaped biology as a science. A wide variety of databases for different purposes are now available. DNA databases like EMBL [1], and GenBank [2] are some of the primary databases hosting sequencing data. Gene Expression Omnibus (GEO) [3] provides curated microarray/assay data-sets. miRBase [4] is a database of microRNA's. DisProt, PROSITE, Protein Data Bank, etc., contains data on protein sequence and their

M. Singh et al. (Eds.): ICACDS 2023, CCIS 1848, pp. 122–135, 2023.
https://doi.org/10.1007/978-3-031-37940-6_11

structures. NCBI [3] hosts a wide variety of biological data. This is certainly not an exhaustive list of databases or even the type of databases. [5] provides a comprehensive list of Biological Databases useful for Human research.

In the present paper we explore the power of the ILP and NMF [6] solutions for the creation of perfect phylogenies, more specifically focusing on the minimal character removal problem. The main contributions of the article are,

- Easy and intuitive presentation of the perfect phylogeny.
- A beginner-friendly nudge in Python and Matlab to solve the Minimum Character removal Problem (MCR) and Maximum character clique problem (MCCP).
- Proffers usage of NMF for the perfect phylogeny problem and shows its significant performance boost.

Recognizing such evolutionary relationships is imperative to comprehend the grand scheme of life, and individual species as well. One of the ways of obtaining these relationships is through a perfect phylogeny mode. Historically, the Steiner tree problem is a closely related, well-studied problem. However, in the current article, we shall explore the perfect phylogeny problem using Integer Linear Programming Formulation, and solve a resultant max-clique problem using Non-Negative Matrix factorization.

The time complexity makes it impossible for probabilistic or brute force methods like for generic phylogenetic tree construction techniques to construct and verify the conformational requirements of perfect phylogenies. Even methods that manage to solve this problem on a similar definition do so in polynomial time, this is indeed excellent but their application on real datasets is comparatively time-consuming so to speak. Additionally, unique perfect phylogenies are NP-hard and intractable [7]. At any rate, the use of creative techniques to find smart solutions for traditional computational bioinformatics problems in an updated, efficient, and eloquent manner are always welcome.

At this point, we are endowed to extend our thanks to Dan Gusfield for his contributions to solving bioinformatics problems with ILP methods. The work done in the paper is inspired by his book, Integer Linear Programming in Computational and Systems Biology [8].

2   Biology Concepts

2.1   Phylogenetics

Each taxon has its set of discrete properties, these properties are biologically referred to as characters or traits, and the prefix discrete is used to denote that these properties can take on a finite number of states or variants. Consider the species of the emperor penguin (*Aptenodytes forsteri*) to be the taxa under consideration and its characteristic to be able to fly or not, the character of flight for any object can take one to two states (Can Fly or Cannot Fly). Similarly, the nucleotide (A, T, C, or G) present at a particular site of a DNA sequence is a four-state character. For the sake of brevity, the taxa under consideration shall be consistently referred to as instances in the discussions to be followed.

The above section will serve as the foundation for the input of the Minimal Character Removal Problem. A matrix M of dimension mxn should be created such that m instances and n columns depict the n characters of each instance. The i, j^{th} element of matrix M (where i & j ϵ $\{1, 2, ..., m\}$) will be the jth character of the ith instance in M. if we let jth character take on some r number of discrete states specified by a set D. Another set E (set of integers), with the same number of elements as D is created, $E = \{1, 2, ..., r\}$, this is done to allow mathematical representations of states. However, we will omit these conversions as our primary goal is to obtain a matrix similar to Fig. 1.

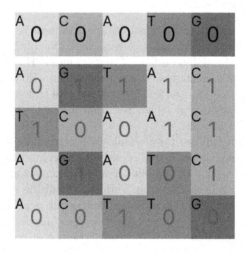

Fig. 1. Binary Mutation Matrix M_A. Shadowed by the original character matrix.

2.2   Phylogenetic Trees

The outputs (in this case going to be M') that are generated will be conventionally and best visualized using the construction of phylogenetic trees (T). Phylogenetic trees are branching diagrams that are created to model the evolutionary history of a set of objects based on the similarity and differences of their characters. Each instance of M in a phylogenetic tree is represented by a node and each node will identify by associating with the characters of the instance. The earliest/oldest-time instance will form the root r of T, subsequent instances will derive from corresponding parent instances with their own mutations in characters, and finally, the present-time instances which will have no further child nodes will be called a leaf l of T.

The Trees are traversed from r to l. From this, we can also state that the character that labels a node v in T is obtained from the root's characters by applying the mutations along the unique path in the tree from the root to u. This also means that if node u' is the unique parent of node u, then the sequence labeling u is derived from the sequence labeling u', by changing the state of every mutating character written on the edge (u', u), this also holds for any parent and child pair in T say r and l.

Concerning the MCR problem, the variation of M used will not be populated with set E rather the mutations of characters of every instance along the phylogenetic tree in comparison to the root's characters will be represented using 0 for no mutation, and 1 for the occurrence of mutation forming a binary set F = {0, 1}, which will be used to populate M. Consequently, the input mutation matrix M will be a binary matrix.

2.3 Perfect Characters

A special class of characters called many names perfect characters, ideal characters, and monophyletic characters are classified by the fact that they mutate only once in a phylogenetic tree. To explain consider a character C' to be part of a Phylogenetic Tree T, then if the character first mutated at any node u, if C' is a perfect character, we will be able to deduce the following things,

– Every descendant of node u also must have the exact same mutation (same state) in C'.
– All other instances in T other than u and its descendants will have the original state of C'.

Now subjecting the same with respect to the MCR problem, the mutation of character C' from an all-zero root sequence will lead to the mutation in C' of u, and the subtree/descendants starting from u take the value 1 for C and all other instances have 0 for C'.

2.4 Character Compatibility

The perfect phylogeny theorem - A binary matrix M is derivable on a perfect phylogeny with the all-zero root sequence, if and only if every pair of columns in M is compatible. Alternatively, consider a tree T with an all-zero root sequence that derives the data in matrix M and every character is perfect, i.e., mutating only once in T, then T is called a perfect phylogeny for M, and M is said to be derivable on a perfect phylogeny.

To accommodate for the unordered nature of instances in M it is ensured that pairwise character compatibility holds, so all characters will remain perfect in the tree representation. Consider only two characters namely $C1$ and $C2$, from the all-zero root of the tree the character pair $(C1, C2)$ by default will have (0, 0) value from which successive mutations are derived. Now the attainable binary pairs for all combinations of mutations of $C1$ and $C2$ will be (0, 1), (1, 0), (1, 1). Provided there are sufficient instances in M (m $\geq$ 3) and if all three of the binary pairs (0, 1), (1, 0), and (1, 1) are present in the columns of $C1$ and $C2$, where it can be observed that in any permutation either $C1$ or $C2$ will have to have mutated twice. This violates the Perfect Character definition, so the two columns are said to be incompatible.

To summarise, two characters in M are said to be incompatible if their associated columns in M contain all three of the binary pairs $(0, 1)$, $(1, 0)$, and $(1, 1)$. Contrary to incompatible characters, if two characters are not incompatible, they are said to be compatible. We should note that saying $C1$ is compatible with $C2$ is the same as saying $C2$ is compatible with $C1$, compatibility is unordered property of characters. Different character removal combinations can be the optimal/ close optimal solutions, but an interesting observation is that when M can be derived on a perfect phylogeny with the all-zero root sequence, the perfect phylogeny is unique.

3   ILP Concepts

Integer Linear Programming is a popular branch of optimization, where objective and constraint functions are linear and variables take only integer values. ILP has been and is being used in various domains including Computer Vision, NLP, Robotics, Signal Processing, Power reduction, etc. [9–12]. Typically it is of the form

$$
\begin{aligned}
\underset{x}{\text{maximize}} \quad & c^T x \\
\text{subject to} \quad & Ax \leq b \\
& x \geq 0 \\
& x \in \mathbb{Z}^n
\end{aligned}
\tag{1}
$$

Depending on the relaxation on x, ILPs can be

1. Pure Integer Linear Programming, where all variables should be integers
2. Mixed Integer Linear Programming (MILP), Where some of the variables are restricted to integers, and others can be any real value
3. Binary Integer Linear Programming, where variables can take only 0 or 1 value.

Computationally, ILP problems are NP-Complete. Cutting plane methods and Branch and Bound algorithms are some of the commonly used techniques to solve an ILP formulation. However, specialized solvers like Gurobi, Simplex, etc. can be used to solve ILP problems.

In general, many problems can be classified as optimization problems or decision problems. An Optimization problem aims to find *best possible solution*, where *best* is measured using some metric. A decision problem on other hand tries to answer yes or no, That is, in terms of a Boolean variable. Problems can be widely classified in terms of their complexity as

1. **P:** Problems with a solution algorithm that runs in polynomial time or less in a deterministic machine

2. **NP:** Decision problems that cannot be solved in polynomial time in a deterministic machine. If it is known the problem can be solved and a solution is at hand, the fact can be checked in polynomial time.

NP-hard problems are subsets of NP problems, which are at least as hard as any other NP problems. Problems that are both NP and NP-hard constitute the set of NP-Complete problems. One polynomial-time solution to one NP-Complete problem implies an immediate solution to all the problems in the set.

4   Problem Definition

Up to this point, there is clarity on the conformations that the models must abide to be classified into perfect phylogeny, but one can also infer that not all real-world datasets will do so. Call them Imperfect phylogenies, now if the cause of this imperfection can be removed (removing the incompatible characters) then the result can be a perfect phylogeny constructed from Imperfect phylogeny. The validation of the results from the actual dataset is done by retaining as much original information from the imperfect phylogeny data as possible. This way of validating the results works both ways (with respect to the solution and input). In other words, remove only as many characters as required to convert the imperfect phylogeny to a perfect one. Remarkably this is an optimization problem hence the ILP approach is suitable.

4.1   MCR - Minimum Character Removal Problem

For a given binary input matrix M of dimension mxn, The expected output is a binary vector C of dimension $1xn$ where the elements of C represent the inclusion/Retention and exclusion/Removal of the columns. This is to achieve a matrix M' that can be derived on a perfect phylogeny.

4.2   Algorithm-1

Algorithm-1 is used to identify all pairs of incompatible characters by comparing the columns of M. It is guided by the perfect phylogeny theorem i.e., it traverses each instance for every given pair of characters from M and keeps track of the occurrences of binary pairs $(0, 1)$, $(1, 0)$, and $(1, 1)$ using Algorithm-1 variables. If all 3 of the binary pairs are found to be present for the character pair in consideration, then the characters are marked as incompatible, and their indices are kept track in an array. These arrays then can be comfortably used to declare constraint set 2 for the ILP formulation.

1. Obtain the dimensions of input matrix M and store them in variables say m, n
2. The use of dynamic data storage methods is essential as the number of incompatible character pairs present in M is unknown prior to the implementation of the algorithm.

3. Next, identification of the pair of elements of every column of M with every other column in M is performed, here identification refers to comparing the element pairs of the columns under consideration with binary pairs $(0, 1)$, $(1, 0)$, and $(1, 1)$. The characters that are considered are kept on track for element pair comparisons (on which the identification process is conducted).
4. To perform identification, initialization of 3 variables(corresponding to each identification) to zero for every pair of columns considered for comparison is required.
5. (a) If any of the considered element pairs matches any of the binary pairs during the comparison the corresponding identification variable is set to 1.
   (b) Following the identification constraints evaluation for every element pair, checking if all 3 identification variables have a value of 1 is needed as in the case where all 3 identification variables have a value of 1, the successive element comparisons for the current character pair can be skipped, which allows the algorithm to move on to the next character pair.
   (5.b) is done to reduce the number of element pair comparisons.
6. Finally appending the character pairs indices to the dynamic data storage if all 3 identified variables have a value of 1. The set of such character pairs will be referred to as set I.

4.3 MCCP - Maximal Character Clique Problem

The MCR problem focuses on the removal of the minimum number of characters from M. From another perspective, the use of the ILP solution of the Maximum Clique problem and solve the MCR problem by modifying the interpretation of input and approaching it from the dual form of MCR. This is the equivalent solution for MCR via the use of the ILP formulation of the Maximal Clique Problem.

Firstly, the modification of input nature is required to achieve compatibility with the maximal clique problem which requires the creation of a graph G from M. G will have n nodes representing each character in M, and connections will be forged using undirected edges for only those nodes of graph G that are pairwise compatible in M. The prerequisite is the compatible characters from M, which is obtained by filtering the characters that are incompatible using Algorithm-1. From the Maximal Clique perspective on G it can be seen that the nodes corresponding to the incompatible characters will not be connected. So, the optimal maximum clique will accommodate during the solving process for the combinations of successive ignorance (removals in the MCR perspective) making some of the initial incompatible pairs be given a chance to become globally compatible if the local incompatibility is ignored in the specific configuration.

Secondly, supplying the graph G as input to the ILP solution guided by the formulation for the maximal-clique problem. The optimal result will be the maximum-sized clique in G called K which represents the character set that can derive a perfect phylogeny from an all-zero root sequence. Every pair of nodes in K will be connected by the definition of a clique, which also means these

Input : Binary Matrix(M)
Output: Incompatible Column Pairs

$m \leftarrow 2$ *number of rows of M*;
$n \leftarrow 2$ *number of columns of M*;
combs $\leftarrow$ `Combinations`$(n,2)$;
lencombs $\leftarrow$ *number of elements in combs*;
incombs $\leftarrow$ *[]*;

for $i \leftarrow 1$ **to** *lencombs* **do**
    $col1 \leftarrow 1^{st}$ *column of ith pair in combs*;
    $col2 \leftarrow 2^{nd}$ *column of ith pair in combs*;
    for $j \leftarrow 1$ **to** m **do**
        if *(col1[j]==0) and (col2[j]==0)* **then**
        | flag00 = 1
        end
        if *(col1[j]==) and (col2[j]==0)* **then**
        | flag10 = 1
        end
        if *(col1[j]==0) and (col2[j]==1)* **then**
        | flag01 = 1
        end
        if *(flag00==1) and (flag01==1) and (flag10==1)* **then**
            // Append i^{th} *pair of combs to incombs incombs* $\leftarrow$ *incombs +*
            combs[i];
            Break;
        end
    end
end
Return incombs;

Algorithm 1: Algorithm-1

characters will be compatible. Particularly, the nodes that are present in G but not K will be the minimum set of characters that have to be removed to form M' from the MCR perspective.

4.4 Algorithm-2

In the previous section, requirements to modify the input matrix M into an adjacency matrix of a graph that represents the connection between the nodes were highlighted. Correspondingly this can be done for any input matrix M using the algorithm described,

1. Initialize G to be a matrix of ones of dimension $n \times n$
2. Get the previously computed indices of incompatible character pairs from Algorithm-1 say for any character pair $(C(i), C(j))$ and corresponding use it to set the $G(i, j)$ and $G(j, i)$ to 0

Input   : Binary Matrix(M) and Incompatible Column Pairs
Output: Maxclique Compatible Adjacency matrix of MCR

$n \leftarrow 2$ *number of columns of* M;
$GK \leftarrow \texttt{Ones}(n,n)$;
lenincombs $\leftarrow$ *number of incompatible column pairs*;

for $i \leftarrow 1$ **to** *lencombs* **do**
$\quad$ $x \leftarrow 1^{st}$ *column index of* i^{th} *pair in combs*;
$\quad$ $y \leftarrow 2^{nd}$ *column index of* i^{th} *pair in combs*;
$\quad$ $GK(x,y) \leftarrow 0$;
$\quad$ $GK(y,x) \leftarrow 0$;
end
$G \leftarrow GK$;
Return G;

Algorithm 2: Algorithm-2

5   MCR ILP Formulation

A binary variable, C_i can be used to represent if column i is selected in the compatible set or not.

$$C_i = \begin{cases} 1, \text{ If node i is in compatible set} \\ 0, \text{ If node i is not in compatible set} \end{cases} \quad (2)$$

The objective function for the problem can be put simply as the maximization of the sum of the binary variables.

$$\sum_{i=1}^{n} C(i) \quad (3)$$

Maximizing the sum of C_i's ensures that the minimal number of columns are removed. Integer variables can be forced to behave as binary variables by setting the upper and lower bounds of the variable. That is,

$$0 \leq C(i) \leq 1 \quad \text{(Constraint set 1)}$$

Hence, any standard ILP solver can be used to solve binary ILP by adding the above constraints.

The next set of constraint definitions will be to specify the incompatible columns that have been identified using Algorithm-1. These constraints are modeled in a way to convey that the selected incompatible column pair must not occur together for the sake of preserving the compatibility of the resulting combination. let set I contain the incompatible pairs identified by Algorithm-1.

$$C(j) + C(k) \leq 1 \forall (j,k) \in I \quad \text{(Constraint set 2)}$$

In essence,

$$\underset{C}{\text{maximize}} \quad \sum_{i=1}^{n} C_i \tag{4}$$
$$\text{subject to} \quad C_j + C_k \leq 1, \forall (j,k) \in I$$

6   MCCP ILP Formulation

The maximum Character compatibility problem essentially boils down to finding the maximal clique from the graph G obtained from the application of Algorithm-2 on matrix M, following the procedure described in Algorithm-2. This section will discuss the ILP formulation of the maximal clique problem in order to solve the MCCP Problem. Further, subsections will discuss the maximal clique problem and its ILP formulation.

6.1   Maximal Clique Problem

Given a graph $G(E, V)$, maximal clique problem [13] is to find the largest possible complete subgraph in G. That is, If a clique $K(V', E')$ is found in G, Each vertex in K, is connected to every other vertex in K. K is said to be maximal clique if there is no other clique $K'(V'', E'')$, such that $\|V''\| > \|V'\|$.

ILP Formulation. As mentioned above, it is natural for us to motivate a binary integer variable, to indicate the presence of an edge. Here, since we are worried about cliques, let a binary integer variable C_i denote if node i is present in a clique K.

$$C_i = \begin{cases} 1, \text{ If node i is in clique K} \\ 0, \text{ If node i is not in clique K} \end{cases} \tag{5}$$

By definition, it is trivial that if two nodes, $n_j, n_k \in V$ are part of a clique K, then there must be an edge between them, i.e., $n_j, n_k \in K \implies (n_j, n_k) \in E$. Equivalently, If there is no edge between nodes n_i and n_j they cannot be part of the same clique. Mathematically, it can be written as

$$C_j + C_k \leq 1 \tag{6}$$

when C_j and C_k are binary variables it can take only the values 0 or 1. So the above constraints tell that C_j and C_k cannot be simultaneously 1. That is, it mathematically asserts that if n_j and n_k do not have an edge between them, then they cannot be part of the same clique.

To find a maximal clique, we would need the maximum amount of C_j's to be one for $j = 1, 2, \cdots |V|$. Hence, the maximal clique problem can be formulated as an ILP,

$$\underset{C}{\text{maximize}} \quad \sum_{i=1}^{|V|} C_i \tag{7}$$

$$\text{subject to} \quad C_j + C_k \leq 1, \forall (j,k) \notin E$$

The constraints from the above equation can be put compactly in a matrix form as $E \cdot c \leq e$ (where $\mathbf{e}$ is a vector of ones). For a constraint e_k, $k \in [1, 2 \cdots P]$ where P is the total number of constraints. Let this constraint correspond to $(n_j, n_k) \notin E$. Then constraint matrix E is defined as

$$E = \begin{cases} e_{k_j} = 1, e_{k_k} = 1 \, if \, , (n_j, n_k) \notin E \\ e_{k_j} = 0, e_{k_k} = 0 \, if \, , (n_j, n_k) \in E \end{cases} \tag{8}$$

NMF Formulation [14]. Introduced an approximate algorithm to solve the maximum clique problem. This study uses their formulation and observes significant performance improvements over ILP. We refer the reader to [14] for a detailed description of the algorithm.

7   Results and Discussion

Table 1. Rutime and Graph Details for MCR (ILP), MCCP (ILP), MCCP (NMF), MCCP (MP), and MCCP (MD) in Matlab

Dim	MCR (ILP)	Nodes	Edges	MCCP (ILP)	MCCP (NMF)	MCCP (MP)	MCCP (MD)
5×50	1.23092	50	783	0.32209	0.05486	0.04619	0.01751
10×50	0.20693	50	914	0.11857	0.01033	0.00694	0.00934
20×50	0.30064	50	13	0.19552	0.00875	0.00886	0.00774
50×50	0.28240	50	10	0.16225	0.01506	0.00829	0.00795
5×100	0.44223	100	4706	0.28601	0.04117	0.01823	0.02330
5×200	68.26072	200	10823	69.53437	0.02288	0.01262	0.03074
5×300	12.51716	300	30560	11.60544	0.03153	0.01164	0.03034
5×400	10.90460	400	76045	10.30237	0.03332	0.01442	0.01695
5×500	70.07120	500	105592	74.34117	0.03956	0.01307	0.01758
5×600	0.32563	600	179076	0.27061	0.04030	0.01642	0.02474
5×700	20.85694	700	210066	20.83842	0.07725	0.02781	0.03300
5×800	1203.09553	800	216477	1200.27734	0.09219	0.01524	0.08226
5×900	1822.04935	900	279682	1896.96825	0.11291	0.01871	0.02660
5×1000	228.16656	1000	477415	1576.97888	0.14909	0.02643	0.02427

The primary focus will be on increasing the number of characters or equivalently columns of M as the major computation of the algorithms and optimizations are focused on the comparisons between columns of M and row comparisons are

Table 2. Rutime for MCR (ILP), MCCP (ILP) in Python

Dim	MCR (ILP)	MCCP (ILP)
5×50	0.04000	0.03854
10×50	0.03521	0.05100
20×50	0.05202	0.05052
50×50	0.04704	0.05054
5×100	0.03202	0.03352
5×200	0.33572	0.40374
5×300	0.57519	0.56690
5×400	0.19655	0.23463
5×500	0.66475	0.63258
5×600	0.05529	0.07904
5×700	1.47360	1.57076
5×800	9.73567	5.73996
5×900	14.05600	11.29155
5×1000	2.64179	2.97423

Table 3. Solutions obtained for MCR (ILP), MCCP (ILP), MCCP (NMF), MCCP (MP), and MCCP (MD)

Dim	Real solution	MCR (ILP)	MCCP (ILP)	MCCP (NMF)	MCCP (MP)	MCCP (MD)
5×50	24	24	24	24	24	24
10×50	22	22	22	22	19	18
20×50	2	2	2	2	2	2
50×50	2	2	2	1	2	2
5×100	79	79	79	78	78	78
5×200	64	64	64	64	57	64
5×300	143	143	143	143	96	143
5×400	315	315	315	315	315	315
5×500	318	318	318	318	270	270
5×600	567	567	567	567	566	566
5×700	461	461	461	461	391	391
5×800	357	357	357	355	254	333
5×900	409	409	409	409	298	298
5×1000	787	787	787	787	735	735

skipped after all three identification variables are achieved, this provides some degree of efficiency in row comparisons. The probability of all 3 identification variables being achieved during comparison increases with an increase in the number of instances, this increase will result in an increase in the number of columns being incompatible, the same is correspondingly reflected in Table 1

(Reduction in number of Edges with an increase in the number of rows). When this happens more columns will need to be removed to form a matrix from which a perfect phylogeny can be derived, this makes them harder to validate the result C. Furthermore driven by the fact that current-day research on sequencing technologies is motivated towards finding easier techniques that allow sequence analysis of molecular data of longer lengths.

Mutation matrices of various sizes were generated and tested with MCR(ILP), MCCP(ILP), MCCP(NMF) [14], MCCP(MP), and MCCP(MD). The resultant compatibility is verified for each of the methods. From the results of codes running on an artificially generated dataset Tables 1, 2, and 3 are documented from which the following robust observations can be stated,

– From Table 1 and 2 the implemented ILP methods run significantly faster in python.
– Table 3 presents the optimal solution count of the characters retained from the solutions produced from the implemented methods in comparison to the optimal solution. From Table 3, the ILP methods(both in Matlab and Python) always produce the optimal solution, the NFM method almost always produces the optimal solution and the MP and MD struggle to produce optimal solutions on substantial data sizes.
– From Table 1, with an increase in the number of edges the runtimes of MCCP methods increase.
– On observing the runtimes of the MCCP solutions in Table 1, the runtimes of MP, MD, and NMF methods are outstandingly faster than the ILP methods. MP and MD are marginally faster than NMF. The runtimes of the MP method are consistently negligibly faster than MD for graphs with more than 100 nodes.

8   Conclusion

Logically it is clear that MCR and MCCP are optimally accurate ways of generating data required for the construction of perfect phylogenies. Successively both were implemented with the use of computationally effective techniques like ILP and NMF with the aid of highly researched and efficient open-source solvers. The techniques were verified on synthesized data of varying dimensions to substantiate both correctness and solution speed. Further, this acts as the base for progressing onto complex relaxed and unrestricted phylogeny problems like the weighted MCR, Forward Convergent phylogeny (FCP), and Full maximum parsimony problem (MPP) along with their reconstructions. Inevitably small improvements through research will compound our knowledge in bioinformatics.

References

1. Kanz, C., et al.: The EMBL nucleotide sequence database. Nucleic Acids Res. **33**, D29–D33 (2005). ISSN 0305-1048. https://doi.org/10.1093/nar/gki098, https://www.ncbi.nlm.nih.gov/pmc/articles/PMC540052/. Accessed 01 Dec 2022

2. Clark, K., et al.: GenBank. Nucleic Acids Res. **44**, pp. D67–D72 (2016). ISSN 0305-1048. https://doi.org/10.1093/nar/gkv1276, https://www.ncbi.nlm.nih.gov/pmc/articles/PMC4702903/. Accessed 01 Dec 2022

3. NCBI Geo: archive for functional genomics data sets—update—Nucleic Acids Research—Oxford Academic. https://academic.oup.com/nar/article/41/D1/D991/1067995. Accessed 01 Dec 2022

4. miRBase: microRNA sequences, targets and gene nomenclature. https://www.ncbi.nlm.nih.gov/pmc/articles/PMC1347474/. Accessed 01 Dec 2022

5. Zou, D., et al.: Biological databases for human research. Genom. Proteomics Bioinform. **13**(1), 55–63 (2015). ISSN 1672-0229. https://doi.org/10.1016/j.gpb.2015.01.006, https://www.ncbi.nlm.nih.gov/pmc/articles/PMC4411498/. Accessed 01 Nov 2022

6. Wang, Y.-X., Zhang, Y.-J.: Nonnegative matrix factorization: a comprehensive review. IEEE Trans. Knowl. Data Eng. **25**(6), 1336–1353 (2013). https://doi.org/10.1109/TKDE.2012.51

7. Habib, M., Stacho, J.: Unique perfect phylogeny is intractable. Theoret. Comput. Sci. **476**, 47–66 (2013)

8. Gusfield, D.: Integer Linear Programming in Computational and Systems Biology: An Entry-Level Text and Course. Cambridge University Press, Cambridge (2019). https://doi.org/10.1017/9781108377737

9. Anbuudayasankar, S., Mohandas, K.: Mixed-integer linear programming for vehicle routing problem with simultaneous delivery and PickUp with maximum route-length. Undefined (2008)

10. Kesav, R.S., Premjith, B., Soman, K.P.: Dependency parser for Hindi using integer linear programming. In: Singh, M., Tyagi, V., Gupta, P.K., Flusser, J., Ören, T., Sonawane, V.R. (eds.) ICACDS 2021. CCIS, vol. 1441, pp. 42–51. Springer, Cham (2021). https://doi.org/10.1007/978-3-030-88244-0_5. ISBN 978-3-030-88243-3

11. Nayak, S.: Chapter eight - integer programming. In: Nayak, S. (ed.) Fundamentals of Optimization Techniques with Algorithms, pp. 223–252. Academic Press (2020). ISBN 978-0-12-821126-7. https://doi.org/10.1016/B978-0-12-821126-7.00008-5, https://www.sciencedirect.com/science/article/pii/B9780128211267000085. Accessed 08 June 2022

12. Aparnna, T., Raji, P.G., Soman, K.P.: Integer linear programming approach to dependency parsing for MALAYALAM. In: 2010 International Conference on Recent Trends in Information, Telecommunication and Computing, pp. 324–326 (2010). https://doi.org/10.1109/ITC.2010.97

13. Bomze, I.M., Budinich, M., Pardalos, P.M., Pelillo, M.: The maximum clique problem. In: Du, D.Z., Pardalos, P.M. (eds.) Handbook of Combinatorial Optimization, pp. 1–74. Springer, Boston (1999). https://doi.org/10.1007/978-1-4757-3023-4_1

14. Belachew, M.T.: NMF-based algorithms for data mining and analysis: feature extraction, clustering, and maximum clique finding. Ph.D thesis (2014)

A Novel Method for Near-Duplicate Image Detection Using Global Features

Kunj Bihari Meena and Vipin Tyagi[(✉)]

Jaypee University of Engineering and Technology, Raghogarh, Guna – MP, India

Abstract. The rapid growth of digital multimedia content has led to an increase in near-duplicate images. Near-duplicate image detection is a critical task in the field of multimedia forensics, which aims to detect and identify illegally distributed copies of an original image. Gaussian Hermite Moments (GHM) have been proven to be an effective global feature for image representation and analysis in various computer vision tasks, including image forgery detection. In this paper, we propose a novel near-duplicate image detection method that utilizes GHM as the global feature descriptor. We conducted experiments on the CoMoFoD image dataset. The experimental results demonstrate that the proposed method outperforms existing methods for near-duplicate image detection under various post-processing operations and geometric transformations, particularly scaling and rotation. Moreover, the proposed method is significantly faster than existing methods for near-duplicate image detection and it can potentially make the proposed method more practical for real-world applications.

Keywords: near-duplicate image detection · global features · digital forensics · Gaussian Hermite moments · CoMoFoD dataset · Image hashing · image copy detection

1 Introduction

Several image processing software are available that can manipulate original images, which is why it is common to come across numerous unauthorized duplicate copies (near-duplicate images) of an original image [23]. Near-duplicate images are a common issue in image processing and can occur due to various reasons, such as resizing or cropping an image, applying different compression algorithms or filters, or taking multiple shots of the same scene with slightly different camera angles or lighting conditions [24]. Although these near-duplicate images can contain the visually similar information as the original image, however the digital content of such images may be different. An original image along with its near-duplicate images is shown in Fig. 2. It has become effortless to duplicate images, and ensuring the protection of image copyright has become crucial [3]. In many cases, it is necessary to verify the authenticity of an image before accepting it. Image authentication methods can be used to establish that an image to be investigated is an original which is captured using digital camera and it is not a near-duplicate image that may be post-processed. This leads us to the domain of image forensics [14]–[16], which employs various methods not only to verify the authenticity of an image but also to identify unauthorized copies i.e. near-duplicate images of an original image.

© The Author(s), under exclusive license to Springer Nature Switzerland AG 2023
M. Singh et al. (Eds.): ICACDS 2023, CCIS 1848, pp. 136–149, 2023.
https://doi.org/10.1007/978-3-031-37940-6_12

The process of detecting near-duplicate images involves identifying images that have been altered slightly but still resemble the original image [12]. This task has significant importance in *various* applications like image retrieval [25], detecting copyright infringement, and identifying spam [25]. It's quite usual for individuals to take several pictures of a scene to capture the moment precisely. However, keeping these similar images can create challenges in managing storage, whether it's cloud-based storage or physical hard disk-based storage. This is because storing redundant or near-duplicate images can take up valuable storage space and make it difficult to efficiently manage and organize image collections. Hence, the detection and removal of the near-duplicate image is an important task in several applications. Overall, near-duplicate image detection is a challenging problem due to the vast number of images available on the internet and the high degree of variability in image content and quality [7]. Due to this near-duplicate image detection has become an active area of research in the past few years.

Several efforts were made to detect near-duplicate images in the past [3, 22, 23]. The existing near-duplicate image detection methods can be categorized as local feature-based near-duplicate image detection and global feature-based near-duplicate image detection. Local feature-based near-duplicate image detection methods aim to identify the similarity between local patches or regions of images [2, 29, 31]. On the contrary, the global feature-based near-duplicate detection methods compare the entire image to identify near-duplicates [1, 8]. Local methods, such as SIFT [2] and SURF [13], are better at capturing the unique features of an image and can detect near-duplicate images with more subtle differences. They are also more robust to image transformations and occlusions, as they focus on smaller image patches rather than the entire image. However, local methods can be computationally expensive, especially when processing large datasets, and may not be as effective in identifying near-duplicate images with larger-scale changes. Moreover, these methods are not robust for rotation and scaling. Hence, to overcome the imitation of local methods, alternatively, global methods were explored for near-duplicate image detection [1, 10]. In general, these methods are computationally efficient and can quickly compare large datasets. They are also robust to geometric transformations, such as scaling and rotation, which can help identify near-duplicates that are not exact copies. Therefore, this paper proposes a global feature-based method using Gaussian Hermite Moments (GHM) which produces rotation and scale-invariant features.

The remaining sections of the paper are arranged as follows: Sect. 2 presents a literature survey related to near-duplicate image detection. Section 3, introduces the concept of Gaussian Hermite Moments. The steps of the proposed method are highlighted in Sect. 4. Section 5, reports the experimental results of the proposed method in comparison with state-of-the-art methods. At last in Sect. 6, the conclusions of the paper are provided.

2 Related Works

In this section, different advanced techniques that are associated with identifying near-duplicate images are evaluated and discussed. Wavelet-based methods use wavelet transforms to decompose an image into its frequency components. These frequency components are then used to identify similar regions between images. Wavelet-based methods are particularly effective in detecting near-duplicates that have undergone geometric transformations. For example, a detection of near-duplicate images was proposed by Venkatesan et al. [26] using a technique that relies on Discrete Wavelet Transform (DWT). However, the method was not properly evaluated under challenging situations. Hence, can not be considered a reliable method.

Perceptual hashing (pHash) [11, 31] is a method that generates a unique hash for each image based on its perceptual content. It involves extracting low-level features from the image, such as color and texture and then uses a hashing algorithm to generate a hash value. This hash value can then be used to compare images for near-duplicate detection as done in [8].

Histogram-based methods, [19, 29] compare the histograms of two images to determine their similarity. These methods work well for detecting near-duplicates that have undergone slight modifications, such as cropping or resizing. Mehta et al. [19] proposed a histogram-based method for identifying near-duplicate images. In this approach, feature vectors with a vector length of 91 were used. According to the authors, the SVM classifier can deliver a precision rate of 100% along with a selectivity value greater than 97%. Furthermore, the proposed method achieves an overall accuracy rate of 99%.

CNNs or Convolutional Neural Networks are complex machine learning models that can be taught to categorize images. They have also been used for near-duplicate detection by training them to learn the features of near-duplicate images [4]. Once trained, these models can be used to detect near-duplicate images [20]. Hu et al. [6] introduce a technique for detecting near-duplicate images that involves a deep-constrained siamese hash coding neural network and deep feature learning. The neural network proposed by them has the ability to extract efficient features that aid in identifying near-duplicate images. These features are then utilized to create an index based on Locality Sensitive Hashing (LSH).

Kozat et al. [11] proposed a technique that employed Singular Value Decomposition (SVD) which is robust under various geometric transformations. Huang et al. [8] investigated an approach to detect the near-duplicate image. To identify texture changes, this technique utilizes the Gray-Level Co-occurrence Matrix (GLCM) to capture the overall statistical characteristics. Kin et al. [10] introduced an approach by using a concept of Gist-PCA (Principal Component Analysis) hashing and nearest neighbor search. Zhou et al. [32] discuss a solution to detect near-duplicate images based on hybrid features by combining SIFT and overlapping region-based global context descriptors. Wang et al. [28] investigated a method using the radon transform. This method was robust under rotation, scaling, and translation. Hsiao et al. [5] introduced a concept of extended feature set to find the near-duplicate images. Further, the authors used machine learning technique support vector machine to classify an image as near-duplicate or different.

3   The Gaussian Hermite Moments

A group of moments called Gaussian-Hermite Moments (GHM) employs Gaussian-Hermite polynomials as the foundation for calculating them. In 1997, J. Shen [21] introduced the concept of GHM. GHM has applications in various fields such as moving object recognition [30], and license plate character recognition [17], copy-move forgery detection [17, 18]. GHM is a dependable approach for feature extraction from images as it is immune to changes in scaling, translation, and rotation. These invariants are essential for the reliable detection of near-duplicate images. Moreover, the GHM features are less sensitive to noise.

The Hermite Polynomial (HP) of an mth degree is calculated using the equation given below:

$$G_m(t) = (-1)^m e^{t^2} \frac{d^m}{dt^m} e^{-t^2} \tag{1}$$

Another way to express Eq. (1) is:

$$G_m(t) = \sum_{i=0}^{\frac{m}{2}} \frac{(-1)^i m!}{i!(m-2i)!} (2t)^{m-2i} \tag{2}$$

The orthogonal property of HP for the weight function $e^{-t^2}\, e^{-t^2}$ is expressed as:

$$\int_{-\infty}^{\infty} e^{-t^2} G_m(t) G_m(t) dt = 2^m m! \sqrt{\pi} \delta_{mn} \tag{3}$$

where δ_{mn} is the Kronecker delta

$$\delta_{mn} = \begin{cases} 1; & m = n \\ 0; & \text{otherwise} \end{cases} \tag{4}$$

The following formula represents an orthonormal form of HP:

$$\hat{G}_m(t) = \left(2^m m! \sqrt{\pi}\right)^{-\frac{1}{2}} e^{(-t^2)/2} G_m(t) \tag{5}$$

which also maintains the orthonormal property:

$$\int_{-\infty}^{\infty} \hat{G}_m(t) \hat{G}_n(t) dt = \delta_{mn} \tag{6}$$

Gaussian-Hermite polynomial with a scale-invariant property can be obtained by setting $t = t/\sigma$ in Eq. (5):

$$\hat{G}_m\left(\frac{t}{\sigma}\right) = \left(2^m m! \sqrt{\pi}\right)^{-\frac{1}{2}} e^{((-\frac{t}{\sigma})^2)/2} G_m\left(\frac{t}{\sigma}\right) \tag{7}$$

In the above equation, σ represents the standard deviation of the Gaussian envelope.

For a digital image $f(x, y)$, the 2D GHM order (m,n) $m, n \in Z$ is obtained by considering Eq. 7 as a basis functions as follows:

$$M_{mn} = \iint^f (x, y) \hat{G}_m\left(\frac{x}{\sigma}\right) \hat{G}_n\left(\frac{y}{\sigma}\right) dx dy \tag{8}$$

Using GHM of order $(0,0)$ to (k,k), it is possible to reconstruct the image. as:

$$f'(x, y) = \sum_{m=0}^{k} \sum_{n=0}^{k} \mu_{mn} \hat{G}_m\left(\frac{x}{\sigma}\right) \hat{G}_n\left(\frac{y}{\sigma}\right) \tag{9}$$

4 Proposed Method

All the steps of the proposed method are highlighted in Fig. 1. The details of each of the steps are as follows:

4.1 RGB to Grayscale Conversion

Processing a three-channel RGB image is a time-consuming process. However, if color-based features are not considered then the equivalent grayscale image can also be used to represent image statistics without compromising the results. As the grayscale image is a single-channel image, the time complexity can be reduced if a grayscale image is used instead of an RGB image. Motivated by this, the proposed method first converts the RGB image into a grayscale image using:

$$Y = 0.299R + 0.587G + 0.114B \tag{10}$$

The color elements which correspond to red, green, and blue are denoted by R, G, and B respectively. In short, the resulting grayscale image represents the same visual information as the original color image, but in a simplified form that is easier to process and analyze.

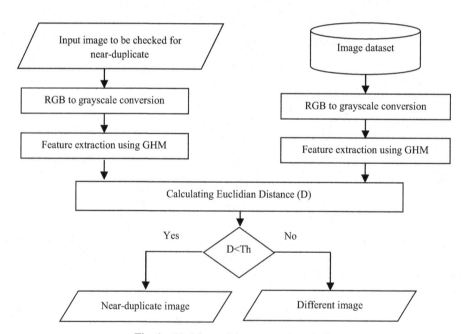

Fig. 1. Workflow of the proposed method

4.2 Feature Extraction Using GHM

Feature extraction step plays an important role in any near-duplicate image detection process. Features that are invariant to different post-processing operations/ geometric transformations are of more importance. Therefore, the proposed method exploits GHM-based global features to represent an image which are rotation and scale invariance. At the same time, the GHM-based features remain almost unchanged even when the image is post-processed by some of the common operations such as JPEG compression, color reduction, image blurring, etc.

In the proposed method, N features are computed for an input image f(x, y) using Eq. 8. Feature with a higher value of N can represent an image in a better way, however it increases computational cost. Experimentally, the value of N is fixed to 32 for all the experiments conducted in Sect. 5.

4.3 Similarity Matching

To find the similarity between two feature vectors that are obtained for the input image and target image, the proposed method employs a widely acceptable feature matching method Euclidian distance. The following is the formula to calculate the Euclidian distance (D) between two feature vectors U and V:

$$D(U, V) = \sqrt{\sum_{k=1}^{N}(U_k - V_k)^2} \tag{11}$$

where N is the number of GHM features calculated for an image. If $D(U, V) < Th$ then the input image is considered as near-duplicate; otherwise, it is considered a different image. The optimal value for the similarity threshold Th, which produces the best outcomes for the proposed approach, is determined to be 12 after conducting a range of experiments.

5 Experimental Results

The MATLAB programming language is used to simulate the proposed method. All the experiments are performed using MATLAB2016a tool on a personal computer. The computer is configured with 8 GB RAM, an Intell Core-i3 processor, and Windows 7 operating system. Section 5.1 explains the image dataset. Further, Sect. 5.2 provides the details of the evaluation parameters used for validating the proposed method.

5.1 Image Dataset Description

According to [23] CoMoFoD image dataset comprises the most challenging near-duplicate image cases as compared to other publicly available image datasets such as Columbia NDI dataset [23], California-ND Dataset [9]. Further, the authors in [23] emphasized that if a method can perform well on the CoMoFoD dataset it can be considered a robust near-duplicate image detection method. Hence, to evaluate the performance of the proposed method, we have created a dataset of 1768 images from the images taken from the CoMoFoD dataset. The CoMoFoD dataset primarily comprises images related to copy-move forgery along with their corresponding authentic images. All the forged images in the CoMoFoD dataset were created from 34 original images. Six post-processing operations viz. Brightness change, contrast adjustment, color reduction, image blurring, noise addition, and JPEG compression are applied to each of the images to create near-duplicate images. Hence, including all these six operations along with 34 original images, a total of 850 images are available in the dataset that can be considered near-duplicate. As there were no images related to rotation and scaling in the CoMoFoD dataset, we have created four copies of each original image by applying four scaling factors [25%, 50%, 150%, and 200%]. Hence, a total of $34 \times 4 = 136$ images were created related to scaling attack. We also created image copies by rotating each of the original images using different angles [15°, 30°, 45°, 60°, 75°, 90°, 105°, 120°, 135°, 150°, 165°, 180°, 195°, 210°, 225°, 240°, 255°, 270°, 285°, 300°, 315°, 330°, and 345°]. Hence, a total of $34 \times 23 = 782$ images were created related to the rotation attack. Figure 2 shows a sample image along with its near-duplicate images from the CoMoFoD dataset.

5.2 Evaluation Parameters Used

Various evaluation parameters are used in the past to evaluate the performance of near-duplicate image detection [22]. However, the True Positive Rate (TPR), and False Positive Rate (FPR) are the most suitable and widely acceptable evaluation parameters in the literature [22]. Hence, the proposed method was also evaluated using these two parameters. These parameters are formulated in Eqs. 12 and 13.

$$\text{TPR} = \frac{n1}{N2} \tag{12}$$

$$\text{FPR} = \frac{n2}{N1} \tag{13}$$

where $n1$ number of actual duplicate images that are detected as near-duplicate images; $n2$ is the number of different (non-near-duplicate) images detected as near-duplicate images; $N1$ is the total number of different (non-near-duplicate) images; $N2$ is the total number of near-duplicate images. A method is considered better if it shows a higher TPR value and at the same time lower FPR value.

5.3 Results of the Proposed Method

We have performed eight sets of experiments that include the evaluation under eight different attacks (post-processing operations / geometric transformations) viz. Brightness change, contrast adjustment, color reduction, image blurring, noise addition, JPEG compression, scaling, and rotation. The experimental results of the proposed method are highlighted in Table 1. The results mentioned in the Table indicate that the proposed method achieves TPR $= 1$ and FPR $= 0$ for color reduction, image blurring, and JPEG compression. That means that the proposed method performs ideally under these three cases. The results are also promising under rotation and scaling, as the proposed method

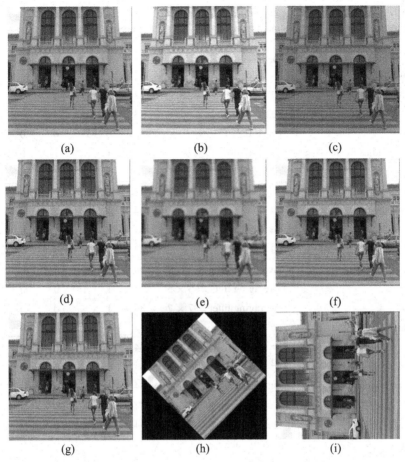

Fig. 2. (a) Original image, (b) Image with brightness change by (0.01, 0.8), (c) Image with contrast adjustment by (0.01, 0.8), (d) Image with the color reduction to 32, (e) Image with blurring by 7 × 7 average filter, (f) Image with noise addition of $\sigma^2 = 0.009$, (g) JPEG compression by quality factor = 20, (h) Image with rotation of 45°, (i) Image with rotation of 90°

obtains TPR = 1 and FPR = 0 in all the cases of scaling and rotation except the scaling of 25%, and rotation of 30° and 300°. In each of these three cases also only one image was not predicted as near-duplicate from 34 images. Hence, TPR value is 0.9705 for these three cases (scaling of 25%, and rotation of 30° and 300°). However, under these situations also the proposed method can maintain FPR = 0.

Under the brightness changes experiment, the results indicate that for lower brightness levels (0.01–0.8), the TPR drops significantly, while the FPR remains at zero. Under contrast adjustment, the results indicate that the TPR = 1 for all levels of contrast, while the FPR increases slightly with decreasing contrast. Under noise addition, the results indicate that for higher levels of noise ($\sigma^2 = 0.005$), the TPR remains at 1, while for the highest level of noise ($\sigma^2 = 0.009$), the TPR drops to 0, and the FPR remains zero.

Table 1. Experimental results of proposed method when subjected to different geometric transformations or post-processing operations

Post-processing/geometric transformation	Details	$n1$	TPR	$n2$	FPR
Brightness change	(0.01, 0.95)	34	1	0	0
	(0.01, 0.9)	33	0.9705	0	0
	(0.01, 0.8)	6	0.1764	0	0
Contrast adjustment	(0.01, 0.95)	34	1	1	0.029
	(0.01, 0.9)	34	1	1	0.029
	(0.01, 0.8)	20	0.5882	0	0
Color reduction	128	34	1	0	0
	64	34	1	0	0
	32	34	1	0	0
Image blurring	Averaging filter $= 3 \times 3$	34	1	0	0
	Averaging filter $= 5 \times 5$	34	1	0	0
	Averaging filter $= 7 \times 7$	34	1	0	0
Noise addition	$\sigma^2 = 0.0005$	34	1	0	0
	$\sigma^2 = 0.005$	34	1	0	0
	$\sigma^2 = 0.009$	0	0	0	0
JPEG compression	Quality factor $= 20$	34	1	0	0
	Quality factor $= 30$	34	1	0	0
	Quality factor $= 40$	34	1	0	0
	Quality factor $= 50$	34	1	0	0
Scaling	25%	33	0.9705	0	0
	50%	34	1	0	0
	150%	34	1	0	0
	200%	34	1	0	0
Rotation	Rotation angle $= 30°$	33	0.9705	0	0
	Rotation angle $= 45°$	34	1	0	0
	Rotation angle $= 60°$	34	1	0	0
	Rotation angle $= 90°$	34	1	0	0
	Rotation angle $= 135°$	34	1	0	0
	Rotation angle $= 180°$	34	1	0	0
	Rotation angle $= 225°$	34	1	0	0
	Rotation angle $= 300°$	33	0.9705	0	0

(continued)

Table 1. (*continued*)

Post-processing/geometric transformation	Details	n1	TPR	n2	FPR
	Rotation angle = 345°	34	1	0	0

Overall, the proposed method appears to be effective to most of the post-processing operations/ geometric transformations tested, except for lower levels of brightness changes and higher levels of noise addition.

5.4 Results Comparison and Analysis

Table 3 compares the experimental results of the proposed approach with existing techniques SVD-based [11], DWT-based [26], and pHash-based [27] for near-duplicate image detection under different post-processing operations/ geometric transformations. The post-processing operations considered include brightness change, contrast adjustment, color reduction, image blurring, noise addition, and JPEG compression, whereas geometric transformations include scaling and rotation.

Under scaling and rotation, the proposed method obtains TPR = 1 and FPR = 0 which is best among all the methods included for comparisons. Although, the DWT-based method obtains TPR = 1 it shows a poor FPR of 0.0882 for rotation and scaling both. This indicates that the DWT-based method predicts some of the visually different images as near-duplicate images. The performance of SVD-based and pHash-based methods is unsatisfactory. This is because both of these methods use features that are not rotation and scale invariant in nature. On the other hand, the proposed method shows the best result due to the rotation and scale-invariant nature of the features obtained from GHM.

Under color reduction, all methods achieves a TPR = 1 and an FPR = 0 except DWT-based method which shows an FPR of 0.0882. Further, it can be noticed from the results shown in Table 3 that none of the method including the proposed one can detect near-duplicate instance if it is corrupted by noise addition. Unfortunately, the proposed method performs poorly under brightness change operation among all. Under brightness change, pHash-based method is the best performer.

Overall, the proposed method showed competitive performance with existing methods and demonstrated robustness under various post-processing operations and geometric transformations (Table 2).

Table 3 highlights the computation time comparisons of the proposed method with existing methods related to near-duplicate image detection. The table lists four methods, including the proposed method, along with the computation time in seconds required by each method for processing the input images of size 512×512 pixels. The computation time for SVD-based is 0.6893 s. The computation time for DWT-based is 0.0736 s. The computation time for pHash-based is 2.4820 s. The computation time for the proposed method is 0.04997 s. Overall, the results in Table 3 shows that the proposed method is much faster than existing methods for near-duplicate image detection, which is a

Table 2. Performance comparison with existing methods

Method	Post-processing operation/ geometric transformations	Details	TPR	FPR
SVD-based [11]	Brightness change	Range [0.01,0.8]	0.8823	0
	Contrast adjustment	Contrast range [0.01, 0.8]	1	0
	Color reduction	Number of colors per channel = 32	1	0
	Image blurring	Averaging filter = 3×3)	0.9411	0
	Noise addition	$\sigma^2 = 0.009$	0	0
	JPEG compression	Quality Factor = 20	1	0
	Scaling	200%	0.0294	0
	Rotation	Rotation angle = 45°	0.0294	0
DWT-based [26]	Brightness change	Range [0.01,0.8]	1	0.3235
	Contrast adjustment	Contrast range [0.01, 0.8]	1	0.0294
	Color reduction	Number of colors per channel = 32	1	0.0882
	Image blurring	Averaging filter = 3×3)	1	0.0882
	Noise addition	$\sigma^2 = 0.09$	0	0
	JPEG compression	Quality Factor = 20	1	0.0882
	Scaling	200%	1	0.0882
	Rotation	Rotation angle = 45°	1	0.0882
pHash-based [27]	Brightness change	Range [0.01,0.8]	1	0
	Contrast adjustment	Contrast range [0.01, 0.8]	0.9117	0
	Color reduction	Number of colors per channel = 32	1	0
	Image blurring	Averaging filter = 3×3)	0.5588	0
	Noise addition	$\sigma^2 = 0.009$	0	0
	JPEG compression	Quality Factor = 20	1	0
	Scaling	200%	0.0882	0
	Rotation	Rotation angle = 45°	0.0882	0
Proposed method	**Brightness change**	**Range [0.01,0.8]**	**0.1764**	**0**
	Contrast adjustment	**Range [0.01, 0.8]**	**0.5882**	**0**

(*continued*)

Table 2. (*continued*)

Method	Post-processing operation/ geometric transformations	Details	TPR	FPR
	Color reduction	**Number of colors per channel = 32**	**1**	**0**
	Image blurring	**Averaging filter = 3 × 3**	**1**	**0**
	Noise addition	$\sigma^2 = 0.09$	**0**	**0**
	JPEG compression	**Quality Factor = 20**	**1**	**0**
	Scaling	**200%**	**1**	**0**
	Rotation	**Rotation angle = 45°**	**1**	**0**

significant advantage in practical applications where large numbers of images need to be processed quickly.

Table 3. Computation time comparisons with existing methods

Method	Time in seconds
SVD-based [11]	0.6893
DWT-based [26]	0.0736
pHash-based [27]	2.4820
Proposed method	**0.04997**

6 Conclusion

The paper introduces a new approach for detecting near-duplicate images based on the global feature extraction mechanism Gaussian Hermite Moments (GHM). The proposed method is designed to address the limitations of existing local feature based methods by providing rotation and scaling invariance while also being computationally efficient. The experimental results indicate that the proposed method obtains a True Positive Rate of 1 and a False Positive Rate of 0 under scaling and rotation, which is better than all the other methods included for comparison. Moreover, the proposed method requires only 0.04997 s to process an input image of size 512×512 pixels, which is much faster than SVD-based, DWT-based, and pHash-based methods. However, the proposed method has limitations of producing low TPR under certain post-processing operations, such as brightness change and noise addition. Overall, the results suggest that the proposed method based on GHM is a promising approach for near-duplicate image detection and has the potential to be useful in practical applications where both accuracy and efficiency are critical.

References

1. An, S., Huang, Z., Weng, D., Chen, Y.: Near duplicate product image detection based on binary hashing. In: ACM International Conference Proceeding Series, pp. 75–80 (2017)
2. Cao, Y., Zhang, H., Gao, Y., Guo, J.: An efficient duplicate image detection method based on Affine-SIFT feature. In: Proceedings – 2010 3rd IEEE International Conference on Broadband Network and Multimedia Technology, IC-BNMT2010, pp. 794–797 (2010)
3. Chen, L., Stentiford, F.W.M.: Comparison of near-duplicate image matching. In: IET Conference Publications, pp. 38–42 (2006)
4. Elaskily, M.A., Alkinani, M.H., Sedik, A., Dessouky, M.M.: Deep learning based algorithm (ConvLSTM) for Copy Move Forgery Detection. J. Intell. Fuzzy Syst. **40**(3), 4385–4405 (2021)
5. Hsiao, J., Chen, C., Chien, L., Chen, M.: A new approach to image copy detection based on extended feature sets. IEEE Trans. Image Process. **16**(8), 2069–2079 (2007)
6. Hu, W., Fan, Y., Xing, J., Maybank, S.: Deep constrained Siamese hash coding network and load-balanced locality-sensitive hashing for near duplicate image detection. IEEE Trans. Image Process. **27**, 4452–4464 (2018)
7. Hu, Y., Cheng, X., Chia, L.T., Xie, X., Rajan, D., Tan, A.H.: Coherent phrase model for efficient image near-duplicate retrieval. IEEE Trans. Multimedia **11**(8), 1434–1445 (2009)
8. Huang, Z., Liu, S.: Perceptual image hashing with texture and invariant vector distance for copy detection. IEEE Trans. Multimed. **9210**, 1–14 (2020)
9. Jinda-Apiraksa, A., Vonikakis, V., Winkler, S.: California-ND: an annotated dataset for near-duplicate detection in personal photo collections. In: 2013 Fifth International Workshop on Quality of Multimedia Experience (QoMEX), pp. 142–147 (2013)
10. Kim, H., Sohn, S., Kim, J.: Revisiting Gist-PCA hashing for near duplicate image detection. J. Signal Process. Syst. **91**(6), 575–586 (2018). https://doi.org/10.1007/s11265-018-1360-0
11. Kozat, S.S., Venkatesan, R., Mihçak, M.K.: Robust perceptual image hashing via matrix invariants. In: Proceedings – International Conference on Image Processing, ICIP, vol. 2, pp. 3443–3446 (2004)
12. Liu, L., Lu, Y., Suen, C.Y.: Variable-length signature for near-duplicate image matching. IEEE Trans. Image Process. **24**(4), 1282–1296 (2015)
13. Ahmed, S.M.H.M.A., Hasan, A.S.M.Z., Hossain, M.A.: Image copy detection using scale invariant feature transform (SURF). Int. J. Comp. Appl. **112**(16) (2015)
14. Meena, K.B., Tyagi, V.: Image splicing forgery detection using noise level estimation. Multimed. Tools Appl. **82**, 1–18 (2021). https://doi.org/10.1007/s11042-021-11483-x
15. Meena, K.B., Tyagi, V.: Distinguishing computer-generated images from photographic images using two-stream convolutional neural network. Appl. Soft Comput. **100**, 107025 (2021)
16. Meena, K.B., Tyagi, V.: A copy-move image forgery detection technique based on tetrolet transform. J. Inf. Security Appl. **52**, 102481 (2020)
17. Meena, K.B., Tyagi, V.: A copy-move image forgery detection technique based on Gaussian-Hermite moments. Multimed. Tools Appl. **78**(23), 33505–33526 (2019). https://doi.org/10.1007/s11042-019-08082-2
18. Meena, K.B., Tyagi, V.: Image forgery detection: survey and future directions. In: Data, Engineering and Applications. Springer Singapore, pp. 163–194 (2019)
19. Mehta, P., Tripathi, R.K.: Near-duplicate detection for LCD screen acquired images using edge histogram descriptor. Multimed. Tools Appl. **81**, 30977–30995 (2022). https://doi.org/10.1007/s11042-022-12637-1
20. Morra, L., Lamberti, F.: Benchmarking unsupervised near-duplicate image detection. Expert Syst. Appl. **135**, 313–326 (2019)

21. Shen, J.: Orthogonal Gaussian–Hermite moments for image characterization. In: SPIE Intelligent Robots Computer Vision XVI, Pittsburgh, pp. 224–233 (1997)
22. Srivastava, M., Siddiqui, J., Ali, M.A.: A review of hashing based image copy detection techniques. Cybernet. Inf. Technol. **19**(2), 3–27 (2019)
23. Thyagharajan, K.K., Kalaiarasi, G.: A review on near-duplicate detection of images using computer vision techniques. Arch. Comput. Meth. Eng. **28**(3), 897–916 (2020). https://doi.org/10.1007/s11831-020-09400-w
24. Tyagi, V.: Understanding Digital Image Processing. CRC Press (2018)
25. Tyagi, V.: Content-Based Image Retrieval: Ideas, Influences, and Current Trends. Springer (2018)
26. Venkatesan, R., Koon, S., Jakubowski, M.H., Moulin, P., Inst, B., Ave, N.M.: Robust image hashing. In: Proceedings 2000 International Conference on Image Processing, pp. 1–3 (2000)
27. Wang, X., Pang, K., Zhou, X., Zhou, Y., Li, L., Xue, J.: A visual model-based perceptual image hash for content authentication. IEEE Trans. Inf. Forensic Security **10**(7), 1336–1349 (2015)
28. Wang, Y., Lei, Y., Huang, J.: An image copy detection scheme based on radon transform. In: 2010 IEEE 17th International Conference on Image Processing, pp. 1009–1012 (2010)
29. Wang, Y., Hou, Z.J., Leman, K.: Keypoint-based near-duplicate images detection using affine invariant feature and color matching. In: ICASSP, IEEE International Conference on Acoustics, Speech and Signal Processing – Proceedings, no. 1, pp. 1209–1212 (2011)
30. Wu, Y., Shen, J.: Moving object detection using orthogonal Gaussian-Hermite moments. Vis. Commun. Image Process. **5308**, 841–849 (2004)
31. Zhao, Y., Wang, S., Zhang, X., Yao, H.: Robust hashing for image authentication using Zernike moments and local features. IEEE Trans. Inf. Forensics Secur. **8**(1), 55–63 (2013)
32. Zhou, Z., Wang, Y., Wu, Q.M.J., Yang, C., Sun, X.: Effective and efficient global context verification for image copy detection. IEEE Trans. Inf. Forensics Secur. **12**, 48–63 (2016)

VSMAS2HN: Verifiably Secure Mutual Authentication Scheme for Smart Healthcare Network

Shivangi Batra[ID], Bhawna Narwal[(✉)][ID], and Amar Kumar Mohapatra[ID]

Department of IT, Indira Gandhi Delhi Technical University for Women (IGDTUW), Delhi, India
{shivangi014mtit21,bhawnanarwal,akmohapatra}@igdtuw.ac.in

Abstract. Digitization has reformed and revamped the healthcare industry and ameliorated patient-clinician interaction. Technology has coherently modified conventional medical workflow and successfully decentralized and stabilized communication between caregivers and patients, making medical services more affordable and accessible to patients. The Internet of Medical Things (IoMT) has efficaciously unified the healthcare ecosystem and made remote patient monitoring more adaptable and reachable. Due to the accessible and economical attributes, incorporating cloud technology has been effective for smart healthcare infrastructure. Nonetheless, the security of sensitive patient data from cyber threats is the key concern for cyber experts. Securing the IoMT network is a pressing priority thereby this paper proposes a VSMAS2HN: A Verifiably Secure Mutual Authentication Scheme for Smart Healthcare Networks. The proposed VSMAS2HN has been tested "SAFE" over AVISPA and BAN Logic has also been presented to support the secure nature of the session key of the suggested scheme.

Keywords: IoMT · Mutual Authentication · BAN Logic · AVISPA

1 Introduction

Technological innovation has brought a positive disruption in varying domains, including the dynamics of business, military, and healthcare. The technical upswing has aggrandized the healthcare sector, making medical services readily available in a trouble-free and more reasonable, and engaging manner. The Internet of Things (IoT) has made a breakthrough in the health industry by creating a web of medical devices and digital health systems together known as the IoMT. IoMT enables both the interconnection and intra-connection of healthcare devices with the outside world and within the medical ecosystem.

The 2019 pandemic, introduced and necessitated the idea of contactless interaction, summarizing the impact and role of technology in methodically actualizing this concept of remote communication. Technology has always been the backbone to expand the healthcare industry beyond the constraints of territorial boundaries and clinical isolations. During the pandemic, it was challenging to remotely monitor the patients where patients were not only geographically restricted but also medically quarantined, not

M. Singh et al. (Eds.): ICACDS 2023, CCIS 1848, pp. 150–160, 2023.
https://doi.org/10.1007/978-3-031-37940-6_13

being in direct contact with the medical staff. Geographic constraints bound the patients in remote areas to visit the primary health centers and restricted them from seeking specialized health advice. Clinical isolation (quarantine) on the other hand was the most fundamental step to prevent the spread of the disease.

IoMT has empowered patients and caregivers to distantly and securely interact with one another. Medical devices such as body sensors and smart wearables enable patients to record their vitals; whereas, the established IoMT network makes it possible for this patient data to be remotely exchanged and accessed by medical clinicians [1, 2]. In such remote and wireless scenarios, it becomes pertinent to secure sensitive patient data from various adversarial attacks.

Cloud facilities such as cost, storage, accessibility, and data privacy make it a suitable component for medical infrastructure. Despite the viability of cloud technology, ensuring the security of data both at rest and during the transmission is paramount. It is thereby peremptory to design a secure, lightweight, and robust scheme to mutually authenticate the communicating entities in the healthcare environment. This work proposes one such scheme to secure communication between medical clinicians, cloud servers, and patients. The security attributes of Elliptic curve cryptography (ECC) have been adopted in the proposed VSMAS2HN. Moreover, biometrics have been utilized to escalate the security of the proposed VSMAS2HN and to maintain data confidentiality.

1.1 Problem Formulation and Contribution

With the expansion of the IoMT network, it is essential to secure the cloud servers from adversarial threats and also the sensitive patient data from various passive and active attacks. Furthermore, it is fundamentally imperative to perform mutual authentication between the communicating entities. Based on the comprehensive understanding of prior literature on cloud-assisted healthcare infrastructure we have proposed a verifiably secure and robust mutual authentication scheme for smart healthcare networks. Fuzzy extractor has been used to enhance the security of the proposed VSMAS2HN. The proposed VSMAS2HN has been tested over AVISPA and BAN Logic has been presented to verify the security of the session key.

1.2 Paper Organization

Succeeding the sections of introduction, problem formulation, and contribution, the upcoming segments of this paper discuss the literature review, network model, and threat model. After the description of the proposed VSMAS2HN, the results of the AVISPA simulation have been laid down along with BAN Logic verification. Finally, the conclusion of the work has been presented.

2 Literature Review

In 2022 Chakraborty and Kishore [3] designed a machine learning (ML) based classification algorithm to speculate heart diseases. To reduce latency, their architecture incorporated fog layer along with cloud technology. They tested their model over several

ML algorithms and analyzed its accuracy using various performance metrics. Although they achieved higher accuracy compared to counterpart schemes, they did not cover the security aspect and domain vulnerabilities. Jeyaraj and Nadar [4] devised a neural network-based patient monitoring system. Smart sensors were an integral component of their proposed system. The accuracy and reliability of their model were significantly high. Seminal contributions were made by Gaurav et al. [5], who discussed the security threats associated with cloud technology in the IoMT environment and also described the existing solutions. Parallel to the works of [5], Rahimi et al. [6] provided a comprehensive study on the merits and demerits of cloud-based medical services. Various parameters including quality of service and applied techniques were considered by the researchers to present their study. Authors of [17] presented a review on the prominence of the burgeoning field of artificial intelligence (AI) in enhancing and upgrading the health-care industry. They discussed the role of AI in predictive analysis, disease detection, and in the precision achieved in AI-assisted surgery techniques. Further, they empha-sized the predominance of developing Point-Of-Care (POC) devices for the futuristic healthcare ecosystem. Pelekoudas-Oikonomou et al. [18] discussed the significance of utilizing blockchain for securing the smart healthcare network. They presented an elab-orative study on available blockchain platforms and the importance of integrating edge technology for optimizing data flow in the healthcare sector.

In 2023, Mageshkumar and Lakshmanan [7] drafted a classification model for a cloud-connected medical system. They utilized the emerging technologies of deep trans-fer learning, fuzzy-c means amongst more. They validated their claims using various simulations and mathematical computations. Jangra and Mangla [8] propounded an optimal load-balancing algorithm for resource scheduling in cloud-based healthcare transmissions using reinforcement learning. They also presented a comparison of their performance parameters (throughput and latency) with that of other pre-existing models. Further, Menon et al. [9] emphasized the prominence of using cloud technology in the healthcare domain. They also discussed the security and accessibility of cloud-linked health services. In the same year, Chenam and Ali [10] presented a keyword-searching scheme for a multi-user and cloud-based medical IoT environment. Their proposed scheme was executed using Raspberry Pi and was also found to be cost-effective. Using the concept of natural language processing, Pulido et al. [11] presented a cloud-based smart navigation robotic platform for aged people. In the same year, Nancy et al. [12] worked in the field of disease prognosis using a predictive deep-learning model. Their model attained an appreciable accuracy percentage of more than 98%. Authors of [13] discussed the security features of an electronic medical record system maintained over a blockchain-based cloud-assisted network along with associated challenges. They also studied the scope of quantum-aware blockchain for the healthcare ecosystem. Al-hajjar et al. [19] presented a focused study on IoMT-based, automated epilepsy seizure detection along with data analysis of patient electroencephalography (EEG) data. They also dis-cussed feature extraction techniques and compared various ML algorithms best suited for clinical employment. Karar et al. [20] furnished an IoMT architecture in clinical diagnosis using Computed Tomography(CT) scans of the patients. They also compared their proffered method with other counterpart approaches. Bhushan et al. [21] discussed the security vulnerabilities and concerns in IoMT. They also studied and explored the sustainable nature of IoMT architecture. Ravikumar et al. [22] propounded the use of neural computing along with fog technology for remote patient monitoring. Their proposed model attained 95% accuracy which was claimed to be significantly higher than related schemes.

3   Network Model

The proposed VSMAS2HN has three communicating entities namely Patient, Clinician, and Cloud Server. Furthermore, sensors are attached to the patient to collect health vitals which are later transmitted over the IoMT network and made available to the sequestered clinician. Remote patient monitoring is achieved through the cloud-assisted IoMT network. Withal, ECC has been used to initialize the proposed VSMAS2HN and generate random primitives, a complex prime number, and a generator. Moreover, biometrics have been incorporated through the fuzzy extractor.

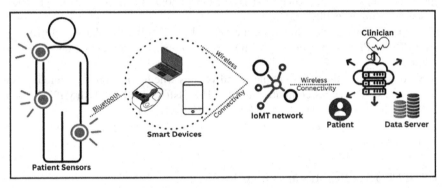

Fig. 1. Network model of VSMAS2HN

Generator and Reproducer are the two fundamental functions of fuzzy extractor, denoted as Gen() and Rep() respectively. Biometrics are affected by noise but the private key remains unchanged when extracted from the imprint of the identity. Whereas, when the imprint is given to Gen(), the function generates a public key and a private key as output. The noisy biometric and obtained public key are inputted to Rep() to recover the private key which will be used to authenticate the identity [15, 16]. Figure 1 illustrates the network architecture for VSMAS2HN.

4   Threat Model

Dolev-Yao threat model [14] has been adopted in this paper, pursuant to which the transmission channel is open to the adversary "V" and here V has the ability to execute prolific passive and active attacks. As per this adversary model, V can probably attempt to anticipate the identity of either the medical clinician or patient.

5 Proposed Scheme

The proposed VSMAS2HN has 3 phases: Initialization, Registration, and Login-Authentication-key exchange. Table 1 describes the various notations used while explaining the proposed VSMAS2HN.

Initialization Phase:

CSR selects an elliptic curve Ej over a finite field Fj. CSR also picks a security parameter sp. N is the point on the elliptic curve. It is worth noting the security attribute of initializing using ECC. Elliptic Curve Discrete Logarithmic problem (ECDLP) is computationally hard. Further, CSR publishes the following parameters: {Ej, Fj, N, sp}.

Registration Phase:

Patient Registration with CSR:

Prior to actual registration, CSR provides a unique and random token number to the patient.

1. While registering, PID and PWD, and patient biometric PBIO are used. Patient computes (PPRk, PPUk) = Gen(PBIO). (Later during login P enters its PBIO*, PPRk is obtained using Rep(PBIO*, PPUk)).
2. The patient also computes his masked id as MPID = Hsf(PID||PPRk) $\oplus$ Hsf(PPWD||PPRk) and sends <MPID, token number> as a registration request to CSR.
3. CSR on receiving the patient request, and verifies whether P is pre-registered or is a new request. Then CSR computes: a. A1 = Hsf(Hsf(MPID||N) $\oplus$ Hsf(MPID||sp)), b. A2 = Hsf(A1 $\oplus$ token number) and c. A3 = Hsf(A1||A2)
4. CSR registers P and sends <A3, token number> to P. P saves the received parameters in its memory.

Medical Clinician Registration with CSR:

Prior to actual registration, CSR provides a unique sequence clinician number (SCN) to the medical clinician.

1. While registering, MCID and MCPWD, and medical clinician biometric MCBIO are used. Patient computes (MCPRk, MCPUk) = Gen(MCBIO). MC also computes his masked id as MPID = Hsf(MCID||MCPRk) $\oplus$ Hsf(MCPWD||MCPRk) and sends <MCID, SCN> as a registration request to CSR.
2. CSR on receiving the request, and verifies whether MC is pre-registered or is a new request. Then CSR computes: a. B1 = Hsf(Hsf(MCID||N) $\oplus$ Hsf(MCID||sp)), b. B2 = Hsf(B1 $\oplus$ SCN) and c. B3 = Hsf(B1||B2)
3. CSR registers MC and sends <B3, SCN> to MC. MC saves the received parameters in its memory.

Patient Login-Authentication and Key Exchange Phase:

The health record of the patient is forwarded to CSR and P gets the prescription from MC through CSR.

1. MPID is computed and checked against the saved MPID when the PID, PPWD, and PBIO are entered by the patient. Corresponding to the MPID, the medical record

is inserted by P as PMR = (MPID, MR). PMR is further signed with the patient's private key as PHMR = PPRk(PMR) which is further encrypted using the public key of MC as UPR = MCPUk(PHMR).

2. P further computes a. $A4 = Hsf(MPID\|A3\|T1) \oplus R1$, b. $A5 = Hsf(A4\|T1\|R1)$ and c. $A6 = A4 \oplus A5$.

3. P sends $<A4, A6, UPR, T1>$ to CSR.

4. CSR on receiving the message verifies the freshness of the message and then computes a. $R1^* = A4 \oplus Hsf(MPID\|A3\|T1)$, b. $A5^* = Hsf(A4\|T1\|R1^*)$ and $A6^* = A4 \oplus A5^*$. The computed A6 is matched with the received A6. If A6 matches then P is authenticated by CSR.

5. Next, CSR computes a. $A7 = MPID \oplus Hsf(SCN\|T2) \oplus R1$, b. $A8 = A7 \oplus Hsf(MPID\|T2\|A6)$ and c. $A9 = A7 \oplus A8$ and sends $<A7, A9, T2, UPR, A4, A6, MPID>$ to MC.

6. MC on receiving the message checks the freshness of the message and computes a. $R1^* = MPID \oplus Hsf(SCN\|T2) \oplus A7$, b. $A8^* = A7 \oplus Hsf(MPID\|T2\|A4\|A6)$ and c. $A9^* = A7 \oplus A8^*$. It then matches the received A9 with the computed A9. If so then communicating entities are authenticated by MC.

7. Further, MC obtains the PMR by first using MCPRk and then using PPUk. After analyzing the medical records, MC enters the associated prescription as MCP = (MCID, MP). MCP is further signed with the clinician's private key as MCHP = MCPRk(MCP) which is further encrypted using the patient's public key as UMCR = PPUk(MCHP).

8. MC further computes a. $B4 = MCID \oplus Hsf(R1^*\|T3\|MPID) \oplus R2$, b. $B5 = Hsf(R2\|N\|T3)$ and c. $B6 = B5 \oplus B6$. It also computes the sessionKey = $Hsf(MPID\|MCID\|R1^*\|R2\|T3)$ and $B7 = PPUk(Hsf(sessionKey))$. MC sends $<UMCR, B7, B4, B6, T3>$ to CSR.

9. CSR on receiving the message checks the freshness of the message and computes a. $R2^* = B4 \oplus MCID \oplus Hsf(R1\|T3\|MPID)$, b. $B5^* = Hsf(R2^*\|N\|T3)$ and c. $B6^* = B4 \oplus B5^*$. Computed B6 is matched with received B6 to authenticate the communicating entities. Then, $B8 = B6^* \oplus A3$. CSR sends $<T4, UMCR, B7, B4, B6, B8, T3, T4, MCID>$ to P.

10. P after checking the freshness of the received message computes a. $R2^* = B4 \oplus MCID \oplus Hsf(R1\|T3\|MPID)$, b. $B5^* = Hsf(R2^*\|N\|T3)$, c. $B6^* = B4 \oplus B5^*$ and $B8^* = B6^* \oplus A3$. B6 and B8 are used by P to authenticate the communicating parameters.

11. P decrypts UMCR by first using MCPUk and then using its own private key. B7 is also decrypted by P using PPRk and sessionKey is thence obtained.

6 AVISPA

The suggested VSMAS2HN has been tested SAFE under AVISPA backends. This guarantees VSMAS2HN is safe against Men in the Middle Attacks, impersonation attacks, and replay attacks. Figs. 2 and 3 presents the AVISPA simulation results.

Table 1. Notation Table

Notations	Description
CSR	Cloud Server
MC	Medical Clinician
P	Patient
PPRk, MCPRk	Private keys of P and MC respectively
PPUk, MCPUk	Public keys of P and MC respectively
PID, PWD, PBIO	Identity, password and biometric of P
MPID	Masked identity of P
MMCID	Masked identity of MC
MCID, MCPWD, MCBIO	Identity, password and biometric of MC
A1, A2, … A9, B1, B2, … B8	Helping parameters
PMR	Patient medical record
MR	Medical record
T1, T2, T3, T4	timestamps
R1, R2	Random nonces
Hsf(.)	Hash Function
⊕	Xor Operation

```
% OFMC
% Version of 2006/02/13
SUMMARY
  SAFE
DETAILS
  BOUNDED_NUMBER_OF_SESSIONS
PROTOCOL
  /home/span/span/testsuite/results/4Conf.if
GOAL
  as_specified
BACKEND
  OFMC
COMMENTS
STATISTICS
  parseTime: 0.00s
  searchTime: 1.30s
  visitedNodes: 9 nodes
  depth: 2 plies
```

```
SUMMARY
  SAFE
DETAILS
  BOUNDED_NUMBER_OF_SESSIONS
  TYPED_MODEL
PROTOCOL
  /home/span/span/testsuite/results/4Conf.if
GOAL
  As Specified
BACKEND
  CL-AtSe
STATISTICS
  Analysed    : 0 states
  Reachable   : 0 states
  Translation: 0.07 seconds
  Computation: 0.00 seconds
```

Fig. 2. SAFE over OFMC AVISPA backend   **Fig. 3.** SAFE over CLA-AtSe AVISPA backend

7 Ban Logic

A formal analysis of VSMAS2HN has been presented with BAN Logic. Notations and rules are fundamental components of BAN Logic which are taken from [14]. BAN Logic verifies the security of the session key. The following goals to be validated are:

Goals: Following BAN Logic directives, needs to confirm the below-mentioned goals/aim(s):

$G1: M C| \equiv (P \xleftrightarrow{UPR,UMCR} MC)$

$G2: M C| \equiv P| \equiv \left(P \xleftrightarrow{UPR,UMCR} MC\right)$

$G3: P| \equiv (P \xleftrightarrow{sessionKey} MC)$

$G4: P| \equiv MC| \equiv (P \xleftrightarrow{sessionKey} MC)$

Idealized Form:

P to MC: Message1: $(P \xleftrightarrow{UPR,UMCR} MC, MCID, R1, R2, Ts)_{P \xleftrightarrow{MPID} MC}$

MC to P: Message2: $(P \xleftrightarrow{MPID} MC, R1, R2, MCID, Ts, P \xleftrightarrow{sessionKey} MC)_{P \xleftrightarrow{MPID} MC}$.

Assumptions: assumptions constitute the shared keys, and fresh values along with trusted propositions. Following assumptions are considered for

$A1 : MC| \equiv \left(P \xleftrightarrow{MPID,MCID,R1,R2,} MC\right)$

$A2 : MC| \equiv \#(Ts), P| \equiv \#(Ts)$

$A3 : MC| \equiv P \Rightarrow P| (P \xleftrightarrow{UPR,UMCR} MC)$

$A4 : P| \equiv \left(P \xleftrightarrow{MPID,MCID,R1,R2,} MC\right)$

$A5 : MC| \equiv \#(R1), P| \equiv \#(R2), MC| \equiv \#(MCID)$

$A6 : P| \equiv MC \Rightarrow MC| \sim (P \xleftrightarrow{sessionKey} MC)$

Analysis:
From message1:

- From A1 and Message meaning rule, we obtain, Eq. 1:

$$\frac{MC| \equiv \left(P \xleftrightarrow{MPID} MC\right), MC(P \xleftrightarrow{UPR,UMCR} MC, v, Ts)_{P \xleftrightarrow{MPID} MC}}{MC| \equiv SN| \sim \left(P \xleftrightarrow{UPR,UMCR} MC, v, Ts\right)} \quad (1)$$

- From A2, Eq. 1 and freshness-conjuncatenation rule, we obtain,

$$\frac{MC| \equiv \#(Ts)}{MC| \equiv \#(P \xleftrightarrow{UPR,UMCR} MC, v, Ts)} \quad (2)$$

- From Eq. 1, Eq. 2 and nonce verification rule, we obtain, Eq. 3:

$$\frac{MC|\equiv P|\sim \left(P \overset{UPR,UMCR}{\longleftrightarrow} MC, MCID, R1, R2, Ts\right), MC| \equiv \#(P \overset{UPR,UMCR}{\longleftrightarrow} MC, MCID, R1, R2, Ts)}{MC|\equiv P| \equiv \left(P \overset{UPR,UMCR}{\longleftrightarrow} MC, v, Ts\right)} \tag{3}$$

- From Eq. 3 and belief rule, we obtain,

$$\frac{MC|\equiv P| \equiv \left(P \overset{UPR,UMCR}{\longleftrightarrow} MC, MCID, R1, R2, Ts\right)}{MC|\equiv P| \equiv \left(P \overset{UPR,UMCR}{\longleftrightarrow} MC\right)} \tag{4}$$

G2 is validated

- From Eq. 4, A3, and jurisdiction rule, we obtain,

$$\frac{MC|\equiv P| \equiv \left(P \overset{UPR,UMCR}{\longleftrightarrow} MC\right),\ MC| \equiv P \Rightarrow P| \sim (P \overset{UPR,UMCR}{\longleftrightarrow} MC)}{MC| \equiv \left(P \overset{UPR,UMCR}{\longleftrightarrow} MC\right)} \tag{5}$$

G1 is validated

From message2:

- From A4 and message meaning rule, we obtain, Eq. 6:

$$\frac{P| \equiv \left(P \overset{MPID,MCID,R1,R2,}{\longleftrightarrow} MC\right),\ P(P \overset{UPR,UMCR}{\longleftrightarrow} MC, v, r_1, w, P \overset{sessionKey}{\longleftrightarrow} MC)_{P\overset{MPID}{\longleftrightarrow}MC}}{SN| \equiv MC| \sim (SN \overset{UPR,UMCR}{\longleftrightarrow} MC, MCID, R1, R2, P \overset{sessionKey}{\longleftrightarrow} MC)} \tag{6}$$

- From A5, Eq. 6, and freshness-conjuncatenation rule, we obtain,

$$\frac{SN| \equiv \#(< MCID, R1, R2, >)}{P| \equiv \#(P \overset{UPR,UMCR}{\longleftrightarrow} MC, v, r_1, w, P \overset{sessionKey}{\longleftrightarrow} MC)} \tag{7}$$

- From Eq. 6, Eq. 7, and nonce verification rule, we obtain, Eq. 8:

$$\frac{P| \equiv MC|\sim (P \overset{UPR,UMCR}{\longleftrightarrow} MC, v, r_1, w, P \overset{sessionKey}{\longleftrightarrow} MC),\ P| \equiv \#(P \overset{UPR,UMCR}{\longleftrightarrow} MC, v, r_1, w, P \overset{sessionKey}{\longleftrightarrow} MC)}{P| \equiv MC| \equiv (P \overset{UPR,UMCR}{\longleftrightarrow} MC, v, r_1, w, P \overset{sessionKey}{\longleftrightarrow} MC)} \tag{8}$$

- From Eq. 8 and belief rule, we obtain,

$$\frac{P| \equiv MC| \equiv (P \overset{UPR,UMCR}{\longleftrightarrow} MC, v, r_1, w, P \overset{sessionKey}{\longleftrightarrow} MC)}{P|\equiv MC| \equiv (P \overset{sessionKey}{\longleftrightarrow} MC)} \tag{9}$$

G4 is validated

- From Eq. 9, A6, and jurisdiction rule, we obtain,

$$\frac{P|\equiv MC| \equiv (P \overset{sessionKey}{\longleftrightarrow} MC),\ P| \equiv MC \Rightarrow MC| \sim (P \overset{sessionKey}{\longleftrightarrow} MC)}{P| \equiv (P \overset{sessionKey}{\longleftrightarrow} MC)} \tag{10}$$

G3 is validated

G1, G2, G3 and G4 have been validated and thereby the security of sessionKey in VSMAS2HN is also verified.

8 Conclusion

This paper proposed a verifiably secure mutual authentication scheme for smart healthcare networks. To develop a secure healthcare ecosystem, a biometric system was suggested in VSMAS2HN. Security of VSMAS2HN was proven using AVISPA. VSMAS2HN has been tested safe over AVISPA, proving it to be "SAFE" and resistant against replay, impersonation, and Men in the Middle attacks. Moreover, BAN Logic verification has also been presented to support the session key security of VSMAS2HN. In the future, these healthcare entities can be linked over blockchain networks.

References

1. Abdussami, M., Amin, R., Vollala, S., et al.: Provably secured lightweight authenticated key agreement protocol for modern health industry. Ad Hoc Netw. **141**, 103094 (2023)
2. Kumar, V., Mahmoud, M.S., Alkhayyat, A., Srinivas, J., Ahmad, M., Kumari, A.: RAPCHI: robust authentication protocol for IoMT-based cloud-healthcare infrastructure. J. Supercomput. **78**(14), 16167–16196 (2022)
3. Chakraborty, C., Kishor, A.: Real-time cloud-based patient-centric monitoring using computational health systems. IEEE Trans. Comput. Soc. Syst. **9**(6), 1613–1623 (2022)
4. Rajan Jeyaraj, P., Nadar, E.R.S.: Smart-monitor: patient monitoring system for IoT-based healthcare system using deep learning. IETE J. Res. **68**(2), 1435–1442 (2022)
5. Gaurav, A., Psannis, K., Peraković, D.: Security of cloud-based medical internet of things (miots): a survey. Int. J. Softw. Sci. Comput. Intell. **14**(1), 1–16 (2022)
6. Rahimi, M., Navimipour, N.J., Hosseinzadeh, M., Moattar, M.H., Darwesh, A.: Cloud healthcare services: a comprehensive and systematic literature review. Trans. Emerg. Telecommun. Technol. **33**(7), e4473 (2022)
7. Mageshkumar, N., Lakshmanan, L.: Intelligent data deduplication with deep transfer learning enabled classification model for cloud-based healthcare system. Expert Syst. Appl. **215**, 119257 (2023)
8. Jangra, A., Mangla, N.: An efficient load balancing framework for deploying resource scheduling in cloud based communication in healthcare. Measurement: Sensors **25**, 100584 (2023)
9. Menon, S.C., Umayal, V.R., Soni, G., Baig, A., Tyagi, A.K.: A visual framework for an iot-based healthcare system based on cloud computing. In: Using Multimedia Systems, Tools, and Technologies for Smart Healthcare Services, pp. 83–95. IGI Global (2023)
10. Chenam, V.B., Ali, S.T.: A designated tester-based certificateless public key encryption with conjunctive keyword search for cloud-based MIoT in dynamic multi-user environment. J. Inform. Secur. Appl. **72**, 103377 (2023)

11. Pavón-Pulido, N., Blasco-García, J.D., López-Riquelme, J.A., Feliu-Batlle, J., Oterino-Bono, R., Herrero, M.T.: JUNO project: deployment and validation of a low-cost cloud-based robotic platform for reliable smart navigation and natural interaction with humans in an elderly institution. Sensors **23**(1), 483 (2023)

12. Nancy, A.A., Ravindran, D., Raj Vincent, P.D., Srinivasan, K., Gutierrez Reina, D.: Iot-cloud-based smart healthcare monitoring system for heart disease prediction via deep learning. Electronics **11**(15), 2292 (2022)

13. Mahajan, H.B., et al.: Integration of Healthcare 4.0 and blockchain into secure cloud-based electronic health records systems. Appl. Nanosci. **13**, 2329–2342 (2022)

14. Narwal, B., Mohapatra, A.K.: SAMAKA: secure and anonymous mutual authentication and key agreement scheme for wireless body area networks. Arab. J. Sci. Eng. **46**(9), 9197–9219 (2021)

15. Qi, M., Chen, J., Chen, Y.: A secure biometrics-based authentication key exchange protocol for multi-server TMIS using ECC. Comput. Methods Programs Biomed. **164**, 101–109 (2018)

16. Lei, C.L., Chuang, Y.H.: Privacy protection for telecare medicine information systems with multiple servers using a biometric-based authenticated key agreement scheme. IEEE Access **7**, 186480–186490 (2019)

17. Manickam, P., Mariappan, S.A., Murugesan, S.M., Hansda, S., Kaushik, A., Shinde, R., Thipperudraswamy, S.P.: Artificial intelligence (AI) and internet of medical things (IoMT) assisted biomedical systems for intelligent healthcare. Biosensors **12**(8), 562 (2022)

18. Pelekoudas-Oikonomou, F., Zachos, G., Papaioannou, M., de Ree, M., Ribeiro, J.C., Mantas, G., Rodriguez, J.: Blockchain-based security mechanisms for IoMT Edge networks in IoMT-based healthcare monitoring systems. Sensors **22**(7), 2449 (2022)

19. Al-hajjar, A.L.N., Al-Qurabat, A.K.M.: An overview of machine learning methods in enabling IoMT-based epileptic seizure detection. J. Supercomputing (2023)

20. Karar, M.E., Khan, Z.F., Alshahrani, H., Reyad, O.: Smart IOMT-based segmentation of coronavirus infections using lung CT scans. Alexandria Eng. J. **69**, 571–583 (2023)

21. Bhushan, B., Kumar, A., Agarwal, A.K., Kumar, A., Bhattacharya, P., Kumar, A.: Towards a secure and sustainable internet of medical things (IoMT): requirements, design challenges, security techniques, and future trends. Sustainability **15**(7), 6177 (2023)

22. Ravikumar, G., Venkatachalam, K., AlZain, M.A., Masud, M., Abouhawwash, M.: Neural cryptography with fog computing network for health monitoring using IoMT. Comput. Syst. Sci. Eng. **44**(1), 945–959 (2023)

Optimal KAZE and AKAZE Features for Facial Similarity Matching

A. Vinay$^{(\boxtimes)}$ ⓘ, Kishan Athirala Vasu$^{(\boxtimes)}$ ⓘ, Pranav Yogi Lodha$^{(\boxtimes)}$ ⓘ, S. Natarajan ⓘ, and T. S. B. Sudarshan ⓘ

Center for Pattern Recognition and Machine Intelligence, PES University, Bengaluru 560085, India
{a.vinay,natarajan,sudarshan}@pes.edu, kishanavasu@gmail.com, lodhapranav2@gmail.com

Abstract. Face Recognition is one of the premier disciplines in the vast field of computer vision and image analysis. A popular method is the Gaussian scale space analysis which limits the performance by smoothing both the noise and natural boundaries in the same proportion. In order to prevent the loss of natural boundaries to smoothing we tap into nonlinear scale space techniques such as KAZE and Accelerated KAZE. KAZE is a multistage 2-D feature detection and description algorithm. It makes use of AOS schemes to develop the nonlinear scale space for analysis. Though the results are satisfactory, it is computationally intense as they solve a humongous module of linear equations. To ascertain the mentioned limitation of KAZE we make use of Accelerated-KAZE, which uses pyramidal structure with Fast Explicit Diffusion incorporated in it, thus minimizing the computation in the step of feature detection of nonlinear scale space. Also, with the use of M-LDB (Modified-Local Difference Binary) descriptor the problem of rotation is solved. Usage of RANSAC after processing in two methods had some disadvantages. It gave bad results when the inliers ratio in the dataset is low. Thus, Optimal RANSAC is employed which works well even when the inliers ratio is as low as 5%. The proposed methods are tested on many standard datasets and various performance parameters.

Keywords: Face recognition · nonlinear scale space · KAZE · AOS schemes · AKAZE · Fast Explicit · Diffusion · RANSAC · Optimal RANSAC

1 Introduction

One of the most crucial uses of biometrics, facial similarity matching, has been widely used in both commercial and law-enforcement contexts. Even after its constant presence, it is plagued by a slew of difficulties that result in a decline in recognition accuracy when there is an occurrence of numerous restrictions such as position, scale, expression, lighting, and translation [2]. The effectiveness of the FR system will be significantly influenced by the feature detector choice.

For creating the scale space form of an input picture, the Gaussian kernel is a simple substitute. Gaussian blurring also equally smoothes noise and details at all scale levels.

© The Author(s), under exclusive license to Springer Nature Switzerland AG 2023
M. Singh et al. (Eds.): ICACDS 2023, CCIS 1848, pp. 161–177, 2023.
https://doi.org/10.1007/978-3-031-37940-6_14

Considering the problem mentioned we make use of nonlinear scale space analysis by making use of blurring which are locally adaptive to the image data. It preserves object boundaries continuing to blur small details [8]. KAZE and AKAZE are the two, 2-dimensional feature detector and descriptor algorithms in nonlinear scale spaces [7, 29]. In comparison to SIFT and SURF which are the most renowned approaches for multiscale feature identification and description, KAZE increases distinctiveness and repeatability by using nonlinear diffusion filtering. The skeleton of this method has a basis on AOS (Additive Operator Splitting) technique. For any step size in this method, nonlinear scale spaces are obtained in a very efficient way. Though the results are impressive, it requires handling a larger system of linear equations for obtaining a solution. This approach is computationally intensive. In order to solve this problem AKAZE (Accelerated – KAZE) is used. In place of using the AOS strategy, the method builds the nonlinear space using Fast Explicit Diffusion (FED). This FED method used is much faster, easier to implement and more accurate than the AOS scheme used earlier. To maintain minimal computational demand and storage, an effective Modified-Local Difference Binary (M-LDB) descriptor is presented.

[13] According to Ondrej Chum and Jırı´ Matas, RANSAC is a very reliable estimating method that eliminates outliers in order to fit any given model, from the data collection. It classifies a given set of points as either inliers or outliers in the model and results in reduction of the duplicate interest point matches. It is difficult to use RANSAC because the inliers present for the same pair of images vary every time. [37] Thus, we make use of OptimalRANSAC, where sets which have a very low inliers ratio as small as 5% are also handled. This is important as the standard RANSAC did not perform well when the inliers ratio was less than 50%. To achieve this, it makes use of re-sampling and re-scoring algorithms for better solutions.

2 Literature Survey

[8] Alcantarilla (2012) proposes a novel feature detection and description algorithm that is both scale-invariant and highly robust to viewpoint changes and image noise. The authors evaluate the performance of the KAZE algorithm on a number of standard datasets, including the Oxford and Mikolajczyk datasets, and compares it to several state-of-the-art feature detection and description methods, such as SIFT and SURF. Overall, the KAZE algorithm proposed in this paper represents a significant contribution to the field of computer vision and has since been widely adopted in both academia and industry for a variety of applications, including object recognition, image retrieval, and motion estimation [10]. An incremental RANSAC-based approach was used by Kondo (2006), for online relocation in large dynamic environments. The method selects a set of robustly matched feature points using RANSAC, and incrementally adds new feature points to the set based on their consistency with the existing set. The approach is shown to be effective in large and dynamic environments, and outperforms traditional RANSAC-based methods [13] Ondrej Chum and Jırı´ Matas proposed Optimal Randomized RANSAC (OR-RANSAC) in their 2005 paper "Optimal Randomized RANSAC". The authors show that RANSAC, a widely used algorithm for robust parameter estimation, is inefficient when applied to problems with a large number of hypotheses, since it requires

many iterations to find the correct solution. OR-RANSAC addresses this problem by introducing a new sampling scheme that reduces the number of iterations needed to find the optimal solution. The authors also provide a theoretical analysis of the algorithm, showing that it achieves the optimal trade-off between the number of iterations and the probability of finding the correct solution. Overall, the paper provides an important contribution to the field of robust parameter estimation and has been widely cited and applied in computer vision and related areas. [14–16] Weickert, terHaarRomeny, and Viergever present their efficient and reliable schemes for nonlinear diffusion filtering in the IEEE Transactions on Image Processing in 1998. They benefit of nonlinear diffusion filtering for image processing, including noise removal, edge enhancement, and feature preservation. They then describe their proposed schemes for implementing this type of filtering, including both isotropic and anisotropic diffusion models.

3   KAZE Features

The word 'KAZE' has a Japanese origin meaning wind and is multi-scale 2-D feature detection and description algorithm. The image is first filtered using a function that changes with time or scale to create the scale space. Gaussian Kernel is then used to create a scale space representation of an input picture. Using this, the input image gets transformed with the use of Gaussian Kernel, whose standard deviation increases. The advantage of using this method is to emphasize the prominent features in the image and reduce the noise. But the image is smoothed using the Gaussian scale space approach i.e., both the noise and the details at all the scales to the same degree. This results in loss of natural boundaries of the image, thus losing on localization.

Non-linear diffusion yields better results and performs well compared to linear methods when we consider some of the operations like segmentation and de-noising of the image. The most probable reason why non-linear method is less used in practical and real time applications is because of its poor efficiency while using some of its approaches. All these methods make use of forward Euler scheme, which uses smaller steps. Hence it takes a longer time to converge and thus higher computational complexity. The application of Additive Operator Splitting (AOS) acts as a backbone in achieving a more stable non-linear scale space despite using varying step size. The main issue with the use of Additive Operator Splitting is to answer the tri-diagonal system of linear equations. This is overcome by using Thomas Theorem, a special type of Gaussian method.

3.1   Nonlinear Diffusion and AOS Schemes

Let us consider a non-linear diffusion formulation:

$$\frac{\partial L}{\partial t} = (c(x, y, t), \Delta L) \tag{1}$$

where divergence (differentiation) is represented by div, gradient by Δ and conductivity function by c. To reduce diffusion at edges, a function got proposed by Perona and Malik named c defined as follows,

$$c(x, y, t) = g(|\Delta L\sigma(x, y, t)|) \tag{2}$$

This ensures the smoothing inside a region preventing it across boundaries. Variations in conductivity function g are given by g1 and g2. g1 works well in high contrast edges whereas g2 performs better in wider regions.

$$g1 = exp\left(\frac{-|\Delta L|^2}{k^2}\right) \tag{3}$$

$$g2 = \frac{1}{1 + \frac{|\Delta L|^2}{k^2}} \tag{4}$$

where k is the contrast factor. Later Weickert proposed a variant of diffusion function which is given by:

$$g3 = \begin{cases} 1, & |\Delta L|^2 = 0 \\ 1 - exp\left\{\frac{-3.315}{\left(\frac{|\Delta L|}{k}\right)^2}\right\}, & |\Delta L|^2 > 0 \end{cases} \tag{5}$$

Here smoothing on either sides of the edges are very strong when compared across it.

Since there is no analytical solution for the non-linear diffusion, we need to discretize it. This is done using the semi-implicit scheme also referred to as linear-implicit scheme.

$$\frac{(L^{i+1} - L^i)}{\tau} = \sum_{l=1}^{m} A_l(L^i)L^{i+1} \tag{6}$$

where A_l matrix reflects the conductivities of the images in each dimension. The solution to L^{i+1} is given by:

$$L^{i+1} = (1 - \tau \sum_{l=1}^{m} A_l(L^i))^{-1}L^i \tag{7}$$

3.2 Computation of Nonlinear Scale Space

We discretize the scale space in logarithmic steps, which is similar to SIFT. A series of O octaves and S sub-levels are created by layering them. Mapping of indices to corresponding scales is done using:

$$\sigma_i(o, s) = \sigma_0 \cdot \frac{o+s}{2^2} \tag{8}$$

such that, o belongs to [0,.........O − 1]
    s belongs to [0,.........S − 1] and
    i belongs to [0,.........N]

where the indices, the octave, and the sub-threshold, respectively, are denoted by i, o and s, 'σ_0' represents the level of base scale, N represents the total count of processed (filtered) images.

This conversion from scale space to time space is done by using the following mapping $\sigma_i \rightarrow t_i$:

$$t_i = \frac{1}{2}\sigma_i^2 \tag{9}$$

where i = {0,......N}.

The mapping is done to get a set of evaluation times enabling us to form the nonlinear scale space. Also, by setting the value of the diffusion function g to 1, the model is also consistent with the Gaussian scale space.

When an input image is fed, it is transformed with the help of Gaussian kernel with standard deviation σ_0 and hence it reduces the image artifacts and noise. Subsequently, the image gradient histogram is calculated along with the contrast parameter k. After the evolution times t_i and the contrast parameter k is obtained, The AOS techniques are used to generate nonlinear scale space. This is as follows:

$$L^{i+1} = (1 - (t_{i+1} - t_i) \cdot \sum_{l=1}^{m} A_l \cdot (L^i))^{-1} L^i \tag{10}$$

3.3 Feature Detection

Scale-normalized determinant is determined for detecting the interest points at different scale levels. For multiscale feature identification, the collection of differential operators also must be normalised.

$$L_{Hessian} = \sigma^2 (L_{xx} \cdot L_{yy} - L_{xy^2}) \tag{11}$$

where L_{xx} represents the second order horizontal derivative, L_{yy} represents the respective vertical derivative and L_{xy} represents the second order cross derivative.

The response at various scale levels is calculated once filtered pictures are produced from the nonlinear scale space L^i, i.e., σ_i. Then the maxima is searched which in scale and special location except for the values where i = 0 and i = N. Every extrema is looked over a rectangular window of size $\sigma_i \times \sigma_i$, depending on the present magnitude of i, lower i − 1 and upper i + 1 filtered images. For removing the non-maxima responses, values are calculated over a window of dimension 3×3 pixels. Interest point groups results invariant features which are described by Brown, is used to find the position of key point with the subpixel accuracy.

Also derivatives of first and second order are computed with the usage of [3×3] Scharr filters with steps of size [σ_i]. The presence of consecutive filters in the desired coordinates is used in estimating the second order derivatives. These filters perform better than the Sobel filter or the Standard Central differences, when it comes to rotation invariance (Fig. 1).

3.4 Feature Description and Development of the Descriptor

The first advantage that we described is rotation invariance. For obtaining this feature, principal position in a local area centered at the interest point location is important. The dominant position is computed in a circular radius $6\sigma_i$, where sampling step is of size σ_i. This is similar to the method adopted in SURF. With the use of a Gaussian that is centred at the critical point, Lx and Ly are weighted for each sample that is utilised. The dominating position is calculated by summing the derivative answers that are distributed in a sliding circle that covers an area with an angle of $\frac{\pi}{3}$. The orientation is determined using the longest vector from all the vectors obtained.

Fig. 1. Keypoint Detection from the original image (Lena.jpg) using KAZE

For the purpose of descriptor, we employ M-SURF descriptor which is adapted to the non-linear scale space. Once the feature is detected at σ_i, the derivatives L_x and L_y are computed over a (24×24) σ_i^2 rectangular area. This rectangular area is divided into four equal divisions of area (9×9) σ_i^2 having an overlap of $2 \sigma_i$. The four divisions are weighted with a Gaussian defined over a mask of 4×4 after its responses are obtained and added into a descriptor vector d_v. Then this process is repeated where each of the samples is rotated as per the dominant rotation and the derivatives are recomputed accordingly. Lastly, the vector d_v whose length is 64 gets normalized to form a unit vector (Fig. 2).

4 AKAZE Features

The KAZE features which were introduced earlier increases the computational complexity by increasing the repeatability and distinctiveness in contrast to the conventional SIFT and SURF methodologies. This is achieved with the help of nonlinear scape space analysis. Numerical methods are needed to summarize the solutions in KAZE because of the absence of analytical solutions which would simplify nonlinear diffusion equations. This is achieved using AOS scheme which is described earlier. This requires solving a large system of equations and hence higher computational cost. Hence this cannot be used real time applications. This is overcome with the introduction of AKAZE feature detection. In this approach we employ Fast Explicit Diffusion (FED) instead of the AOS scheme for building the nonlinear scale space. This FED method used is much faster and much easier to implement becoming accurate than the previous AOS scheme. We embedded this FED scheme in a pyramidal framework and speed up the process of feature detection process. Hence, it is termed as Accelerated-KAZE (AKAZE). Modified-Local Difference Binary (M-LDB) descriptor is used to preserve low computation cost.

4.1 Fast Explicit Diffusion (FED)

To address these drawbacks, Fast Explicit Diffusion incorporates both explicit and semi-implicit techniques. These FED schemes received motivation from the decomposition of the box filters. It's main aim lies in to perform M cycles consisting of diffusion steps n using different step sizes τ_i, which originated from box filter factorization:

$$\tau_i = \frac{\tau_{max}}{2cos^2(\frac{\pi(2j+1)}{4n+2})} \tag{12}$$

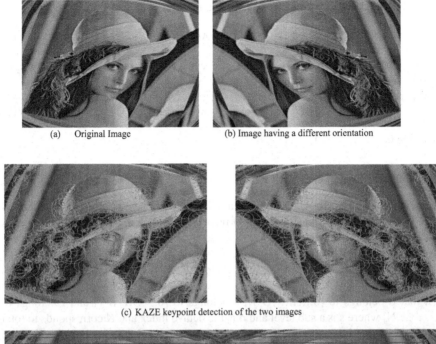

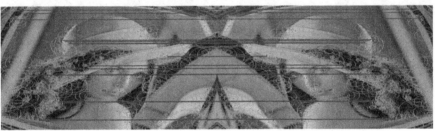

(d) KAZE matching of the two images

Fig. 2. Shows the original image (a) and the identical image (b), but with a different orientation. The key points of Images (a) and (b) are shown, respectively, in Images (c) and (d).Image (e) shows the KAZE matching which shows rotation invariance feature.

where τ_{max} is the maximum step size assuming it does not contravene the stability condition. Stoppage time θ_n of is expressed as:

$$\theta_n = \sum_{j=0}^{n-1} \tau_j = \tau_{max} \frac{n^2 + n}{3} \tag{13}$$

The discretization of the nonlinear diffusion equation involving explicit scheme is represented as:

$$\frac{(L^{i+1} - L^i)}{\tau} = A(L^i) \cdot L^i \tag{14}$$

where $A(L^i)$ constitutes the encoding matrix of the image and τ is a stride length of time less than τ_{max}. This is done to ensure that the stability condition holds true. The term

L^{i+1} is computed directly using the below equation:

$$L^{i+1} = (1 + \tau A(L^i))L^i \tag{15}$$

where I is the identity matrix. A FED cycle consists of n different stride size τ_j is determined using:

$$L^{i+1,j+1} = (1 + \tau_j A(L^i))L^{i+1,j} \tag{16}$$

where $j = 0\ldots\ldots, n - 1$

4.2 Nonlinear Scale Space with FED

Similar to KAZE, here the scale space is a definite sequence consisting of 'O' octaves and 'S' sub-levels. Also transformation of respective indices to corresponding scales is done using the following approach:

$$\sigma_i(o, s) = 2^{\frac{o+s}{s}} \tag{17}$$

such that o belongs to $[0,\ldots\ldots.O\text{-}1]$, s belongs to $[0,\ldots\ldots.S\text{-}1]$ and i belongs to $[0,\ldots\ldots.N]$ where s is a sub level and o is the octave index and N corresponds to total number of filtered images. The conversion of the discretized units to time units is done similar to the KAZE approach. Here let us have a look at the algorithms for:

1. Pyramidal Fast Explicit Diffusion approach
2. Fast Explicit Diffusion inner cycle:

Algorithm 1

Input: Image K^0, contrast parameter λ, τ_{max} and evolution times t_i
Output: Filtered images K^i, i = 0...N

while (i = 0 to N-1) do
    Calculation of diffusivity matrix $A(K^i)$

    Set the value for FED outer cycle time $\tau = t_{i+1}\text{-}t_i$
    Calculate the number of inner steps n, i.e. of FED

    Calculate the step sizes τ_j
    Set the value of Prior $K^{i+1,0} = K^i$

    Then, compute $K^{i+1} = $ **INNERCYCLEFED** $(K^{i+1,0}, A(K^i), \tau_j)$
    if $(o_{i+1} > o_i)$

        Down sample K^{i+1} with mask $(\frac{1}{4}, \frac{1}{2}, \frac{1}{4})$
        $\lambda = \lambda X 0.75$

    end
end

Algorithm 2

Function: INNERCYCLEFED $(K^{i+1,0}, A(K^i), \tau_j)$
    while (j = 0 to n-1) **do**

$$K_{i+1,j+1} = (I + \tau_j A(K_i))\, K_{i+1,j}$$

    end
    return $K_{i+1, n}$
end

Here, for nonlinear diffusion filtering, Algorithm 1 uses a pyramidal FED method and Algorithm 2 is FED inner Cycle function that is called in the first algorithm.

4.3 AKAZE Feature Detection and Description

In AKAZE, for each filtered images the determinant of the Hessian is calculated in the nonlinear space for feature detection. This is similar to the one used in KAZE for feature detection.

We employ Modified-Local Difference Binary (M-LDB), which makes use of gradient and intensity data from the nonlinear scale space, for feature definition. The Local Difference Binary which follows the same principle of BRIEF, was introduced by X.Yang and K.T.Cheng. But in contrast it does the binary tests over the mean of areas neglecting the contribution of single pixels adding to the robustness in the technique. Also along with intensity values, the average of the derivatives in vertical and horizontal sections being analyzed is taken resulting in three bits per comparison. LBD makes use of various grids of 2×2, 3×3, 4×4, etc. by dividing them into finer steps. When considering the rotation of the key points, visiting all the key points of the integral images is costly in computation. AKAZE achieves invariance in rotation by approximation of the main position of KAZE. After that the rotation of LDB grid is performed. Then the grids are sub sampled in steps which is a function of σ in contrast of the usage of the mean of pixels located in each subdivision. Then M-LDB employs the derivatives which are computed during the feature detection and thus reduces the total steps for constructing the descriptor. Making use of methods like PCA for reducing the length of the descriptor will reduce the computation cost without affecting the performance of the system (Fig. 3).

5 RANSAC and Optimal-RANSAC

RANSAC is a very reliable estimating method that eliminates outliers from the data set in order to suit any specified model. The demerit with standard RANSAC is it's non repeatable property. Hence it becomes difficult in using it for the same pair of images as the set of inliers. It can vary for each run. Another drawback is that the typical RANSAC stops the run when the chance of finding more inliers hits a predetermined level rather than looking for the best collection of inliers. This results in ill performance when the set is highly contaminated i.e. when the outlier ratio is high. The stopping criteria used in RANSAC is completely based on the statistical hypothesis. This hypothesis might indicate that the consensus is not achieved even after the determination of almost all the

Fig. 3. Keypoint Detection from the original image (Lena.jpg) using AKAZE

inliers. Because of all these reasons we use Optimal RANSAC here as a post processing procedure after KAZE and AKAZE. The advantage of Optimal RANSAC is that it supports extremely low inlier handle ratio settings, which can be as small as 5%. This is important as the standard RANSAC performed badly below the threshold of the inliers ratio of 50%.

Optimal RANSAC is one which randomly samples the bare minimum points which in the collection of related pairings is necessary. It uses resampling method to search for tentative. We also make sure to handle exceptions when one more inliers are found.

5.1 The Stopping Criteria

A reliable stopping criterion is obtained by using Optimal-RANSAC. In case of standard RANSAC the stopping criteria was based on a hypothesis and would stop when the probability of finding the inliers had reached a threshold value. Experimental research revealed that, up until and unless the ideal set is discovered, Optimal-RANSAC would seldom reach the same consensus twice. When the number of tentative inliers is tiny, especially fewer than 30, the likelihood of discovering a non-optimal set twice grows significantly. The possible solution to this is the use of two equal sets such that the final one is the most optimal for such a case. Another problem arises when two distinct optimal sets are found, like for example if set S_a contains more inliers than S_b. If the two sets are re-scored, the very same sets are obtained back. One of the possible solutions might be trimming the set to a lower tolerance ε' which forces the method to search for similar S_b set almost every time.

5.2 Resampling

Resampling is done when the number of inliers obtained is more than 5. The number 5 was likely chosen as the threshold for resampling because it strikes a balance between computational efficiency and accuracy. Selecting too few inliers (e.g., 2 or 3) may result in a less robust solution that is highly sensitive to noise or outliers, while selecting too many inliers (e.g., 10 or more) may be computationally expensive and not significantly improve the accuracy of the solution. This is a process to get the most optimal best solution for a given set of points. There can be many lines than can be fitted for a given points. But there can only one optimally fitted line which is the best solution. Resampling is done to optimize the result obtained. The methodology applied for resampling is that it resamples the points using one-fourth of tentative inliers. The collection is then iteratively rescored if there are more inliers present than there were previously. But in case the set obtained turns out to be larger than the one before, the algorithm is restarted.

5.3 Rescoring

Rescoring is used to get the optimal set by faster convergence. This repeatedly re-estimates the model and do rescore the set, till that time when the set remains the same without any change. In order to avoid entering into an endless loop we restrict the iteration to either 20 or 30. Here 20 is used as an example. The choice of 20 iterations for loop control in KAZE and AKAZE facial recognition have been based on previous experimentations or observations of the algorithm's behavior. The value of 20 has been found to be sufficient for achieving convergence in a reasonable amount of time, while avoiding the risk of getting stuck in an endless loop.

If the number of iterations is increased to 100, the algorithm may take longer to converge and require more computational resources. However, it may result in a more accurate solution in some cases. Conversely, if the number of iterations is reduced to 5, the algorithm may converge too quickly and not reach the optimal solution, resulting in suboptimal performance.

5.4 Pruning

Once the tentative inlier set is obtained, we can further prune it by removing the most extreme tentative inliers. Then the remaining inliers are further employed to re-estimate the model. Until all the tentative inliers fall inside the tolerance ε', the process is repeated. Every time a tentative inlier is eliminated from the set using this method, the best model is guaranteed to fit. This technique assures that the best model is fitted every time we remove a tentative inlier from the set. The algorithm for Optimal-RANSAC uses the functions **random_sampling**, **re_sampling**, **pruning_set** which find its definition in Alcantarilla et al.

The model is then re-estimated using the remaining inliers.

6 Results

The results obtained for KAZE are tabulated in Table 1. Some of the factors that might affect how well the algorithm performed were evaluated on various datasets. In Fig. 4, the suggested method employing KAZE is depicted.

Image (a) on application of KAZE yields 324 key points whereas Image (b) yields 259 key points. The total number of inliers obtained after the application of Optimal-RANSAC is 40 and KAZE feature extraction time was found to be 629.417 ms.

As evident from the above table the accuracy of KAZE is maximum in case of Grimace Database. The values of Recall and Precision are quite high with KAZE method in almost all the databases. The table also indicates that Yale Database shows least

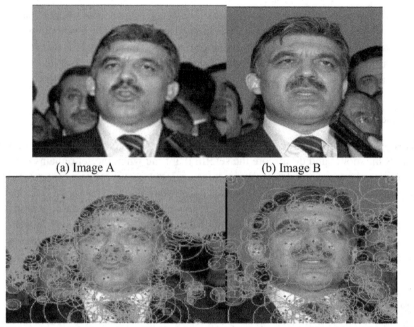

(a) Image A　　　　　　　　　(b) Image B

(c) KAZE keypoint detection of Image (a)　(d) KAZE keypoint detection of Image (b)

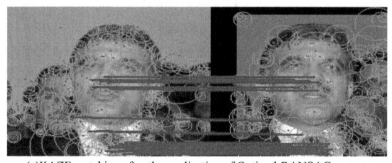

(e)KAZE matching after the application of Optimal-RANSAC

Fig. 4. Image (a) has the photo of a person and Image (b) has the same person's photo in a different background and varied resolution. The keypoints of Images (a) and (b) are shown, respectively, in Images (c) and (d). Image (e) shows the KAZE matching of the person's face from both the images after the application of Optimal-RANSAC

accuracy, specificity but yet high FRR results. TER and FAR values in case of ORL Database is the least obtained.

The results obtained for AKAZE are tabulated in Table 2 [9]. The algorithm was tested over the same datasets and the parameters for the performance of the algorithm were determined similar to KAZE. The proposed method using AKAZE is depicted in Fig. 5.

Table 1. Tabulation of KAZE results for FACE RECOGNITION by using various parameters and testing it on a wide set of Standard DATASETS

KAZE	ORL Database	Grimace Database	Face 95 Database	Face 96 Database	Yale Database
Sensitivity	93.50	95.20	94.31	95.67	94.79
Specificity	83.33	73.91	59.75	54.54	22.22
Precision	97.39	96.36	95.02	96.14	92.85
Accuracy	92.17	92.63	90.53	92.47	88.57
Recall	93.50	95.20	94.31	95.67	94.79
F1 Score	95.40	95.78	94.66	95.91	93.81
FAR	0.16	0.26	0.40	0.45	0.77
FRR	0.06	0.04	0.05	0.04	0.05
HTER	0.11	0.15	0.22	0.24	0.41
TER	0.18	0.20	0.28	0.29	0.46

Image (a) on application of AKAZE yields 191 key points whereas Image (b) yields 156 key points. The total number of inliers obtained after the application of Optimal-RANSAC is 14 and AKAZE feature extraction time was found to be 149.356 ms. This analysis proves that matching in AKAZE is better than KAZE.

As evident from the above table the accuracy of AKAZE is maximum in case of Grimace Database and least in the case of Yale Database. The values of Recall and Precision are quite high with the present AKAZE method in almost all the databases. The table also indicates that Yale Database shows least accuracy, specificity in contrast to high FRR results. TER and FAR values in case of ORL Database is the least obtained.

Comparing the results of KAZE and AKAZE methods, it is very much evident that AKAZE method is better than KAZE methodology. Also, AKAZE feature extraction outperforms KAZE feature extraction in terms of speed.

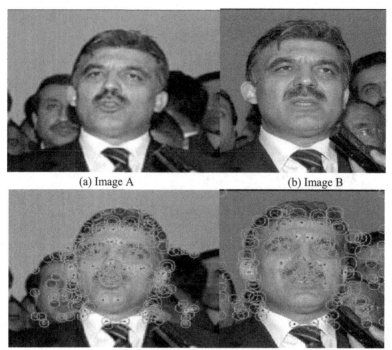

(a) Image A   (b) Image B

(c) AKAZE keypoint detection of Image (a)   (d) AKAZE keypoi   nt detection of Image (b)

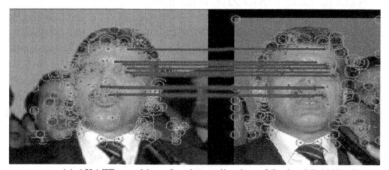

(e) AKAZE matching after the application of Optimal-RANSAC

Fig. 5. Image (a) has the photo of a person and Image (b) has the same person's photo in a different background and varied resolution. Image (c) and (d) shows the key points of Images (a) and (b) respectively. Image (e) shows the AKAZE matching of the person's face from both the images after the application of Optimal-RANSAC

Table 2. Tabulation of AKAZE results for FACE RECOGNITION by using various parameters and testing it on a wide set of Standard DATASETS

AKAZE	ORL Database	Grimace Database	Face 95 Database	Face 96 Database	Yale Database
Sensitivity	96.00	97.59	95.80	97.48	95.78
Specificity	83.33	75.00	62.19	57.85	40.00
Precision	97.46	96.42	95.38	96.48	93.81
Accuracy	94.34	94.73	92.13	94.40	90.47
Recall	96.00	97.59	95.80	97.48	95.78
F1 Score	96.72	97.00	95.59	96.98	94.79
FAR	0.16	0.25	0.37	0.42	0.60
FRR	0.04	0.02	0.04	0.02	0.04
HTER	0.10	0.13	0.20	0.22	0.32
TER	0.14	0.16	0.25	0.24	0.36

7 Conclusion

Here we have made use of the 2-dimensional non-linear scale space analysis for Face Recognition. KAZE and AKAZE are the most famous of them. KAZE eliminates the disadvantages that the previous methods like SIFT and SURF which depended on Gaussian scale space for analysis. KAZE employs AOS schemes which shows better results than the conventional methods. One of the disadvantage of KAZE was it had high computations. This was overcome by the use of AKAZE technique for Face Recognition. This makes use of FED scheme instead of the AOS scheme as in the case of KAZE methodology. This reduces the computation costs spent on building the non-linear scale space. A modification to the LDB descriptor in the case of KAZE methodology also made it more efficient. AKAZE results also prove that it is very much faster and gave better results than the earlier used KAZE method. The use of Optimal-RANSAC after the KAZE and AKAZE methods gave us the optimal set of inliers in almost all the cases. It also demonstrated good results even when the inlier ration was very less.

References

1. Best-Rowden, L., et al.: Unconstrained face recognition: establishing baseline human performance via crowdsourcing. In: 2014 IEEE International Joint Conference on Biometrics (IJCB). IEEE (2014)
2. Heisele, B., Ho, P., Wu, J., Poggio, T.: Face recognition: component-based versus Global Approaches. Comput. Vis. Image Underst. **91**(1–2), 6–21 (2003)
3. Endres, F, Hess, J., Engelhard, N., Sturm, J., Cremers, D., Burgard, W.: An evaluation of the RGB-D SLAM system. In: 2012 IEEE International Conference on Robotics and Automation (ICRA), pp. 1691–1696. IEEE (2012)

4. Lowe, D.G.: Distinctive image features from scale-invariant key points. Int. J. Comput. Vision **60**(2), 91–110 (2004)
5. Liu, C., Yuen, J., Torralba, A.: Dense scene alignment using SIFT flow for object recognition. In: IEEE Confernce on Computer Vision and Pattern Recognition, CVPR (2009)
6. Dreuw, P., Steingrube, P., Hanselmann, H., Ney, H., Aachen, G.: SURF-face: face recognition under viewpoint consistency constraints. In: BMVC, pp. 1–11 (2009)
7. Bay, H., Ess, A., Tuytelaars, T., Van Gool, L.: SURF: speeded up robust features. Comput. Vis. Image Underst. **110**, 346–359 (2008)
8. Alcantarilla, P.F., Bartoli, A., Davison, A.J.: KAZE features. In: Fitzgibbon, A., Lazebnik, S., Perona, P., Sato, Y., Schmid, C. (eds.) ECCV 2012. LNCS, vol. 7577, pp. 214–227. Springer, Heidelberg (2012). https://doi.org/10.1007/978-3-642-33783-3_16
9. Lazebnik, S., Schmid, C., Ponce, J.: Beyond bags of features: Spatial pyramid matching for recognizing natural scene categories. In: IEEE Conference on Computer Vision and Pattern Recognition, CVPR (2006)
10. Tanaka, K., Kondo, E.: Incremental RANSAC for online relocation in large dynamic environments. In: IEEE International Conference on Robotics and Automation, pp. 1025–1030 (2006)
11. Lindeberg, T.: Feature detection with automatic scale selection. Intl. J. Comput. Vision **30**, 77–116 (1998). Duits, R., Florack, L., De Graaf, J., terHaarRomeny, B.: On the axioms of scale space theory. J. Math. Imaging Vis. **20**, 267–298 (2004)
12. Perona, P., Malik, J.: Scale-space and edge detection using anisotropic diffusion. IEEE Trans. Pattern Anal. Mach. Intell. **12**(7), 629–639 (1990). https://doi.org/10.1109/34.56205
13. Chum, O., Matas, J.: Optimal randomized RANSAC. IEEE Trans. Pattern Anal. Mach. Intell. **30**(8), 1472–1482 (2008). https://doi.org/10.1109/TPAMI.2007.70844
14. Weickert, J., Romeny, B.M.T.H., Viergever, M.A.: Efficient and reliable schemes for nonlinear diffusion filtering. IEEE Trans. Image Process. **7**(3), 398–410 (1998)
15. ter Haar Romeny, B.M.: Front-End Vision and Multi-Scale Image Analysis. Multi-Scale Computer Vision Theory and Applications, written in Mathematica. Kluwer Academic Publishers (2003)
16. Weickert, J.: Efficient image segmentation using partial differential equations and morphology. Pattern Recogn. **34**, 1813–1824 (2001)
17. Qiu, Z., Yang, L., Lu, W.: A new feature-preserving nonlinear anisotropic diffusion method for image denoising. In: British Machine Vision Conference, BMVC, Dundee, UK (2011)
18. Mikolajczyk, K., Schmid, C.: A performance evaluation of local descriptors. IEEE Trans. Pattern Anal. Machine Intell. **27**, 1615–1630 (2005)
19. Weickert, J., Ishikawa, S., Imiya, A.: Linear scale-space has first been proposed in Japan. J. Math. Imaging Vision **10**(3), 237–252 (1999)
20. Viola, P., Jones, M.J.: Robust real-time face detection. Intl. J. Comput. Vision **57**, 137–154 (2004)
21. Brown, M., Lowe, D.: Invariant features from interest point groups. In: British Machine Vision Conference. BMVC, Cardiff, UK (2002)
22. Weickert, J., Scharr, H.: A scheme for coherence-enhancing diffusion filtering with optimized rotation invariance. J. Vis. Commun. Image Represent. **13**, 103–118 (2002)
23. Calonder, M., Lepetit, V., Özuysal, M., Trzinski, T., Strecha, C., Fua, P.: BRIEF: Computing a local binary descriptor very fast. IEEE Trans. Pattern Anal. Machine Intell. **34**(7), 1281–1298 (2011)
24. Cao, X., Wei, Y., Wen, F., Sun, J.: Face alignment by explicit shape regression. In: IEEE Conf. on Computer Vision and Pattern Recognition (CVPR), pp. 2887–2894 (2012)
25. Grewenig, S., Weickert, J., Bruhn, A.: From box filtering to fast explicit diffusion. In: Goesele, M., Roth, S., Kuijper, A., Schiele, B., Schindler, K. (eds.) DAGM 2010. LNCS, vol. 6376, pp. 533–542. Springer, Heidelberg (2010). https://doi.org/10.1007/978-3-642-15986-2_54

26. Heinly, J., Dunn, E., Frahm, J.-M.: Comparative evaluation of binary features. In: Fitzgibbon, A., Lazebnik, S., Perona, P., Sato, Y., Schmid, C. (eds.) ECCV 2012. LNCS, pp. 759–773. Springer, Heidelberg (2012). https://doi.org/10.1007/978-3-642-33709-3_54
27. Lenc, J., Gulshan, V., Vedaldi, A.: VLBenchmarks. http://www.vlfeat.org/benchmarks/ (2012)
28. Leutenegger, S., Chli, M., Siegwart, R.Y.: BRISK: Binary robust invariant scalable key points. In: IEEE International Conference on Computer Vision (ICCV), pp. 2548–2555 (2011)
29. Rublee, D., Rabaud, V., Konolige, K., Bradski, G.: ORB: an efficient alternative to SIFT or SURF. In: IEEE International Conferene on Computer Vision (ICCV), pp. 2564–2571 (2011)
30. Yang, X., Cheng, K.T.: LDB: An ultra-fast feature for scalable augmented reality. In: IEEE and ACM International Symposium on Mixed and Augmented Reality (ISMAR) (2012)
31. Zimmermann, K., Svoboda, T., Matas, J.: Tracking by an optimal sequence of linear predictors. IEEE Trans. Pattern Anal. Machine Intell. **31**(4), 677–692 (2009)
32. Fischler, M.A., Bolles, R.C.: Random sample consensus: a paradigm for model fitting with applications to image analysis and automated cartography. Commun. ACM **24**(6), 381–395 (1981)
33. Okabe, T., Sato, Y.: Object recognition based on photometric alignment using RANSAC. In: Computer Vision and Pattern Recognition, 2003. Proceedings. 2003 IEEE Computer Society Conference on, vol. 1, pp. I–221. IEEE (2003)
34. Wang, Y., Huang, H., Dong, Z., Manqing, W.: Modified RANSAC for SIFT-Based InSAR Image Registration. Prog. Electromagnet. Res. M **37**, 73–82 (2014)
35. Calonder, M., Lepetit, V., Strecha, C., Fua, P.: BRIEF: binary robust independent elementary features. In: Daniilidis, K., Maragos, P., Paragios, N. (eds.) ECCV 2010. LNCS, vol. 6314, pp. 778–792. Springer, Heidelberg (2010). https://doi.org/10.1007/978-3-642-15561-1_56
36. Hast, A., Nysjö, J., Marchetti, A.: Optimal RANSAC-Towards a Repeatable Algorithm for Finding the Optimal Set (2013)
37. Choi, S., Kim, T., Wonpil, Y.: Performance evaluation of RANSAC family. J. Comput. Vision **24**(3), 271–300 (1997)
38. Chum, O., Matas, J., Kittler, J.: Locally optimized RANSAC. In: The Annual Pattern Recognition Symposium of the German Association for Pattern Recognition, pp. 236–243 (2003)
39. Chum, O., Matas J., Obdrzalek S.: Enhancing RANSAC by generalized model optimization. In: Asian Conference on Computer Vision (2004)

Modified InceptionV3 Using Soft Attention for the Grading of Diabetic Retinopathy

Shroddha Goswami⬤, K Ashwini$^{(\boxtimes)}$⬤, and Ratnakar Dash⬤

National Institute of Technology, Rourkela, Rourkela, India
ashwinikamakshi@gmail.com, ratnakar@nitrkl.ac.in

Abstract. Diabetic retinopathy (DR) is an eye ailment affecting retinal blood vessels. Many deep learning methods for detecting DR have been presented as manual diagnosis is time-consuming and inconvenient. InceptionV3 architecture was modified using a soft attention module in this proposed framework. The attention technique's basic concept is to concentrate on specific relevant parts by assigning the weights accordingly. Contrast Limited Adaptive Histogram Equalization (CLAHE) is the pre-processing method applied initially to the fundus images to improve the contrast level. Along with this, augmentation has been done to increase the number of images, which are then trained and validated using a modified InceptionV3 model. The experimental results show that the suggested model better diagnoses all stages of DR than existing techniques and outperforms the existing model on the IDRiD and DDR datasets.

Keywords: Diabetic Retinopathy · deep learning · CLAHE · InceptionV3

1 Introduction

Diabetic Retinopathy (DR) is an eye condition that can develop in people having Diabetes Mellitus (DM) - type 1 or type 2. DR is caused due to the damage caused to the blood vessels and light-sensitive tissues at the back of the eyes due to high blood sugar levels. At first, DR might cause no or only mild vision problems, but if not detected at early stages, it can cause permanent blindness in people. This disease is irreversible, so the only way to retain eyesight is the prevention of DR. Apart from the manual diagnosis, there is a need for a Computer aided-diagnosis system that classifies DR according to its grade. This disease is still regarded as the leading cause of blindness in diabetic patients all over the globe, especially in developing countries.

According to a paper released by the National Library of Medicine [1], the International Diabetes Foundation (IDF) approximates the global population with DM to be 463 million in 2019 and estimated the number to go up to 700

M. Singh et al. (Eds.): ICACDS 2023, CCIS 1848, pp. 178–188, 2023.
https://doi.org/10.1007/978-3-031-37940-6_15

million in 2045. DR was discovered to be the most typical side effect of the same and a major contributor to avoidable blindness in working adults in the modern world. The paper also stated that the number of adults found to have DR worldwide was 103.12 million, anticipated to rise to around 160.50 million by 2045.

In India, a survey was conducted to understand the effect DM and DR have, and it was considered an emerging disease with significant public health impact [2]. The study found that 16.9% of diabetics experienced DR, mild retinopathy was 11.8%, and Sight-threatening Diabetic Retinopathy (STDR) was 3.6%. In a paper published in the Indian Journal of Ophthalmology [3], a survey was conducted on people with DR in at least one eye for 9 years. It was seen that people, especially males, in the age group of 30–50 years were in the risk zone of getting DR. It was also observed that every fourth person having DM developed diabetic retinopathy. Apart from this, every second person is having diabetic retinopathy STDR, which is a number not to be taken casually.

According to the International Standard for severity scale, there are generally five stages of DR according to Fig. 1. The presence of abnormalities determines this classification, and their magnitude defines the intensity of the disease. The five stages of DR are healthy retina or No DR, Mild Non-proliferative DR, Moderate Non-proliferative DR, Severe Non-proliferative DR and Proliferative DR. The stages mentioned are in increasing order of the severity of the disease, starting from no abnormality till complete blindness.

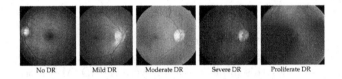

No DR   Mild DR   Moderate DR   Severe DR   Proliferate DR

Fig. 1. Stages of DR

Convolutional Neural Network (CNN) has been commonly used to make significant contributions in identifying and grading fundus images according to their phases. CNN is a network architecture for deep learning that can accept an image as input and attribute importance (learnable weights and biases) to different image characteristics. A lot of research has already been conducted in this field to grade DR and has achieved good accuracy.

In this paper, some of the works done previously in the grading of DR have been reviewed. Some pre-existing models have been implemented to determine the best suited for our data set using Transfer learning. It has been found that InceptionV3 performed better among all the pre-existing models; thus, it has been modified using soft attention to classify DR. This model is then studied with some state-of-the-art models. The rest part of the paper consists of some related works in Sect. 2, the work that has been done for this paper has been explained in Sect. 3, succeeded by the results obtained in Sect. 4 and the conclusion in Sect. 5.

2   Related Works

Extensive work has been done in the multi-class classification of DR, achieving good results. Harry Pratt et al. [4] proposed a network that involves CNN architecture and data augmentation that can recognize the minute features involved in the classification task, such as hemorrhages and micro-aneurysms on the retina, and provide a diagnosis automatically. Their developed model achieved an accuracy of 75% and a sensitivity (correctly identifying who has the disease) of 95%. Li et al. [5] used ResNet architecture to develop a cross-disease attention network (CANet)to evaluate the internal relationships between DR and Diabetic Macular Edema (DME) while utilising simply image-level supervision. For this purpose, they have used the Messidor and the IDRiD data sets for training and testing purposes. Their model received a joint accuracy of 85.1%.

Another deep learning framework from ResNet was proposed by Doaa K. Elswah et al. in their paper [6]. In this, the framework is divided into three stages - the pre-processing and augmentation of fundus images, the pre-processed images are passed through a ResNet CNN. Finally, grading, the images are according to their classes. IDRiD data set was used for the implementation of the model. The pre-processing in this model was done in various stages and combinations to get a good result. The accuracy of the framework came up to 83.90%.

Gayathri et al. [7] created a customized CNN model specifically for feature extraction. The output features from the CNN model are then used as input for various classifiers such as Random Forest, Support Vector Machine, etc. Data sets that were used for this implementation were IDRiD, Messidor, and Kaggle Datasets. For multi-class classification, the model can provide an accuracy of 99.59%.

Pour et al., in their paper [8], used CLAHE to reduce the noise by limiting the contrast of the images. The authors have used the EfficientNet architecture of CNN to implement their model. The model was trained on the Messidor data set and tested on the IDRiD data set. The images' scaling of all three dimensions (depth, width, and resolution) are done by a factor. The model was able to achieve a sensitivity of 92%.

Pao et al. proposed a model [9] of using a bi-channel convolutional neural network to identify and grade DR images. In the proposed model, images from the publicly available Kaggle Diabetic Retinopathy were taken. Then a bi-channel CNN was constructed to simultaneously process the luminance of the gray image and the green channel extracted from the image. The accuracy came out to be 87.83%, sensitivity to be 77.81%, and specificity to be 93.88%.

A course-to-fine network in diabetic retinopathy has been proposed by Wu et al. [10], which gave a reliable method to grade fundus images. The course network is pre-trained with attention modules. It performs classification to classify images as DR and no DR. The fine network is also pre-trained and performs the four-class classification. This network classifies images into mild NPDR, moderate NPDR, severe NPDR, and PDR. Image normalization and data augmentation have been done on the images as a part of preprocessing their images. The IDRiD data set was used to get the results. The model's accuracy, sensitivity, and specificity are 56.19%, 64.21%, and 87.39%. He et al. [11] proposed a CABNet architecture, an attention model. It is used to capture small, detailed lesions equally and more precisely.

Shaikh et al. [12] proposed a Lesion-aware architecture. It is another attention-based architecture that jointly trains the autoencoder and neural vector support machine to obtain lesion-specific attention features. This model has been implemented on APTOS and IDRiD data sets. Cherukuri et al. [13] proposed the Hinge attention network (HA-Net). A convolutional LSTM layer is used to prioritize the essential feature maps before passing them onto the Hinge attention model that had been proposed.

Though the existing works give good results, most cannot strike a balance between all the performance parameters. Ensuring a balance is essential, as it will give an optimum result. An attention technique is helpful as the attention techniques help the model give importance to the parts in the images that are more relevant for detecting the stage of diabetic retinopathy. In this paper, one such attention mechanism called soft attention, has been used.

3   Proposed Model

The model proposed in this paper uses soft attention at its crux. Soft attention is a kind of attention mechanism that helps to focus more on the lesions, microaneurysms, and blood vessels of the fundus images. It gives more numerical weight to those pixels in the image. Transfer learning has been used to find the model that will be best for the data set used here, which is found to be InceptionV3. InceptionV3 is combined with the soft attention model to give rise to the proposed modified InceptionV3 model. Two data sets have been used for the execution of the proposed models: IDRiD and DDR.

3.1   Preprocessing

The preprocessing module consists of resizing the data set so that all the images are of the same size, increasing the contrast of the images to get a clearer view of the images' small details, and increasing the number of images using augmentation. In medical data sets, the quality and number of images usually fall short in helping the model make a reasonable interpretation, which is why preprocessing plays a crucial role.

Image data augmentation is done on the data set to improve the model's performance. New images are created in this using various operations like flipping, rotating, changing brightness, and zooming in or out. For the data set used here, augmentation tasks done on the images are flipping horizontally, vertically, and rotating by 90°. The images are also resized to 299 × 299 pixels. The input image size for InceptionV3 is 299 × 299 pixels, which is why this value was chosen. Also, resizing gives uniformity to all the images.

Apart from the augmentation, the contrast in images is increased. Adaptive histogram equalization (AHE) is a process by which the contrast of images is increased. In contrast to conventional histogram equalisation, the adaptive technique computes many histograms, each corresponding to a different image section. The brightness levels of the image are then distributed using them.

Since the histogram in these places is heavily concentrated, conventional AHE tends to over-amplify the contrast in regions of the image that are nearly constant. AHE might thus lead to noise amplification in almost constant locations. Contrast Limited AHE (CLAHE), a form of adaptive histogram equalization, limits contrast amplification to lessen the issue of noise amplification. CLAHE works with discrete sections, called tiles, instead of processing the entire image. Combining the adjacent tiles using bi-linear interpolation eliminates the artificially created borders.

Here, CLAHE has been applied to identify the blood vessels and lesions of the fundus image. The RGB channel of the fundus image has been converted to LAB color space, where the L channel is for brightness, and the other channels are for color. All three channels have been separated, and CLAHE has only been applied to the L channel. After the preprocessing, this enhanced L channel has been merged back with the A and B channels. The change that CLAHE brings about in images is shown in Fig. 2.

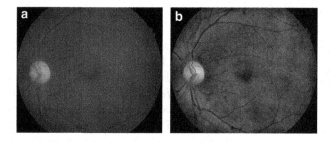

Fig. 2. (a) Before CLAHE (b) After CLAHE

3.2   Selection of Deep Learning Model

Transfer learning uses a pre-trained model, usually on extensive data sets, so the models don't have to be trained from scratch. The weights obtained from the training are then used to set the weights of the new model initially. Transfer learning poses quite a few advantages for the users. The models are trained on the ImageNet dataset by default. The ImageNet data set contains over 1000 classes, a sizable collection of human annotations and pictures developed by researchers.

Transfer learning helps researchers experiment to see which pre-defined model is best suited for them.

For carrying out the experiments, the pre-trained model used for this report is InceptionV3 [14], InceptionResNetV2 [15], Xception [16], ResNet50 [17], DenseNet169 [18], DenseNet201 [18], and DenseNet121 [18]. Out of these, InceptionV3 gave the best results, followed by Xception and DenseNet169. The results for the same are mentioned in Table 1.

3.3   Proposed Modified InceptionV3 Model

As mentioned earlier, InceptionV3 gave the best result using Transfer Learning. The original architecture of InceptionV3 has been shown in Fig. 3.

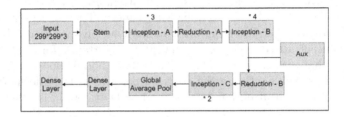

Fig. 3. InceptionV3 end-to-end Original Architecture

The IDRiD and DDR data sets have been used for this proposed framework for implementation purposes. The IDRiD data set (Indian Diabetic Retinopathy Image Dataset) is the pioneer in representing the fundus images of the Indian population [19]. A total of 516 images are present in this data set. It is the only set of data that includes both regular lesions caused by DR and normal retinal structures that have been image-level annotated. The DDR data set [20] is a database constituted from using the fundus images in Chinese hospitals. It contains 13,673 fundus images of 9598 from 147 hospitals and covers 23 provinces in China.

The two data sets used the soft attention unit to classify the images. Since it was observed that InceptionV3 gave the best results for our data set, this is the model that has been modified further. It was experimentally observed when the soft attention layer was applied between a few blocks; the best results were obtained. Only the useful features are multiplied with a higher weight, and the irrelevant features are given a lower weight. Apart from this, transfer learning was used for the first few blocks of the InceptionV3. That is, till Reduction-A, transfer learning has been used, and then the model has implemented the soft attention model. Using this combination helped incorporate the advantages of transfer learning and the main soft attention block. This helps to give more attention to the essential features and increases the model's performance. Figure 4 shows the end-to-end architecture of the same.

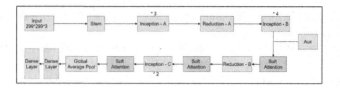

Fig. 4. Modified InceptionV3 end-to-end architecture

The total number of parameters for this model was 27,076,181. Out of this, 27,041,749 are trainable, and 34,432 are non-trainable parameters. The architecture of InceptionV3 is 48 layers deep.

Soft Attention. Datta et al. [21] used soft attention to improve the performance of a CNN network for treating skin cancer. The concept of selective attention in cognitive science describes how we focus on specific nearby things. It helps us concentrate so that we can tune out unimportant information and pay attention to what matters most. The soft-attention module boosts efficiency by concentrating predominantly on the essential portions of the information. The architecture of the soft attention unit, as used by Datta et al., is given below in Fig. 5.

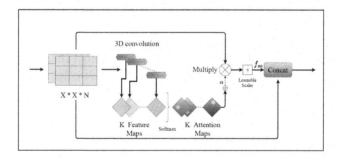

Fig. 5. Soft Attention architecture as used by Datta et al. [21]

This architecture was originally proposed by Shaikh et al. [22] in 2020. It was used to verify handwritten text. It uses a 3D convolution to identify the most important features. The feature tensor (t) travels down the deep neural network and is the structure's input.

$$f_{sa} = \gamma t(\sum_{i=1}^{K} softmax(W_i * t)) \tag{1}$$

A 3D convolution layer having weights, $W_i \in R^{h*w*d*K}$, where K is the number of 3D weights, receives this feature tensor, $t \in R^{h*w*d*K}$. To produce

K = 16 attention maps, the output of this convolution is normalized using the softmax function. These attention maps are combined to create a single attention map that serves as a weighting function α, as seen in Fig. 5. The prominent feature values are then carefully scaled by first multiplying this α by t, and then by γ, a learnable scalar. Finally, a residual branch is formed by joining the carefully scaled features (f_{sa}) with the original feature t. For the network to moderately learn to regulate the amount of attention it needs, we initialize γ with 0.01 during training.

4   Results

To check the effectiveness of the proposed model, the entire data sets have been split up into training and testing sets in the ratio of 80:20. Tensorflow framework has been used to implement the same. The learning rate used for the model was 0.002, and the number of epochs was 200. Adam optimizer has been selected, as experimentally, it gave the best results. Since the data sets used in this paper have multiple classes, the loss function used for checking the loss was categorical cross-entropy which is used for multi-class classification. The hardware used for the experiments, Intel(R) Xeon(R) Silver 4210 CPU @220 GHz with 64GB RAM and 16GB GPU running on Windows 10 Pro, has been used. NVIDIA RTX A4000, version 522.06, was used as the GPU to compute the proposed model.

Transfer learning was used to find the model that would give us the best result, as mentioned in the previous segment. It was found that InceptionV3 gave the best accuracy of 56% among all models. Considering validation accuracy as the primary performance metric, it was selected as the model to continue further work with. The validation accuracy of all the considered models is mentioned in Table 1.

Table 1. Comparison of the pre-defined model using transfer learning

Model	Validation Accuracy	Validation Loss
InceptionV3	0.56	2.63
Xception	0.54	2.51
DenseNet169	0.51	2.13
DenseNet121	0.50	2.43
DenseNet201	0.47	2.43
InceptionResNetV2	0.47	2.48
ResNet50	0.31	1.48

InceptionV3 was used to classify the images in the IDRiD and DDR data sets. The results of the original architecture are compared with the proposed

Table 2. Performance Metrics on IDRiD data set

Model	Accuracy	Loss	Sensitivity	Specificity	Precision
Original InceptionV3	0.79	2.01	0.79	0.95	0.80
Modified InceptionV3 + SA	0.82	0.82	0.82	0.96	0.82

model. Table 2 gives the result using the IDRiD data set. It is observed that the proposed model gave an accuracy of 82% on the IDRiD data set.

Table 3 gives the result using the DDR data set. The accuracy of the proposed model was found to be 79% on the DDR data set. Both the results show a significant improvement on the original models.

Table 3. Performance Metrics of DDR data set

Model	Accuracy	Loss	Sensitivity	Specificity	Precision
Original InceptionV3	0.75	1.79	0.74	0.93	0.75
Modified InceptionV3 + SA	0.79	1.34	0.79	0.94	0.79

The Modified InceptionV3 is also compared with some state-of-the-art architectures. The proposed model has been compared with some already proposed attention models on the respective data sets. On IDRiD, the models by Shaik et al. [12], Li et al. [5], and Cherukuri et al. [13] have been compared. On the DDR data set, the proposal by He et al. [11] was considered for comparison purposes. It was seen that the performance metrics of the proposed model were better than that of the state-of-the-art architecture considering the accuracy in both the IDRiD and DDR data sets. The outcome is shown in Tables 4 and 5.

Table 4. Comparison of Performance Metrics with Existing Attention Models on IDRiD data set

Model Name	Accuracy (%)
LA-NSVM [12]	63.24
CAN-Net [5]	65.1
HA-Net [13]	66.41
Modified Inceptionv3 + SA Model	**82.13**

Table 5. Comparison of Performance Metrics with Existing Attention Model on DDR data set

Model Name	Accuracy (%)
CAB-Net [11]	78.13
Modified Inceptionv3 + SA Model	**79.03**

5   Conclusion

Since DR is one of the major contributors to blindness, it is vital to bring methods for accurate and fast identification and grading of this problem. The soft attention module and InceptionV3 are used in this work to give a five-stage rating of DR. At first, the images are pre-processed using CLAHE, and then augmentation is done. Then, the images are classified using the modified InceptionV3 model. The outcomes obtained are better when compared with the state-of-the-art methods. The results show that the model obtained considerable sensitivity and specificity, important metrics for medical image processing. This demonstrates the efficacy of the attention-based model for the classification of images.

References

1. Teo, Z.L., et al.: Global prevalence of diabetic retinopathy and projection of burden through 2045: systematic review and meta-analysis. Ophthalmology **128**(11), 1580–1591 (2021)
2. Vashist, P., et al.: Prevalence of diabetic retinopathy in India: results from the national survey 2015–19. Indian J. Ophthalmol. **69**(11), 3087 (2021)
3. Das, A.V., Prashanthi, G.S., Das, T., Narayanan, R., Rani, P.K.: Clinical profile and magnitude of diabetic retinopathy: an electronic medical record-driven big data analytics from an eye care network in India. Indian J. Ophthalmol. **69**(11), 3110 (2021)
4. Pratt, H., Coenen, F., Broadbent, D.M., Harding, S.P., Zheng, Y.: Convolutional neural networks for diabetic retinopathy. Procedia Comput. Sci. **90**, 200–205 (2016)
5. Li, X., Hu, X., Yu, L., Zhu, L., Fu, C.W., Heng, P.A.: CANet: cross-disease attention network for joint diabetic retinopathy and diabetic macular edema grading. IEEE Trans. Med. Imaging **39**(5), 1483–1493 (2019)
6. Elswah, D.K., Elnakib, A.A., Moustafa, H.E.D.: Automated diabetic retinopathy grading using ResNet. In: 2020 37th National Radio Science Conference (NRSC), pp. 248–254. IEEE (2020)
7. Gayathri, S., Gopi, V.P., Palanisamy, P.: A lightweight CNN for diabetic retinopathy classification from fundus images. Biomed. Signal Process. Control **62**, 102–115 (2020)
8. Pour, A.M., Seyedarabi, H., Jahromi, S.H.A., Javadzadeh, A.: Automatic detection and monitoring of diabetic retinopathy using efficient convolutional neural networks and contrast limited adaptive histogram equalization. IEEE Access **8**, 136668–136673 (2020)

9. Pao, S.-I., Lin, H.-Z., Chien, K.-H., Tai, M.-C., Chen, J.-T., Lin, G.-M.: Detection of diabetic retinopathy using bichannel convolutional neural network. J. Ophthalmol. (2020)
10. Wu, Z., et al.: Coarse-to-fine classification for diabetic retinopathy grading using convolutional neural network. Artif. Intell. Med. **108**, 101936 (2020)
11. He, A., Li, T., Li, N., Wang, K., Fu, H.: CABNet: category attention block for imbalanced diabetic retinopathy grading. IEEE Trans. Med. Imaging **40**(1), 143–153 (2020)
12. Shaik, N.S., Cherukuri, T.K.: Lesion-aware attention with neural support vector machine for retinopathy diagnosis. Mach. Vis. Appl. **32**(6), 1–13 (2021)
13. Shaik, N.S., Cherukuri, T.K.: Hinge attention network: a joint model for diabetic retinopathy severity grading. Appl. Intell. **52**(13), 15105–15121 (2022)
14. Szegedy, C., Vanhoucke, V., Ioffe, S., Shlens, J., Wojna, Z.: Rethinking the inception architecture for computer vision. In: Proceedings of the IEEE Conference on Computer Vision and Pattern Recognition, pp. 2818–2826 (2016)
15. Szegedy, C., Ioffe, S., Vanhoucke, V., Alemi, A.A.: Inception-v4, inception-ResNet and the impact of residual connections on learning. In: Thirty-first AAAI Conference on Artificial Intelligence (2016)
16. Chollet, F.: Xception: deep learning with depthwise separable convolutions. Proceedings of the IEEE Conference on Computer Vision and Pattern Recognition, pp. 1251–1258 (2017)
17. He, K., Zhang, X., Ren, S., Sun, J.: Deep residual learning for image recognition. In: Proceedings of the IEEE Conference on Computer Vision and Pattern Recognition, pp. 770–778 (2016)
18. Huang, G., Liu, Z., Van Der Maaten, L., Weinberger, K.Q.: Densely connected convolutional networks. In: Proceedings of the IEEE Conference on Computer Vision and Pattern Recognition, pp. 4700–4708 (2017)
19. Prasanna, p., et al.: Indian diabetic retinopathy image dataset (IDRiD). IEEE Dataport (2018)
20. Li, T., Gao, Y., Wang, K., Guo, S., Liu, H., Kang, H.: Diagnostic assessment of deep learning algorithms for diabetic retinopathy screening. Inf. Sci. **501**, 511–522 (2019)
21. Datta, S.K., Shaikh, M.A., Srihari, S.N., Gao, M.: Soft attention improves skin cancer classification performance. In: Reyes, M., et al. (eds.) IMIMIC/TDA4MedicalData -2021. LNCS, vol. 12929, pp. 13–23. Springer, Cham (2021). https://doi.org/10.1007/978-3-030-87444-5_2
22. Shaikh, M.A., Duan, T., Chauhan, M., Srihari, S.N.: Attention based writer independent verification. In: 2020 17th International Conference on Frontiers in Handwriting Recognition (ICFHR), pp. 373–379. IEEE (2020)

Comprehensive Study of Cyber Security in AI Based Smart Grid

Priyansh Sanghavi$^{(\boxtimes)}$ ⓘ, Riya Solanki ⓘ, Viral Parmar ⓘ, and Kaushal Shah ⓘ

School of Technology, Pandit Deendayal Energy University, Gandhinagar, Gujarat 382007, India
Priyansh4299@gmail.com

Abstract. One of the most beneficial purposes of artificial intelligence is the smart grid. The whole globe is rapidly migrating from traditional grid systems to smart grids driven by AI. One of the most difficult jobs is protecting the smart grid from cyber assaults. Any cyber-attack may compromise the smart grid's confidentiality, integrity and availability. Significant research is being performed in industry, government, and academia to improve smart grid security in order to secure the smart grid from intruders. In this paper, we explore the history of cyber-attacks on smart grids throughout the globe, various cyber-attacks, and countermeasures in smart grid cyber-security. This research examined the difficulties and potential solutions of such AI-based smart grids. We highlight cyberattack types and conduct an in-depth study of the smart grid's cyber security. Emphasis is placed on the discussion & investigations of security breaches, assault countermeasures, and security requirements. We aim to develop a comprehensive understanding of cyber-security threats and countermeasures in the context of smart grid technologies, as well as a road strategy for further cyber-security studies in this area.

Keywords: Smart Grid · Cyber Attack · Artificial Intelligence · Cyber Threat · SCADA Security · Machine Learning

1 Introduction

Modern technology has been added to the old electricity infrastructure, resulting in the construction of smart grids. Some examples of energy measures include smart metres and home sensors placed at the customer location, production metre, renewable power generation system, smart inverters & energy saving technology installed on the grid's point. Despite the intermittent nature of renewable energy and its sensitivity to several environmental elements, including environmental conditions such as temperature, moisture, air velocity, and position, sustainable power sources reduce power costs because the cost of creating clean and renewable energy is zero.

Several methodologies exist for estimating solar energy, wind energy and battery condition to absorb renewable energy effectively and expeditiously. The intelligent grid provides full-duplex communication between scattered grids and sensors. The smart grid's has numerous benefits Compared to conventional grids, such as enhanced electricity reliability, self-healing, affordability with the incorporation of renewable power,

M. Singh et al. (Eds.): ICACDS 2023, CCIS 1848, pp. 189–202, 2023.
https://doi.org/10.1007/978-3-031-37940-6_16

flexible power generation, more ecologically sustainable procedure, the consolidation of scattered energy resources, client side real time power usage monitoring, automation of functions via the incorporation of Artificial Intelligence designs, remote energy monitoring, Rapid fault reaction, virtual fusing & remote metering. Fixing these vulnerabilities becomes more challenging when smart grid information is hosted in the cloud. Cyber security is gradually becoming a vital component to ensure that the smart grid is always safe and reliable. By manipulating energy use, these types of attacks may cause widespread power outages [1, 2].

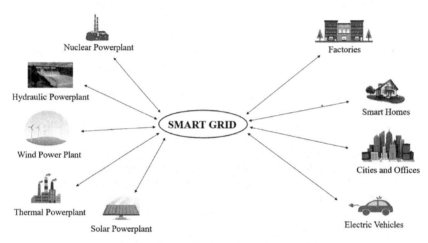

Fig. 1. Application of Smart Grid

Figure 1 shows a simplified view for understanding the process of smart grids. Aside from physical security, cyber security is becoming an important the likelihood of problems grows as the complexity of the grid increases[3]. For example, if hundreds of sensors are placed and one of them begins relaying inaccurate data even though there is no defect in the production equipment, The whole of the grid system's functioning might be destabilised [4]. The 2nd concern is security, particularly information exchange among gadgets and the grid. The challenges of smart grid's communications routes may impede the protection of smart grid's information & cyberattacks may result in physical damage to the smart grids. We did a comprehensive investigating analysis of current smart grid's vulnerabilities and mitigation strategies, which is the principal contribution of this study. Information interchange between machines and the grid poses a second security risk. Cyberattacks may cause physical harm to the smart grid, and the complexity of smart grid communication pathways might affect the security of smart grid data [5]. The primary contribution of this paper is an in-depth investigation of existing grid risks and Its solutions. Any cyber-attack targets the network communication, the workers who run the network, or the consumers who use the network. We provide approaches for reducing the likelihood of any cyber-attack at any point.

2 Literature Review

An Overview of Cyberattacks on Smart Grids as Well as Their Blackouts:

2.1 Cyberattack on Saudi Aramco – Saudi Arabian Oil Company, 2017

In 2017, malicious software took advantage of a vulnerability in the company's safety system at one of its petrochemical facilities, making Saudi Aramco the target of cyber-attacks. It's the first-time malicious software has been used to protect computers at a factory from a potential tragedy.

An official from the power facility was quoted in the Independent[6] as saying that the purpose of the assault was not simply to disable equipment or delete data but also to convey a political message.

2.2 Cyber-Attack on Ukraine Power Grid, 2015 and 2016

Cyber-attacks on the energy industry are becoming more frequent and pose a growing threat to reliability and security. Successful attacks on the Ukrainian electricity grid in 2015 & 2016 demonstrate this threat. As a result of an attack on the company's computers and SCADA systems, thirty substations were knocked down for three hours. As a consequence of these incidents, attackers could obtain access to distributing network operator consoles and remotely shut off circuit breakers, which led to power disruptions in the affected areas. As a result of this assault, thirty substations were rendered inoperable, which impacted around 230,000 people. It was the very first cyber assault on smart grids that was successful. Attackers can infiltrate communication networks and change information or overload highly interconnected networks with internet traffic. It restricts the ability of network operators to monitor and operate the network.

According to the BBC[7], the power outage in 2016 equated to one-fifth of Kyiv's total usage that night. Some experts believe was intended to physically harm the electrical infrastructure.

2.3 Blackout Affecting Parts of Arizona and Southern California, 2011

According to officials, the outage was most likely triggered by an employee removing a piece of monitoring equipment that was creating difficulties at a power substation in southwest Arizona. A final tally of 2.7 million citizens was without power due to the blackout that occurred in Arizona and Southern California on September 8, 2011. During a particularly hot day, peak-hour demand rises as well as a single high-voltage line has a failure owing to a flaw. As a result, carrying energy to the San Diego area becomes necessary. Within a few minutes after the power was redistributed, other lines and transformer failures occurred, disconnecting San Diego from the remainder of Western Connections. This disparity between demand and supply in the vicinity of San Diego resulted in overloads and power outages [8].

2.4 Cyber-Attack on Iran Nuclear Facility, 2010

A dangerous Stuxnet computer worm targeted Iran Nuclear Plant. Several centrifugal pumps at Iran's nuclear uranium enrichment facility reportedly exploded due to Stuxnet. Stuxnet was designed to disrupt and destroy Iran's nuclear program. Nonetheless, it demonstrated the capacity to harm critical infrastructure significantly by infecting computerized controllers and SCADA systems that regulate industrial equipment [9].

2.5 The United States and Canada Were also Affected by a Blackout in 2003

The accident was caused on August 14, 2003, as a high voltage transmission line traversed northern Ohio, it encountered several overgrown trees. The line has deteriorated due to the powerful current that flows through this. The fault was supposed to raise an alarm at the monitoring centre nonetheless, the warning system has failure. Later, 3 more power lines fell into trees and collapsed, adding stress on the surviving power lines. A few hours later, they too went down owing to overload, precipitating a cascading of failed throughout south-eastern Canada & eight north-eastern states. The greatest blackout in North American history left Up to two days without power for 50 million people [10].

3 Vulnerabilities in the Smart Grid

A network for a smart grid may have a vulnerability if there is a point where an attacker may exploit a weakness in the system to obtain access to the network and harm the system. Because it can connect with so many different domains through so many other protocols, the smart grid is vulnerable to many different kinds of assaults. In this part, we look at the factors that might make the grid more vulnerable to cyber infiltration. But first, let's go through the many forms of cyber-attacks. There are two different types of attacks: passive and active. Figure 2 shows the basic concept of passive and active assaults. Passive assaults are those in which no damage is done to the information & the attacker observes it [11].

In contrast, active attacks are more damaging because the attacker alters or stops the recipient from obtaining the information. For example, in Fig. 2(a) The sender has sent a message to the receiver through the internet. During the transmission of a message, a hacker will perform a kind of attack through which the attacker can only read or observe the message. The term for this type of attack is passive attack. In Fig. 2(b) sender has sent a message to the receiver through the internet. During the transmission of a message, a hacker will perform a kind of attack through which the attacker can edit the message. This type of assault is known as an active assault.

Passive attacks are categorized as eavesdropping and traffic analysis. The active attacks are classified as masquerade assaults, replay attacks, false data attacks, Dos attacks & DDoS attacks.

The following is an explanation of the most important factors contributing to the smart grid's vulnerability to cyberattacks.

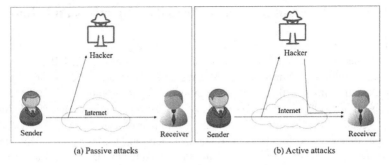

(a) Passive attacks   (b) Active attacks

Fig. 2. Active & Passive Attack

3.1 A Rise in the Prevalence of Smart Electronic Gadgets being Installed

The number of devices inside a connection correlate directly to the number of access points available to malicious users. Even if the security of only a specific point is breached, the integrity of the whole network infrastructure is put at risk.

3.2 Installation of Components Provided by a Third Party

Specialists do not recommend using third party's parts in the system. It made the system more susceptible to assault. There is a possibility that these computers are infected with trojans, which might subsequently spread the infection to other devices connected to the network.

3.3 Insufficient Staff Training

Each technological tool must be used with the proper training. When employees lack proper training, Phishing scams may take advantage of such vulnerabilities more easily.

3.4 Utilizing the Internet Protocols

Most data transfer protocols are not secure. Several protocols use a format that does not include encryption when they transfer data. Consequently, they are excellent targets for assaults known as "man in the middle" that steal data [1].

3.5 Maintenance

Although ensuring smooth operations is presumably the primary goal of maintenance, this process may sometimes become a channel for hackers. During routine maintenance, operators often disable the security equipment in order to conduct the review. In 2015, the Eastern European electric utility industry reported a similar incident.

4 Smart Grid's Primary Cyber Security Objectives and Goals

The NIST (National Institute of Standards and Technology) has created a model that may be used by the medical field to improve the safety of smart grids. The NIST proposes the following 19 smart grid requirements[12]: Awareness Training, Access Control, keeping a Good Account of Everything, Authorization and Evaluation of the Safety Measures, Management of Configurations, Assurance of the Continuity of Operations, Methods of Identifying and Authenticating Individuals, Administration of Information and Documentation, Reactions to Incidents, The Development and Maintenance of Information Systems for Smart Grids, Protection of the Press, Safety in the Physical and Environmental Environment, Organizational Structure, Administration of the Security Program, Safety Procedures for Employees, Risk Assessment and Management, Acquiring Information Systems and Services Relating to Smart Grids, Information System and Communication Protection for Smart Grid, The Information System for Smart Grids and the Integrity of the Information.

Each security need must have a unique identity, category, requirement, supplementary advice, enhancement, extra consideration, and impact level. The following are the five key goals of implementing cybersecurity measures in smart grids: Authentication, Authorization, Confidentiality, Integrity, and Availability are the Five as of Information Security [12].

In the following table, Table 1, we explain the many types of attacks that can be made, as well as which security goals are compromised by each form of assault.

5 Techniques for Detecting and Mitigation Cyber Attacks

Because of the involvement of numerous stakeholders, managing smart grid information, especially from smart metres, becomes a difficult task. A framework provides principles for integrating privacy and security across multiple domains for better smart meter security and privacy protection. Three areas comprise the security framework: transmission safety, protected computer safety, and system management safety. Communication security includes crypto-system safety, router safety, & network privacy. An effective organizational system, End to End (E2E) cryptography & multi-hop routing may help accomplish communication security goals [13].

Furthermore, one of the key functions of smart meters is to record the quantity of energy utilized and variables such as voltage and frequency. Diverse methods have been proposed in the literature to handle the cybersecurity aspects of the smart grid. Moreover, When the sophistication and inclusion of artificial intelligence increase, more research will be performed to improve the grid's dependability [14].

5.1 Attack Detection and Prevention Based on Machine Learning

As the smart grid replaces the old grid, hundreds more sensors are being integrated into the infrastructure of the smart grid. Regularly observing the condition of the equipment to which they are attached, these sensors generate massive amounts of information in event logs or time series information. Before being transferred to the servers, the data

Table 1. Under the attack, security objectives are compromised

Attack Category	Security Goal Compromised	Description
Flooding Attack	Availability	preventing people from using resources
Denial of Service, Distributed Denial of Service	Availability	Stop fulfilling the user's request
Jamming	Availability	Network Interference or try to Jamming the Network
Buffer Overflow	Availability, Confidentiality	Overwriting the buffer's memory
Man in the middle	Confidentiality	Obtaining packet information between the sender and the recipient
Packet Sniffing	Confidentiality	Examining the package
Social Engineering Attack	Confidentiality, Integrity	Attacking people rather than equipment or networks
Session hijacking attack	Confidentiality, Integrity	preventing the user from accessing resources for a set period of time
False Data Injection	Integrity	tampering with real-world data
Replay Attack	Integrity	Send data again and repeated
Data Manipulation	Integrity	Data tempering

collected from the sensors is kept on a server and pre-processed. Storing the data on a local server is as secure as feasible. When data is stored on the cloud, the person has far greater control over how that information is utilized since the data may be streamed to a computer remotely using the GETS instruction. Based on the information, the model distinguished between three scenarios: attacks, natural disasters, and no event. To locate an FDIA, researchers used an ensemble-based ML model technique. Using the IEEE 14 transport system, the model was evaluated. Using multi-layered perceptron, the impact of FDIA on the Artificial-Intelligence based intelligent grids is investigated in detail. The results of this study suggest that even if just 20% of an information is fake, it may reduce the performance of machine learning algorithms by 15%, which might have serious consequences for the smart grids' decision-making process. In the event of "data poisoning," for instance, if there is a problem, as well as the model, failed to foresee it due to inaccurate information, the grid might become unsustainable, which could have disastrous consequences. Using the artificial neural networks & support-vector computer methods, the performance of the model was compared. The fundamental goal of a DDoS attack is to overwhelm the communications servers with a bogus request, preventing it from functioning as a communication server. Approximately 700,000 packages and 49 characteristics were used during model training. The publicly accessible UNSW-NB15 data collection was utilized to develop the model [15].

5.2 Cloud-Based Detection and Mitigation

The usage of cloud computing might improve the smart grid more resistant to DDoS attacks if they occur. DDoS assaults on the smart grid might now be thwarted thanks to the creation of a firewall that is hosted in the cloud. The experiment required the generation of 250-Gbps of data to mimic a distributed denial of service assault. The calculations demonstrate that the grid Open flow Protocol firewall has an exceptionally low amount of delay. An attribute based online/offline cryptography using public-keys system was created for smart grid apps that use the cloud in order to safeguard information access for certified persons. These applications make use of the cloud. Safe HAN (Home Area Network) that relies on the cloud, which may be compromised by brute force, replay, capture, and other attacks. A deep belief network (DBN) consisting of many RBMs as well as a BP neural network was proposed as a means of assessing the security of a smart grid. The following five categories of potential security breaches were investigated: risks associated with policy and organisation, risks associated with general technology, hazards associated with SaaS, PaaS, and IaaS [14, 16].

5.3 A Detection and Countermeasures System Based on Blockchain

Blockchain is a distributed ledger technology, has lately emerged to be one of the most valuable technologies in a variety of businesses due to the security it provides. In a blockchain, each block has an indexing, timestamp, immediately preceding hashing, hash, and data in addition to its own hash. It is believed that the blockchain security can be attributed to the use of hashing. If an attacker attempts to alter the hash information of a block, they are required to also modify the hash information of every block that came before it. This requires a significant amount of processing power when the quantity of blocks in a chain is enormous[16].

Some researchers have suggested a blockchain-based architecture to protect against FDIA by facilitating the transfer of information between self-driving systems and their human operators. The model may be broken down into three distinct layers: the information layer, the recognition layer as well as the block-chain layer. The data layer is charged with information collecting, which is subsequently transmitted to the qualitative data was obtained for community surveillance. The block-chain layer is responsible for the protection of the transaction record and the community detection. In another paper, a secure data forwarding system for smart metres & service providers that is based on blockchain technology was proposed. In another piece of research, the authors presented a secure data transmission system for smart meter and service providers that was built on blockchain technology. The solution eliminates the potential for FDIA on the intelligent meter's side. During this inquiry, smart metres will be used to start each operation, while the supplier will take on the role of the master node. Transaction information is routinely audited and broadcasted, which allows for it to be confirmed and shared throughout the network. Connecting service providers makes use of peer- to-peer (P2P) networking. Verification of the consensus is necessary before a new transaction or block can be added, and the subsequent block can only be added once verification has been completed. Every transaction creates its own key by utilising the SHA-256 hashing technique. The authors of this study proved, with the use of a framework based on blockchain technology, that data may be transmitted between peers inside a Peer-to-peer service provider network. The research led to the development of a decentralised security paradigm, which was considering the cloud infrastructure & smart contracts found in the block-chain ecosystem. This approach is broken up into stages that include things like registrations, scheduling, authentication, and paying [17, 18].

5.4 Hardware-Based Security

The IoT is an integral component of the smart grid network. In a physical attack, the attacker attempts to circumvent the authentication mechanism. Inside channel analysis, an adversary predicts the cryptographic keys by analysing the profile of variables including current, voltage, and frequency. The main goals of hardware Trojans are unauthorised access to confidential information, circuit tampering, and decreased circuit reliability. Nevertheless, with the development of ML, which is extraordinarily adept in predicting the behaviour based on historical performance, the behaviour of PUFs is also 95% predictable precision. Researchers created a customizable tristate PUF to defend PUFs against ML-based assaults that utilise an XOR based technique to obscure the link in between question and response. Due to this uncertainty, the ML model is incapable of identifying any pattern between the question and its answer. This research found that when CTPUF was used as a reference, machine learning techniques such as the supported vector machine, the artificial neural network, and the logistic regression model achieved an accuracy of around 60%. Another study demonstrated the constraints of voltage based scaling authentication, demonstrating that it is vulnerable to abuse by means of machine learning models. In the study, a VOS strategy resistant to ML was proposed by combining previous problems with keys. Following the application of CSoS (challenging self-obfuscation structure), the findings revealed that the accuracy of the ML model was roughly 51.2% [19].

5.5 Detections of Fraudulent Metres Using AI

To perform a "successful" attack, the adversary must advance farther as the specified state variable grows. Therefore, it is natural to develop a defence system that can spot an assault with the fewest compromised metres possible. FDI assaults can be detected in instantaneously by observing the evolution of data over time. The Kullback-Leibler divergence (KLD) gauges the gap between the historical and current rate of measurement change. The measurement sample contains the FDI attack if the Kullback-Leibler divergence result for the instantaneously measurement sample is higher than the threshold which is based on historical measurements. Once the assault has been identified, an AI-based load evaluator is utilised to anticipate the current sample's load using reliable historical samples. This locates the compromised measuring sensors and warns the system administrator. The entire procedure for discovering and detecting FDI assaults [2, 20, 21].

5.6 Assault Utilizing a False Data Injection (FDI)

The online world is beset by cyber-attacks, with the insertion of bogus data regarded as one of the most harmful types of assault. An assault known as a false data injection assault involves the purposeful injection of fictitious information or data into a network. This misleading information makes it difficult to comprehend. The Internet of Things devices that are connected to a network that uses a smart grid are susceptible to attacks on the communication network since a smart grid's is a combination of a 2-way network of communication with a conventional grid. There are three subcategories that fall under the category of false data injection: randomised FDIA, guided constraint FDIA, & targeted unregulated FDIA [22].

The attackers will inject data packets at random during in the random FDIA. This will be done regardless of the patterns of the information that are being utilised. In planned constrained FDIA, the attacker examines the information before the start of the assault and then violates it by including a predefined number of mistakes, such that it gives the impression that the data is accurate even when it is not. In order to compromise particular variables in the uncontrolled FDIA that is the focus of the adversary's attack, the adversary gives each variable a value, nonetheless, the compromising of these because that leads to the penetration of additional variables. The attacker can inject packets of data at random during the random FDIA, independent of the distributions of the values that are really being used. In targeted limited FDIA, the attacker reviews the data in advance of the attack and then corrupts it by adding a predetermined quantity of errors, making it appear as though the data is accurate. The adversary compromises certain variables in the targeted unregulated FDIA by assigning them a value, but the compromise of these variables leads to the compromise of more variables [22].

In the following table, Table 2, we explore the several types of technologies that can be used to detect cyberattacks and that can assist us in the prevention of cyberattacks. Additionally, we demonstrate some of the benefits and drawbacks of using these technologies. This technology will assist us in warding off any potential cyber-attacks.

Table 2. Techniques for Cyber-attack Detection

Techniques for Detecting & Mitigating Cyber-attacks	Advantage	Disadvantages
Machine-Learning- Based	1. High precision 2. Models that are simple to deploy 3. Automation of tasks 4. Constant and adaptable learning	1. To achieve the best results, records in bulk must be used to train the model 2. It takes a long time and is computationally expensive to train a model 3. There is a risk of overfitting or underfitting if hyperparameters are not tweaked

(continued)

Table 2. (*continued*)

Techniques for Detecting & Mitigating Cyber-attacks	Advantage	Disadvantages
Blockchain-Based	1. In general, highly secure 2. Dispersed data storage 3. Smart contracts are unchangeable 4. Every transaction is encrypted	1. There has been few research on blockchain-based cyber security 2. Excessive energy use to power all the nodes 3. Private blockchains lack security
Cloud-Computing-Based	1. Extremely safe 2. Not computationally costly 3. Very low latency	1. High bandwidth availability 2. Relying on a cloud service provider

In the following table, Table 3, we describe the many types of cyberattacks and the techniques that can be used to identify them. These techniques will help us detect the various types of cyberattacks and establish protection strategies. This table also demonstrates the systems being attacked and presents the proposed solutions.

Table 3. Attack types with detection & preventive strategies

Attack Category	Detection/Mitigation Technique Type	Proposed Solution/Research performed	Target of Attack
Flooding attack	Time measurement of flooded packets	Detection Using Bait Messages	Communication network
Denial of service (DoS)	-	The impact of DoS on the AMI network	AMI Network
Denial of service (DoS)	-	The influence of DoS on load frequency control	Load frequency controller
Distributed Denial of service (DDos)	Cloud computing	Cloud computing capability	Communication network
Distributed Denial of service (DDos)	Cloud computing	Cloud Based firewall Available	Communication network
Data manipulation	Cloud computing	Online/offline searchable encryption technique based on attributes	AMI & SCADA network

(*continued*)

Table 3. (*continued*)

Attack Category	Detection/Mitigation Technique Type	Proposed Solution/Research performed	Target of Attack
Data Manipulation, Social Engineering and Session Hijacking	Cloud computing	DBN (Deep Belief Network)	SCADA network
Social Engineering Attack (SEA)	-	The effect of SEA on industrial control system security is measured by calculating the mean time to compromise under assault	Humans at organizations
Social Engineering Attack (SEA)	-	37 intrusion detection and prevention systems were studied, and relevant IDPS were recommended	SCADA & AMI Network
FDIA	Deep – Machine Learning	Conditional Deep Belief Network (CDBN)	SCADA network
FDIA	Machine Learning	Ensemble Based Learning	AMI Network
FDIA	Machine Learning	A multilayer neural network will be used to investigate the influence of FDIA on an artificial intelligence-based smart grid	Communication Network
FDIA	Machine Learning	An artificial neural network model is being developed to anticipate the presence of a cyber assault	Communication network to poison PV generation data
FDIA	Blockchain	Secure message transfer mechanism based on blockchain for smart metres and service providers	Smart Meters

6 Conclusion

This study overviewed the growing problems with cybersecurity in smart grids. Recent research projects have devised many ways to protect the smart grid from cyber-attacks.

The shift from a traditional grid to a smart grid adds a different level of difficulty, and innovation carries inherent dangers. In addition to the enormous problem of developing, managing, and keeping a stable physical structure for the smart grids, it is also quite difficult to construct an architecture for the communications infrastructure and ensure that it is kept up to date. This study carried out a detailed investigation of the communications network of the smart grid. Additionally, it conducted a comprehensive analysis of potential cyberattacks and solutions for mitigating their effects. There is no meaningless assault, even the tiniest hit might have severe repercussions.

We genuinely think that the communication infrastructure and personnel using or maintaining it is susceptible to cyber-attacks and might become the attacker's first option if they do not regulate the assaults securely. Therefore, a remedy was suggested to construct a powerful smart grid infrastructure aimed at safeguarding buyers, the smart grid's communications networks & its employees. This was done because we believe that the communication infrastructure and the young folk who use or organize it are vulnerable to cyber-attacks.

Acknowledgment. We would like to acknowledge Pandit Deendayal Energy University, Gandhinagar, Gujarat, India and Center for Cyber and Information Security, ComExpo Cyber Security Foundation, Ahmedabad, Gujarat, India for their support during this study.

References

1. Ali, S.S., Choi, B.J.: State-of-the-art artificial intelligence techniques for distributed smart grids: a review. Electronics (Switzerland) **9**(6), 1–28 (2020). https://doi.org/10.3390/electronics9061030
2. Bakkar, M., Bogarra, S., Córcoles, F., Aboelhassan, A., Wang, S., Iglesias, J.: Artificial intelligence-based protection for smart grids. Energies **15**(13), 4933 (2022). https://doi.org/10.3390/en15134933
3. Bhattacharya, S., et al.: Incentive mechanisms for smart grid: state of the art, challenges, open issues, future directions. Big Data Cogn. Comput. **6**(2), 47 (2022). https://doi.org/10.3390/bdcc6020047
4. Tufail, S., Parvez, I., Batool, S., Sarwat, A.: A survey on cybersecurity challenges, detection, and mitigation techniques for the smart grid. Energies **14**(18), 5894 (2021). https://doi.org/10.3390/en14185894
5. Khoei, T.T., Ould Slimane, H., Kaabouch, N.: A comprehensive survey on the cyber-security of smart grids: cyber-attacks, detection, countermeasure techniques, and future directions
6. A cyber attack in Saudi Arabia failed to cause carnage, but the next attempt could be deadly | The Independent | The Independent. https://www.independent.co.uk/news/long_reads/cyber-warfare-saudi-arabia-petrochemical-security-america-a8258636.html. Accessed 27 Aug 2022
7. Ukraine power cut 'was cyber-attack' – BBC News. https://www.bbc.com/news/technology-38573074. Accessed 27 Aug 2022

8. Zhang-Kennedy, L., Rocheleau, J., Mohamed, R., Baig, K., Chiasson, S., Assal, H.: {Black-IoT}: {IoT} Botnet of High Wattage Devices Can Disrupt the Power Grid (2018). Accessed 29 Mar 2023
9. Kenney, M.: Cyber-terrorism in a post-stuxnet world. Orbis **59**(1), 111–128 (2015). https://doi.org/10.1016/j.orbis.2014.11.009
10. Final Report on the Blackout in the United States and Canada: Causes and Recommendations (2003)
11. Mohammadi, F.: Emerging challenges in smart grid cybersecurity enhancement: a review. Energies **14**(5), 1380 (2021). https://doi.org/10.3390/en14051380
12. Guidelines for smart grid cybersecurity. Gaithersburg, MD (2014). https://doi.org/10.6028/NIST.IR.7628r1
13. Du, D., Li, X., Li, W., Chen, R., Fei, M., Wu, L.: ADMM-based distributed state estimation of smart grid under data deception and denial of service attacks. IEEE Trans. Syst. Man Cybern. Syst. **49**(8), 1698–1711 (2019). https://doi.org/10.1109/TSMC.2019.2896292
14. Gunduz, M.Z., Das, R.: Cyber-security on smart grid: threats and potential solutions. Comput. Netw. **169**, 107094 (2020). https://doi.org/10.1016/j.comnet.2019.107094
15. Kumari, A., et al.: AI-empowered attack detection and prevention scheme for smart grid system. Mathematics **10**(16), 2852 (2022). https://doi.org/10.3390/math10162852
16. Zhang, H., Wang, J., Ding, Y.: Blockchain-based decentralized and secure keyless signature scheme for smart grid. Energy **180**, 955–967 (2019). https://doi.org/10.1016/j.energy.2019.05.127
17. Agung, A.A.G., Handayani, R.: Blockchain for smart grid. J. King Saud Univ. – Comput. Inform. Sci. **34**(3), 666–675 (2022). https://doi.org/10.1016/j.jksuci.2020.01.002
18. Mehta, P., Gupta, R., Tanwar, S.: Blockchain envisioned UAV networks: challenges, solutions, and comparisons. Comput. Commun. **151**, 518–538 (2020). https://doi.org/10.1016/j.comcom.2020.01.023
19. Moslemi, R., Mesbahi, A., Velni, J.M.: A fast, decentralized covariance selection-based approach to detect cyber attacks in smart grids. IEEE Trans. Smart Grid **9**(5), 4930–4941 (2018). https://doi.org/10.1109/TSG.2017.2675960
20. Bose, B.K.: Artificial intelligence techniques in smart grid and renewable energy systems – some example applications. Proc. IEEE **105**(11), 2262–2273 (2017). https://doi.org/10.1109/JPROC.2017.2756596
21. Khanna, K., Panigrahi, B.K., Joshi, A.: AI-based approach to identify compromised meters in data integrity attacks on smart grid. IET Gener. Transm. Distrib. **12**(5), 1052–1066 (2018). https://doi.org/10.1049/iet-gtd.2017.0455
22. Tufail, S., Batool, S., Sarwat, A.I.: False data injection impact analysis in AI-based smart grid. In: Conference Proceedings – IEEE SOUTHEASTCON, Institute of Electrical and Electronics Engineers Inc. (2021). https://doi.org/10.1109/SoutheastCon45413.2021.9401940

Sentiment Classification of Diabetes-Related Tweets Using Transformer-Based Deep Learning Approach

V. S. Anoop(⊠) ⓘ

School of Digital Sciences, Kerala University of Digital Sciences,
Innovation and Technology, Thiruvananthapuram, India
anoop.vs@duk.ac.in

Abstract. It is estimated that 1 out of 77 people in India are formally diagnosed with diabetes, which is alarming. Nowadays, people use social media platforms such as Facebook and Twitter to express their beliefs, feelings, and concerns on any topic, including healthcare. It is interesting and highly useful to analyze this humongous data to unearth latent patterns for further analysis and decision-making. This study aimed at (1) analyzing the sentiment of the diabetes-related tweets from India and (2) unearthing latent themes of discussion about diabetes in India from Twitter. We collected tweets from India using the keywords - *diabetes*, *diabetes mellitus*, and *hyperglycemia*. We identified the sentiments of retrieved tweets using a deep learning sentiment classifier followed by a topic modeling approach to analyze the major themes of discussion. The results show that the majority of the sentiments were positive, but still, there are concerns regarding diabetes risk and the alarming rate of diabetes in India. Our topic modeling results showed some interesting patterns from the discussions, such as *diabetes advisory*, *diabetes risk*, *diabetes warning*, and *diabetes control*.

Keywords: Diabetes · Sentiment analysis · Text mining · Topic modeling · Natural language processing · Deep learning · Infodemiology · Public health

1 Introduction

Recent statistics published by the World Health Organization estimated that approximately 422 million people across the globe with diabetes[1]. Most of these people live in low or middle-income countries having limited access to effective healthcare resources. It is estimated that annually around 1.5 million deaths are reported globally due to diabetes, which may grow exponentially in the coming years. The major cause of blindness, kidney failure, heart attacks, and stroke

---

[1] https://www.who.int/health-topics/diabetes.

© The Author(s), under exclusive license to Springer Nature Switzerland AG 2023
M. Singh et al. (Eds.): ICACDS 2023, CCIS 1848, pp. 203–214, 2023.
https://doi.org/10.1007/978-3-031-37940-6_17

is diabetes. India, the second most diabetes-affected country in the world, has approximately 77 million people with diabetes diagnosed, making 1 in every 77 individuals. This is an alarming rate that is expected to be growing exponentially over the coming years. It is the need of the hour to devise and implement better surveillance and remedial interventions [28,29]. It is an undeniable fact that every diabetes patient should get affordable access to treatment, including the availability of insulin, to help reduce complications and also to control the death rate [30].

There are several studies reported that use social media for public health surveillance [9,16,33]. The pandemic, such as COVID-19, stressed the need to analyze user-generated health content for improving public health initiatives and interventions [1–5,8,13,15,17]. People often post their opinions, concerns, and views on many healthcare topics on social media platforms like Twitter and Facebook [22,31]. They share information such as symptoms, medications, and adverse drug reactions on such platforms, and analyzing such content could be useful for prioritizing public health strategies for healthcare stakeholders [20, 24,27]. A significant amount of research studies on the same directions have been reported using data from Twitter[2], which is a micro-blogging and social networking service [7,26,34]. The present study also assessed Indian tweets to (1) analyze people's views and sentiments toward diabetes and (2) leverage major themes of these discussions.

To identify public opinion toward diabetes in India, such as how they are responding to new medication for diabetes, we first need to assign polarities of negative, positive, and neutral to the retrieved tweets. This study uses existing sentiment analysis models for this purpose and then trains a diabetes tweets classifier using Transformers [32], a deep learning technique that uses an"attention" mechanism. We use BERT [10] for text encoding. The present study is one of the early attempts to analyze Indian tweets to find the sentiments and themes of discussion concerning diabetes. The major contributions of this work may be summarized as below:

1. Discusses recent prominent approaches that use social media, such as Twitter, for public health surveillance.
2. Analyze Indian tweets related to diabetes to discover people's sentiments and opinions.
3. Models topics from the tweets to leverage latent themes concerning diabetes in India

Organization of this Paper: Section 2 discusses some of the recent and prominent approaches that use social media for health surveillance with a focus on diabetes; Sect. 3 presents the proposed methodology, and Sect. 4 discusses the results. In Sect. 5, the authors conclude the paper.

---

[2] https://twitter.com/home.

2   Related Studies

Twitter has been extensively used for many public sentiment analysis and health surveillance studies [18,35]. A thematic analysis of how leading medical centers are using Twitter as a platform was conducted by taking a case study of medical centers [21]. This study reported that out of all the themes leveraged, the diet and cancer themes were the most significant and consistent, but diabetes was the least used topic in the tweets. This study thus recommended amendments in the content strategies for healthcare institutions. An approach for diabetes tweets classification using a capsule network and gravitational search algorithm was proposed by Diviya et al. [11]. The authors have performed sentiment classification of tweets related to health conditions, producing comparable results to other state-of-the-art approaches.

A social media analytics and reachability analysis approach was proposed by Karmegam et al., in which the sentiment analysis was done by a lexicon-based approach. The second phase of the work was a reachability study that was conducted based on the number of tweets, re-tweets, and favorite counts. The authors have chosen a total of 1840 tweets for this study and the insights from the study were that the major proportion of tweets are positive [19]. A method for sentiment analysis on diabetic diagnosis healthcare using machine learning technique has been reported in the literature recently [25]. Their approach used support vector machines for classifying user sentiments using medical opinions on the diagnosis. The authors have also used Gaussian distribution to find the probability of getting diabetes based on patient conditions.

A study that analyzed data from Twitter for sentiment analysis of diabetes was reported by Gabarron et al. [14]. They have used larger Twitter data from the Norwegian Diabetes Association for sentiment analysis. An interesting study that analyzed social media usage in the diabetes communities was published [12]. The authors have extracted and curated a dataset of 29.6 billion tweets. Their analyses leverage interesting insights, such as the number of tweets that will be increased in November due to World Diabetes Day. A gap analysis study on diabetes was conducted [23]. Their study suggested that using a diabetic online community is beneficial and will be an active area of research.

There are only a few studies reported from India using Twitter data to analyze public sentiment and opinions on diabetes-related tweets. These studies use limited data and shallow-learning approaches to classify tweets into positive, negative, and neutral polarities. This proposed work uses a larger dataset to classify the tweets. Also, the thematic analysis of the tweets on diabetes is carried out to unearth the latent themes prevailing in these discussions. The findings from this study may be beneficial for health policymakers and other healthcare stakeholders to get informative insights useful for data-driven decision-making.

3   Proposed Methodology

This section details the methodology used in this study for analyzing the public sentiment and themes of discussion on diabetes-related tweets. Figure 1 shows the overall workflow of the proposed approach.

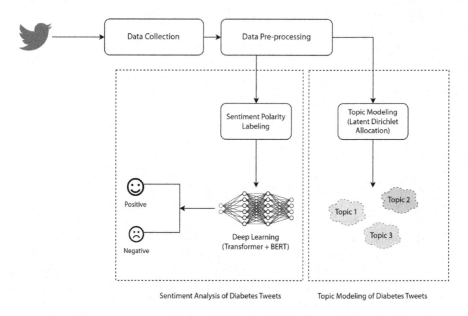

Fig. 1. Workflow of the methodology used for sentiment analysis and topic modeling

3.1   Data Acquisition and Pre-processing

Python programming language library called "snscrape" was used to collect diabetes-related tweets, and we have used *diabetes*, *diabetes mellitus*, and *hyperglycemia* as search terms. As this work only focuses on analyzing the tweets from India, we have made constraints by mentioning the above search keywords with "#India". The tweets were collected from 2012 to 2022, and a plot showing the tweet distribution for these years is shown in Fig. 2. Twitter data contains noises such as special characters, URLs, and emojis. We have done an extensive pre-processing of the tweets using the regular expression (re) library in Python, and for removing the emojis, *demoji* library available at https://pypi.org/project/demoji/ has been used.

3.2   Sentiment Classification

First, we have split the dataset into train, test, and validation to fine-tune and train the BERT pre-trained model and to classify the tweets using Transformer

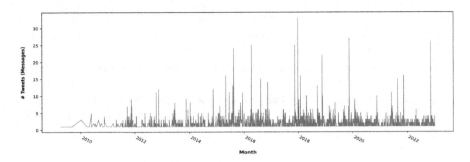

Fig. 2. Coverage of tweets used for this study

architecture. After splitting the dataset, we used *vader lexicon* from Natural Language Toolkit (NLTK)[3] to assign polarity values (1 for positive and 0 for negative). Some of the representative tweets labeled according to the polarity values are shown in Table 1. All the collected tweets are labeled using *vaderSentiment* library available at https://github.com/cjhutto/vaderSentiment and made the experiment-ready copy of the dataset. This work uses Transformer architecture with BERT, a pre-trained model to build the classifier. We have used *torch* implementation of the Transformer with *bert-large-uncased* as the pre-trained model for classification. For this experiment, we have used the number of epochs = 10, training batch size = 16, evaluation batch size = 64, warm-up steps = 500, and weight decay = 0.01.

Table 1. A snapshot of the tweets with sentiment polarity

Sl. No.	Tweet	Sentiment polarity
1	Medical Nutrition Therapy can bring down your HBA1C by 2%	1 (Positive)
2	Indians fare better than almost all other subcontinental countries, when it comes to eye problems arising due to diabetes	1 (Positive)
3	Records estimate 77 million individuals had diabetes which is expected to rise to over 134 million by the year 2045	0 (Negative)
4	Sad to see india in this list! We're country of yoga and aadiyogi..why we not wake up!	0 (Negative)
5	Not only is #India the #diabetes capital of the world, it is witnessing an alarming increase amongst the young	0 (Negative)

3.3 Topic Modeling

Topic models are text-understanding algorithms that generate hidden themes from a large collection of unstructured text documents. Many topic modeling algorithms are available in machine learning and differ in their assumptions

---

[3] https://www.nltk.org/.

about leveraging the "topics". LDA [6] is one of the most widely employed algorithms in the text mining and machine learning domains. This work also employs the LDA algorithm to analyze diabetes-related tweets to find the major topics of discussion. We use the Gensim[4] library and the implementation of the LDA algorithm available at https://pypi.org/project/gensim/. We removed the user mentions and other noises from tweets using the "re" library of Python and stopwords using NLTK's stopword list. The number of topics has been set as ten and the number of iterations as 300, and these parameters are finalized using a trial and error approach.

4   Results and Discussion

This section presents the results obtained on implementing the methodology discussed in Sect. 3. The coverage of positive and negative tweets considered in the train and test set of the dataset is shown in Fig. 3. The training and validation loss for the sentiment classification algorithm for different steps from 500 to 2500 is shown in Fig. 4. The precision, recall, and accuracy for the trained classifier model are shown in Fig. 5. and the precision, recall, and f-measure comparison is shown in Fig. 6. In this work, we have only used *bert-large-uncased*. Our primary intention was not to compare the classification accuracy for different pre-trained models. Various BERT models are trained on generic data such as Wikipedia articles and specific data from social media such as Twitter. A comparative analysis of the classification accuracy values for diabetes tweets classification may be attempted in future work. A word cloud showing major keywords found in tweets related to diabetes is shown in Fig. 7. We can observe words such as *diabetes*, *india*, *health*, *obesity*, and *insulin* are prominent in the tweets.

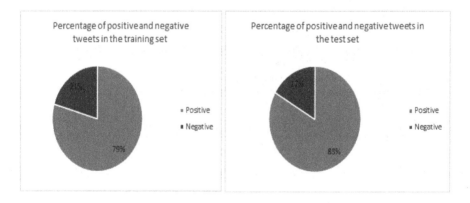

Fig. 3. Percentage of positive and negative comments in the train and test set

The topic modeling analysis generated some interesting insights on the major themes of discussion in diabetes-related tweets. We have taken topics, which are

---

[4] https://pypi.org/project/gensim/.

collections of words, and manually generated topic labels. These topic labels are assigned to topics for better interpretation. The result of the topic words to label assignment is given in Table 2. We could leverage topic labels such as *diabetes advisory, diabetes risk, diabetes warning, diabetes control, diabetes cases*, and *medical tourism* to map to topic words. The topic-level analysis of the dataset shows that people often discuss how diabetes can be prevented by adopting a better lifestyle and exercises such as yoga. We also found that organizations such as DiabetesIndia (@diabetesindia22), Diabetes India Youth in Action (@DIYAbetesINDIA), and National Diabetes Obesity and Cholesterol Foundation (@NDOCfoundation) regularly update Twitter using their handles on diabetes risks and preventive measures (Fig. 8).

Fig. 4. The training and validation loss for the sentiment classifier algorithm

Table 2. Major topics and corresponding labels

Sl. No	Topic words	Topic Label
1	diabetes, health, diet, plan, diabetic, exercise, healthy, yoga	Diabetes advisory
2	disease, million, people, risk, study, health, obesity, ncdc	Diabetes risk
3	health, medical, care, tourism, travel, medicaltourism, india	Medical tourism
4	obesity, hypertension, covid, death, people, diabetes, india	Diabetes warning
5	health, lifestyle, news, thanks, insulin, latest, obesity, free	Diabetes control
6	diabetes, india, world, people, capital, million, china, diabetic	Diabetic cases

Our topic modeling analysis further strengthens these findings using the topic labels *diabetes advisory, diabetes warning*, and *diabetes control*. The other topics (for instance, *medical tourism*) may be related to promotional posts from other organizations, but we may need detailed analysis to confirm the same. For this study, we did not differentiate tweets based on their characteristics, such as the user or promotional tweets, which may be attempted in future analysis.

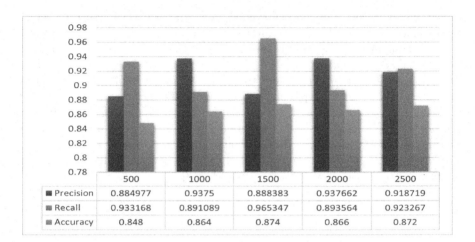

Fig. 5. Precision, recall, and accuracy comparison for the sentiment classifier for steps from 500 to 2500

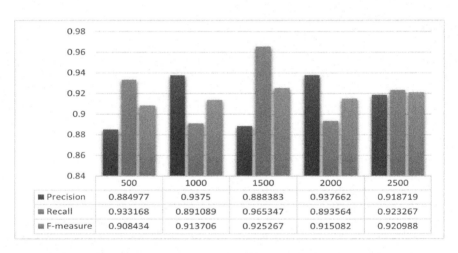

Fig. 6. Precision, recall, and f-measure comparison for the sentiment classifier for steps from 500 to 2500

Fig. 7. A word cloud showing major keywords from all the tweets

Fig. 8. World clouds for topic #0, topic #2, topic #3, and topic #7 (representative)

5   Conclusion

In this study, we analyzed 5000 tweets from India to analyze the sentiments and topics of discussion related to diabetes. We labeled tweets using the Vader Sentiment analyzer, and the Transformer model with BERT was used for building the classifier. The best-recorded f-measure was 92.52% which is comparable with state-of-the-art deep learning models. We also implemented topic modeling, a natural language processing technique to leverage discussion topics from Twitter on diabetes. Our modeling revealed interesting patterns and themes, such as *diabetes risk*, *diabetes advisory*, *diabetes warning*, and *diabetes control*. These findings may be of immense use to public health practitioners, community health workers, and health policymakers to devise impactful, data-driven interventions to improve public health.

Limitations of this Study: First, we used a dataset with only 5000 tweets for analysis as our search keywords returned limited tweets. This may be due to the fewer users using Twitter as a medium in India compared to other geographical locations around the globe. Also, a comprehensive comparison of different deep learning approaches and pre-trained models may be another interesting research dimension that can be attempted to enhance this work.

References

1. Anoop, V.S., Asharaf, S.: Aspect-oriented sentiment analysis: a topic modeling-powered approach. J. Intel. Syst. **29**(1), 1166–1178 (2018)
2. Anoop, V.S.: Text mining and natural language processing for health informatics: recent trends and the way forward. Machine Learning and Data Analytics for Predicting, Managing, and Monitoring Disease, pp. 111–121 (2021)
3. Anoop, V., Sreelakshmi, S.: Public discourse and sentiment during Mpox outbreak: an analysis using natural language processing. Public Health **218**, 114–120 (2023)
4. Anoop, V.S., Thekkiniath, J., Govindarajan, U.H.: We chased COVID-19; did we forget measles? - public discourse and sentiment analysis on spiking measles cases using natural language processing. In: Morusupalli, R., Dandibhotla, T.S., Atluri, V.V., Windridge, D., Lingras, P., Komati, V.R. (eds.) MIWAI 2023. LNCS, vol. 14078, pp. 147–158. Springer, Cham (2023). https://doi.org/10.1007/978-3-031-36402-0_13
5. Banerjee, D., Meena, K.: Covid-19 as an "infodemic" in public health: critical role of the social media. Frontiers in Public Health **9**, 610623 (2021)
6. Blei, D.M., Ng, A.Y., Jordan, M.I.: Latent Dirichlet allocation. J. Mach. Learn. Res. **3**, 993–1022 (2003)
7. Cano-Marin, E., Mora-Cantallops, M., Sánchez-Alonso, S.: Twitter as a predictive system: a systematic literature review. J. Bus. Res. **157**, 113561 (2023)
8. Chandran, N.V., Anoop, V.S., Asharaf, S.: A topic modeling-guided framework for aspect-oriented sentiment analysis on social media. In: Handbook of Research on Opinion Mining and Text Analytics on Literary Works and Social Media, pp. 132–146. IGI Global (2022)
9. Chaurasia, R., Ghose, U.: Social media and the COVID-19 pandemic: boons and banes. Digital Innovation Pandemics, 183–223 (2023)
10. Devlin, J., Chang, M.W., Lee, K., Toutanova, K.: Bert: pre-training of deep bidirectional transformers for language understanding. arXiv preprint arXiv:1810.04805 (2018)
11. Diviya Prabha, V., Rathipriya, R.: Diabetes twitter classification using hybrid GSA. In: Nayak, J., Das, A.K., Naik, B., Meher, S.K., Brahnam, S. (eds.) Nature-Inspired Optimization Methodologies in Biomedical and Healthcare. Intelligent Systems Reference Library, vol. 233, pp. 195–219. Springer, Cham (2023). https://doi.org/10.1007/978-3-031-17544-2_9
12. Fagherazzi, G., Ravaud, P.: Digital diabetes: perspectives for diabetes prevention, management and research. Diabetes Metab. **45**(4), 322–329 (2019)

13. Fung, I.C.H., Tse, Z.T.H., Fu, K.W.: The use of social media in public health surveillance. West. Pac. Surveill. Response J: WPSAR **6**(2), 3 (2015)

14. Gabarron, E., Larbi, D., Dorronzoro, E., Hasvold, P.E., Wynn, R., Årsand, E.: Factors engaging users of diabetes social media channels on Facebook, Twitter, and Instagram: observational study. J. Med. Internet Res. **22**(9), e21204 (2020)

15. Jickson, S., Anoop, V.S., Asharaf, S.: Machine learning approaches for detecting signs of depression from social media. In: Anwar, S., Ullah, A., Rocha, Á., Sousa, M.J. (eds.) Proceedings of International Conference on Information Technology and Applications. LNNS, vol. 614, pp. 201–214. Springer, Singapore (2023). https:// doi.org/10.1007/978-981-19-9331-2_17

16. Jing, F., Li, Z., Qiao, S., Zhang, J., Olatosi, B., Li, X.: Using geospatial social media data for infectious disease studies: a systematic review. Int. J. Digital Earth **16**(1), 130–157 (2023)

17. John, R., Anoop, V.S., Asharaf, S.: Health mention classification from user-generated reviews using machine learning techniques. In: Anwar, S., Ullah, A., Rocha, Á., Sousa, M.J. (eds.) Proceedings of International Conference on Information Technology and Applications. LNNS, vol. 614, pp. 175–188. Springer, Singapore (2023). https://doi.org/10.1007/978-981-19-9331-2_15

18. Jordan, S.E., Hovet, S.E., Fung, I.C.H., Liang, H., Fu, K.W., Tse, Z.T.H.: Using twitter for public health surveillance from monitoring and prediction to public response. Data **4**(1), 6 (2018)

19. Karmegam, D., Mappillairaju, B.: Social media analytics and reachability evaluation-# diabetes. Diabetes Metab. Syndr. Clin. Res. Rev. **16**(1), 102359 (2022)

20. Khan, P.I., Razzak, I., Dengel, A., Ahmed, S.: Performance comparison of transformer-based models on twitter health mention classification. IEEE Trans. Comput. Soc. Syst. **10**, 1140–1149 (2022)

21. Kordzadeh, N.: Health promotion via twitter: a case study of three medical centers in the USA. Health Promot. Int. **37**(2), daab126 (2022)

22. Lekshmi, S., Anoop, V.: Sentiment analysis on COVID-19 news videos using machine learning techniques. In: Basu, S., Kole, D.K., Maji, A.K., Plewczynski, D., Bhattacharjee, D. (eds.) Proceedings of International Conference on Frontiers in Computing and Systems: COMSYS 2021, vol. 404, pp. 551–560. Springer, Singapore (2022). https://doi.org/10.1007/978-981-19-0105-8_54

23. Litchman, M.L.: State of the science: a scoping review and gap analysis of diabetes online communities. J. Diabetes Sci. Technol. **13**(3), 466–492 (2019)

24. Martínez-Martínez, F., Roldán-Álvarez, D., Martín, E., Hoppe, H.U.: An analytics approach to health and healthcare in citizen science communications on twitter. Digital Health **9**, 20552076221145348 (2023)

25. Nagaraj, P., Deepalakshmi, P., Muneeswaran, V., Muthamil Sudar, K.: Sentiment analysis on diabetes diagnosis health care using machine learning technique. In: Saraswat, M., Sharma, H., Balachandran, K., Kim, J.H., Bansal, J.C. (eds.) Congress on Intelligent Systems. LNDECT, vol. 114, pp. 491–502. Springer, Singapore (2022). https://doi.org/10.1007/978-981-16-9416-5_35

26. Naik, D., Ramesh, D., Gorojanam, N.B.: Enhanced link prediction using sentiment attribute and community detection. J. Ambient Intell. Humanized Comput. **14**, 1–18 (2022)

27. Naseem, U., Kim, J., Khushi, M., Dunn, A.G.: Identification of disease or symptom terms in reddit to improve health mention classification. In: Proceedings of the ACM Web Conference 2022, pp. 2573–2581 (2022)

28. Rajput, D.S., Basha, S.M., Xin, Q., Gadekallu, T.R., Kaluri, R., Lakshmanna, K., Maddikunta, P.K.R.: Providing diagnosis on diabetes using cloud computing environment to the people living in rural areas of India. J. Ambient. Intell. Humaniz. Comput. **13**(5), 2829–2840 (2022)

29. Sachdev, M., Misra, A.: Heterogeneity of dietary practices in India: current status and implications for the prevention and control of type 2 diabetes. Eur. J. Clin. Nutr. **77**, 1–11 (2022)

30. Sathyanath, S., et al.: An economic evaluation of diabetes mellitus in India: a systematic review. Diabetes Metab. Syndr. Clin. Res. Rev. **16**, 102641 (2022)

31. Varghese, M., Anoop, V.: Deep learning-based sentiment analysis on COVID-19 news videos. In: Ullah, A., Anwar, S., Rocha, Á., Gill, S. (eds.) Proceedings of International Conference on Information Technology and Applications. LNNS, vol. 350, pp. 229–238. Springer, Singapore (2022). https://doi.org/10.1007/978-981-16-7618-5_20

32. Vaswani, A., Shazeer, N., Parmar, N., Uszkoreit, J., Jones, L., Gomez, A.N., Kaiser, Ł., Polosukhin, I.: Attention is all you need. In: Advances in Neural Information Processing Systems, vol. 30 (2017)

33. Velasco, E., Agheneza, T., Denecke, K., Kirchner, G., Eckmanns, T.: Social media and internet-based data in global systems for public health surveillance: a systematic review. Milbank Q. **92**(1), 7–33 (2014)

34. Verma, R., Chhabra, A., Gupta, A.: A statistical analysis of tweets on COVID-19 vaccine hesitancy utilizing opinion mining: an Indian perspective. Soc. Netw. Anal. Min. **13**(1), 1–12 (2023)

35. Yepes, A.J., MacKinlay, A., Han, B.: Investigating public health surveillance using twitter. In: Proceedings of BioNLP 15, pp. 164–170 (2015)

Assessment and Prediction of a Cyclonic Event: A Deep Learning Model

Susmita Biswas[1]($^{(\boxtimes)}$) (iD) and Mourani Sinha[2] (iD)

[1] Department of Cyber Science and Technology, Brainware University, West Bengal, Kolkata 700125, India
dsb.cst@brainwareuniversity.ac.in

[2] Department of Mathematics, Techno India University, West Bengal, Kolkata 700091, India

Abstract. Deep learning models based on long short-term memory (LSTM) and bi-directional short-term memory (BLSTM) networks are applied in this present study to predict ocean wave heights for different lead times in the Arabian Sea, during tropical cyclones (TC). The BLSTM model can serve as a computationally effective prediction technique for significant wave height data with higher accuracy in combination with the conventional numerical wave models. In the Arabian Sea, two grids namely G1 (72°E, 9°N) and G2 (71°E, 15.5°N) have been chosen through which OCKHI (29 November-06 December 2017) and NISARGA (01–04 June 2020) cyclones passed and the model computed wave heights are generated. The wave heights are subjected to prediction during extreme conditions using the deep learning models and the accuracy is compared in terms of root mean square error. For grid G1 a time-series is considered from 1979–2017 and for grid G2 a shorter time-series from 2010–2020. The LSTM and BLSTM models are trained with 80% data and tested with various hidden units and epoch values to achieve a lesser error. For the training set, the longer time-series including or not including the TC conditions gave lesser error compared to the shorter time-series. For the 20% testing set, the shorter time-series including the cyclonic data gave more error. Further predictions with different delays or lead times are carried out resulting in an increase in error. This is related to the fact that the power of the model diminutes as the forecast horizon increases.

Keywords: Deep learning · LSTM · BLSTM

1 Introduction

The tropical wave of the Arabian Sea (AS) has strong wind speed during the summer monsoon season, resulting in high wave heights. In recent years, the frequency and severity of tropical storms have increased due to the warmth and coldness of the seas, and the number of storms has also increased. The distribution of the tropical cyclone wave spectrum is complex and difficult to predict. A third-generation wave spectral ocean wave model needs better parameterizations to simulate the asymmetry of the ocean fields during a cyclonic event. In this study, deep learning architecture based on LSTM and BLSTM networks are proposed in this work with various lead times to predict ocean

M. Singh et al. (Eds.): ICACDS 2023, CCIS 1848, pp. 215–227, 2023.
https://doi.org/10.1007/978-3-031-37940-6_18

wave heights during cyclonic periods. Biswas and Sinha (2022) have focused on shallow learning model and deep learning model for estimation of sea surface temperature in the two different grid point of Bay of Bengal and Arabian sea region. Recursive neural networks with LSTM (which will briefly refer to as LSTM) have appeared as effective and scalable models for various learning problems connected to sequential data (Greff et al. 2017). LSTM network architecture perfectly captures temporary features and long-term dependencies from historical data (Patel et al. 2018). Data intelligent algorithms designed to predict SWH of coastal waves in a relatively short period of time in coastal areas can generate important information on increasing renewable energy production (Ali and Prasad 2019). Realistic prediction of ocean wave during a cyclone is generally carried out using third-generation spectral ocean wave models like WAVEWATCH III (Tolman 1991) which includes propagation generation and dissipation of wind waves for varied depth. Cyclone is one of the most devastating natural calamities in the earth (Singh et al., 2005). This study involves time-series data of wave height parameter during OCKHI cyclone and NISARGA cyclone in the AS. A similar study is conducted by Biswas and Sinha (2020) to predict Indian Ocean wind speed data with reasonable accuracy. Bethel et al (2022) reproduced accurate observed hurricane induced wave heights (nowcasts) in the Caribbean Sea rather than predicted ones (forecasts) using LSTM neural networks. Adytia et al (2022) introduced wave heights with accuracy in the Java Sea, Indonesia using a combination of numerical wave models and deep learning models. SWH of southwestern Atlantic Ocean was estimated using deep learning LSTM model by Minuzzi and Farina (2023). Tao Song et al (2023) indicates for developing the latest deep learning algorithm to estimate reliable SWH. Domala and Lee (2022) have estimated the wave parameters with upgraded machine learning and deep learning techniques. Afzal et al (2023) developed a machine learning (ML) technique with generalized extreme value (GEV) theory and applied it to uttermost wave analysis in terms of estimating significant wave heights. For that reason, integrating observational data is essential to prepare accurate predictions of significant wave height. In this study, the LSTM and BLSTM models are used for estimation of cyclonic ocean wave heights for different lead times and the accuracy of the predictions are evaluated in view of root mean square error analysis.

2   Data and Methodology

In this work, the ERA5 hourly data estimates are considered, a fifth generation ECMWF (European Centre for Medium-Range Weather Forecasts) atmospheric reanalysis data, at 50 km spatial resolution. Significant wave height data for the two grid points G1 (72°E and 9°N) and G2 (71°E, 15.5°N) in the AS region are downloaded (https://cds.climate. copernicus.eu/cdsapp#!/dataset/reanalysis-era5-single-levels?tab=form) and processed through which the OCKHI and the NISARGA cyclones passed. For the OCKHI cyclone grid 06, 12, 18- and 24-h data are taken into account from 1979 to 2017 and for the NISARGA grid from 2010 and 2020. The OCKHI cyclone time-series data is referred to as the long time-series data while the NISARGA cyclone time-series data is referred to as the short time-series data. Four separate experiments are conducted using the above time-series data. First the cyclonic data is not included and then included for the above mentioned two cyclones. LSTM and BLSTM models using the deep learning approach are trained to estimate the wave height parameter in the two grid points of the AS region. In this regard, 06 hourly wave height data are chosen as input variables for the development of the models. Two different types of deep learning models (LSTM and BLSTM) have been trained with the available wave measurements. 80% of the data set is applied to train the models, and 20% remaining data as testing data. The performance of the two different models during the training and testing period are evaluated. For predicting the cyclonic and non-cyclonic wave heights, BLSTM model produced more accurate results than the conventional LSTM model. For each model, four different types of time series data set are used (data with and without OCKHI period and data with and without NISARGA period). Next the BLSTM model having better accuracy is subjected to prediction for different lead times 1, 2, 3, 4, 8, 12 corresponding to 06, 12, 18, 24, 48 and 72 h (H) respectively. The mean square error (MSE) and root mean square error (RMSE) are used to measure and evaluate the model performances.

3 Results and Discussion

The next instantaneous wave height forecast based on current and previous values is estimated using deep learning models. The present research is an explorative study to predict the significant wave-height data typically dominated by deep learning models in the AS region. LSTM and BLSTM models are introduced for predicting wave height data for cyclonic and non-cyclonic time-series data for two different grids, G1 (72°E and 9°N) and G2 (71°E and 15.5°N), through which OCKHI and NISARGA cyclones passed respectively. Time-series data at G1 grid from 1979 to 2017 (39 years) and G2 grid from 2010 to 2020 (11 years) are utilized for both the deep learning models. The data sets consist of 55520 (1979–2016), 56980 (1979–2017), 14608 (2010–2019) and 15336 (2010–2020) points for 38 years, 39 years, 10 years, 11 years respectively. The 38- and 10-years data represent the phases without the given two cyclones and the other two (39 and 11 years) corresponds to the periods including the given cyclones. Figure 1a and b depict the time-series plots for G1 and G2 grids of the AS region. The peak of more than 6 m during the OCKHI cyclone is well depicted in the figure. For the given grid G1 (72°E and 9°N) the significant wave height was 6.5 m on 02 December 2017, 06 h. For the G2 grid the significant wave height was 4.2 m on 02 June 2020, 12 h. Performances of LSTM and BLSTM models for various inputs have been analyzed and the results are discussed. BLSTM model is also used to forecast significant wave height data for various lead times 1, 2, 3, 4, 8, 12 corresponding to 06, 12, 18, 24, 48 and 72 h. Separate experiments are conducted for the given time-series using LSTM and BLSTM models. The time series data are divided into (80%) training and (20%) testing set. Different epochs and hidden units are tested in the present study. Considering LSTM network model, 250 hidden nodes and 0.005 initial run rate are chosen for the respective time-series data. For this state-of-the-art network model, the optimal number of epochs are found to be 1000 for the periods 1979–2016 and 1979–2017 using the LSTM model. For the periods 2010–2019 and 2010–2020 it is found to be 2500. These generated results having higher accuracies. For the BLSTM models by varying the number of two hidden units and epochs, the optimal values are searched to be 250, 250 for the hidden units and 500 and 1000 for the epochs respectively for the long length and short length time-series data. Several experimental setups are prepared. First training and testing are accomplished to predict a long growth curve starting from 1979 to 2020. Training and testing are executed various times to abate the bias due to random initialization. Meanwhile quantitative evaluation and comparison is done using RMSE. It computes the differences or residuals between observed and predicted values. Results of time-series prediction of various wave heights are compared in terms of graphs visualization and RMSE.

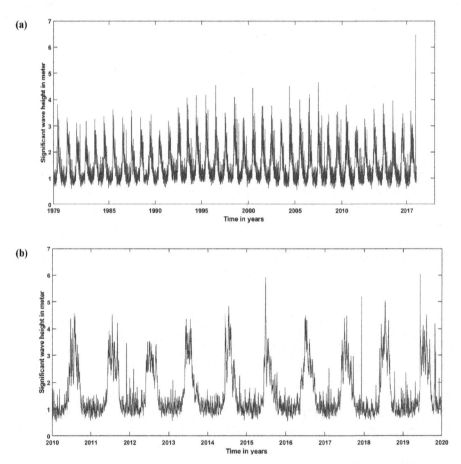

Fig. 1. (a) ERA5 significant wave height at G1 grid (72°E, 9°N) from 1979–2017. (b) ERA5 significant wave height of G2 grid (71°E, 15.5°N) from 2010–2020

Figure 2 shows comparison between training forecasted and training observed result using the LSTM model for the G1 grid. The RMSE value increased from 0.07 to 0.3 m on including the cyclonic data.

Figure 3 gives similar comparisons for the G2 grid using the LSTM model. For the short time-series the error increased from 0.12 to 0.26 m on including the cyclonic data.

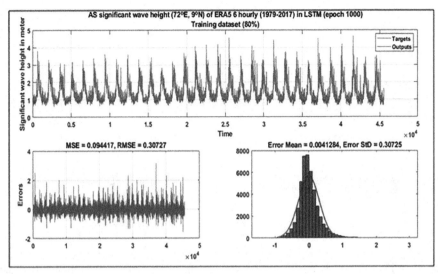

Fig. 2. LSTM network used for significant wave height at G1 (72°E, 9°N) from 1979–2017 (training dataset)

Figure 4 shows the comparisons for the training set for the G1 grid using the BLSTM model and Fig. 5 similarly for the G2 grid. The errors reduced significantly for the cyclonic data sets with the BLSTM model. Next the testing data sets are considered with the LSTM and BLSTM models. Figure 6 and 7 gives the comparison between testing forecasted and testing observed data using the LSTM model for the G1 and G2 grids respectively. For both the grids with the testing data set the error increased from 0.1 to 0.3 m on including the cyclonic wave heights. Figure 8 and 9 gives similar comparisons and error analysis using the BLSTM model. Although the error is less in comparison to the LSTM model it is more for the short time-series than the long one. Thus, deep learning models can be applied reasonably to predict significant wave height data for both non-cyclonic and cyclonic conditions.

In a later experiment BLSTM model is applied to the given four time series data and the wave height is well predicted with lead times 1, 2, 3, 4, 8, 12 corresponding to 06, 12, 18, 24, 48 and 72 h respectively. Advanced forecasting is useful for certain operational plans and the signature of significant wave height change plays a major role. Delay 1, 2, 3, 4, 8, 12, and epoch 1000 for long length time-series and epoch 2500 for short length time-series have been used to evaluate the forecast for the next 72 h using the BLSTM model. These models show lower accuracy with increasing lead times as expected. By increasing the lead time, the differences, obtained from BLSTM, between the measured and predicted value are evident. However, the general patterns of the variations of wave height are still reasonably captured by the BLSTM model. In addition, when the lead time reaches 12 (72 h), shifts between the measured and predicted wave height time series by BLSTM model can be easily noted. It is easily notified that the shifts obviously increase as the lead time arises. Table 1 gives the RMSE values estimated for the training data sets using the LSTM model for the AS grids. Table 2 gives similar RMSE values

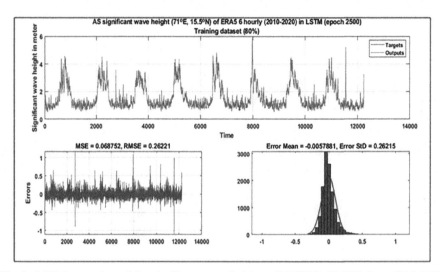

Fig. 3. LSTM network used for significant wave height at G2 (71°E, 15.5°N) from 2010–2020 (training dataset)

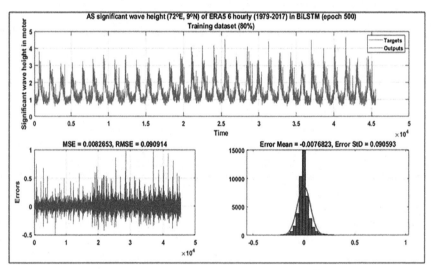

Fig. 4. BLSTM network used for significant wave height at G1 (72°E, 9°N) from 1979–2017 training dataset)

using the BLSTM model and the training data sets. The results of the testing data sets are given in Tables 3 and 4. The tables gives the RMSE values for the different epochs and hidden units for the different time-series data sets. The error values which are minimum are given in bold. Tables 5 and 6 give the RMSE values for the training and testing data sets using the BLSTM model for different lead times. The performance decreases gradually as the lead time increases. For the training set the error reaches 0.3 m for the

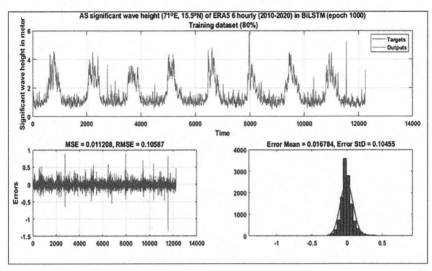

Fig. 5. BLSTM network used for significant wave height at G2 (71°E, 15.5°N) from 2010–2020 (training dataset)

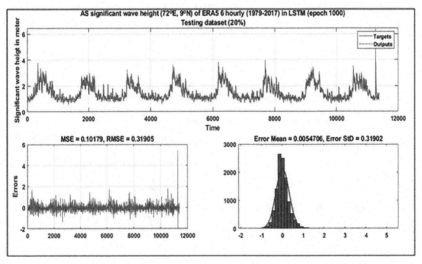

Fig. 6. LSTM network used for significant wave height at G1 (72°E, 9°N) from 1979–2017 (testing dataset)

long data set and 0.4 m for the short data set for 72 h forecast time, including the cyclonic wave heights. For the testing set the error increased more for the short time-series. Thus, for all the models and the time-series data sets the errors are depicted in the plots and the tables given below. The BLTM model performed better with higher accuracy than the LSTM model in all the cases in respect of RMSE values.

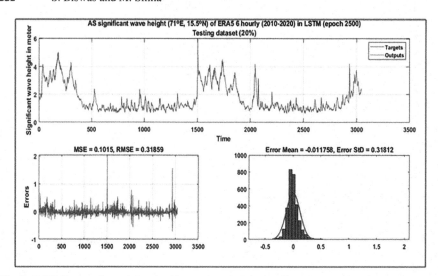

Fig. 7. LSTM network used for significant wave height at G2 (71°E, 15.5°N) from 2010–2020 (testing dataset)

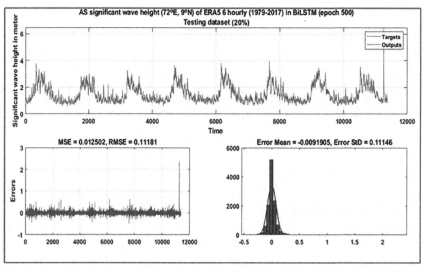

Fig. 8. BLSTM network used for significant wave height at G1 (72°E, 9°N) from 1979–2017 (testing dataset)

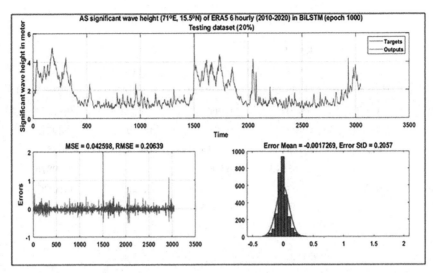

Fig. 9. BLSTM network used for significant wave height at G2 (71°E, 15.5°N) from 2010–2020 (testing dataset)

Table 1. RMSE values for LSTM training set

Epoch	Hidden Unit	RMSE			
		G1 (72°E, 9°N)		G2 (71°E, 15.5°N)	
		(1979–2016)	(1979–2017)	(2010–2019)	(2010–2020)
50	250	0.07658	0.34543	0.12343	0.23098
100	250	0.87945	0.35432	0.13214	0.28342
250	250	0.98765	0.38879	0.22012	0.27912
500	250	0.88997	0.34543	0.13259	0.26999
1000	**250**	**0.073991**	**0.30727**	0.15043	0.28976
2500	**250**	0.80897	0.32223	**0.12085**	**0.26221**

Table 2. RMSE values for BLSTM training set

Epoch	Hidden Unit	RMSE			
		G1 (72°E, 9°N)		G2 (71°E, 15.5°N)	
		(1979–2016)	(1979–2017)	(2010–2019)	(2010–2020)
50	250,250	0.07075	0.06644	0.09125	0.11254
100	250,250	0.06677	0.07078	0.08975	0.89715
250	250,250	0.72315	0.64147	0.92910	0.08594
500	**250,250**	**0.065543**	**0.09091**	0.10349	0.09876
1000	**250,250**	0.76856	0.87651	**0.92791**	**0.10587**
2500	250,250	0.09890	0.07563	0.09987	0.10875

Table 3. RMSE values for LSTM testing set

Epoch	Hidden Unit	RMSE			
		G1 (72°E, 9°N)		G2 (71°E, 15.5°N)	
		(1979–2016)	(1979–2017)	(2010–2019)	(2010–2020)
50	250	0.10914	0.29876	0.10354	0.30189
100	250	0.09247	0.29601	0.11405	0.29186
250	250	0.09421	0.30172	0.12098	0.38976
500	250	0.09131	0.30172	0.15031	0.32796
1000	**250**	**0.08899**	**0.31905**	0.14019	0.31987
2500	**250**	0.91234	0.34119	**0.13541**	**0.31859**

Table 4. RMSE values for BLSTM testing set

Epoch	Hidden Units	RMSE			
		G1 (72°E, 9°N)		G2 (71°E, 15.5°N)	
		(1979–2016)	(1979–2017)	(2010–2019)	(2010–2020)
50	250,250	0.07659	0.07856	0.12181	0.10651
100	250,250	0.07402	0.08768	0.10987	0.12518
250	250,250	0.08461	0.077191	0.11834	0.20789
500	**250,250**	**0.07280**	**0.11181**	0.10986	0.26987
1000	**250,250**	0.09897	0.98076	**0.10731**	**0.20639**
2500	250,250	0.07987	0.99826	0.11297	0.21686

Table 5. RMSE values for BLSTM training set with different lead times

Delay	Hidden Units	RMSE			
		G1 (72°E, 9°N)		G2 (71°E, 15.5°N)	
		(1979–2016)	(1979–2017)	(2010–2019)	(2010–2020)
		Epoch 1000	Epoch 1000	Epoch 2500	Epoch 2500
1 (6 H)	**250, 250**	0.06554	0.09091	0.09298	0.10879
2 (12 H)	**250, 250**	0.12151	0.11332	0.18441	0.19996
3 (18H)	**250, 250**	0.14658	0.14636	0.20129	0.21982
4 (24H)	**250, 250**	0.17559	0.17897	0.22850	0.22574
8 (48H)	**250, 250**	0.20019	0.26261	0.32608	0.33872
12(72H)	**250, 250**	0.30838	0.31032	0.39162	0.39332

Table 6. RMSE values for BLSTM testing set with different lead times

Delay	Hidden Units	RMSE			
		G1 (72°E, 9°N)		G2 (71°E, 15.5°N)	
		(1979–2016)	(1979–2017)	(2010–2019)	(2010–2020)
		Epoch 1000	Epoch 1000	Epoch 2500	Epoch 2500
1 (6 H)	**250, 250**	0.07280	0.11181	0.10731	0.20639
2 (12 H)	**250, 250**	0.12662	0.13462	0.20642	0.21987
3 (18H)	**250, 250**	0.15404	0.16142	0.22056	0.23376
4 (24H)	**250, 250**	0.18146	0.19363	0.25916	0.26761
8 (48H)	**250, 250**	0.26047	0.27417	0.37525	0.40045
12(72H)	**250, 250**	0.30914	0.31678	0.45496	0.46164

4   Conclusions

This study involves the estimation of significant wave height in the AS region using deep learning models during cyclonic conditions. To appraise the increase in error using different lead times is a prior assignment for wave height forecasting. To investigate the performances of the forecasting models quantitatively, different error measures including the MSE, RMSE, error mean and error standard are employed as judgement criteria. These predictions are based on qualitative analysis and mathematical modeling. Statistical assessment of the quality of the forecast leads to significant results. MSE compares the goodness of fit between the observed and predicted data. A low value of MSE indicates high efficiency of the model. Details of the MSE, RMSE, error mean and error standard of wave height predictions are studied and summarized in this paper using deep learning models. Error measures of LSTM and BLSTM models are also contrasted along with training and testing errors with lead times. In case of advance 72 h forecast, for BLSTM model, the RMSE is lower than 0.5 m, and error standard is 0.4. Trial and error method are considered to compare model output with desired output in respect of root mean square error. BLSTM models show very less error in forecasted wave heights. LSTM shows constant trend for non-cyclonic period but BLSTM provides better result in cyclonic and non-cyclonic time. By analyzing the graphs and tables in the above study it is concluded that BLSTM model performed better than LSTM. Thus, deep network architectures can be developed for uttermost wave prediction having wide application in the domains of ocean engineering. A major drawback of the present study is that the time series data involved is having less variability and less dynamics. In future deep network models are to be tested with wave parameters having more variability and more dynamics, especially during extreme conditions and it can introduce different variables, such as rainfall, snowfall, Sea surface temperature etc., for the implementation of combining LSTM-CNN architecture.

Conflict of Interest:.   On behalf of all authors, the corresponding author states that there is no conflict of interest.

References

Adytia, D., Saepudin, D., Pudjaprasetya, S.R., Husrin, S., Sopaheluwakan, A.A.: Deep learning approach for wave forecasting based on spatially correlated wind features, with a case study in the java sea, Indonesia. Fluids **7**, 39 (2022). https://doi.org/10.3390/fluids7010039

Bethel, B.J., Sun, W., Dong, C., Wang, D.: Forecasting hurricane-forced significant wave heights using a long short-term memory network in the Caribbean Sea. Ocean Sci. **18**, 419–436 (2022). https://doi.org/10.5194/os-18-419-2022

Biswas, S., Sinha, M.: Performances of deep learning models for Indian Ocean wind speed prediction. Model. Earth Syst. Environ. **7**(2), 809–831 (2020). https://doi.org/10.1007/s40808-020-00974-9

Biswas, S., Sinha, M.: Assessment of shallow and deep learning models for prediction of sea surface temperature. In: Sk, A.A., Turki, T., Ghosh, T.K., Joardar, S., Barman, S. (eds.) Artificial Intelligence: First International Symposium, ISAI 2022, Haldia, India, February 17–22, 2022, Revised Selected Papers, pp. 145–154. Springer Nature Switzerland, Cham (2022). https://doi.org/10.1007/978-3-031-22485-0_14

Domala, V., Lee, W.: Wave data prediction with optimized machine learning and deep learning techniques. J. Comput. Des. Eng. **9**, 1107–1122 (2022). https://doi.org/10.1093/jcde/qwac048

Greff, K., Srivastava, R.K., Koutnik, J., Steunebrink, B.R., Schmidhuber, J.: LSTM: a search space odyssey. IEEE Trans. Neural Netw. Learning Syst. **28**(10), 2222–2232 (2017). https://doi.org/10.1109/TNNLS.2016.2582924

Minuzzi, F.C., Farina, L.: A deep learning approach to predict significant wave height using long short-term memory. Ocean Model. **181**, 102151 (2023). https://doi.org/10.1016/j.ocemod.2022.102151

Patel, M., Patel, A., Ghosh, R.: Precipitation Nowcasting: Leveraging bidirectional LSTM and 1D CNN. arXiv:1810.10485 (cs) (2018)

Ali, M., Prasad, R.: Significant wave height forecasting via an extreme learning machine model integrated with improved complete ensemble empirical mode decomposition. Renew. Sustain. Energy Rev. **104**, 281–295 (2019). https://doi.org/10.1016/j.rser.2019.01.014

Song, T., Wang, J., Huo, J., Wei, W.: Prediction of significant wave height based on EEMD and deep learning. Front. Mar. Sci. **10**, 1089357 (2023). https://doi.org/10.3389/fmars.2023.1089357

Singh, O.P., Khan, T.M.A., Sazedur Rahman, M.: Has the frequency of intense tropical cyclones increased in the north Indian Ocean. Current Sci. **80**(4), 575–580 (2005). https://doi.org/10.1016/j.neunet.2005.06.042

Tolman, H.: A third-generation model for wind waves on slowly varying, unsteady, and inhomogeneous depths and currents. J. Phys. Oceanogr. **21**(6), 782–797 (1991)

Afzal, M.S., Kumar, L., Chugh, V., et al.: Prediction of significant wave height using machine learning and its application to extreme wave analysis. J. Earth Syst. Sci. **132**, 51 (2023). https://doi.org/10.1007/s12040-023-02058-5

Technology Enabled Self-directed Learning: A Review and Framework

Sarika Sharma[1]([✉]) [iD], Vipin Tyagi[2] [iD], and Anagha Vaidya[1,2] [iD]

[1] Symbiosis Institute of Computer Studies and Research, Symbiosis International (Deemed University), Atur Centre, Model Colony, Pune 411016, India
sarika4@gmail.com

[2] Jaypee University of Engineering and Technology, Raghogarh, Guna 473226, MP, India

Abstract. The use of technology to support self-directed in education sector has been on the rise in recent years. Higher educational institutions have utilized technology to enhance teaching and learning in various ways. Advancements in technology have made laptops, tablets, and smartphones more accessible and replaced personal computers as the primary technology in classrooms, giving teachers more options for supporting their students. Researchers have also explored how technology can be used to accommodate the diverse needs of learners and promote autonomous, self-directed learning. With the outbreak of COVID-19, online classes became more prevalent, highlighting the need to further investigate technology-enabled self-directed learning. This paper reviews the existing literature on the topic and presents it in a framework.

Keywords: Self-directed learning · Higher education · Technology · learning algorithms

1 Introduction

In today's ever evolving world where technology is considered to be the backbone of every sector, the education sector is one of the most prominent ones. Before the COVID-19 pandemic, the most common means of conducting the teaching-learning activities was in physical mode with face-to-face methods. Technology was used only to some extent and only to add-on to the traditional teaching methods. Distance learning and lifelong learning models used technology generously in their models. However, the declaration of pandemic by the World Health Organization, has forced the educational institute to embrace technology to maintain continuity in their teaching-learning methods. Due to COVID-19 [1], as per a report by UNESCO, during April 2020 most of the schools as well as the higher educational institutes were severely affected and that led to the disruption followed by their closure. The number identified was 1.5 billion youngsters in 195 countries. The only way to keep the ball rolling was the adoption of Technology based platforms such as flipped classrooms, WebEx, google meets, and Microsoft teams to name a few. The use of technology gave a boost to self-directed learning where learners can take their own initiative for learning [2], pushing students towards self-directed learning.

M. Singh et al. (Eds.): ICACDS 2023, CCIS 1848, pp. 228–240, 2023.
https://doi.org/10.1007/978-3-031-37940-6_19

1.1 Theoretical Foundations of Self-directed Learning

Self-directed learning cab be seen as a process as well as the objective for adult-learning which leads to the successful learning activities [3]. It is also an approach through which learners involved as motivated to take their own responsibilities and control through self-management and self-monitoring [4]. In the context of SDL, studies have highlighted the need for nurturing the skills for self-direction in a virtual environment, which can lead to improvements in the achievement of academic outcomes [5]. These researchers proposed an SDL model specifically in the context of online education where they added a dimension of learning context representing the learning environment. Learners in the virtual setup can be benefited from online learning through the adoption of self-directed learning [6, 7].

1.2 Technology-Enabled Self-directed Learning

Researchers have begun to study the affect technology creates on learner-directed learning. According to the definition of learner autonomy by [8] it is the proficiency to cultivate a self-determined and responsible learner in educational environments for empowerment and transformations in the educational environments. The five phases of progression toward autonomy, as outlined by [9] include awareness, involvement, intervention, creation, and transcendence. These researchers argue that ICT can play a positive role in promoting the development of learner autonomy. Additionally, [10] have examined how the relationship between learner autonomy and technology has evolved over the past 20 years, identifying five key themes in technology-mediated autonomy research including training and strategies for learners, language advising and self-access, autonomy for teachers, social technologies for learning, and tele-collaboration. There is immense scope for further research studies on technology-enabled self-directed learning.

Self-directed learning can be seen as an automated approach where learners drive more ownership of their learning process, and technology plays a key role in supporting this approach. The advancements in technology such as online learning opportunities, mobile device accessibility, and pedagogical shifts have made it possible for learners to have more control over when and where they learn. This has resulted in a change in the traditional role of the teacher, who now serves more as a facilitator and guide. Innovations such as Khan Academy and Flipped Classroom also have driven towards the changes in the traditional classroom settings and highlighted the imperative role of self-directed learning. With these approaches, learners can access learning materials and resources on their own and teachers can focus on addressing individual needs and questions.

1.3 Research Gap and Objectives

During the pandemic for almost two years, the teaching was conducted online mode. Therefore as the pandemic is over and academic institutions are heading towards their usual mode of teaching, post-pandemic, there is a need to understand the impacts and implications on student learning and self-direction. Self-directed learning is a context that is well developed and well addressed in the previous body of research, but the effects of online learning remain unclear [11, 12]. Literature surveys revealed that the studies in self-directed learning either focused on life-long learning or mostly the case of language education [13, 14]. [15] a study on the readiness of self-directed learning in emergency situations like COVID-19 revealed that self-directed learning skills were improved.

The present study addresses this imperative issue of technology-enabled self-directed learning in adults by compiling relevant research studies through a systematic literature review. The objectives are:

Objective 1: To assimilate and create existent knowledge on to Self-directed learning (SDL).
Objective 2: To propose a research framework
Objective 3: To identify open research issues on technology-enabled SDL.

2   Systematic Literature Review: The Methodology

A systematic literature review is a methodical process that follows a specific procedure. There are various guidelines and steps that can be followed when conducting a systematic literature review. In this study, we followed the steps outlined by [16], which is a widely accepted method in scientific research. The five steps we adopted and followed are as follows. For clarity, the procedure is also represented in a pictorial form (Fig. 1).

Step 1: Setting of Research objectives
This study aims to present a systematic literature review of technology-enabled self-directed learning. The three research objectives are framed and presented in the previous section.

Step 2: Database search with a search string
The next step in the systematic literature review process is to identify a database for searching for relevant articles using a search string developed by the researchers. This study explores the research done on technology-enabled self-directed learning. The Web of Science (WoS) database was selected due to its suitability for technology-related studies and its reputation for providing high-quality research papers and a wide range of studies, especially in emerging fields. As a result, WoS is often a top choice for literature review-based studies.

The formation of search string was done with the help of literature, where first, the keywords are identified from the literature review of the SDL. Then with the additional words like technology and computer, a search string is formulated as:
   (Technology OR Computer) AND "self-directed learning"
The databases were searched using the above string in January 2023, which generated a result of 47 research articles.

Step 3: Inclusion Exclusion criteria for further assessment
Authors visited and analyzed the 47 research articles for the abstracts of each and every article for suitability for inclusion in the analysis. The articles which were not related to educational learning were excluded. Hence the final included studies are 38 for the analysis, as per the scope of the study. Initially, all the authors conducted this exercise separately and there concluded together. The deviations were addressed through the meetings amongst the co-authors to form a common conscience. To understand the

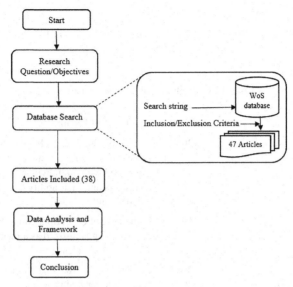

Fig. 1. Steps followed for the Systematic review process

timeline of the research on technology-enabled self-directed learning a graph is generated for the year-wise number of studies on this topic (Fig. 2). It can be observed that the field of technology-enabled SDL is evolving since the last decade as 33 articles are contributed to the same. It can be noted that the year 2022 has witnessed 7 studies.

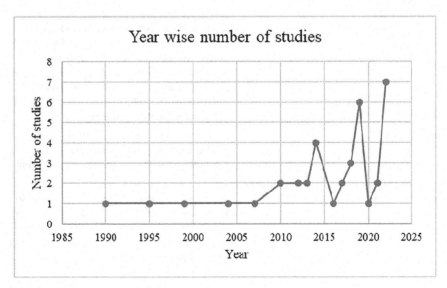

Fig. 2. Year-wise number of studies

Step 4: Analysis and Compilation

The 38 articles which were finally selected for the inclusion in the literature review are compiled and presented in Table 1, with their major findings.

Table 1. Compilation of the research articles

Author(s)	Year	Highlights of Findings
Lee, C; et al. [17]	2017	The study found that self-management, a desire for learning, and self-control are all positively associated with an individual's use of computers and their ability to learn independently
Mason, R [18]	1988	In the field of education, computer conferencing is utilized as a means to promote self-direction in learners
Demir, O et al. [19] Tzeng, SY et al. [20]	2014 2022	Research has shown that there is a strong correlation between students' attitudes toward computers and the level of their engagement in the activity of self-directed learning using technology, particularly for younger students
Mentz, E; Van Zyl, S [21]	2018	An investigation was conducted to examine the connection between the application of cooperative learning approaches and students' ability to direct their own learning
Radoyska, P [22]	2012	A web portal was proposed as a means of facilitating self-directed learning. The key features and functionalities required for the portal were outlined. The design and infrastructure of the system were described. An example of the portal's implementation as a virtual research lab focused on electronics and computer technologies was provided
Vas, R et al. [23]	2018	Paper utilizes the semantic technologies to realize connectivism through information exploration principle as an important.measure
Barett, H. [24] Shinkareva, ON; Benson, AD [25]	1990 2007	Findings state that self-directed learning is associated with higher levels of proficiency in using personal computers. It was observed that the variability in preferred learning strategies was more pronounced based on the type of user interface preferred than on the individual's learning style

(continued)

Table 1. (*continued*)

Author(s)	Year	Highlights of Findings
Muench, ST; Mahoney, JP [26]	2004	The early response to the concept of the Pavement Guide has been positive, suggesting that the development of self-directed learning (SDL) products in pavement training is a promising area to explore further
Sumuer, E [27] Pan, XQ [28] An, FH et al. [29]	2018 2020 2022	Authors have indicated that technology usage in education is a crucial factor in determining how well students can engage in self-directed learning through technology, as it impacts their online communication self-efficacy and computer self-efficacy
Mishra, P et al. [30]	2013	The use of technology in a meaningful way in learning environments can have a significant impact on promoting self-directed learning
Wong, HKM; Mok, IAC [31] Lao, ACC et al. [32]	2019 2017	A Technology Self-Directed Learning (TSDL) package was developed, which aimed to strengthen students' understanding through the use of five-step Self-Directed Learning Indicators
Geng, S et al. [33] Lopez-Ubeda, R et al. [34]	2019 2022	This study involves students and evaluated the effects of technology readiness, self-directed learning, and learning motivation
Akerlind, GS; Trevitt, AC [35]	1999	Factors that can either exacerbate or alleviate resistance to computer-facilitated, problem-based courses are discussed, with examples provided
Zheng, BB [36]	2022	The goal of this study was to investigate which technologies students adopt for self-directed learning. Key factors that influence students' self-directed technology use were identified
Bullock, SM [37]	2013	The study suggests that self-directed learning experiences in teacher education are valuable, but experiences that focus specifically on digital technologies may inadvertently reinforce a bias toward using technology in teaching
Chau, KY et al. [38]	2021	This study examines the correlations to understand the combined effects of presence in synchronous learning environments and provides educators with insight into the future trends of learning and instructional strategies in online teaching

(*continued*)

Table 1. (*continued*)

Author(s)	Year	Highlights of Findings
Lee, K et al. [39]	2014	The study's results confirmed the validity of the four-factor structure model and found that students who reported engaging in self-directed learning (SDL) and cooperative learning (CL) in face-to-face settings also participated in these forms of learning in technology-supported contexts
Rashid, T; Asghar, HM [40]	2016	The results suggest that the use of technology has a direct and positive impact on students' engagement and self-directed learning. The findings indicate the intricate interplay between students' technology use, engagement, self-directed learning, and academic performance
Jou, M; Wu, YS [41]	2012	An interactive learning system was created to aid self-directed learning of microfabrication technology by taking both the technical and functional perspectives into account and integrating it with a web-based learning system
Markheva, OJ [42] Seleke, B et al. [43]	2010 2019	Methods to enhance the effectiveness of self-directed learning are identified. The ratio of class hours to self-directed learning in the cycle of teacher vocational training is outlined
Wu, CM [44]	2014	This study investigates the impact of a new teaching model (combining classroom instruction with web-based instruction) and found that it has a significant effect on learners' learning strategies, which can enhance their self-directed learning
Khosravi, S et al. [45]	2022	This study examines the utility of using psychophysiological eye-tracking data collected from participants in response to visual stimuli
Teo, T et al. [46]	2010	Developed and validated a Self-Directed Learning with Technology Scale (SDLTS) for young students

(*continued*)

Table 1. (*continued*)

Author(s)	Year	Highlights of Findings
Toth, J et al. [47]	2021	This study evaluated changes in self-directed learning (SDL) among students in a Pharmacists' Patient Care Process (PPCP) course that uses adaptive learning technology (ALT) as a teaching method
Avsec, S; Savec, VF [48]	2022	The study suggests that for successful teaching and learning of science, technology, and design thinking in teacher education, educators and course designers should consider adjusting their implementation of design thinking
Briede, B [49]	2019	Analyzed the crucial elements of the Strategy-Driven Learning (SDL) approach, including the use of electronic tools to support the implementation of learning strategies and the creation of adaptable learning environments
Morris, TH; Rohs, M [50]		Investigated the possibility of utilizing digital technology to assist students in implementing the Strategy-Driven Learning (SDL) approach and utilizing digital tools for educational purposes
Mentz, E; Bailey, R [51]	2019	The findings also showed that the use of technology to facilitate cooperative learning can effectively support self-directed learning
Choy, D; Cheung, YL [52]	2022	It was determined that technology can aid in the cultivation of self-directed and collaborative learning skills among students through the use of ICT-enhanced learning activities
Hemrungrote, S; Aunsri, N [53]	2014	This research designed and developed an e-learning system to address the challenge of promoting self-directed learning (SDL) through the use of induction module lessons. The goal of the system is to encourage students to take initiative in organizing their learning
Curran, V et al. [54]	2019	The growing utilization and reliance on digital and mobile technologies have significant implications for the development of organizational and workplace policies that can facilitate efficient self-directed learning in a digital era

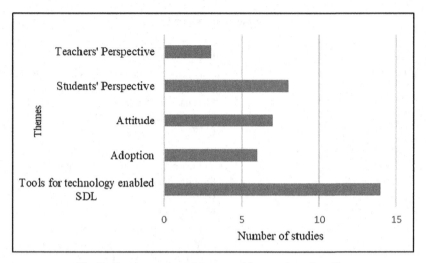

Fig. 3. Representation of themes with a number of studies

Step 5: Qualitative analysis and proposed framework
All the studies included were visited for initial overall understanding. A qualitative research approach was adopted for the thematic analysis using the inductive reasoning method. A research framework was proposed based on the themes generated in these studies, also represented by the number of studies (Fig. 3). The framework is presented below:

I Tools for technology enabled SDL
II Implementation

II (A) Adoption
II (B) Attitude

III Stakeholders' perspective

III (A) Students
III (B) Teachers

3   Results and Discussions

Self-directed learning is imperative and an emerging field of research where there is immense scope for further contributions to existing literature. Through this systematic literature review, authors identified the three major streams which are, tools and techniques related to self-directed learning, implementation aspects from the adoption and attitude of the users, and stakeholder's perspective. Stakeholders are the teachers who will adopt the self-directed learning tools as well as the students who are the ultimate users and the subjects who will be benefited from the self-directed learning using technology.

4 Conclusion and Future Scope

This study makes a valuable contribution to the area of technology-enabled self-directed learning, which is an emerging area of research. After the pandemic, there is more and more adoption of technology for teaching activities. SDL is here to stay and the future is bright. The authors have compiled the work done in this area through a systematic literature review approach. This compilation is presented through graphs as well as a framework. Certainly, there are a few limitations also. First of all the database used is the web of science. There can be the inclusion of other databases such as Scopus and EBSCO for a holistic approach to the study. Some future research areas and open issues are identified through the studies which are included in this. These are:

- Personalized learning: using technology to tailor the learning experience to the individual student's needs and preferences.
- Gamification: using game design elements in non-game contexts, such as education, to increase engagement and motivation.
- Adaptive learning: using technology to adjust the learning experience in real time based on the student's performance.
- Learning analytics: using data and analytics to understand and improve the learning process.
- Mobile learning: using mobile devices, such as smartphones and tablets, to deliver educational content and facilitate learning.
- Virtual and augmented reality: using VR and AR technology

References

1. UNESCO (2020). https://en.unesco.org/news/13-billion-learners-are-still-affectedschool-uni versity-closures-educational-institutions
2. Knowles, M.: Self-directed Learning. The Adult Education Company, New York (1975)
3. Song, Y., Hill, J.R.: A conceptual model for understanding self-directed learning in online environments. J. Interact. Online Learn. **6**(1), 27–42 (2007)
4. Garrison, D.R.: Self-directed learning: toward a comprehensive model. Adult Educ. Q. **48**(1), 18–33 (1997)
5. Carson, E.H.: Self-directed learning and academic achievement in secondary online students (dissertation), Chattanooga: University of Tennessee at Chattanooga (2012). http://pqdtopen. proquest.com/doc/1038364482.html?FMT=ABS
6. Beach, P.: Self-directed online learning: a theoretical model for understanding elementary teachers' online learning experiences. Teach. Teach. Educ. **61**, 60–72 (2017)
7. LaTour, K.A., Noel, H.N.: Self-directed learning online: an opportunity to binge. J. Mark. Educ. **4**(2), 174–188 (2021)
8. Jiménez, M., Lamb, T., Vieira, F.: Mapping Autonomy in Language Education: A Framework for Learner and Teacher Development. Peter Lang, Frankfurt (2017)
9. Chitashvili, N.: The concept of autonomy in second language learning. Georgian Electron. Sci. J.: Educ. Sci. Psychol. **2**(11), 17–22 (2007)
10. Reinders, H., White, C.: 20 years of autonomy and technology: how far have we come and where to next?. Lang. Learn. Technol. **20**(20), 143–154 (2016)

11. Coman, C., Ru, L.G., Mesean-Schmitz, L., Stanciu, C., Bularca, M.C.: Online teaching and learning in higher education during the coronavirus pandemic: students' perspective. Sustainability **12**(24), 1–24 (2020)
12. Dama, C., Langford, M., Dan, U.: Teachers' agency and online education in times of crisis. Comput. Hum. Behav. **121**, 1–16 (2021)
13. Basereh, N., Pishkar, K.: Self-directed learning and self-efficacy belief among Iranian EFL learners at the advanced level of language proficiency. J. Appl. Linguist. Lang. Res. **3**(1), 232–240 (2016)
14. Zainuddin, Z., Perera, C.J.: Supporting students' self-directed learning in the flipped classroom through the LMS TES BlendSpace. On the Horiz. **26**(4), 281–290 (2018)
15. Watson, M.K., Barrella, E., Skenes, K.: Self-directed learning readiness among engineering students during emergency online instruction. In: Proceedings of 2021 IEEE Frontiers in Education Conference (FIE), pp. 1–5 (2021)
16. Khan, K.S., Kunz, R., Kleijnen, J., Antes, G.: Five steps to conducting a systematic review. J. R. Soc. Med. **96**(3), 118–121 (2003)
17. Lee, C., Yeung, A.S., Ip, T.: University English language learners' readiness to use computer technology for self-directed learning. System **67**, 99–110 (2017)
18. Mason, R.: Computer conferencing - a contribution to self-directed learning. Br. J. Edu. Technol. **19**(1), 28–41 (1988)
19. Demir, Ö., Yaşar, S., Sert, G., Yurdugül, H.: Examination of the relationship between students' attitudes towards computer and self-directed learning with technology. Educ. Sci. **39**, 257–266 (2014)
20. Tzeng, S.Y., Lin, K.Y., Lee, C.Y.: Predicting college students' adoption of technology for self-directed learning: a model based on the theory of planned behavior with self-evaluation as an intermediate variable. Front. Psychol. **13**, 1–10 (2022)
21. Mentz, E., Van Zyl, S.: The impact of cooperative learning on self-directed learning abilities in the computer applications technology class. Int. J. Lifelong Educ. **37**(4), 482–494 (2018)
22. Radoyska, P.: Web portal for self-directed learning, experimentation and collaboration on projects in electronics and computer technology. In: Conference Name: 6th International Technology, Education and Development Conference, pp. 1160–1169 (2012)
23. Vas, R., Gadot, M., Weber, C., Gkoumas, D.: Implementing connectivism by semantic technologies for self-directed learning. Int. J. Manpow. **39**(8), 1032–1046 (2018)
24. Barrett, H.: Adult Self-Directed Learning, Personal-Computer Competence, and Learning Style (1991). Unpublished dissertation. https://citeseerx.ist.psu.edu/document?repid=rep1&type=pdf&doi=37dec66a367d34a20a34c87934cbcaf82aa46874
25. Shinkareva, O., Benson, A.: The relationship between adult students' instructional technology competency and self-directed learning ability in an online course. Hum. Resour. Dev. Int. **10**, 417–435 (2007)
26. Muench, S.T., Mahoney, J.P.: Computer-based multimedia pavement training tool for self-directed learning. Transp. Res. Rec. **1896**(1), 3–12 (2004)
27. Sumuer, E.: Factors related to college students' self-directed learning with technology. Australas. J. Educ. Technol. **34**(4), 29–43 (2018)
28. Pan, X.Q.: Technology acceptance, technological self-efficacy, and attitude toward technology-based self-directed learning: learning motivation as a mediator. Front. Psychol. **11**, 1–10 (2022)
29. An, F., Xi, L., Yu, J., Zhang, M.: Relationship between technology acceptance and self-directed learning: mediation role of positive emotions and technological self-efficacy. Sustainability **14**(10390), 1–13 (2022)
30. Mishra, P., Fahnoe, C., Henriksen, D.: Creativity, self-directed learning and the architecture of technology rich environments. TechTrends **57**, 10–13 (2013)

31. Wong, H.K.M., Mok, I.A.C.: Students' mathematics experience of the technology self-directed learning (TSDL) pedagogy. In: Conference Name: 12th Annual International Conference of Education, Research and Innovation, pp. 1614–1623 (2019)

32. Lao, A.C.-C., Cheng, H.N.H., Huang, M.C.L., Ku, O., Chan, T.-W.: Examining motivational orientation and learning strategies in computer-supported self-directed learning (CS-SDL) for mathematics: the perspective of intrinsic and extrinsic goals. J. Educ. Comput. Res. **54**(8), 1168–1188 (2017)

33. Geng, S., Law, K.M.Y., Niu, B.: Investigating self-directed learning and technology readiness in blending learning environment. Int. J. Educ. Technol. High. Educ. **16**(1), 1–22 (2019). https://doi.org/10.1186/s41239-019-0147-0

34. López-Úbeda, R., García-Vázquez, F.A.: Self-directed learning using computer simulations to study veterinary physiology: comparing individual and collaborative learning approaches. Vet. Rec. **191**(8), e1732, 1–9 2022)

35. Åkerlind, G.S., Trevitt, A.C.: Enhancing self-directed learning through educational technology: when students resist the change. Innov. Educ. Train. Int. **36**(2), 96–105 (1999)

36. Zheng, B.: Medical students' technology use for self-directed learning: contributing and constraining factors. Med. Sci. Educ. **32**(1), 149–156 (2021). https://doi.org/10.1007/s40670-021-01497-3

37. Bullock, S.M.: Using digital technologies to support self-directed learning for preservice teacher education. Curriculum J. **24**(1), 103–120 (2013)

38. Chau, K.Y., Law, K.M.Y., Tang, Y.M.: Impact of self-directed learning and educational technology readiness on synchronous e-learning. J. Organ. End User Comput. **33**(6), 1–20 (2021)

39. Lee, K., Tsai, P.S., Chai, C.S., Koh, J.H.L.: Students' perceptions of self-directed learning and collaborative learning with and without technology. J. Comput. Assist. Learn. **30**(5), 425–437 (2014)

40. Rashid, T., Asghar, H.M.: Technology use, self-directed learning, student engagement and academic performance: examining the interrelations. Comput. Hum. Behav. **63**, 604–612 (2016). https://doi.org/10.1016/j.chb.2016.05.084

41. Jou, M., Wu, Y.S.: Development of a web-based system to support self-directed learning of microfabrication technologies. Educ. Technol. Soc. **15**, 205–213 (2012)

42. Markheva, O.Y.: Organization of self-directed learning of the future foreign languages teachers powered by information technologies. Inf. Technol. Learn. Tools **16**(2), 1–11 (2010)

43. Seleke, B., Havenga, M., Beer, J.D.: The enhancement of self-directed learning through the engagement in problem-based learning activities during a professional development programme on Indigenous Knowledge for Technology teachers. In: Proceedings of Teaching and Education Conferences 9612049, International Institute of Social and Economic Sciences, pp. 362–371 (2019)

44. Wu, C.M.: Research on application of information technology in promoting learners' translation competence and self-directed learning. Advanced Development of Engineering Science. Advanced Material Research. IV

45. Khosravi, S., Khan, A.R., Zoha, A., Ghannam, R.: Self-directed learning using eye-tracking: a comparison between wearable head-worn and webcam-based technologies. In: IEEE Global Engineering Education Conference (EDUCON2022), Tunis, Tunisia, 28–31 March 2022, pp. 640–643 (2022)

46. Teo, T., et al.: The self-directed learning with technology scale (SDLTS) for young students: an initial development and validation. Comput. Educ. **55**(4), 1764–1771 (2010)

47. Toth, J., Rosenthal, M., Pate, K.: Use of adaptive learning technology to promote self-directed learning in a pharmacists' patient care process course. Am. J. Pharm. Educ. **85**(1), 28–33 (2021)

48. Avsec, S., Savec, V.F.: Mapping the relationships between self-directed learning and design thinking in pre-service science and technology teachers. Sustainability. **14**(14), 1–28 (2022)

49. Briede, B.: Students' self-directed learning in the context of industrial challenges: Latvia University of life sciences and technologies case. In 5th International Conference on Higher Education Advances (HEAd 2019) Universitat Politècnica de València, València, pp. 685–694 (2019)

50. Morris, T.H., Rohs, M.: The potential for digital technology to support self-directed learning in formal education of children: a scoping review. Interact. Learn. Environ. **31**(4), 1974–1987 (2021). https://doi.org/10.1080/10494820.2020.1870501

51. Mentz, E., Bailey, R.: A systematic review of research on the use of technology-supported cooperative learning to enhance self-directed learning. In: Self-Directed Learning for the 21st Century: Implications for Higher Education. NWU Self-Directed Learning Series (2019)

52. Choy, D., Cheung, Y.L.: Comparison of primary four students' perceptions towards self-directed learning and collaborative learning with technology in their English writing lessons. J. Comput. Educ. **9**, 783–806 (2022). https://doi.org/10.1007/s40692-022-00220-4

53. Hemrungrote, S., Aunsri, N.: E-learning development to support self-directed learning via induction module lessons: a case study of Introduction to information technology course. In: 2014 Asia-Pacific Signal and Information Processing Association Annual Summit and Conference, APSIPA 2014, pp. 1–7 (2014)

54. Curran, V., et al.: Adult learners' perceptions of self-directed learning and digital technology usage in continuing professional education: an update for the digital age. J. Adult Continuing Educ. **25**(1), 74–93 (2019)

Detecting Toxic Comments Using FastText, CNN, and LSTM Models

Hetvi Gandhi[(⊠)] ⓘ, Rounak Bachwani ⓘ, and Archana Nanade ⓘ

Computer Engineering Department, Mukesh Patel School of Technology Management, and Engineering, NMIMS University, Mumbai, India
hetvisg@gmail.com

Abstract. The use of social media has become a necessary daily activity. It provides a platform to share news, information, and social interaction. However, many people now take social media platforms for granted, using them to harass and threaten others, which results in cyberbullying. Toxic comments are online remarks that are insulting, abusive, or inappropriate, and frequently cause other users to quit a debate. People are unable to openly express their thoughts owing to cyberbullying and harassment. Identifying and classifying such remarks by hand is a time-consuming, inefficient, and unreliable operation. To solve this issue, this research article focuses on developing a deep learning system to analyze toxicity and produce efficient results in order to restrict its negative consequences, which will aid institutions to put the necessary measures into practice. Our proposed model uses Long Short-Term Memory (LSTM) along with FastText word embedding, resulting in a model with high performance. To make the social networking experience better, this model tries to improve the detection of different sorts of toxicity. Toxic, Severe Toxic, Obscene, Threat, Insult, and Identity-hate are the six categories that our methodology divides such comments into. Multi-Label Classification aids us in providing an automatic answer to the problem of poisonous remarks that were faced.

Keywords: Convolutional Neural Network (CNN) · FastText · Long Short-Term Memory (LSTM) · Natural language processing (NLP) · Text classification · Toxic comments

1 Introduction

The rise of social media has resulted in an increase in online forum remarks. It was due to the fact that it allows users to keep in touch with other users, creates a sense of belonging, and provides a forum for users to share their views and thoughts.

According to Fig. 1, the statistics say that the use of social media is increasing every year. In today's modern world, social media has an undeniable impact on our culture, economy, and overall worldview. Although it has provided many benefits, it has also brought a negative impact on society. Everyone is seeking acceptance and their emotions are getting attached to the likes, comments, and tags they receive on social media. People posting on social media get both good and bad comments. The impact of

© The Author(s), under exclusive license to Springer Nature Switzerland AG 2023
M. Singh et al. (Eds.): ICACDS 2023, CCIS 1848, pp. 241–252, 2023.
https://doi.org/10.1007/978-3-031-37940-6_20

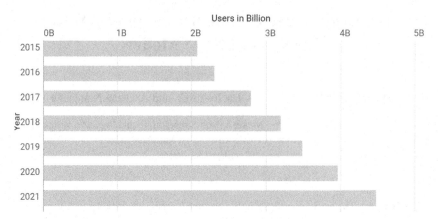

Fig. 1. Growth of social networking year by year [1]

hate and abusive comments affect the users in a lot of ways than we think. One-fourth of children have experienced cyberbullying, as per a study undertaken by the National Children's Home [3] in the United Kingdom. These issues have left teenagers with serious mental scars and even led to teenage suicide. Greta Thunberg, a Swedish climate change activist faced a lot of hateful abuse online. Her social media manager received threatening calls after Greta's tweet [2] supporting farmers' protests in India, went viral on Twitter.

In July 2020, the Pew Research Center [4] conducted a study of US adults. According to the survey, over two-thirds of Americans believe that social media has an adverse influence on current events in the country. Only one out of every ten Americans appears to believe that social media platforms ideally have a positive influence on how things are progressing, whereas a third of Americans believe that social media has no influence on society. Those who are critical of the impact of social media point to misinformation as well as online abuse on the platform. They are also concerned that people will believe everything they come across or that they won't know what to believe.

Researchers are exploring whether Deep Learning (DL) approaches can be used for comment classification tasks. Self-regulatory measures include anti-bullying teaching resources, monitoring systems, blocking, and masking software; human or computerized systems such as supervised machine learning, are used to implement cyberbullying policies.

Twitter has developed a safety center [5] where individuals can learn about being safe online, with sections tailored to teen, parent, and educational audiences. It also announced online safety cooperation with the mental health organization Cycle Against Suicide.

Comment sections are a perk, not a must. Comment sections should not be the rule, but rather a thoughtful gift to readers on an article-by-article basis. That implies readers should be grateful for the ability to remark directly on the site rather than feeling violated when it isn't available. Many people, including myself, will not read a comment section if it is not regulated; in fact, many people will not read an entire site, including the articles, if it is not constrained. "Unmoderated comments appear to have a minor, but real negative influence on readers' opinion of the sites on which they appear," writes Adam Felder of The Atlantic.

Identifying harassment in social media discussions is vital, as is determining the number of unpleasant sentiments conveyed inside the comment. Organizations can conserve time and effort in regulating these platforms by adopting an automated way of classifying comments. This study primarily focuses on determining the toxicity levels of derogatory comments from various social media websites. The proposed model could aid in identifying those who abuse individuals. As a result, it will eventually assist in the execution of policies and the penalty of individuals who do not obey, which can subsequently be used to bring down the level of toxicity in online forums. This study represents a methodology to build a multi-label classifier using LSTM, CNN, and Fast-Text that separates comments into six groups according to their toxicity level: toxic, severe toxic, obscene, threat, insult, and identity-hate.

The contributions of our paper are as follows:

- Proposing a model for the detection of toxic comments using FastText word embedding, LSTM, and CNN architecture.
- A detailed analysis and implications of the CNN and LSTM model.

The key takeaways from our paper are to guide deep learning researchers to adopt a suitable deep learning model. From our detailed experimentation and findings, it is found that the LSTM model is outperforming the CNN.

The paper has the following sections - Sect. 2 has the related work, where we have listed several existing methods in toxic comment detection. Section 3 has the proposed system and methodology. Section 4 has a comparative analysis of the proposed models. Section 5 concludes our study with implications.

2 Related Work

Deep Learning has led to plenty of major breakthroughs around the world [6] and has become a recent trend. In Natural Language Processing, deep learning has generated an impact that smoothly and efficiently exceeds the normal techniques. Many researchers have devoted their work to the study of comments in recent years (Table 1).

3 Proposed System and Methodology

Our proposed model performs multi-label categorization of toxic comments using three models namely LSTM, LSTM with FastText, and LSTM-CNN with FastText. In this proposed system, social networks will provide input in the form of comments, which will then be processed before being sent to the word embedding phase. In this phase, texts are broken up into words and vectorized, which are subsequently processed by the proposed model. The subsequent toxicity labels will be predicted and illustrated after the trained model has evaluated the comment.

Strings (plain text) cannot be processed by machine or deep learning techniques. Such strings cannot be processed by these algorithms as the primary input. Word embedding solves this problem by converting raw text into a vectorized format that the model can

Table 1. Related work on detecting toxic comments.

Citation	Dataset	Algorithm	Embedding	Research gap
[13]	Toxic comment classifier jigsaw dataset	Logistic regression – 91.48%	None	Accuracy can be improved using deep learning techniques
[14]	Toxic comment classifier jigsaw dataset	CNN – Accuracy not given	FastText	Metrics not given
[15]	Toxic comment classifier jigsaw dataset	Logistic Regression - 89.46%	None	Accuracy can be improved using deep learning techniques
[16]	Toxic comment classifier jigsaw dataset	RVVC – 97%	For word embedding, GloVe is used, and FastText is used for sub- word embedding	–
[17]	Movie Review Data	LSTM-CNN (Filter Size: 5x600) – 87.31%	Word embedding used but not mentioned which one	Accuracy can be improved
[18]	Toxic comment classifier jigsaw dataset	Support Vector Machine -86.5%	None	Accuracy can be improved using deep learning techniques
[19]	Not mentioned	LSTM– 94.94%	SpaCy	–
[20]	Dataset is self-made along with Kaggle's toxic comment	Logistic Regression – 92%	None	Word embeddings could be used to increase efficiency
[21]	Toxic comment classifier jigsaw dataset	CNN – 98.05%	FastText	LSTM can be used to make it more efficient

interpret. By computing the difference of the two vectors commonly known as embedding space, this format can be utilized to identify semantic relations between related words. In the proposed framework, FastText is used as a word embedding method.

Facebook Research's FastText [7] is a library for quickly learning word embedding and text categorization. It is an expansion of the word2vec approach, in which each word is represented as an n-gram of characters rather than directly training word vectors. For instance, the word "Machine" (Fig. 2) is represented in FastText, with n = 3 is <ma, mac, ach, chi, hin, ine, ne>, the angle brackets designate the word's beginning and ending. This enables the embeddings to recognize suffixes and prefixes, which aids in the comprehension of shorter words. After the representation of words using character n-grams, a skip-gram model is developed to determine the word embeddings. The model is termed as bag of words with a sliding window over a word since the internal structure of the word is not taken into account. It doesn't concern what sequence the n-grams are in if the n-grams fall within this window FastText performs well even with rare words Despite a word being missed throughout the training phase, FastText [8] can identify its embeddings by splitting it into n-grams.

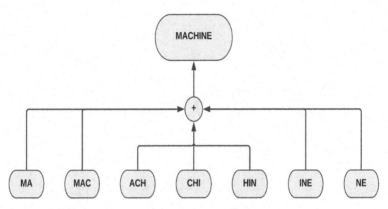

Fig. 2. A visual guide to FastText Word Embedding

Convolutional Neural Network (CNN)

CNN is a type of deep learning algorithm commonly used for image recognition and computer vision tasks. Unlike traditional neural networks, which process data in a linear fashion, CNNs use convolutional layers to extract features from the input image. These features are then combined in fully connected layers to make predictions about the contents of the image. CNNs have achieved impressive results in a wide range of applications, including object detection, face recognition, and medical image analysis. Their success can be attributed to their ability to automatically learn hierarchical representations of the input data, which enables them to recognize complex patterns and structures in images (Fig. 3).

Long Short-Term Memory

Variable-length sequence processing will prominently be a principal method in RNNs. On top of the RNN model, the LSTM model is improved and constructed. Inclusion of a memory cell and controlled gates, LSTM helps historical information to be preserved (Figs. 4 and 5).

- **Comments from the dataset:** The model was built using Kaggle's Toxic Comment Classification Dataset on Wikipedia's Talk page edits.
- **Data Analysis:** Our study uses exploratory data visualizations to better understand the problem in order to find a solution.

Input

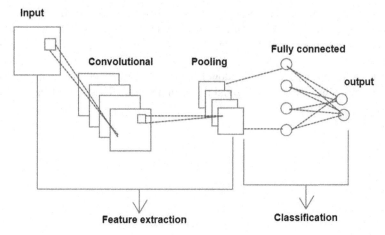

Feature extraction   Classification

Fig. 3. Basic CNN architecture [9]

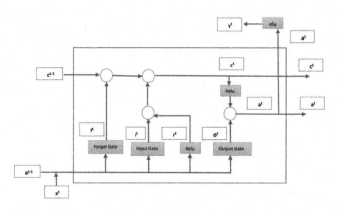

Fig. 4. The repeating module in an LSTM contains four interacting layers [10]

Figure 6 depicts that the dataset contains comments ranging in length between 0 and 1400 characters. It is evident that we have fewer lengthy comments and the most common length is from 0 to 200.

The labels in the dataset are seen in Fig. 7. As a result, we can deduce that the number of toxic comments in our dataset is significantly higher. Identity Hate comments, on the other hand, are the lowest.

- **Data Pre-processing:** Data pre-processing is done at a high standard in order to prepare it for the model. Multiple morphological techniques were used in the pre-processing, including text normalization, lemmatization, stopwords detection and removal, and tokenization.

  - **Text Normalization** refers to the process of deleting characters from between lines of text. Repetitive characters, punctuations, needless white spaces between words,

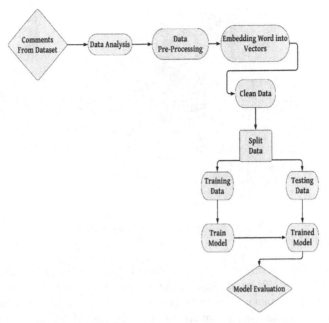

Fig. 5. The flow of the proposed methodology

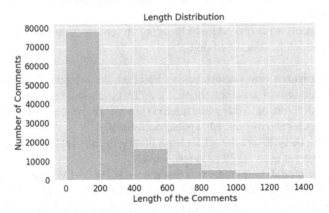

Fig. 6. Visualization of the data

and non-English characters should all be removed. as well as "n" Lower-case data conversion

- **Lemmatization** is a procedure of determining a word's normalized form. It unites a word's multiple inflected forms into a single, analytical unit called a lemma [11]. A word's meaning is usually determined by its use in the text; likewise, different forms of words convey related meanings, such as "bag" and "bags," which have the same meaning. A search for "bag" and a search for "bags "would most likely get

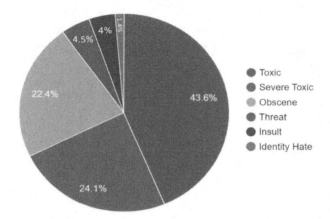

Fig. 7. Visualization of labeled data

the same results. This type of contrast between distinct forms of words is known as "inflection," [12] but it causes a variety of difficulties in interpreting inquiries.

- **Stop-words** contain conjunctions, pronouns, assisting verbs, and other terms like is, am, are, he, she, and it, that appear in the dataset in large numbers yet have no importance. After reviewing the dataset, we noticed some stopwords and removed them as well. The vocabulary size is considerably reduced after the removal of stop words.
- **Tokenization** is the procedure of breaking up sentences into independent tokens, each of which is a one-word unit.

- **Embedding words into vectors:** Using FastText, the pre-processed comments are converted into vectorized representation. FastText is chosen because it considers the underlying structure of words while interpreting word representations, which may be advantageous for words that only appear occasionally.
- **Splitting the Data:** Using the sci-kit learn library, the dataset is split into training set and validation set in an 80:20 ratio respectively.
- **Training the model:** Our model was first trained using LSTM. The architecture is as follows, GlobalMaxPoolong layer, followed by a GlobalMaxPooling layer, dropouts, and dense layers. The architecture of LSTM-CNN with FastText model is as follows, an Embedding layer, an input layer, LSTM, Convo1D, MaxPooling1D, GlobalMax-Pooling1D, Batch Normalization, Dense Layers, and Dropouts. The architecture of the LSTM with FastText model is as follows, an Embedding layer, followed by an input layer, LSTM, GlobalMaxPooling1D, Dense Layers, and Dropouts in this model. We trained all the above three models for 2 epochs.
- **Model Evaluation:** Loaded the model with unseen data, i.e. validation set, to make predictions and to measure the performance. Detailed evaluation and analysis are given in the Comparative Analysis section of this paper.

4   Comparative Analysis

The accuracy, loss, validation accuracy, and validation loss were noted while training the model in order to compare and conclude the best model which is shown in Table 2. Our proposed model was able to achieve high accuracy majorly because of two things, one is FastText word embedding which performs better than most of the other word embeddings. Another factor is hyperparameter tuning, which includes the length of sentences in the dataset, total vocabulary, neural network architecture, and the number of neurons that were optimized to a great extent. The LSTM model was for our personal experimental purpose of understanding FastText word embedding.

Table 2. Comparative analysis of LSTM, LSTM with FastText, and LSTM-CNN with FastText.

Proposed models	Accuracy ↑	Loss ↓	Val Accuracy ↑	Val Loss ↓
LSTM	98.88%	4.64%	99.46%	4.95%
LSTM (FastText)	**99.08%**	**4.87%**	99.32%	**4.59%**
LSTM-CNN (FastText)	*98.31%*	5.39%	**99.46%**	4.88%

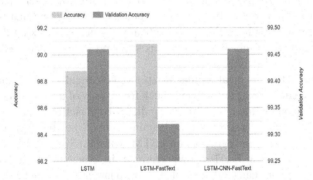

Fig. 8. A visual comparison of accuracy and validation accuracy

The above Fig. 8 and Fig. 9 compares the loss and accuracy of all three proposed models. Low loss and high accuracy are desirable. Comparing the models, our study concludes that LSTM with FastText word embedding outperforms the other models.

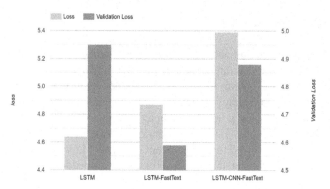

Fig. 9. A visual comparison of loss and validation loss

5   Discussion and Conclusion

Social media has become an integral part of our lives. Some use it for entertainment purposes whilst some use it as a platform to showcase their skills and earn money. It has given people the liberty to express their views on various issues and topics. A lot of times, people post atrocious comments that spread negativity, hate, and toxicity [22] among users which may psychologically and mentally affect the user. Hence it is very important to make it a safe platform for everybody. It is something that is concerning and needs to be dealt with quickly and practically. Manually checking the comments is impractical and sometimes inaccurate. This study proposes a methodology that can accurately detect toxicity in the comments.

A model would be a practical approach that can automatically detect toxic comments. It can be embedded into the platform and check for toxicity of a comment before it goes onto the public. In our opinion, there must be policies in the platform that allows blocking toxic comments, giving warnings to even banning users if they try to post such hateful comments. This study mainly focuses on LSTM with FastText and LSTM-CNN with FastText. The LSTM with FastText outperformed the other model. In the future, the dataset is something one can work on such as using a larger dataset and even working on removing the skewness or the imbalance that exists. Furthermore, advanced, and more efficient architecture like transformer-based, BERT, can be used to make the model more reliable and accurate. In future we would like to integrate our model with real social media platforms using parallel processing techniques for faster and reliable detection of comments.

References

1. Website. https://backlinko.com/social-media-users. Accessed 19 Mar 2023
2. Jung, J., Petkanic, P., Nan, D., Kim, J.H.: When a girl awakened the world: a user and social message analysis of Greta Thunberg. Sustainability **12**, 2707 (2020). https://doi.org/10.3390/su12072707
3. Amedie, J.: The Impact of Social Media on Society (2015)
4. Duggan, M.: Online harassment. Pew Research Center (2014)

5. Konikoff, D.: Gatekeepers of toxicity: reconceptualizing Twitter's abuse and hate speech policies. Policy Internet **13**, 502–521 (2021). https://doi.org/10.1002/poi3.265
6. Minar, M.R., Naher, J.: Recent Advances in Deep Learning: An Overview (2018)
7. Di, W., Bhardwaj, A., Wei, J.: Deep Learning Essentials, January 2018
8. Santos, I., Nedjah, N., de Macedo Mourelle, L.: Sentiment analysis using convolutional neural network with fastText embeddings. In: 2017 IEEE Latin American Conference on Computational Intelligence (LA-CCI), pp. 1–5 (2017). https://doi.org/10.1109/LA-CCI.2017.828 5683
9. Website. https://medium.com/techiepedia/binary-image-classifier-cnn-usingtensorflow-a3f 5d6746697. Accessed 19 Mar 2023
10. Website. https://colah.github.io/posts/2015-08-Understanding-LSTMs/. Accessed 19 Mar 2023
11. Manning, C.D., Raghavan, P., Schütze, H.: Introduction to Information Retrieval. Cambridge University Press (2008)
12. Plisson, J., Lavrac, N., Mladenic, D.: A rule based approach to word lemmatization. In: Proceedings of IS, vol. 3 (2004)
13. Husnain, M., Khalid, A., Shafi, N.: A novel preprocessing technique for toxic comment classification. In: 2021 International Conference on Artificial Intelligence (ICAI), pp. 22–27 (2021). https://doi.org/10.1109/ICAI52203.2021.9445252
14. Mestry, S., Singh, H., Chauhan, R., Bisht, V., Tiwari, K.: Automation in social networking comments with the help of robust fastText and CNN. In: 2019 1st International Conference on Innovations in Information and Communication Technology (ICIICT), pp. 1–4 (2019). https://doi.org/10.1109/ICIICT1.2019.8741503
15. Rahul, Kajla, H., Hooda, J., Saini, G.: Classification of online toxic comments using machine learning algorithms. In: 2020 4th International Conference on Intelligent Computing and Control Systems (ICICCS), pp. 1119–1123 (2020). https://doi.org/10.1109/ICICCS48265. 2020.9120939
16. Rupapara, V., Rustam, F., Shahzad, H.F., Mehmood, A., Ashraf, I., Choi, G.S.: Impact of SMOTE on imbalanced text features for toxic comments classification using RVVC model. IEEE Access **9**, 78621–78634 (2021). https://doi.org/10.1109/ACCESS.2021.3083638
17. Zhang, J., Li, Y., Tian, J., Li, T.: LSTM-CNN hybrid model for text classification. In: 2018 IEEE 3rd Advanced Information Technology, Electronic and Automation Control Conference (IAEAC), pp. 1675–1680 (2018). https://doi.org/10.1109/IAEAC.2018.8577620
18. Sumanth, P., Samiuddin, S., Jamal, K., Domakonda, S., Shivani, P.: Toxic speech classification using machine learning algorithms. In: 2022 International Conference on Electronic Systems and Intelligent Computing (ICESIC), pp. 257–263 (2022). https://doi.org/10.1109/ICESIC 53714.2022.9783475
19. Dubey, K., Nair, R., Khan, M.U., Shaikh, P.S.: Toxic comment detection using LSTM. In: 2020 Third International Conference on Advances in Electronics, Computers and Communications (ICAECC), pp. 1–8 (2020). https://doi.org/10.1109/ICAECC50550.2020.9339521
20. Vichare, M., Thorat, S., Uberoi, C.S., Khedekar, S., Jaikar, S.: Toxic comment analysis for online learning. In: 2021 2nd International Conference on Advances in Computing, Communication, Embedded and Secure Systems (ACCESS), pp. 130–135 (2021). https://doi.org/10. 1109/ACCESS51619.2021.9563344

21. Pavel, M.I., Razzak, R., Sengupta, K., Niloy, M.D.K., Muqith, M.B., Tan, S.Y.: Toxic comment classification implementing CNN combining word embedding technique. In: Smys, S., Balas, V.E., Kamel, K.A., Lafata, P. (eds.) Inventive Computation and Information Technologies. LNNS, vol. 173, pp. 897–909. Springer, Singapore (2021). https://doi.org/10.1007/978-981-33-4305-4_65

22. Varma, R., Verma, Y., Vijayvargiya, P., Churi, P.P.: A systematic survey on deep learning and machine learning approaches of fake news detection in the pre- and post- COVID-19 pandemic. Int. J. Intell. Comput. Cybern. **14**(4), 617–646 (2021). https://doi.org/10.1108/IJICC-04-2021-0069

UEye: Insights on User Interface Design Using Eye Movement Visualizations

S. Akshay$^{(\boxtimes)}$ ⓘ, Anupam Shukla ⓘ, and Vishnu K Raman ⓘ

Department of Computer Science, School of Computing,
Amrita Vishwa Vidyapeetham, Mysuru, Karnataka, India
`s_akshay@my.amrita.edu`

Abstract. The internet has become a primary need in the present, and understanding the importance of a well-designed User Interfaces(UI) is a prerequisite. Eye-tracking can help design appropriate UI that can have a positive impact on any business. Eye Tracking is the method of knowing where a person is looking. This study presents a novel eye-tracking system that uses a high-resolution webcam to record raw eye gaze data. The system includes an initial calibration process that adapts to the user's screen size and lighting conditions to improve tracking accuracy. After calibration, the system records eye movement data during a user's interaction with a designated test subject. Preprocessing is applied to ensure data reliability, and fixations are calculated based on Euclidean distance, time differences, and velocity thresholds. The resulting fixations are used to generate visual heat maps and scan paths that reveal user behavior on the test subject. Further analysis, such as region-wise attraction percentages, helps identify important UI components that users focus on the most. Overall, this eye- tracking system provides valuable insights into user behavior and can inform UI design decisions to enhance user experience. The proposed system is not only cost-effective but also provides decent accuracy of 83%, making it a useful tool for designers and organization to optimize their UI components for user attention.

Keywords: User Interface · Eye tracking · Visualization · Heatmap · Fixation

1 Introduction

The study of eye-tracking has been around for more than 110 years and as time has progressed, we have seen many advancements in the discipline, unlike in the past when they were practiced in unique laboratories where many dimensions had to be taken care of. Eye-tracking consists of acquiring the position of a person's eyes when focusing at a visual [6]. The position of the eyes is calculated multiple times every second and overlaid on a stimulus recording [1, 4, 22]. The generally accepted hypothesis behind eye detection is that of a direct match between where people are looking and where they focus their attention. Thus, by

© The Author(s), under exclusive license to Springer Nature Switzerland AG 2023
M. Singh et al. (Eds.): ICACDS 2023, CCIS 1848, pp. 253–264, 2023.
https://doi.org/10.1007/978-3-031-37940-6_21

looking at the eye movements of people, we can get detailed information about their attention processes and know about what they find important, interesting, or confusing. There are a variety of eye trackers available but they differ in the methods used to find the eye position. In 2001, Tobii Technology started developing advanced IR-based eye trackers. These days not only IR-based eye trackers but also webcam-based eye-trackers are gaining popularity keeping in mind that the data collected from webcam-based eye-trackers may not be as accurate as IR-based trackers. The most common type of eye movement is called fixations and saccades [16,18]. Fixations are points of relatively stable gaze when individuals are expected to pay attention to the stimulus at their point of regard, where the eyes change the focus by jumping from one place on the stimulus to another. These speedy jumps are referred to as saccades. Heatmaps and scan paths can help to collect data on how people interact with your website, so you can use those insights to improve UI and the metrics that matter to your business The user interface is the graphical layout design of an application or website [12]. It includes buttons, text, images [13,26], form fields, and all the other items the user interacts with. Eye-tracking allows you to discover usability problems without disturbing natural user behavior. By analyzing the eye movements, a product designer can get insights and improve the UI design elements. Webcam is easily available as compared to IR-based eye-tracker so collecting data will be easy and one big advantage of the webcam is getting more participation as compared to IR-based eye-tracker as webcam is easily accessible. Considering the fact that the availability of an IR-based eye tracker is limited due to its cost and hence this research focuses on the development of a system that is cost-effective. The proposed system collects data from an inbuilt or external webcam. The data collected is then preprocessed and fixations are calculated on the same using the IVT algorithm. Further fixations are visualized using a heatmap and certain areas are recognized where the focus of visitors is high. The recognized areas can then be used to place certain UI components on which higher user attention is required. The system thus can be used as an analysis tool by designers, web content creators. The article explains the related work in Sect. 2. The proposed system is explained in Sect. 3. The detailed experimentation is presented in Sect. 4. The results obtained and visualizations are reported in Sect. 5. The conclusion and the future scope of the work is presented in Sect. 6.

2   Related Work

Some of the existing technologies for eye tracking are presented by [8,10,17,25]. The work by [24] defines eye tracking as the analysis of eye movements such as fixation and fixation time. Fixation maps are a two-dimensional record of the location of the analyzed fixations. They use a 3D Gaussian function to produce fixation map. The study also mentions how an area of interest can enable gaining insights into a specific region of the stimulus. This becomes a major advantage when the points of interest on a stimulus in high and more information related to a specific region are required. It is known that fixations are an essential part of eye-tracking so to identify them with respect to spatial criteria like velocity Velocity Threshold Identification(IVT) algorithm can be used [7,19]. IVT is based on velocity. IVT begins by calculating point-to-point velocity where each velocity is calculated as the distance between the current point to the next or previous point and it classifies each point as fixation based on a threshold where if the velocity at a point is less than a certain threshold point then it is classified as fixation. I-AOI algorithm identifies fixations that occur within specified target areas. I-AOI starts by connecting the data point with target areas. A point in the target area is called a fixation. This technique is useful to gain deeper insights into a specific region. [11] talk about the number of users that are required for a study to be accurate when compared to a study which has a comparatively high number of users. It is found that a relatively low number of users can produce up to 75% similar results as compared to a study with a greater number of users. Also, it was observed that the required number of users might be higher in cases where a user is browsing through a page. work by [9] discusses the eye gaze tracking problems faced while using a low-cost and convenient web camera in a desktop environment, as compared to gaze tracking techniques that require specific hardware such as high resolution camera and infrared light sources. Also, issues related to eye detection failure in cases where the environmental light is low are highlighted. The experiment achieved an accuracy of 1.28° with no head movements and 2.27° with minor head movements [21]. [10] use an algorithm that uses iris geometry information to decide the area that contains the eye. In the candidate region once it finds the eye region then it selects the symmetry of each eye. The algorithm can work on any complex image so there is no restriction on the foreground and background. To demonstrate the effectiveness and robustness of the algorithm it is tested on people with different eye colors and some wearing glasses. When the eyes are not completely open only then the algorithm reaches a high detection rate. [5]use visualization method to size the spatio-temporal traits of individual element and on the equal time to perceive the rising organization behaviours. It offers the end result set to be record the similarity and exclusive throughout a set observers acting a not unusual place task. Set of rules are shows in story line visualization shows, same group of people viewing the same film clips. It records the what are people looking at dynamic clips gaze pattern. [14] Eye movements are processed in extraordinary stages. However a heat map makes the use of colors to differ the maximum gazed area and much less gazed are with the aid of using attention gaze information. Eye tracking revel remarkable possibilities for enhancing the visualization, which closely is predicate on correctly in the conveying fact true images. General highlight in eye tracking is the unique requirements and traits of eye tracking with inside the context of information visualization. Several eye movement based applications for gender detection [1], blink detection [4], program comprehension [6], online learnig [2], Digital topography [12], motivates us to propose a system that is completely based on eye movement features. The availability of various eye tracking technologies and its applications as mentioned in this review inspired us to come with a eye movement based solution to the effective UI design.

3 Proposed Method

This state-of-the-art study presents a system that records raw eye gaze data using an external CASE-U 1080p webcam with auto low light correction and a 110-degree field of view at 30 frames per second. Although there is no dependency where the specified webcam should be used but it is still recommended to use a 1080p webcam to get more accurate results. Every user will perform an initial calibration such that the system adapts to the current screen size, the lighting and improve its tracking. This calibration is a continuous learning process for the system. After calibration, an image has to be selected as a test subject and then the system starts recording the eye movement of the user. The recorded data is then preprocessed to ensure its reliability. Next fixations are calculated based on the preprocessed data. Euclidean Distance is calculated between two points $(x1, y1)$ and $(x2, y2)$

$$d = \sqrt{(x2 - x1)^2 + (y2 - y1)^2} \tag{1}$$

where $x1$ and $y1$ are the coordinates of the first point and $x2$ and $y2$ are the coordinates of the second point and d is the Euclidean Distance between the two points. Next, difference between the timestamp of the two points is being calculated.

$$t = t1 - t0 \tag{2}$$

where $t0$ is the timestamp of the first point and $t1$ is the timestamp of the second point and $t1$ is the time difference between the two points. Finally, the velocity between the two points is calculated by dividing the calculated Euclidean distance and time. If the calculated velocity is less than the threshold only then the point is considered as a fixation. [3]

$$v = d/t \tag{3}$$

$$v < v_{\text{thr}} \tag{4}$$

Here, v is the calculated velocity and $v/_{\text{thr}}$ the velocity threshold. The same calculation continues for every gaze point and the result will be a list of fixations. The system here relies heavily on fixations for all the analysis that it does. These fixations are then plotted as a heatmap on the test subject. A scan path is also plotted which clearly depicts the position from where a user started looking on the screen and the sequence in which the rest of the UI components were viewed. More analysis such as most viewed region, region-wise attraction percentage, first viewed region, last viewed region, and area of interest is then produced by the system. Here, a region is a section on the UI, the system has by itself divided the UI into 3 columns where the width of the first column, the third column is 20% and the width of the second column is 60% of the whole screen. The reason for these specified 3 columns and the static width ratio is that a columnar approach in building a website is most commonly used method. These metrics help in assessing the participant's behavior on the screen, in knowing the participant's approach in finding a specified option on the screen, the position from where a participant starts looking at the screen, and the path followed thereafter. Region-wise attraction percentage also helps in finding a specific spot on the screen where a participant focuses the most. These identified regions can be used to place certain important components of the UI, the components that the designer or the content owner wants every visiting user to look at [20]. The Fig. 1 below shows a graphical representation of the process. A flow chart depicting the proposed system is given in Fig. 2.

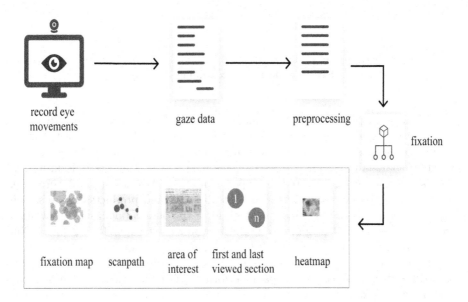

Fig. 1. UEye: The Proposed System

4   Experimentation

Participants of the experiment are people whose age varies in the range of 20–35. As this is the age group that has had access to various social/e-commerce platforms for a considerable time and is familiar with the technology. Any participant who is not familiar with technology will not be considered a right fit for the experimentation. A total of 25 participants were selected for the study, these are technically sound people aged between 20–35. Every participant has signed a consent to allow the collected data to be used for the study. Participants were asked to be well seated in front of the screen and to make sure they are in the webcam's frame and a well-lit surrounding so that the eye tracker can detect their features clearly and accurately. Every participant has to perform calibration of the eye tracker by clicking on 9 various points on the screen 5 times each while continuously looking at that specific point. An accuracy percentage

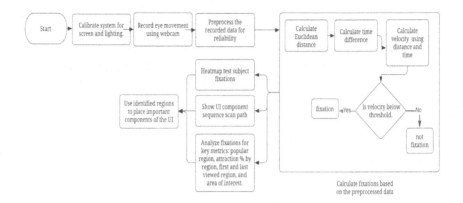

Fig. 2. A flowchart explaining the UEye

will then be displayed on the screen. If accuracy is anywhere below 70% then it is suggested to redo the calibration process so that the quality of the study is not compromised by any means. After calibration is complete the eye tracker starts tracking and storing the eye movements. Online e-commerce desktop website is the test subject selected for experimentation. A series of tasks were given to participants where task 1 was to spot all sponsored products on the screen. This would require a participant to look for a tag on a product card that signifies a specific product is sponsored on the product list page. Task 2 was to find Reynolds 045 pen on the product list page. Each participant had to complete the given tasks one after another and the eye gaze data was continuously being collected for every task the participant performs. After a participant completes all the tasks, they were to answer a questionnaire that focuses on knowing the difficulty that a participant faced while completing each of the given tasks. Results from this questionnaire will help in determining if the results produced by the eye tracker are accurate or not. If a participant decides to quit the experiment midway, then their data was altogether excluded from the study.

5   Results and Discussions

Based on the data collected, the eye tracker then produces certain results such as fixations, most viewed region, region-wise attraction percentage, first viewed region, last viewed region, scan path, and area of interest. Each one of these features will help in understanding the user behavior on a user interface and give deeper insights into the user's experience not only in terms of difficulty in spotting an option on the user interface but much more than that. This can be leveraged by individuals and firms to study their user interface. In the first task, the participant was asked to spot the sponsored products on the products page of an online ecommerce desktop website. Figure 3 below shows the user behavior on the specific interface. A heatmap will help in identifying regions where a user has looked the most and give insights into regions the user has spent more time looking at and a scan path shows the path that the user followed gazing through the various sections of the website. Both heatmap and scan path are being generated from the fixations.

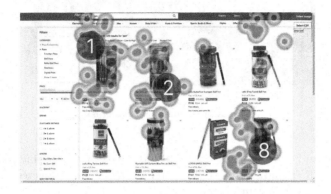

Fig. 3. A heatmap for first task

Figure 4 shows that on the first, second, and eighth product the user has spent a considerable amount of time, and hence we have got multiple fixations around the same product which is also the reason for the dense red color around these three products. The scan path Fig. 3 clearly outlines the path that participant followed and we see that participant started by looking at product 1's product name and gaze moved above after this which is where the advertisement icon is present. We also notice that for the first product participant took some time which created a considerable number of fixations but from the next product the participant was particularly looking at the specific position where the advertisement tag is present. Looking at these results we can assert that the advertisement tag on this specific page is well placed and designed such that a user is able to spot it without having to look around the screen searching for it. We can also see a few fixations on other parts of the UI such as some around the fourth product. Now there can be multiple reasons for these fixations, one being the user has in fact looked at these points for some information and another reason could be the eye tracker's fault. We should always consider the fact that a webcam might lose focus sometimes and during this timeframe the eye tracker might go wrong and store incorrect gaze points.

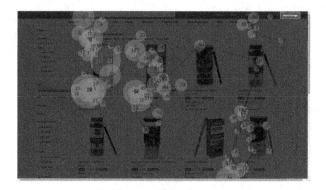

Fig. 4. A scanpath for first task

In the second task, the participant was asked to spot a specific product [23]. From the heatmap and scan path generated from the data collected in this task, we noticed that the first product always attracts more attention than any other region on the UI. Here the participant was asked to spot Reynolds 045 blue pen which is the sixth product. In the Fig. 5 see a considerable number of fixations around the second product. This is because the product name here is much similar to what the participant is looking for and we also see that more than the product name the participant has focused on the image to identify if a specific product is by Reynolds. Once the participant could confirm the brand name by looking at the image only then the gaze moved to the product name. The participant also seems to be less keen on looking at the products on the right side of the UI. In cases where the product to be spotted was on the right side, then the participant would end up going through all the products in the center of the screen first and only then would look at the products listed on the right side of the screen [15] (Fig. 6).

To dig deeper into the ignorance towards products on the right parts of the screen we calculated region-wise gaze attraction percentage based on the fixations for all the above tasks both individually and combined. For this, we split the screen into 3 columns, where the first and last column took up 20% width of the whole screen and the second column took up the rest of 60% width. This way of dividing the screen into 3 columns is based on the modern web design pattern. The columns are split such that most of the websites designed in a columnar pattern will fit right in these columns.

From the attraction percentage of the first and second tasks individually we could spot that about 87% and 100% of the fixations lie in the center region of the screen for both the tasks respectively (Refer to Fig. 7 and 8). Marginal attraction can be seen in regions on both ends of the UI. With this, we can conclude the attraction on either end of the UI is less when compared to the center, and on an ecommerce website, it is found that the first product always gets the most attention when compared to any other region of the website. Any ecommerce website should be using the first product position to place an adver-

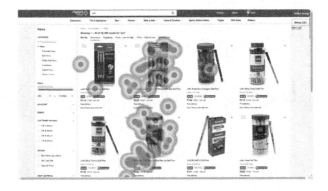

Fig. 5. A heatmap for second task

Fig. 6. A scanpath of task two.

Fig. 7. AOI of first task.

tisement as it will attract the most gaze and will generate better advertisement outcomes. Use cases of the study are not limited to ecommerce platform, any website that contains advertisements will become a potential use case as know-

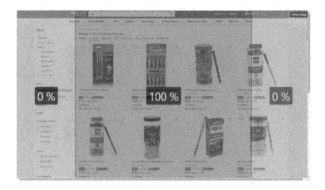

Fig. 8. AOI of second task.

ing the areas on a website where the user gaze is maximum in terms of time spent in looking at that specific region will help the creators decide where to place advertisement on a specific page to gain the maximum user attention and increase user engagement. This can significantly increase the revenue generated from advertisements. Moving away from the advertisement use case, the system can also be used to place the right content on the right position in a website. For instance, if a creator wants to focus on increasing the newsletter sign-up on their website then the same can be leveraged to know the areas on UI where the user attention is highest and that particular region can be used to place the newsletter sign-up component on the website. At a later point in time if the creator wants to focus on increasing visits to blog, then the same region on UI can be used to place snippets of blogs and the previously placed newsletter sign-up component can be temporarily placed somewhere else.

6   Conclusion

In this study, a cost-effective eye-tracking system was developed using a high-resolution webcam to record raw eye gaze data. The system includes an initial calibration process that adapts to the user's screen size and lighting conditions to improve tracking accuracy. The resulting fixations were used to generate visual heatmaps and scan paths that reveal user behavior on the test subject. Further analysis helped identify important UI components that users focus on the most. Overall, this eye-tracking system provides valuable insights into user behavior and can inform UI design decisions to enhance user experience. This significantly increases the accessibility of eye tracking in future to anyone out there and benefiting from the technology for their specific use case. In future, work can be done on improving the systems accuracy in conditions where the environment light is low, alongside study on improving the calibration process.

References

1. Akshay, S., Lakshmipriya, V., Sikha, S.: Analysis of spatial eye gaze data for aiding gender identification. In: 2017 International Conference on Intelligent Computing and Control (I2C2), pp. 1–5. IEEE (2017)
2. Akshay, S., Vasanth, P.: A CNN based model for identification of the level of participation in virtual classrooms using eye movement features. In: 2022 IEEE International Conference on Electronics, Computing and Communication Technologies (CONECCT), pp. 1–6 (2022). https://doi.org/10.1109/CONECCT55679.2022.9865694
3. Al-Rahayfeh, A., Faezipour, M.: Eye tracking and head movement detection: a state-of-art survey. IEEE J. Transl. Eng. Health Med. 1, 2100212–2100212 (2013)
4. Amudha, J., Roja Reddy, S., Supraja Reddy, Y.: Blink analysis using eye gaze tracker. In: ISTA 2016. AISC, vol. 530, pp. 237–244. Springer, Cham (2016). https://doi.org/10.1007/978-3-319-47952-1_18
5. Balint, J.T., Arendt, D., Blaha, L.M.: Storyline visualizations of eye tracking of movie viewing. In: 2016 IEEE Second Workshop on Eye Tracking and Visualization (ETVIS), pp. 35–39. IEEE (2016)
6. Bednarik, R., Tukiainen, M.: An eye-tracking methodology for characterizing program comprehension processes. In: Proceedings of the 2006 Symposium on Eye Tracking Research & Applications, pp. 125–132 (2006)
7. Bojko, A.A.: Informative or misleading? heatmaps deconstructed. In: Jacko, J.A. (ed.) HCI 2009. LNCS, vol. 5610, pp. 30–39. Springer, Heidelberg (2009). https://doi.org/10.1007/978-3-642-02574-7_4
8. Chavda, R., Doshi, A., Barve, M., Karani, R.: Real time eye-tracking using web camera. Int. J. Comput. Sci. Eng. (IJCSE) 3, 64–67 (2015)
9. Cheung, Y.M., Peng, Q.: Eye gaze tracking with a web camera in a desktop environment. IEEE Trans. Hum. Mach. Syst. 45(4), 419–430 (2015)
10. D'Orazio, T., Leo, M., Cicirelli, G., Distante, A.: An algorithm for real time eye detection in face images. In: Proceedings of the 17th International Conference on Pattern Recognition, 2004. ICPR 2004, vol. 3, pp. 278–281. IEEE (2004)
11. Eraslan, S., Yesilada, Y., Harper, S.: Identifying patterns in eyetracking scanpaths in terms of visual elements of web pages. In: Casteleyn, S., Rossi, G., Winckler, M. (eds.) ICWE 2014. LNCS, vol. 8541, pp. 163–180. Springer, Cham (2014). https://doi.org/10.1007/978-3-319-08245-5_10
12. Incoul, A., Ooms, K., De Maeyer, P.: Comparing paper and digital topographic maps using eye tracking. In: Brus, J., Vondrakova, A., Vozenilek, V. (eds.) Modern Trends in Cartography. LNGC, pp. 339–356. Springer, Cham (2015). https://doi.org/10.1007/978-3-319-07926-4_26
13. Jassim, F.A., Qassim, H.E.: Five modulus method for image compression. arXiv preprint arXiv:1211.4591 (2012)
14. Kurzhals, K., Burch, M., Pfeiffer, T., Weiskopf, D.: Eye tracking in computer-based visualization. Comput. Sci. Eng. 17(5), 64–71 (2015)
15. Leckner, S.: Presentation factors affecting reading behaviour in readers of newspaper media: an eye-tracking perspective. Vis. Commun. 11(2), 163–184 (2012)
16. Lupu, R.G., Ungureanu, F.: A survey of eye tracking methods and applications. Buletinul Institutului Politehnic din Iasi, Automatic Control and Computer Science Section 3(1), 72–86 (2013)
17. Papoutsaki, A., Sangkloy, P., Laskey, J., Daskalova, N., Huang, J., Hays, J.: Webgazer: scalable webcam eye tracking using user interactions. In: Proceedings

of the 25th International Joint Conference on Artificial Intelligence (IJCAI), pp. 3839–3845. AAAI (2016)

18. Rosch, J.L., Vogel-Walcutt, J.J.: A review of eye-tracking applications as tools for training. Cogn. Technol. Work **15**(3), 313–327 (2013)

19. Salvucci, D.D., Goldberg, J.H.: Identifying fixations and saccades in eye-tracking protocols. In: Proceedings of the 2000 Symposium on Eye Tracking Research & Applications, pp. 71–78 (2000)

20. Sanchez-Brito, M., Garcia-Hernandez, C.F.: Eyes segmentation as a help to communication process. In: 2017 International Conference on Computing Networking and Informatics (ICCNI), pp. 1–4. IEEE (2017)

21. Schoning, J., Faion, P., Heidemann, G., Krumnack, U.: Eye tracking data in multimedia containers for instantaneous visualizations. In: 2016 IEEE Second Workshop on Eye Tracking and Visualization (ETVIS), pp. 74–78. IEEE (2016)

22. Scott, N., Green, C., Fairley, S.: Investigation of the use of eye tracking to examine tourism advertising effectiveness. Curr. Issue Tour. **19**(7), 634–642 (2016)

23. Špakov, O., Miniotas, D.: Visualization of eye gaze data using heat maps. Elektronika ir elektrotechnika **74**(2), 55–58 (2007)

24. Wooding, D.S.: Fixation maps: quantifying eye-movement traces. In: Proceedings of the 2002 Symposium on Eye Tracking Research & Applications, pp. 31–36 (2002)

25. Zemblys, R.: Eye-movement event detection meets machine learning. Biomed. Eng. 2016, **20**(1) (2017)

26. Zia, M.A., Ansari, U., Jamil, M., Gillani, O., Ayaz, Y.: Face and eye detection in images using skin color segmentation and circular hough transform. In: 2014 International Conference on Robotics and Emerging Allied Technologies in Engineering (iCREATE), pp. 211–213. IEEE (2014)

Code Mixed Information Retrieval for Gujarati Script News Articles

Payal Joshi$^{(\boxtimes)}$ ⓘ and Dhaval Joshi ⓘ

Department of Information and Communication Technology, Veer Narmad South Gujarat University, Surat, Gujarat, India
{pkpandya,dajoshi}@vnsgu.ac.in

Abstract. Large amount of digitized content is being created and made available online in many Indian languages. Searching from such content is difficult because of lack of regional language input tools and interface. Further documents may also contain commonly used English words written using Indian script which we refer as code mixed content. One of the easier methods of query input could be typing query's words using Roman script mixing words of Indian language and English in single script which we refer as code mixed queries. For various Indian languages work is done on single language (code) transliterated search but code mixed information retrieval is very less explored area. Gujarati is an Indian regional language spoken by 55 million people in the world, but very less work is done for code mixed information retrieval and it lacks transliteration pair corpus. To search code mixed contents we need to identifying language of a word and we have to back transliterate it from Roman to Indian language script. Moreover certain words belong to both the languages. All these issues are major challenges in code mixed IR especially for Indian languages. A novel hybrid approach is proposed in this paper for Gujarati words transliteration which combines the use of Roman script Gujarati Soundex based match with alternative use of machine learning based improved CRF with post processing. A huge transliteration pair corpus of Gujarati script is created for words of both languages. Analysis of approaches for language identification and transliteration is presented in this paper along with compared retrieval result for different transliteration approaches and obtained an overall increase of 17% in MAP as compared to other approaches.

Keywords: Language Identification · Transliteration · Information Retrieval

1 Introduction

In India, a large amount of digital content is published online in the form of eBooks and news articles of different regional languages on daily basis. Mixing some commonly used English words in such mono-lingual or mono-script contents is common which we refer as code mixed contents. Making this content searchable for the search engine user

---

http://fire.irsi.res.in.
https://msir2016.github.io/.

© The Author(s), under exclusive license to Springer Nature Switzerland AG 2023
M. Singh et al. (Eds.): ICACDS 2023, CCIS 1848, pp. 265–276, 2023.
https://doi.org/10.1007/978-3-031-37940-6_22

is difficult because of various social and technical reasons. It is a known fact that English is the most popular foreign language and use of transliterated Roman script code mixed query to search for Indian language documents can facilitate user for the same [1]. The process of phonetically representing the words of a language in a non-native script is called transliteration [14]. It is a major challenge to match code mixed Roman script query words with Indian language documents. For many Indian languages like Hindi, Bangla, Punjabi good amount of work is done on mono-lingual transliterated search. Gujarati is one of the regional language spoken by 55 million people. Gujarati is a low resource language and no substantial work is done in code mixed query expansion for cross script (transliterated) information retrieval. Gujarati language also lacks linguistic resources and transliteration pair corpus.

In this paper the problem of Code Mixed Information Retrieval (CMIR) is described with its research issues. Gujarati news corpus search is a crucial requirement for people of Gujarat and for other Gujarati people settled in various other locations around the world. This is the major source of motivation for research on CMIR for Gujarati news corpus.

Rest of the paper is organized as follows: Sect. 2 defines problem, scope and challenges, Sect. 3 describes related work. Section 4 describes details about code mixed query expansion approaches of language identification and back transliteration along with analysis of these approaches and propose a new approach. Section 5 describes methodology used and implementation. Section 6 describes dataset used for experimentation and comparative analysis of results and Sect. 7 concludes the paper.

2   CMIR: Problem Definition and Challenges

In this section, we define the problem, objectives and challenges of code mixed document retrieval with reference to FIRE Gujarati language evaluation dataset and query set.

2.1   Problem Definition

Every language is written in a particular script. E.g. English is written in Roman script. Code mixed means using one script, words of one than one language (code) are mixed. E.g. To search documents of Gujarati language for query યુરો અપનાવાતા દેશો, user can type query in Roman script as "Euro apanavata desho". In this query, words of two languages are mixed using Roman script. Euro is English language word and other two words belong to Gujarati language which we refer as code mixed query.

Transliterated or cross script document retrieval means, using one language script query, we search documents of other language script. E.g. "Euro apanavata desho" query is typed in Roman script but it aims to search documents which contain words યુરો અપનાવાતા દેશોwhich are words written in Gujarati script. So code mixed information retrieval means mixing words of more than one language(code) using a single script, attempt to search documents of different script. Motivation of this problem is taken from FIRE Mixed Script Information Retrieval (MSIR) task. 2 But very less work is done on problem of code mixed information retrieval.

Moreover, lack of research on the problem of Code Mixed search is also mentioned in [1] by P. Gupta.

Observing test queries given in Table 3, challenges observed are: (1) Language identification, (2) Handling words common in both languages and (3) Handling spelling variations and possible multiple transliterations.

3 Related Work

Code mixed information retrieval has attained very less attention in area of information retrieval especially for Gujarati language. In this section we describe work done for transliterated search, MSIR and CMIR in Indian languages.

P Gupta and P Rosso [1] have employed deep learning approach for mixed script information retrieval in Hindi song lyrics search. Shraddha Patel and Vaibhavi Desai [2], in FIRE 2014 in Mixed Script IR task have employed combination of bi-gram and tri-gram for language identification using LIGA and rule based approach for transliteration. But they have used Hindi as base language for transliteration to Gujarati language. Moreover they do not handle words common to both languages. Royal Denzil Sequiera, Shashank S Rao, and Shambavi B R [3] have also used rule based tri-gram approach to identify language and dictionary based approach to back transliterate a word to its native script. Irshad Ahmad Bhat, Vandan Mujadia, Aniruddha Tammewar, Riyaz Ahmad Bhat, and Manish Shrivastava in [4] have used ID3 classifier for language identification and Indic-converter for transliteration of Gujarati language. A. Agarwal [5] has employed CRF approach for Hindi named entity transliteration. S Gella, J Sharma and K Bali [6] have used MaxEnt classifier for language identification and hash function based approach for transliteration from English to Hindi and used Hindi as base language for transliteration to Gujarati and Bangla languages. S Singhal and N Tyagi [7] used hybrid approach for Hindi named entity transliteration. Verulkar, P., Balabantray, R. C., and Chakrapani, R. A [8] have used dictionary lookup based approach for transliterated search for Hindi lyrics. M Dhore, S Dixit and T Sonwalkar [9] trained a statistical machine translation system for transliteration of Hindi to English named entities using CRF-based approach. They showed 85.79% accuracy and showed that CRF is best suited for processing Indian languages. P. Pakray1, P. Bhaskar [10] have adapted joint source channel models for Hindi and Bangla transliteration. S. Banerjee, A. Roy, A. Kulia, S. Naskar, S. Bandyopadhyay. P. Rosso in [11] have used CRF and heuristics for language identification and PB-SMT for transliteration of Bangla words. N Joshi, I Mathur [12] have used rule based scheme for English to Punjabi transliteration. S. Varshney and J. Bajpai in [14] have done English to Hindi cross language information retrieval using transliteration of query terms. U. Barman, A. Das, J. Wagner and J. Foster in [15] and A. Das and B. Gambäck in [16] have worked on language identification of code mixed social media content in Hindi and Bangla languages. A. Das and B. Gambäck. P H Rathod, M L Dhore and R M Dhore in [17] have worked on Hindi and Marathi to English Machine Transliteration using support vector machine.

Most of the work is done for monolingual queries. Moreover, work done in Gujarati language is mostly rule based or is done with Hindi as a base language. Gujarati language also lacks transliterated pairs corpus. In Gujarati language and on FIRE Gujarati language news corpus data, no work for CMIR is done by anyone before. We present results along with comparison of different approaches at each step.

4   Code Mixed Query Expansion Methodology

Generally query expansion performs stop words removal and stemming. But code mixed queries written in Roman script, query expansion also involves language identification and back-transliteration to Gujarati script. In this section tasks and techniques involved in this process are described:

4.1   Stop Words Removal, Abbreviations and Numeric Value Identification

We have a list of 117 stop words of Gujarati and English language. For code mixed query expansion, first we remove words matching in this list. As Gujarati is the main language we are dealing with in the document corpus, we have removed Gujarati stop words and most common English language stop words like is, are, to, am etc.

Moreover, numeric values and abbreviations are also processed. Numeric values are identified using regular expression. Abbreviations are identified using rule based dictionary search. For this, a list of common abbreviations is developed.

4.2   Language Identification

All query keywords are typed in Roman script mixing Gujarati as well as English language words. So it is crucial to identify language of a word. For language identification using machine learning, word n-grams are used as language model features to train the model. For this 9448 labeled n-grams of English and Romanized Gujarati words collection is prepared. Support vector machine and logistic regression models are trained and compared their performance.

4.3   Stemming

Gujarati is morphologically rich language. To handle morphology, we have applied stemming at two levels: before back-transliteration and after back-transliteration. Before back-transliteration, common suffixes are removed and whole words are preserved in order to improve accuracy of language identification and transliteration. And a query may contain English words as well, so morphology of both the languages is handled in stemming process. After transliteration, minor suffixes of Gujarati diacritics are removed.

4.4 Back-Transliteration

In this paper back transliteration refers to conversion of query terms typed in Roman script back to Gujarati script. In this section various techniques of transliteration are discussed.

4.4.1 Dictionary Lookup with Exact Match: For Gujarati and English Language

In this transliteration mining technique, parallel pairs of Romanized words with its transliteration to Gujarati script is used. In this list, query term is searched and exact matching parallel Gujarati script transliterated word is used in query expansion. In this method spelling variations can't be handled.

4.4.2 Dictionary Lookup Using Romanized Gujarati Soundex: For Gujarati and English Language

In this transliteration mining technique also parallel pairs of Romanized words with its transliteration to Gujarati script is used. In this list, query term is searched in list using Roman script Gujarati soundex and matching word's transliteration(s) is/are picked from list and used in query expansion.

No work is done on Romanized Gujarati soundex algorithm. For Gujarati language Romanized words, changes are done in matching similar sounding letters in order to handle Gujarati accent and spelling variations. E.g Transliteration of the word Raja can be done as રજા or રાજા. Moreover Roman transliterations of many Gujarati characters are made up of double characters. Some of the classes of Romanized Gujarati soundex are "oo, u", "I, ee", "jh, j, z", "sh, s", "ksh, ks, x", "v, w".

4.4.3 Transliteration Generation Using CRF: For Gujarati Language

Transliteration generation refers to the method of chunking the word and converting it to its equivalent back-transliterated form and combining it back. This process is mostly based on machine learning approach.

Word chunking for transliteration generation is the process of phonetically dividing words into syllables. Syllables are parts that are made up of a vowel sound with or without a closely combined consonant sound.

Maximum Entropy (MaxEnt) machine learning model is trained for automated word chunking using manually chunked Romanized Gujarati words list as given in [18]. E.g. da-sta-ve-j for the word dastavej, vya-va-stha for the word vyavastha.

After word chunking, chunked word fragments can be transliterated using Conditional Random Fields (CRF) machine learning approach. In machine transliteration CRF can be used to generate the target language word from a source language word. Introduction and use of CRF model for transliteration done by A. Agarwal in [5] and S. Dhore in [9]. E.g. Word dastavej is first chunked using MaxEnt model as da-sta-ve-j and then it is transliterated as દ-સ્તા-વે-જ and chunked fragments are combined as દસ્તાવેજ.

4.4.4 Improved Transliteration Generation Using CRF with Post Processing Heuristic: For Gujarati Language

Transliteration generation using CRF may not generate accurate result every time. E.g. word Abhipray can be transliterated as અભિપ્રયinstead of અભિપ્રાય. To overcome this issue for this problem, a novel approach is developed by applying post processing heuristic as given in [19] which searches CRF result in Gujarati words list using Gujarati script soundex matching with which we can match word અભિપ્રયwith અભિપ્રાયand get the accurate word. For this we have used Gujarati words corpus of FIRE data. R Shah in [13] has also developed Soundex for Gujarati and Hindi script. But in this they have not matched certain characters like ક્ષwith ઝ્ર, ઙwith રwhich are included in this algorithm.

4.5 Proposed Hybrid Approach

In this paper two novel approaches are developed for back transliteration of Gujarati words (I) Romanized soundex as described in Sect. 4.4.2 and (II) Improved transliteration Generation using CRF with post processing heuristic as described in Sect. 4.4.4. As the proposed method, both of them are combined and an algorithmic hybrid approach is developed. As per the algorithm, if no word is matched with Romanized soundex, then improved transliteration generation using CRF with Gujarati script soundex based dictionary lookup approach is used for transliteration. Further, when user types the word, all generated transliterations are shown with web based interface and allow the user to select transliteration of his/her choice to avoid ambiguity of multiple possible transliterations.

Screenshot of web based interface developed is shown below in Fig. 1.

Fig. 1. Web based user selection interface.

5 Resources Developed and Implementation Document Search

For code mixed transliterated query expansion many resources have been created to solve the research problem.

5.1 Resources Developed and Used

For implementation of this methodology following data resources are developed and used:

1. List of 9448 labeled n-grams of English and Romanized Gujarati words.
2. Romanized Gujarati and Gujarati script transliteration pairs for 15,610 Gujarati language words.
3. English and its equivalent Gujarati script transliteration pairs for 3,130 English language words.
4. List of 120 words common to both languages along with their transliterations in Gujarati script in both languages.
5. List of commonly used abbreviations.
6. Romanized Gujarati Transliterated words of manually chunked 2080 words for training automated word chunking with MaxEnt model.
7. Romanized Gujarati and Gujarati Transliterated pairs of manually chunked 2080 words for training of transliteration using CRF. Sample data is shown in Table 8.
8. List of common Gujarati and English stop words in Roman script.
9. List of Romanized Gujarati suffixes to stem transliterated Gujarati query terms.
10. Mapping of English letter and numbers with their equivalent transliterated form in Gujarati script. E.g. A – એ, B- બી, C-સી, D-ડી.

5.2 Implementation

Implementation phase is divided in two parts: (I) Indexing and (II) Query expansion and document retrieval.

5.2.1 Indexing Gujarati Document Corpus

We have developed indexing module which performs common indexing tasks like tokenization, stop words removal and stemming for Gujarati script documents. For this we have used Gujarati language stop words list and suffix list as mentioned in Sect. 5.1

5.2.2 Code Mixed Query Expansion

All query keywords are typed in Roman script mixing Gujarati as well as well as English words. From user query first all stop words are removed and then query level stemming is done. Stemming trims major suffixes of query terms. Like word filmo becomes film. Word 'games' becomes 'game'. This process is followed by following phases:

Phase 1: Words containing numbers or abbreviations are back-transliterated character by character.
Phase 2: Language labeling: For this, SVM machine learning algorithm is used. Model is trained using collection of labeled n-grams (2, 3, 4 and 5 g) of English and Romanized Gujarati words.
Phase 3: For word labeled as English, English-Gujarati transliterated pairs list is matched using Soundex algorithm and query is expanded with its equivalent transliteration(s) in Gujarati script. Also same English word is searched in list of common words in order to cope up with possibility of a word to be common in both languages. E.g. are, same, deep etc.
Phase 4: For word labeled as Romanized Gujarati, Roman-Gujarati transliterated pairs list is searched using Romanized Gujarati Soundex algorithm.

Phase 5: If Gujarati word is not found in Phase 3 transliterated pairs list, then it is back transliterated using proposed hybrid approach as mentioned in Sect. 4.5.

Phase 6: All transliterated terms will be stemmed using Gujarati stemmer removing all diacritics.

5.2.3 Document Retrieval

All expanded query keywords are matched with document index and matching documents are retrieved. Automatically transliterated words are matched with document words. Vector space model similarity model is used for retrieval.

6 Experiments and Results

FIRE benchmark data of Gujarati news articles corpus and test queries are used in this research. Set 50 queries were given to different users and users were asked to type these queries in Romanized transliterated words in web user interface as shown in Fig. 1. These user queries are then taken as input in our system and they are searched against the documents index.

6.1 Dataset

The FIRE dataset comprises of document collection (D), query set (Q) and relevance judgments. The collection (D) contains 3, 13, 163 documents containing Gujarati script news articles from the year 2001 to 2010. Each document represents a news article from "Gujarat Samachar" newspaper. Statistics of the document collection is given in Table 1. From the year 2001 to 2010 Table 2 contains statistics of query collection. Table 3 lists a few examples of queries from Q.

Table 1. Statistics of Gujarati Collection.

Size of Collection	2.7 GB
Number of text Documents	3,13,163

Table 2. Statistics of queries.

Number of queries (number 176 to 225) with title and description	50
Number of Unique Gujarati language words (Stop words and morphological words eliminated)	245
Number of Unique English language words (Stop words and morphological words eliminated)	52
Total number of abbreviations or numeric values	29

Sample transliterated queries entered by user are given in Table 4.

Table 3. Sample of original queries.

ઑસ્ટ્રેલિયન એલચી કચેરી સામે બૉમ્બમારો

પ્રથમ ક્રિકેટર જેણે 700 ટેસ્ટ વિકેટ લીધી

Table 4. Sample of user input Roman transliterated queries.

Austreliyan elachi kacheri same bombmaro

Pratham cricketer jene 700 test wicket lidhi

6.2 Results

Query Terms Language Labeling Accuracy and Comparison. Query tokens are categorized in three categories: (I) English language words, (II) Gujarati language words and (III) Abbreviations and numbers. First numbers and abbreviations from query are identified using rule based approach and then remaining words in query are labeled as English or Gujarati using machine learning approach. Result of language identification accuracy of query words using support vector machine (SVM) and logistic regression (LR) algorithm is given below in Table 5: (Query set 176–225, Title-Description).

Table 5. Result of language labeling accuracy of query terms.

	LR		SVM	
	ENG	GUJ	ENG	GUJ
Precision	0.64	0.93	**1.00**	**0.96**
Recall	0.75	0.89	**0.82**	**1.00**
F-Score	0.69	0.91	**0.90**	**0.98**

For English words 36% improvement is obtained in precision, 7% increase in recall and 21% increase in F-Score. For Gujarati words 3% improvement is obtained in precision, 12% increase in recall and 7% increase in F-Score.

Back-Transliteration Accuracy and Comparison
Result of transliteration of English as well as Romanized Gujarati words is as shown in Table 6. For numbers and abbreviations we have obtained back transliteration accuracy of 96.55%.

Table 6. Result of back-transliteration of Romanized Gujarati query terms.

	LR		SVM			
	Imp. Soundex	Imp. Soundex + Web UI	CRF	Imp. CRF	Roman Guj. Sound-ex	Proposed Hybrid Approach
Precision	0.48	**1.00**		0.88	0.90	**0.96**
Recall	0.96	**0.82**	0.28	0.73	1.00	**1.00**
F-Score	0.64	**0.90**		0.80	0.95	**0.98**

As shown is result comparison of Table 6, proposed approaches for English and Gujarati languages performs best for back-transliteration of English and Romanized Gujarati words back transliteration.

Expanded Queries. Once user inputs query word, all its possible back-transliterations are listed to him/her and user selects the desired word. These transliterations are then expanded with query in order to retrieve documents written in both Gujarati and Roman script. E. g. Query "Euro apanavata desho" is expanded as "યુરો અપનાવાતા દેશો"and it retrieves documents matching these words in Gujarati script.

Gujarati Language Documents Retrieval. To conduct experiments standard adhoc retrieval setup is done and vector space model is used as similarity model. Comparison of retrieval result using each approach for Gujarati word transliteration is as shown in Table 7 with a graph depicting performance enhancement in MAP in Fig. 2.

Table 7. Result of retrieval performance.

Language and Transliteration Methods						
GUJ	ENG	MAP	P@ R0.00	P@ R0.10	P_5	P_10
CRF	*IS	0.10	0.21	0.18	0.10	0.09
Improved CRF	IS	0.12	0.38	0.30	0.20	0.16
Soundex	IS	0.24	0.53	0.47	0.31	0.25
Soundex + Improved CRF	IS	0.24	0.53	0.47	0.31	0.25
(Soundex + ImpCRF) with Web UI	**IS + web UI**	**0.27**	**0.53**	**0.49**	**0.32**	**0.26**

*IS = Improved Soundex

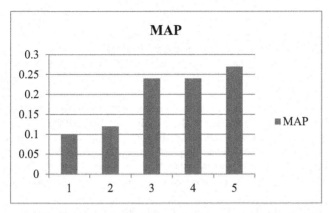

Fig. 2. Comparison MAP result of document retrieval.

6.3   Result Analysis

As shown in Table 7 and Fig. 2, proposed approach with web UI performs best for code mixed Romanized user queries to retrieve Gujarati documents. We obtained results with overall increase of 17% in MAP, 32% increase in P@R0.00 and 22% increase in P_5. We obtained similar performance for soundex and soundex+ imp CRF for Gujarati because all query words are available in transliteration pair corpus because of a huge transliteration pair corpus developed by us. But if word is not found in dictionary then it will be back transliterated using improved CRF and will get result based on accuracy of improved CRF approach as shown in Table 6. Improved CRF performs better than CRF approach so we have combined Romanized soundex with improved CRF.

7   Conclusion

Code mixed cross script document retrieval is a crucial requirement for current search engines and digital libraries, especially in resource scare language like Gujarati. In this paper novel approaches have been developed for transliteration of Gujarati and English language words to Gujarati script and compared their performance. Moreover interactive transliteration selection interface is developed. We have also contributed transliteration corpus for Gujarati and English words in Gujarati script. Proposed approaches are best performing with MAP = 0.27 and overall increase of 17% as compared to other approaches.

This research work will make the scope of search broad for Gujarati language documents by giving user flexibility to search Gujarati documents using code mixed Roman script to match Gujarati script documents. In future enhancements are planned in this work by adding more training data in order to cope up with wide vocabulary of both languages and also going to extend the work by adding more features.

References

1. Gupta, P., Bali, K., Banchs, R.E., Choudhury, M., Rosso, P.: Query expansion for mixed-script information retrieval. In: The 37th International ACM SIGIR Conference on Research and Development in Information Retrieval, SIGIR 2014, Gold Coast, QLD, Australia, pp. 677–686. ACM (2014)
2. Patel, S., Desai, V.: LIGA and syllabification approach for language identification and back transliteration. In: Shared Task Reported by DAIICT in FIRE-2014 (2014)
3. Royal, D., Rao, S., Shambavi, B.: Word-level language identification and back transliteration of romanized text. In: A Shared Task Report by BMSCE in FIRE-2014 (2014)
4. Bhat, I., Mujadia, V., Tammewar, A., Bhat, R., Shrivastava, M.: IIIT-H system submission for FIRE2014 shared task on transliterated search. In: A Shared Task Report in FIRE-2014 (2014)
5. Agarwal, A.: Transliteration involving English and Hindi languages using syllabification approach. M.Tech thesis, Indian Institute of Technology, Bombay (2010)
6. Gella, S., Sharma, J., Bali, K.: Query word labeling and back transliteration for Indian Languages. In: A Shared Task in FIRE-2013 (2013)
7. Singhal, V., Tyagi, N.: A hybrid approach of English – Hindi named – entity transliteration. Int. J. Adv. Technol. Eng. Sci. 3(2), 580–587 (2015). ISSN 2348-7550
8. Verulkar, P., Balabantray, R., Chakrapani, R.: Transliterated search on Hindi lyrics. Int. J. Comput. Appl. 121(1), 32–37 (2015). ISSN 0975-8887
9. Dhore, M., Dixit, S., Sonwalkar, T.: Hindi to English machine transliteration of named entities using conditional random fields. Int. J. Comput. Appl. (0975–8887) 48(23) (2012)
10. Pakray, P., Bhaskar, P.: Transliterated search system for Indian languages. In: The Proceedings of the Fifth Forum for Information Retrieval Evaluation (FIRE 2013) (2013)
11. Banerjee, S., Roy, A., Kulia, A., Naskar, S., Bandyopadhyay, S., Rosso, P.: A hybrid approach for transliterated word-level language identification: CRF with post processing heuristics. In: FIRE-2014 Proceedings of the Forum for Information Retrieval Evaluation. ACM (2014)
12. Bhalla, D., Joshi, N., Mathur, I.: Rule based transliteration scheme for English to Punjabi. Int. J. Nat. Lang. Comput. (IJNLC) 2(2) (2013)
13. Shah, R., Singh, D.: Improvement of Soundex algorithm for Indian language based on phonetic matching. Int. J. Comput. Sci. Eng. Appl. (IJCSEA) 4(3) (2014)
14. Varshney, S., Bajpai, J.: Improving performance of English-Hindi cross language information retrieval using transliteration of query terms. Int. J. Nat. Lang. Comput. (IJNLC) 2(6) (2013)
15. Barman, U., Das, A., Wagner, J., Foster, J.: Code mixing: a challenge for language identification in the language of social media. In: Proceedings of the First Workshop on Computational Approaches to Code Switching, Doha, Qatar, pp. 13–23 (2014)
16. Das, A., Gambäck, B.: Identifying languages at the word level in code-mixed Indian social media text. In: The 11th International Conference on Natural Language Processing (ICON-2014), Goa, India (2014)
17. Rathod, P., Dhore, M., Dhore, R.: Hindi and Marathi to English machine transliteration using SVM. Int. J. Nat. Lang. Comput. (IJNLC) 2(4) (2013). https://doi.org/10.5121/ijnlc.2013.2404
18. Joshi, P.: Syllabification of Romanized Gujarati words using machine learning. Int. J. Res. Electron. Comput. Eng. (IJRECE) 6(4) Version 5 IJRECE_F811 (2018). ISSN 2348-2281 (Online)
19. Joshi, P.: English to Gujarati transliteration using machine learning. Int. J. Res. Electron. Comput. Eng. (IJRECE) 6(4) Version 5 IJRECE_F812 (2018). ISSN 2348-2281 (Online)

Plant Disease Classification Using VGG-19 Based Faster-RCNN

Marriam Nawaz[1] , Tahira Nazir[2] , Muhammad Attique Khan[3]([✉]) ,
Venkatesan Rajinikanth[4] , and Seifedine Kadry[5,6,7]([✉])

[1] Department of Software Engineering, University of Engineering and Technology Taxila,
Taxila 47050, Pakistan
[2] Faculty of Computing, Department of Computer Science, Riphah International University
Gulberg Green Campus, Islamabad, Pakistan
[3] Department of Computer Science, HITEC University, Taxila, Pakistan
`attique.khan@ieee.org`
[4] Department of Computer Science and Engineering, Division of Research and Innovation,
Saveetha School of Engineering, SIMATS, Chennai 602105, India
[5] Department of Applied Data Science, Noroff University College, 4612 Kristiansand, Norway
`Seifedine.kadry@noroff.no`
[6] Artificial Intelligence Research Center (AIRC), Ajman University, Ajman 346,
United Arab Emirates
[7] Department of Electrical and Computer Engineering, Lebanese American University, Byblos,
Lebanon

Abstract. Early plant disease diagnosis can help farmers avoid spending money
on costly crop pesticides and aid them to boost food production. Scientists have
put a lot of effort to classify plant diseases, but it is difficult to quickly locate
and identify different crop abnormalities due to the high degree of resemblance
between the normal and damaged parts of plant leaves. Additionally, the procedure
of detecting plant diseases has been made more challenging due to the extensive
color, size, shape, and intensity variations in the background and foreground of the
plant images. To address the existing difficulties, we have introduced an effective
deep learning (DL) based system called Faster-RCNN o recognize and classify
various types of plant diseases. The suggested method consists of 3 basic steps.
In order to identify the area of interest in investigated samples, we first create
annotations for them which are later used for Faster-RCNN training. The Faster-
RCNN model employs the VGG-19 network to extract the relevant keypoints from
the given images which are later passed to the regressor and classification units to
identify and categorize the various crop diseases using the estimated features. We
evaluated our method on a widely used standard plant sample repository called
the PlantVillage database, and the findings show that our approach is reliable for
classifying plant diseases under a variety of image-capturing scenarios.

Keywords: Crop disease · VGG-19 · Faster-RCNN · Localization ·
Classification

M. Singh et al. (Eds.): ICACDS 2023, CCIS 1848, pp. 277–289, 2023.
https://doi.org/10.1007/978-3-031-37940-6_23

1 Introduction

The Food and Agriculture Organization (FAO) of the United Nations forecasts that by 2050, there may be 9.1 billion people on the planet, which would lead to a rise in food demand [1]. The main obstacles to the expansion of food production are the lack of big planting areas and the dearth of clean water. To meet the demands of a big population, it is vital to accelerate agricultural development in constrained areas. However, the existence of multiple plant pathogens also significantly lowers the amount and quality of food. Since these losses have a negative effect on the farmer's revenue and can lead to high prices of food, it is crucial to recognize these disorders at the earliest phase. In its more serious stages, it can cause hunger, particularly in developing nations around the globe. The majority of agricultural examinations are carried out by people going to the production sites. Visual check, however, is a time-consuming operation that is very reliant on the availability of trained staff personnel. Second, because it is challenging for humans to see each individual crop, physical agricultural assessment is not considered much accurate [2]. Consequently, to increase the amount of food production and spare the farmers from costly pesticide methods, the early diagnosis and classification of plant illnesses is a necessary activity. Therefore, investigators are currently developing automated processes for identifying numerous plant diseases in order to prevent manual examination [3].

The scientific world initially used methods from the fields of cell genetics and microbiology for agricultural disease monitoring [4, 5]. These techniques necessitate an extensive amount of human skill and struggle from significant computing costs. The majority of agricultural production is done by low-income individuals [6], therefore such pricey methods are not practical for them [7]. Researchers have incorporated numerous hand-coded strategies in the sector of agriculture with the emergence of machine learning (ML) [8]. Easy digital data-gathering techniques have made it possible to gather large volumes of information in real-time that are used by ML-based techniques to make smart choices. Support vector machines (SVM) [9], decision trees (DT) [10], Gaussian architectures [11], and KNN are some of the methods being investigated for examining infections in plants. Such ML-based solutions are simpler to comprehend and only need a minimal quantity of data to build models, but they still take a little time and rely heavily on skilled humans. Additionally, the classic ML-based feature computation techniques constantly necessitate a trade-off between computational effort and classification power [12]. DL-based techniques are now being tested primarily to address the shortcomings of ML-based strategies. Recurrent neural networks (RNNs) [13], deep belief networks (DBNs) [14], and other DL techniques [15] are currently highly regarded in plant science. These methods can autonomously pick out the relevant selection of image key-points without the assistance of domain experts. When it comes to performing pattern recognition, deep learning techniques (DL) mirror how the human brain functions by localizing and recognizing a variety of things by looking at their images. Several forms of deep neural networks (DNNs) demonstrate cutting-edge precision over multispectral evaluation, and DL-based architectures exhibit robust results in the field of agricultural research and are well-suited to a variety of applications [16]. A lot of research is being done in farming using methods like GoogLeNet [17], AlexNet [18], VGG [19], and ResNet [20] for activities including estimating food amount, detecting crop heads, quantifying fruits, recognizing and classifying crop diseases, etc. Due to their capacity to make use of the topological information from the image features, these algorithms are able to demonstrate accurate prediction performance while consuming less computing time.

Despite the fact that the experts have done a lot of effort to classify plant diseases, it is still difficult to identify them at the beginning stages since the diseased and healthy plant parts share a lot of common properties [21]. Additionally, the existence of noise and blurring in the input photos, fluctuations in lighting and brightness, and different plant leaf shapes have made the screening procedure more challenging. Thus, there is still an opportunity for improving performance in terms of processing speed as well as reliability in identifying plant diseases. We have introduced a DL-based strategy called Faster-RCNN for crop disease identification and categorization to address the issues at hand. The primary contributions of our study are as follows:

- Fine-tuned an object identification approach called Faster-RCNN to classify numerous diseases of plants which increased the categorization results.
- Accurate identification of the afflicted areas of crop leaves due to the Faster-RCNN model's ability to handle over-fit network training data.
- To demonstrate the efficacy of the suggested model, a thorough investigation of the presented approach against existing cutting-edge crop illness categorization techniques across a public repository called PlantVillage is performed.
- Establishes a method that enables the identification of the infected plant area even in the incidence of noise, blurring, and fluctuations in color, shape, and illumination of images.

The remaining of the manuscript is organized as: Sect. 2 explains the related work, while the presented model is elaborated in Sect. 3. The details of the used database along with attained results are discussed in Sect. 4, and collusion is drawn in Sect. 5.

2 Related Work

Here, a thorough investigation of existing techniques used for the detection and classification of plant leaf diseases is elaborated.

Le et al. [22] introduced a work for the recognition of numerous plant diseases. The work initially performed a pre-processing phase that used the employment of morphology-based open and close approaches to remove the incidence of noisy patterns from the input images. Then, as a next phase the work applied a method called the filtered local binary pattern approach to compute the features of the given samples. Whereas, in the next stage, the calculated keypoints were used with the SVM predictor to accomplish the classification job. The method discussed in [22] is effective to classify various categories of crop abnormalities, however, unable to perform well for samples with intense lightning alterations. Another approach employing the pattern-based feature descriptor was proposed in [23] where a technique called Directional Local Quinary Patterns (DLQP) was introduced to extract the relative set of keypoints. This work used the SVM predictor to accomplish the classification task. The work [23] signifies improved classification results for recognizing the plant diseases, however, unable to tackle the incidence of blurring in images. Sun et al. [24] presented an ML-based approach to categorize the various types of crop abnormalities. In the first phase, an approach named simple linear iterative cluster was utilized for distributing the suspected sample into numerous blocks. While in the next stage, two approaches named the fuzzy salient region contour and Gray Level Co-occurrence Matrix (GLCM) were applied for keypoints calculation from the sample blocks. The work utilized the SVM algorithm to classify the various types of crop diseases. The approach [24] performs better to recognize different types of plant leaf abnormalities, however, requires a huge set of samples for model training. A comparable technique was discussed in [2], where the GrabCut strategy was utilized to split the samples and the LBP approach to obtain the image features. Additionally, the SVM predictor was trained using the computed features to complete the task of classifying crop leaf diseases. The methodology [2] works fine for identifying agricultural diseases but may struggle with noisy photos. Ramesh et al. [25] provided a method to carry out the categorization of leaf abnormalities in crops in which the HOGs technique was employed to estimate the sample keypoints and the RF predictor was utilized to carry out the categorization procedure. This method [25] offers a lightweight model to categorize crop leaf diseases, however, classification results require to be improved. A strategy to identify the crop illness of the turmeric plant was put forward in [26]. Firstly, the input visual sample was segmented using the K-means method. The SVM algorithm was then trained to identify diseases using the keypoints from the images, which were computed using the GLCM approach. Although this method [26] is effective in identifying plant diseases, it is not resistant to significant luminance variations in the input visual samples. For categorizing various infections of crop leaves, an ML-based method was presented in [27]. Various approaches, including SIFT, LBP, and GLCM, were used to calculate the features of the input images. The SVM, KNN, and RF learners were trained using the obtained characteristics in the following phase. With the RF predictor, this strategy [27] produces effective categorization accuracy, but the efficiency still takes effort. Another approach for identifying crop diseases was suggested in [28]. The first phase was extracting the keypoints from the suspected visual samples with a feature vector having a dimension of 172 using the 14 colour spaces. The collected keypoints were also used to train an SVM learner. The methodology in [28] is reliable for classifying crop diseases but is ineffective for data with distortions.

With better outcomes, many scientists have looked into DL-based algorithms for identifying crop diseases [29]. Few-Shot Learning (FSL), a DL strategy for classifying and identifying plant diseases, was developed by Argüesoa et al. [30]. The Inception V3 methodology was applied to compute dense features, and the SVM predictor was utilized for the classification problem. The accuracy of disease categorization is improved in the study [30], but a more comprehensive and challenging data sample is needed to evaluate the technique. To estimate the sample features and carry out the categorization job, [31] presented a CNN model made up of three convolutional layers. This research [31] provides an inexpensive method for the automatic diagnosis of agricultural infections, although it suffers from the issue of framework over-fitting. A mobile phone-deployable, compact method for detecting plant diseases was suggested by Richey et al. [32]. The approach was used to categorize the illness affecting maize production. The calculation and categorization of features were performed using the ResNet50 architecture. Although the approach [32] works well for identifying plant diseases, the processing conditions made it unsuitable for all sorts of smart phones.

Similar research for identifying tomato plant diseases was suggested in [33]. From the investigated images, a dense keypoints detection model called AlexNet was utilized to calculate the trustworthy set of sample features on which the KNN algorithm was trained to carry out the categorization job. The architecture in [33] is reliable for classifying tomato plant illness, although KNN is a slower approach.

Another architecture for classifying leaf spot diseases was also described in [34], using a residual model to retrieve dense keypoints and a CNN-based predictor to categorize plant diseases. Although the leaf disease classification results of the algorithm in [34] is consistent, a higher computation burden results from this. A method called region-based CNN (RCNN) was put out by Dwivedi et al. [35] to identify and classify the various grape plant leaf disorders. The deep keypoints were extracted in the initial phase using the ResNet18 architecture. The RCNN algorithm then grouped the calculated features into distinct classifications. Even while the work [35] shows increased classification results, it does not adapt well to unseen data samples. Multiple architectures, including VGG, ResNet, and DenseNet, were employed in [36] to compute features from the given visual data samples. The work [36] demonstrates strong accuracy using the DenseNet technique, although it has a significant financial cost. Even though the scientific world has done a significant amount of work on the automatic localization and recognition of plant leaf diseases, effectiveness still needs to be improved. The methods in [37, 38] adopted the idea of using object detection approaches for the timely localization and classification of plant leaf diseases and exhibited better classification results. However, these methods are applied to locate the leaf diseases of the only tomato plant.

3   Method

In this study, we present a method for categorizing and diagnosing a variety of crop illnesses from investigated visual samples by using a DL method called the Faster-RCNN. Annotations are first made to identify the impacted area in the input samples, and these annotations are then utilized to train the Faster-RCNN architecture. Lastly, using the test instances, the trained model recognizes and categorizes crop illness. In Fig. 1, the whole process of the suggested solution is depicted. Our evaluation findings demonstrate that Faster-excellent RCNN memory capacity makes it resilient to the classification of plant diseases.

3.1   Data Preparation

It is vital to accurately identify the impacted fraction from the investigated samples in order to have a reliable and precise training approach. To achieve this, we developed annotations of plant photos using the public software LabelImg [26], which allows us to precisely find the region of interest (ROIs). The outcome, which contains the location coordinates of the impacted areas, is stored in a CSV format and afterward given to the Faster-RCNN architecture for training the network together with the input samples.

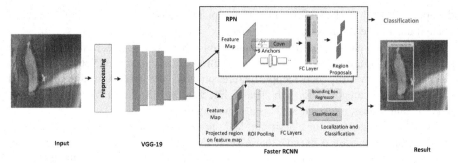

Fig. 1. Model depiction

3.2 Feature Extraction Using Faster-RCNN

For the automatic localization and recognition of various plant illnesses, we have used Faster-RCNN, a deep learning approach. The Faster-RCNN method makes use of several convolution windows, which enables it to assess the plant pattern's architecture and derive a relevant set of keypoints. Due to the computational complexity of the RCNN and Fast-RCNN frameworks, we have chosen the Faster-RCNN architecture for classifying crop diseases. Additionally, RCNN and Fast-RCNN algorithms use manual methods for feature calculation, such as EdgeBox [39] or selective search [40], which makes it impossible to acquire a reliable collection of sample features. On the other hand, the Faster-RCNN method more effectively addresses the challenges of RCNN and Fast-RCNN by adding a new component called the Regional Proposal Network (RPN) for the autonomously extracting keypoints from the source samples.

In the field of plant leaves infection recognition and categorization, locating the ROIs from the source images are struggling from 2 major issues: i) locating the exactly affected region from leaves due to massive shades resemblance in diseased and normal region ii) class associated with each identified region. The RPN component of the employed model has given it the ability to more accurately detect the ROIs and categorize the investigated images since it makes use of data on the shape, color, and structure of unhealthy parts and ensures a greater recall rate by employing a few nominated windows.

The Faster-RCNN model contains potentially 4 phases to locate and recognize the different plant illnesses:

3.2.1 Convolution Layers

The Faster-RCNN framework comprises a total of 16 convolutional, relu, and pooling layers which are used for keypoints extraction from the suspected sample. Then the computed keypoint maps are passed to the next RPN module and related layers.

3.2.2 RPN

The RPN module consists of 3 × 3 convolve layers that produce the anchors and bounding boxes and act as a base for computing the object proposals.

3.2.3 ROI Pooling

The ROI pooling layer takes the output from the convolution layers as well as the RPN component to generate the keypoints maps, which are then supplied as input to all fully linked layers.

3.2.4 Classification

As a last phase, the Faster-RCNN architecture accomplish the detection and recognition step by developing the bounding boxes to recognize the infected portions of plants and regulate the associated category.

4 Experiment and Results

In this section, a detailed description of the dataset used to test the model performance is given. Moreover, the used evaluation measures are described. We have performed a thorough assessment of the proposed approach to show its effectiveness for plant diseases categorization.

4.1 Dataset

For performance analysis of the presented architecture, a standard dataset called PlantVillage [41] is utilized in this work which can be freely accessed and downloaded from the Kaggle website. This data sample comprises a total of 54,306 images belonging from 14 different groups of plants with 12 normal and 26 affected plant categories.

4.2 Evaluation Metrics

The parameters used to test the efficacy of our approach are elaborated in this section. Clearly, we have employed the accuracy, precision, recall, mean average precision (mAP), and intersection over union (IOU), measures to numerically discuss of our results and defined as follows (Figs. 2, 3 and 4).

$$Accuracy = \frac{TP + TN}{TP + FP + TN + FN} \tag{1}$$

$$mAP := \sum_{i=1}^{T} AP(t_i)/T \tag{2}$$

4.3 Detection Performance

An autonomous crop diseases system's primary need is that it be reliable in categorizing the numerous plant leaf diseased regions belonging to different abnormalities. Therefore, we examined the photos from the PlantVillage database to determine the recognition rate of the suggested method, and some of the pictorial results are shown in Fig. 5. The given

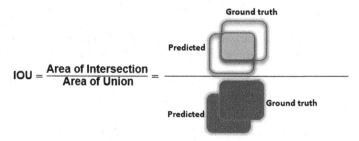

$$IOU = \frac{\text{Area of Intersection}}{\text{Area of Union}} =$$

Fig. 2. Geometrical representation of IOU.

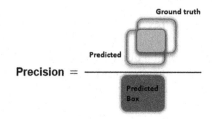

$$\text{Precision} = $$

Fig. 3. Geometrical representation of Precision.

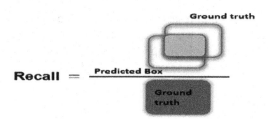

$$\text{Recall} = $$

Fig. 4. Geometrical representation of Recall.

findings in Fig. 5 can be used to determine that the proposed approach is reliable for classifying samples of various classes. Moreover, we have used two measures called mAP and IOU to check the quantitative results of the used framework. In a clearer manner, we attain the mAP score of 0.89 and IOU of 0.92 respectively. It can be concluded from the attained scores and visual findings that our proposed architecture can be reliably applied to recognize and classify various crop abnormalities.

4.4 Comparison with Base Approaches

We have conducted an evaluation to analyze the detection power of our work with different object identification models namely Fast-RCNN [42], Faster-RCNN [43] You Only Look Once (YOLO)[44], and single-shot detector (SSD) [45], and results are discussed in Table 1. It can be seen from the Table that our approach outperforms the other approaches. Clearly, the lowest results are attained by the ResNet-101-based SSD model which is due to the reason that the SSD model employs the one-stage detector for plant disease classification which results in the reduction of computed characteristics and reduces the classification results. The comparable results are achieved by the VGG-16-based Faster-RCNN model, however, comparatively, we obtain a mAP value of 0.87 which is higher than all other methods. The main reason for the efficient performance of our work is due to the more extraction of a more discriminative set of image features through VGG-19 which presents the complex transformation of image samples in viable manners.

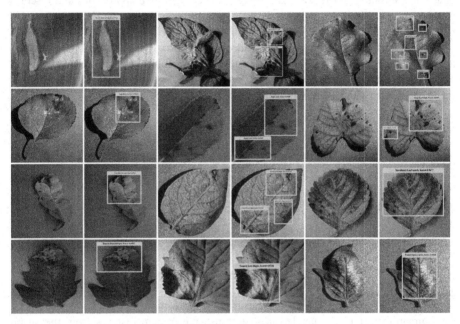

Fig. 5. Model recognition results

Table 1. Performance analysis of the presented model with other object identification approaches.

Models	Base	mAP
Fast-RCNN	VGG-16	0.85
Faster-RCNN	VGG-16	0.88
YOLOv3	DarkNet-53	0.83
SSD	ResNet-101	0.81
Proposed Faster-RCNN	VGG-19	0.87

4.5 Comparison with the Latest Works

To more assess the plant diseased regions classification results of our approach, here, we considered the latest methods and compared our results with them. To numerically discuss the results, we have taken measures namely precision, recall, and accuracy, and the results are shown in Table 2. It can be seen that the works in [46, 47] attain precision values of 99.35% and 99.47% respectively, while the presented technique acquires a precision value of 99.48% which is higher than both works. Similarly, in the case of recall and accuracy, our technique has outperformed the comparative with the values of 99.90% and 99.51% respectively. Clearly, the peer works to attain the average precision value of 97.91%, which is 99.48% for our case and we have provided a performance gain of 1.57%. Similarly, for recall and accuracy, the comparative approaches attain scores of 99.62% and 97.90% which are 99.90%, and 99.51% for our work, and presented performance gains of 0.28%, and 1.60% for the mentioned metrics.

The foundation for the better performance of our approach is that the methods [46, 47] utilize dense model architectures which cause the overfitting problem. Comparatively, our approach utilizes a shallow model structure and computes a more reliable group of image keypoints which results in its effective performance.

Table 2. Performance analysis with the latest techniques.

Model	Precision	Recall	Accuracy
Mohanty et al. [46]	99.35%	99.35%	99.35%
Geetharamani et al. [47]	96.47%	99.89%	96.46%
Proposed	99.48%	99.90%	99.51%

5   Conclusion

The human experience is needed to distinguish the minute features from the suspicious samples of various crops in order to manually diagnose and classify the different plant leaf disorders. Additionally, the process of crop leaf infection categorization has become more difficult by the extreme variances in the size, hue, and architecture of plant leaves. We have designed an effective strategy to address these issues called VGG-19-based Faster-RCNN, and the outcomes are assessed using the PlantVillage data collection which is a standard and challenging dataset. According to both the qualitative and numerical findings the suggested method can successfully substitute manual systems for the diagnosis of plant leaf diseases. As the focus of the current research is on identifying the disease solely in plant leaves, we intend to broaden the application of our method in the future to include other plant components as well.

Conflict of Interest.   All authors declared no conflict of interest in this work.

References

1. Bruinsma, J.: The resource outlook to 2050: by how much do land, water and crop yields need to increase by 2050. In: Expert Meeting on How to Feed the World, vol. 2050, pp. 24–26 (2009)

2. Pantazi, X.E., Moshou, D., Tamouridou, A.A.: Automated leaf disease detection in different crop species through image features analysis and One Class Classifiers. Comput. Electron. Agric. **156**, 96–104 (2019)
3. Wolfenson, K.D.M.: Coping with the food and agriculture challenge: smallholders' agenda. Food Agriculture Organisation of the United Nations, Rome (2013)
4. Sankaran, S., Mishra, A., Ehsani, R., Davis, C.: A review of advanced techniques for detecting plant diseases. Comput. Electron. Agric. **72**(1), 1–13 (2010)
5. Dinh, H.X., Singh, D., Periyannan, S., Park, R.F., Pourkheirandish, M.: Molecular genetics of leaf rust resistance in wheat and barley. Theor. Appl. Genet. **133**(7), 2035–2050 (2020)
6. Patil, S., Chandavale, A.: A survey on methods of plant disease detection. Int. J. Sci. Res. **4**(2), 1392–1396 (2015)
7. Ferentinos, K.P.: Deep learning models for plant disease detection and diagnosis. Comput. Electron. Agric. **145**, 311–318 (2018)
8. Gebbers, R., Adamchuk, V.I.: Precision agriculture and food security. Science **327**(5967), 828–831 (2010)
9. Joachims, T.: Making large-scale SVM learning practical. Technical report (1998)
10. Rokach, L., Maimon, O.: Decision trees. In: Maimon, O., Rokach, L. (eds.) Data mining and knowledge discovery handbook, pp. 165–192. Springer, Boston (2005). https://doi.org/10.1007/0-387-25465-X_9
11. Birgé, L., Massart, P.: Gaussian model selection. J. Eur. Math. Soc. **3**(3), 203–268 (2001)
12. Bello-Cerezo, R., Bianconi, F., Di Maria, F., Napoletano, P., Smeraldi, F.: Comparative evaluation of hand-crafted image descriptors vs. off-the-shelf CNN-based features for colour texture classification under ideal and realistic conditions. Appl. Sci. **9**(4), 738 (2019)
13. Roska, T., Chua, L.O.: The CNN universal machine: an analogic array computer. IEEE Trans. Circ. Syst. II: Analog Digit. Signal Process. **40**(3), 163–173 (1993)
14. Zaremba, W., Sutskever, I., Vinyals, O.: Recurrent neural network regularization. arXiv preprint arXiv:1409.2329 (2014)
15. Salakhutdinov, R., Hinton, G.: Deep boltzmann machines. In: Artificial intelligence and statistics, pp. 448–455. PMLR (2009)
16. Gewali, U.B., Monteiro, S.T., Saber, E.: Machine learning based hyperspectral image analysis: a survey. arXiv preprint arXiv:1802.08701 (2018)
17. Szegedy, C., et al.: Going deeper with convolutions. In: Proceedings of the IEEE Conference on Computer Vision and Pattern Recognition, pp. 1–9 (2015)
18. Yuan, Z.-W., Zhang, J.: Feature extraction and image retrieval based on AlexNet. In: Eighth International Conference on Digital Image Processing (ICDIP 2016), vol. 10033, p. 100330E: International Society for Optics and Photonics (2016)
19. Vedaldi, A., Zisserman, A.: VGG convolutional neural networks practical. Dept. Eng. Sci. Univ. Oxford **2016**, 66 (2016)
20. Thenmozhi, K., Srinivasulu Reddy, U.: Crop pest classification based on deep convolutional neural network and transfer learning. Comput. Electron. Agric. **164**, 104906 (2019)
21. Paul, A., Ghosh, S., Das, A.K., Goswami, S., Das Choudhury, S., Sen, S.: A review on agricultural advancement based on computer vision and machine learning. In: Mandal, J.K., Bhattacharya, D. (eds.) Emerging technology in modelling and graphics. AISC, vol. 937, pp. 567–581. Springer, Singapore (2020). https://doi.org/10.1007/978-981-13-7403-6_50
22. Le, V.N.T., Ahderom, S., Apopei, B., Alameh, K.: A novel method for detecting morphologically similar crops and weeds based on the combination of contour masks and filtered local binary pattern operators. GigaScience **9**(3), giaa017 (2020)
23. Ahmad, W., Shah, S.M., Irtaza, A.: Plants disease phenotyping using quinary patterns as texture descriptor. KSII Trans. Internet Inf. Syst. **14**(8), 3312–3327 (2020)
24. Sun, Y., Jiang, Z., Zhang, L., Dong, W., Rao, Y.: SLIC_SVM based leaf diseases saliency map extraction of tea plant. Comput. Electron. Agric. **157**, 102–109 (2019)

25. Ramesh, S., Hebbar, R., Niveditha, M., Pooja, R., Shashank, N., Vinod, P.V.: Plant disease detection using machine learning. In: 2018 International Conference on Design Innovations for 3Cs Compute Communicate Control (ICDI3C), pp. 41–45. IEEE (2018)
26. Kuricheti, G., Supriya, P.: Computer vision based turmeric leaf disease detection and classification: a step to smart agriculture. In: 2019 3rd International Conference on Trends in Electronics and Informatics (ICOEI), pp. 545–549. IEEE (2019)
27. Kaur, N.: Plant leaf disease detection using ensemble classification and feature extraction. Turk. J. Comput. Math. Educ. **12**(11), 2339–2352 (2021)
28. Shrivastava, V.K., Pradhan, M.K.: Rice plant disease classification using color features: a machine learning paradigm. J. Plant Pathol. **103**(1), 17–26 (2020). https://doi.org/10.1007/s42161-020-00683-3
29. Walter, M.: Is this the end? Machine learning and 2 other threats to radiologys future, p. 13 (2016)
30. Argüeso, D., et al.: Few-Shot Learning approach for plant disease classification using images taken in the field. Comput. Electron. Agric. **175**, 105542 (2020)
31. Agarwal, M., Singh, A., Arjaria, S., Sinha, A., Gupta, S.: ToLeD: tomato leaf disease detection using convolution neural network. Procedia Comput. Sci. **167**, 293–301 (2020)
32. Richey, B., Majumder, S., Shirvaikar, M., Kehtarnavaz, N.: Real-time detection of maize crop disease via a deep learning-based smartphone app. In: Real-Time Image Processing and Deep Learning 2020, vol. 11401, p. 114010A. International Society for Optics and Photonics (2020)
33. Batool, A., Hyder, S.B., Rahim, A., Waheed, N., Asghar, M.D.: Classification and identification of tomato leaf disease using deep neural network. In: 2020 International Conference on Engineering and Emerging Technologies (ICEET), pp. 1–6. IEEE (2020)
34. Karthik, R., Hariharan, M., Anand, S., Mathikshara, P., Johnson, A., Menaka, R.: Attention embedded residual CNN for disease detection in tomato leaves. Appl. Soft Comput. **86**, 105933 (2020)
35. Dwivedi, R., Dey, S., Chakraborty, C., Tiwari, S.: Grape disease detection network based on multi-task learning and attention features. IEEE Sens. J. (2021)
36. Akshai, K.P., Anitha, J.: Plant disease classification using deep learning. In: 2021 3rd International Conference on Signal Processing and Communication (ICPSC), pp. 407–411. IEEE (2021)
37. Nawaz, M., et al.: A robust deep learning approach for tomato plant leaf disease localization and classification. Sci. Rep. **12**(1), 18568 (2022)
38. Albahli, S., Nawaz, M.: DCNet: DenseNet-77-based CornerNet model for the tomato plant leaf disease detection and classification. Front. Plant Sci. **13**, 957961 (2022)
39. Uijlings, J.R.R., Van De Sande, K.E.A., Gevers, T., Smeulders, A.W.M.: Selective search for object recognition. Int. J. Comput. Vision **104**(2), 154–171 (2013)
40. Dollár, P., Zitnick, C.L.: Fast edge detection using structured forests. IEEE Trans. Pattern Anal. Mach. Intell. **37**(8), 1558–1570 (2014)
41. Hughes, D., Salathé, M.: An open access repository of images on plant health to enable the development of mobile disease diagnostics. arXiv preprint arXiv:1511.08060 (2015)
42. Girshick, R.: Fast R-CNN. In: Proceedings of the IEEE International Conference on Computer Vision, pp. 1440–1448 (2015)
43. Ren, S., He, K., Girshick, R., Sun, J.: Faster R-CNN: towards real-time object detection with region proposal networks. IEEE Trans. Pattern Anal. Mach. Intell. **39**(6), 1137–1149 (2016)
44. Redmon, J., Farhadi, A.: YOLOv3: an incremental improvement. arXiv preprint arXiv:1804.02767 (2018)
45. Liu, W., et al.: SSD: single shot multibox detector. In: Leibe, B., Matas, J., Sebe, N., Welling, M. (eds.) ECCV 2016. LNCS, vol. 9905, pp. 21–37. Springer, Cham (2016). https://doi.org/10.1007/978-3-319-46448-0_2

46. Mohanty, S.P., Hughes, D.P., Salathé, M.: Using deep learning for image-based plant disease detection. Front. Plant Sci. **7**, 1419 (2016)
47. Geetharamani, G., Pandian, A.: Identification of plant leaf diseases using a nine-layer deep convolutional neural network. Comput. Electr. Eng. **76**, 323–338 (2019)

Classification of Real and Deepfakes Visual Samples with Pre-trained Deep Learning Models

Marriam Nawaz[1] , Ali Javed[1] , Tahira Nazir[2] , Muhammad Attique Khan[3(✉)] ,
Venkatesan Rajinikanth[4] , and Seifedine Kadry[5,6(✉)]

[1] Department of Software Engineering, UET Taxila, Taxila 47050, Pakistan
[2] Department of Computing, Riphah International University, Islamabad, Pakistan
[3] Department of Computer Science, HITEC University, Taxila, Pakistan
`attique.khan@ieee.org`
[4] Department of Computer Science and Engineering, Division of Research and Innovation,
Saveetha School of Engineering, SIMATS, Chennai 602105, India
[5] Department of Applied Data Science, Noroff University College, 4612 Kristiansand, Norway
`Seifedine.kadry@noroff.no`
[6] Department of Electrical and Computer Engineering, Lebanese American
University, Byblos, Lebanon

Abstract. Serious security and privacy problems have arisen as a result of significant advancements in the creation of deepfakes. Attackers can easily replace a person's face with the target person's face in an image using sophisticated Deep learning (DL) algorithms to spoof their identity. Deepfakes detection algorithms have been proposed in response to the growing concerns about the potential harm caused by deepfakes. However, a reliable deepfakes detector that can keep up with contemporary deepfakes creation techniques is required. In this work, we have proposed an end-to-end methodology for detecting manipulated visual content. We used multiple CNN models i.e., ResNet18, ResNet50, DenseNet65, DenseNet77, and DenseNet100 along with the SVM classifier to compute effective cues from the input facial faces to distinguish between actual and altered content. We have also applied the concept of transfer learning to solve the issue of model overfitting and improve generalizability against different manipulation algorithms. A comparison study is carried out to evaluate the performance of several feature extractors. Through thorough experiments performed using the Deepfakes Detection Challenge dataset, our results demonstrated that DenseNet100 surpasses the other CNN models by better recognizing deepfakes.

Keywords: Deepfakes · SVM · deep learning · CNN

1 Introduction

People's facial appearance has rich attributes that deliver influential biometric signals to distinguish humans. In general, human recognition approaches are heavily being employed in various applications including law implementation institutes, surveillance systems to individual verification on cell phones [1]. Due to the specific characteristics

© The Author(s), under exclusive license to Springer Nature Switzerland AG 2023
M. Singh et al. (Eds.): ICACDS 2023, CCIS 1848, pp. 290–303, 2023.
https://doi.org/10.1007/978-3-031-37940-6_24

of each human face, it has been considered the best way to recognize humans across the globe [2]. Meanwhile, the introduction of easy-to-use tools like FaceApp [3], Adobe Photoshop [4], and Face2Face [5] are enabling people to edit their visual content to make them more appealing. Visual editing can be performed in two ways namely: i) recreational edits where face retouching techniques are used to improve the appearance of a person's look, ii) malicious edits that involve the altering of visual appearance with the intent of spreading fake information. Such alterations always complicate the human recognition process. Nevertheless, forging digital content is not new. Fabrication of audiovisual data has been utilized for several years to create political stress [6] or to create censorship by removing humans from multimedia content. For example, in the twentieth century, political revolutionists were murdered and then removed from snaps via using masking techniques in the Great Terror period of the Soviet Union [7].

Now, the enhancement of deep learning-based methods like generative adversarial networks (GAN) has empowered multimedia fabrication techniques among one such phenomenon as deepfakes [8]. In deepfakes, the facial region of the target person is swapped or morphed according to the source person to depict the target saying or doing something that is done by the source. The great development in GANs has increased the realism of such fake content that it is difficult for the audience to differentiate between real and fake data [9]. Even though, at the start, deepfakes were a source of entertainment for people unless these manipulations appear to propagate political disorder, nonconsensual porn, and character assassination. Moreover, easier internet access has resulted in the drastic growth of such content on social media. According to a report published in [10], the growth rate of such manipulated videos is 100% per year. Though the term deepfakes was first initiated in 2017 on Reddit, however, according to [10] there are about 21,244 highly realistic forged content on cyberspace, which was only 7,964 by December 2018. Such an exponential increase in the growth rate of such altered content is threatening and provoking researchers to take active measures. Initially, the generation of highly realistic deepfakes require extensive training data, therefore, celebrities were the main of such manipulations. However, with the introduction of few-shot learning-based approaches, now the general audience is also not safe from its devastating effects. The prevailing situation requires the development of such frameworks which can be used to check the visual content for its authenticity before using them in processing any legitimate claims.

Several techniques have been introduced to verify the truthfulness of the images and videos. Hadi et al. [11] introduced a framework to detect real and fake content from videos. After the preprocessing step, the 16 Gabor filters were used to compute the features from the input samples. Later, a CNN-based binary classifier was trained on the calculated keypoints to differentiate the original and manipulated samples. The work [11] performs well for deepfakes detection, however, unable to generalize well to unseen examples. Xu et al. [12] introduced a method to locate the forensic manipulations made within the videos. For this reason, the work [12]introduced a supervised contrastive framework for identifying forged and pristine data. The approach used the Xception model for feature computation and to execute the classification task. This [12] method is generalized well to unseen examples, however, performance needs further enhancements. Another technique to recognize the forensic alteration within visual content was presented in [13], where a model namely the fused facial region feature descriptor was presented. The work accompanied the facial features from several conventional machine learning approaches like SIFT, SURF, ORB, etc. to better present the human face attributes. The technique [13] shows better fake faces recognition performance, however, unable to tackle the several adversarial attacks made within altered content. Kolagati et al. [14] performed an analysis to discriminate the real and fake visual content. Initially, the frames were extracted from the video samples. Then, first, the landmarks features were computed from the extracted frames which were passed as input to MLP. Then, in the second step, the entire frae was passed as input to the CNN model to compute the deep features. In the third step, the result from the MLP and the CNN were joined and sent s input to a fully connected layer, which generated the final results. The work [14] performs well for classifying the pristine and manipulated videos, however, unable to perform well for video samples containing more than one person face.

Sun et al. [15] proposed an approach to identify the forensic changes in videos via employing spatial and temporal information. Initially, the landmark features from the video frames were extracted via using the Face2Face and Dlib library, then a two-way RNN framework was used to exploit the temporal features. Finally, both the landmarks and temporal keypoints were used to discriminate the real and fake content. The work presented in [15] showed better deepfakes detection results, however, suffering from high computational costs. Another framework namely the FakeBuster was presented in [16] to locate the real and fake videos of people. The approach [16] is robust to deepfakes detection, however, requires evaluation on a more challenging dataset. Yavuzkilicn et al. [17] introduced a hybrid approach for the recognition of alterations made within visual data. Three deep learning frameworks namely the VGG16, VGG19, and ResNet18 were used for feature computation from the input samples. Then, the features from all three networks were combined and employed for the classifier training to perform the classification task. The approach [17] works well for deepfakes detection, however, at the expenditure of a high computational burden. Masood et al. [8] presented the analysis of several pre-trained deep learning-based architectures to demonstrate their behavior for deepfakes detection. Several approaches like VGG16, VGG19, InceptionV3, and DenseNet-169 were used for feature vector estimation. Then, the SVM classifier was trained for all used feature extractors to identify the pristine and forge faces from the videos. The methodology [8] reports the best results for the DenseNet-69 frameworks, however, the work requires evaluation for the adversarial attacks. Another work was presented in [18] to detect real and forged images. Initially, the landmarks from the input videos were extracted via using OpenCV, which were later passed to the LSTM model to execute the deepfakes classification task. The technique [18] is robust to deepfakes detection and classification, however, requires testing on a more challenging dataset.

Khalid et al. [19] proposed an approach namely OC-FakeDect for locating the alterations of visual samples. The technique uses the one-class Variational Autoencoder (OC-VAE) along with the added encoder structure. The work [19] was trained only for fake samples. The framework shows better deepfakes detection performance, however, accuracy requires further improvements. Another technique was introduced in [20], where both the spatial and temporal features were used to locate the forensic manipulations. Initially, the multi-task cascaded technique was used to detect the faces from the video samples. Then, the EfficientNet was used to compute the spatial features, while the RNN approach was employed for temporal sequence analysis. The work proposed in [20] exhibits an improved deepfakes classification method, however, requires extensive training data. Wang et al. [21] proposed an approach where 3D CNN models termed the 3D, ResNet, ResNeXt, and I3D were used for deepfakes classification. The work [21] shows that the 3D approaches perform well for forensic analysis, however, unable to perform well for noisy samples. Yang et al. [22] introduced a technique for deepfakes recognition from the videos. The work was based on the assumption that the forged content lacked to maintain the 3D head poses. Initially, the landmarks were computed to measure the orientation of head poses, which were then used for the SVM classifier. The technique [22] is robust to deepfakes detection, however, unable to perform well for the compressed video samples.

Several works from history have been presented for the timely detection and classification of real and manipulated visual content, however, there is a requirement for a more accurate system due to the generation of more realistic datasets. In this work, we have presented the analysis of several deep learning approaches for deepfakes detection. The main contributions of the presented work are as follows:

- We have employed the concept of transfer learning to demonstrate the general flow of deepfakes identification to resolve the issue of network over-fitting.
- A comparative analysis of several deep learning approaches namely ResNet18, ResNet50, DenseNet65, DenseNet77, and DenseNet100 models for forensic manipulation detection.
- Extensive evaluations on a complex database named Deepfake Detection Challenge Dataset (DFDC) are performed to exhibit the effectiveness of DL approaches.

2 Method

In this part, we have elaborated on the details of the employed technique. The entire pipeline followed by the proposed approach is presented in Fig. 1. Our framework followed three main steps namely facial region identification, keypoints calculation, and prediction respectively. The details of all steps are discussed in the subsequent sections.

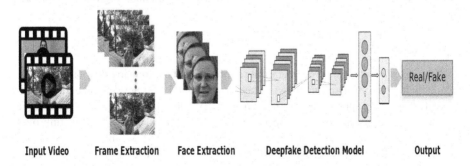

| Input Video | Frame Extraction | Face Extraction | Deepfake Detection Model | Output |

Fig. 1. The generic pipeline of the proposed work

2.1 Facial Region Identification and Extraction

In the first step, we have extracted the faces from the input videos as most of the manipulations are created within the facial regions. So, to locate the faces in video frames, we have used the OpenFace2 [23] toolkit. The basic purpose of selecting the OpenFace2 [23] toolkit for face extraction is that this software utilized 2D and 3D landmarks estimation to locate the faces and empowered to better capture the head orientations, and eye alignments and robustly extract the action units. Furthermore, the OpenFace2 toolkit is invariant to face alignment in video frames and can efficiently detect human faces under the occurrence of intense light changes [24]. Such structural description of the OpenFace2 library allows it to better compute the features of the facial portion. Additionally, to maintain the computational complexity, we have taken 20 frames per second from all visual samples.

2.2 Keypoints Calculation

The second phase of the proposed approach is concerned with the calculation of deep features from the extracted faces which are then used to categorize the original and fake content. To obtain the discriminative set of image features, we have nominated several latest pre-trained deep learning-based approaches like ResNet18, ResNet50, DenseNet65, DenseNet77, and DenseNet100. The reason to employ the pre-trained model for deepfakes detection is that these approaches are already trained on huge datasets like ImageNet and have already gained extensive knowledge which allows them to calculate more distinctive keypoints set from the video frames. Thus, the usage of the pre-trained approaches for a new application like for visual manipulation classification enhances the training procedure by quickly learning the domain characteristics. As these models have already gained significant knowledge and can easily adapt to a new task. This entire procedure is named 'transfer learning'. In the processing of executing the transfer learning process, the early model's layers are concerned to extract lower-level keypoints and the latter is related to learn the application-specific characteristics. A graphic description of transfer learning is depicted in Fig. 2. The used models namely ResNet18, ResNet50, DenseNet65, DenseNet77, and DenseNet100 are capable of learning nominative keypoints from the facial areas i.e. face structure, eyes, nose alignments, lip sizes, etc.

2.2.1 ResNet

ResNet [25] is a well-known deep learning approach that exploits identity shortcut links together with residual mapping between layers to acquire better performance. In conventional deep learning approaches, the input from the previous layer is passed as input to the next layer to compute a dense set of feature vectors [26]. Such a network setting improves the classification performance, however, also results in an intense computational burden. Besides, the intense increase in depth of framework architecture can cause the gradient vanishing problem which can reduce the model recognition performance. To resolve the issues of such approaches, the ResNet framework was presented as it used skip links by surpassing the several layers and founding the basis for residual blocks (RBs). Such a

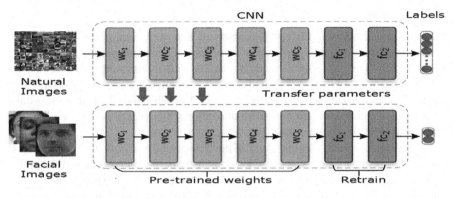

Fig. 2. A visual depiction of transfer learning

network of ResNet models allows it to reemploy feature maps computed by the prior layers and improved the classification results by minimizing the model complexity as well. The RB is the major structural unit of the ResNet framework. We have shown the pictorial representation of the residual block in Fig. 3. An RB consists of Convolution and batch normalization layers. Moreover, the model uses the ReLU activation approach and shortcut links in its architecture.

Within all RBs, the stacked layers launch residual plotting via generating shortcut connections that accomplish mapping (x). The output is produced by combing the results from all stacked layers by using the output residual function $F(x)$. The output of the RB is given by using Eq. 1:

$$Y = F(x) + x \tag{1}$$

where x is denoting the input, F designates the residual method and Y depicts the output. In our work, we have used two variants of the ResNet method with a depth of 18 and 50 for deepfakes classification.

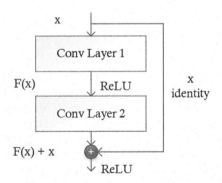

Fig. 3. The structural depiction of Residual Block

2.2.2 DenseNet

DenseNet is an empowered deep learning-based approach comprising various dense blocks (DBs) that are continuously joined with extra-added convolutional and pooling layers [27, 28]. The denseNet network is effective in demonstrating the difficult sample transformations robustly which contributes to resolving the concern of disappeared object location information to some level. Besides, DenseNet aids the features transmission process and supports their redundant utilization which makes the DenseNet model more useful for classifying pristine and altered visual content.

The DBs are the main building block of the DenseNet framework as shown in Fig. 4, in which within all N-1 layers, $n \times n \times m_0$ demonstrates the features maps (FPs) with n and m_0 signifying the FPs length and total channels respectively. To shorten the channel length, a non-linear transformation designated as $H(.)$ is used containing numerous methods like Batch Normalization (BN), and Relu. In Fig. 4, the long-dashed arrow are depicting the dense links which join the earlier and the coming layers via the output calculated by $H(.)$ and $n \times n \times (m_0 + 2m)$ is the final outcome from $N + 1$ layer. The wide dense links enhance FPs largely, so, a transition layer (TL) is proposed to lessen the keypoints dimension than the previous DB. In our work, we have used three variants of the DenseNet model with the depth of 65, 77, and 100 respectively.

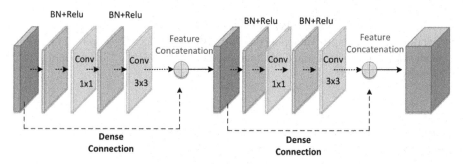

Fig. 4. Visual depiction of DB [29]

2.3 Classification

After the feature extraction, we have performed the classification of deepfakes into two categories i.e., fake, and real. For the said reason, we have applied the binary SVM predictor, which creates the hyperplanes to put the decision margin. The main motivation for choosing the SVM is that it can deal with binary problems more efficiently than the other methods i.e., KNN, Naïve Bayes, etc. Moreover, it has the ability to reduce the total experimental error as well preserving the complexity point of the mapping task. In deepfakes classification, SVM achieves a robust solution and also handles the overfitting during the training process. All of the above qualities of SVM provide us with accurate and robust solutions and also have the ability to recognize unseen data. The total features vector is denoted by N and can be described as $(u(j), v(j)), j = 1,....n$, where $v(j) \in \{1, -$

1] displays the classes into real and fake. SVM creates the hyperplane for every vector that is linearly separating our classes as:

$$D^T \cdot u^{(j)} + b \geq 1 \; if \; v^{(j)} = +1 \tag{2}$$

$$D^T \cdot u^{(j)} + b < 1 \; if \; v^{(j)} = -1 \tag{3}$$

The bias is symbolized by b while the weight vector is denoted by D. The target is to achieve the maximum margin among 2 vectors by decreasing the norm $\|D\|$, as shown below:

min $\|D\|$, such that

$$v^{(j)} \left(D^T \cdot u^{(j)} + b \right) \geq 1 \tag{4}$$

The classification of deepfakes can be found using the given Eq. 5:

$$\begin{cases} original, \; f(u^{(j)}) = +1, \\ deepfake, \; f(u^{(j)}) = -1 \end{cases} \tag{5}$$

where $f(u) = sign \, (D^T \cdot u^{(j)} + b)$ is the discriminant function.

3 Experiments and Results

Here, we have discussed the description of the used dataset, and evaluation measures. Moreover, we have performed a detailed experiment to show the efficacy of the employed deep learning models for deepfakes detection.

3.1 Dataset

To evaluate our work, we have used the DFDC dataset issued by Facebook which is online accessible on the Kaggle competition site [30]. The forged samples of the DFDC database are created by employing two unknown AI approaches. This dataset contains a total of 19,000 pristine video samples along with 100,000 manipulated videos. For our model, we have randomly distributed the DFDC dataset into a 70–30 ratio for the training and testing purpose respectively.

3.2 Evaluation Matrices

Numerous standard evaluation measures namely like (P), recall (R), accuracy (A), true positive rate (TPR), and F1-score. The mathematical demonstration of all used performance measures can be found here:

$$P = \frac{\tau}{(\tau + f)} \tag{6}$$

$$R = \frac{\tau}{(\tau + \eta)} \tag{7}$$

$$A = \frac{(\mho + \jmath)}{(\mho + \jmath + \int + \eta)} \tag{8}$$

$$F1 - score = \frac{2PR}{P + R} \tag{9}$$

Here, $\mho$ is denoting the true positive (true predicted manipulated video samples), and $\jmath$ shows the true negatives (true predicted real samples). Moreover, $\int$ shows false positive (wrong detected forged samples), while η demonstrates false-negative (wrong marked real samples) respectively.

3.3 Implementation Details

We have experimented with all models in Python language with TensorFlow on an Nvidia GTX1070-based GPU system. Furthermore, we have trained the SVM classifier with different feature computations by using the DFDC database with 50 epochs and a 0.001 learning rate.

3.4 Results

In this part, we have performed a detailed evaluation of the employed deep learning models namely ResNet18, ResNet50, DenseNet65, DenseNet77, and DenseNet100 with the SVM classifier with the help of several standard metrics used for model performance measures in the area of deepfakes detection and classification. We have used about 8500 samples for the SVM training, while about 3500 images are used for model verification.

Initially, we have discussed the precision values obtained for the ResNet18, ResNet50, DenseNet65, DenseNet77, and DenseNet100 models respectively, as this metric shows how much a model is capable of discriminating the original content from forged samples. The obtained values are given in Fig. 5. It is quite clear from the values elaborated in Fig. 5 that the employed DL models are quite robust to recognize the pristine samples. In a more distinctive way, the work attained the highest precision value for the DenseNet100 model with a value of 97.50%, while the second-highest score is obtained by the DenseNet77 model with a number of 96.41%. Moreover, the lowermost precision results are acquired by the ResNet18 approach with a score of 94.82%.

Another main evaluation metric for deepfakes detection-based approaches is the recall measure which determines the capability of a model in recognizing the fake samples. As the deepfakes systems are highly employed in processing legal claims, therefore, misclassification of deepfakes samples as real can cause huge damage. Hence, the optimization of the recall metric is an essential task for deepfakes detection approaches. We have discussed the obtained values for all five employed deep learning approaches with the SVM classifier in Fig. 6. The results shown in Fig. 6 are clearly showing the efficacy of the employed approaches in recognizing the manipulated samples from the original content. More descriptively, we have attained the recall values of 0.951, 0.954, 0.955, 0.960, and 0.973 for the ResNet18, ResNet50, DenseNet65, DenseNet77, and DenseNet100 respectively.

Furthermore, we have discussed the F1-Score along with the error rate for all five employed deep learning models along with the SVM classifier as this metric provides

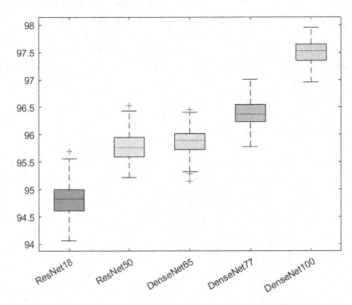

Fig. 5. Precision values for all employed deep learning models with the SVM classifier

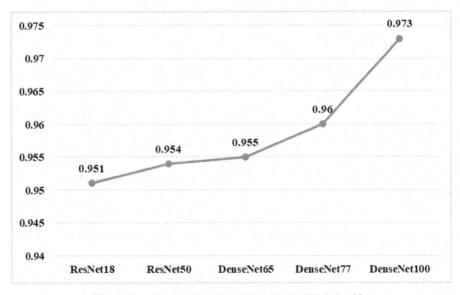

Fig. 6. Recall values for all models with the SVM classifier

an overall examination of the effectiveness of the models. The larger value of the F1-Score depicts a better model performance. The obtained F1-Score values along with the error rates for the ResNet18, ResNet50, DenseNet65, DenseNet77, and DenseNet100 models are shown in Fig. 7. The values reported in Fig. 7 are showing the efficiency of deep learning models with the SVM classifier for forensic manipulation detection. More clearly, we have attained the highest F1-Score value with the minimum error rate for the DenseNet100 model with values of 97.40%, and 2.60% respectively. While the ResNet18 shows the lowest F1-Score with the highest error rate with values of 94.90% and 5.1% respectively.

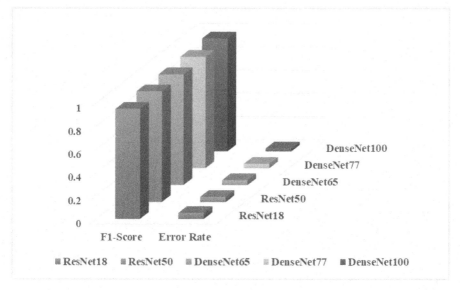

Fig. 7. F1-Scores with error rates for all deep learning models with the SVM classifier

Moreover, we have reported the accuracy values for the ResNet18, ResNet50, DenseNet65, DenseNet77, and DenseNet100 models along with the SVM classifier in Table 1 from where it is quite evident that all models perform well for deepfakes detection. The highest accuracy value is reported by the DenseNet100 model with the SVM classifier with a value of 97.80%, while the DenseNet77 shows the second-largest accuracy value of 97.20%.

Table 1. Accuracy values for all deep learning models with the SVM classifier

Models	Accuracy
ResNet18 + SVM	0.966
ResNet50 + SVM	0.973
DenseNet65 + SVM	0.969
DenseNet77 + SVM	0.972
DenseNet100 + SVM	0.978

From all the conducted analyses, we can conclude that the deep models with the SVM classifiers are robust to detect the forensic changes made within visual samples. Moreover, from all employed models, the DenseNet100 framework is found more efficient for classifying the real and deepfakes samples.

4 Conclusion

With the introduction of new deepfakes generation technologies capable of producing realistic manipulated videos, detecting deepfakes has become a difficult challenge. In this study, we developed an end-to-end framework that can be effectively used to identify visual manipulations. We have also used the concept of transfer learning to improve performance in distinguishing between distinct deepfakes and reduce model overfitting. We used a variety of performance assessment criteria, such as precision, accuracy, recall, and F1-score, to analyze the effectiveness of features extracted by different DL models for the purpose of deepfakes detection. The comparative results reveal that the DenseNet100 model obtains higher overall accuracy than the other models by extracting robust feature vectors. We conclude that our work gives insight into creating a more generalizable model with improved detection performance and serves as a baseline for the research community interested in developing deepfakes detection models.

Acknowledgments. This work was supported by the grant of the Punjab Higher Education Commission (PHEC) of Pakistan via Award No. (PHEC/ARA/PIRCA/20527/21).

Conflict of Interest. All authors declared no conflict of interest in this work.

References

1. Akhtar, Z., Rattani, A., Hadid, A., Tistarelli, M.: Face recognition under ageing effect: a comparative analysis. In: Petrosino, A. (ed.) ICIAP 2013. LNCS, vol. 8157, pp. 309–318. Springer, Heidelberg (2013). https://doi.org/10.1007/978-3-642-41184-7_32
2. Nawaz, M., et al.: Single and multiple regions duplication detections in digital images with applications in image forensic. J. Intell. Fuzzy Syst. **40**(6), 10351–10371 (2021)
3. FaceApp (2022). https://www.faceapp.com/
4. Park, J.S., Chung, M.S., Hwang, S.B., Lee, Y.S., Har, D.-H.: Technical report on semi-automatic segmentation using the Adobe photoshop. J. Digit. Imaging **18**(4), 333–343 (2005)
5. Kohli, A., Gupta, A.: Detecting DeepFake, FaceSwap and Face2Face facial forgeries using frequency CNN. Multimed. Tools Appl. **80**(12), 18461–18478 (2021). https://doi.org/10.1007/s11042-020-10420-8
6. Güera, D., Delp, E.J.: Deepfake video detection using recurrent neural networks. In: 2018 15th IEEE International Conference on Advanced Video and Signal Based Surveillance (AVSS), pp. 1–6. IEEE (2018)
7. Gellately, R.: Lenin, Stalin, and Hitler: The age of social catastrophe. Alfred a Knopf Incorporated (2007)

8. Masood, M., Nawaz, M., Javed, A., Nazir, T., Mehmood, A., Mahum, R.: Classification of deepfake videos using pre-trained convolutional neural networks. In: 2021 International Conference on Digital Futures and Transformative Technologies (ICoDT2), pp. 1–6. IEEE (2021)

9. Nawaz, M., et al.: Image authenticity detection using DWT and circular block-based LTrP features. CMC-Comput. Mater. Continua **69**(2), 1927–1944 (2021)

10. Moudhgalya, N.B., Divi, S., Adithya Ganesan, V., Sharan Sundar, S., Vijayaraghavan, V.: DeepTrace: a generic framework for time series forecasting. In: Rojas, I., Joya, G., Catala, A. (eds.) IWANN 2019. LNCS, vol. 11506, pp. 139–151. Springer, Cham (2019). https://doi.org/10.1007/978-3-030-20521-8_12

11. Jameel, W.J., Kadhem, S.M., Abbas, A.R.: Detecting deepfakes with deep learning and gabor filters. Aro Sci. J. Koya Univ. **10**(1), 18–22 (2022)

12. Xu, Y., Raja, K., Pedersen, M.: Supervised contrastive learning for generalizable and explainable deepfakes detection. In: Proceedings of the IEEE/CVF Winter Conference on Applications of Computer Vision, pp. 379–389 (2022)

13. Wang, G., Jiang, Q., Jin, X., Cui, X.: FFR_FD: effective and fast detection of deepfakes via feature point defects. Inf. Sci. **596**, 472–488 (2022)

14. Kolagati, S., Priyadharshini, T., Rajam, V.M.A.: Exposing deepfakes using a deep multi-layer perceptron–convolutional neural network model. Int. J. Inf. Manage. Data Insights **2**(1), 100054 (2022)

15. Sun, Z., Han, Y., Hua, Z., Ruan, N., Jia, W.: Improving the efficiency and robustness of deepfakes detection through precise geometric features. In: Proceedings of the IEEE/CVF Conference on Computer Vision and Pattern Recognition, pp. 3609–3618 (2021)

16. Mehta, V., Gupta, P., Subramanian, R., Dhall, A.: Fakebuster: a deepfakes detection tool for video conferencing scenarios. In: 26th International Conference on Intelligent User Interfaces-Companion, pp. 61–63 (2021)

17. Yavuzkilic, S., Sengur, A., Akhtar, Z., Siddique, K.: Spotting deepfakes and face manipulations by fusing features from multi-stream CNNs models. Symmetry **13**(8), 1352 (2021)

18. Yasrab, R., Jiang, W., Riaz, A.: Fighting deepfakes using body language analysis. Forecasting **3**(2), 303–321 (2021)

19. Khalid, H., Woo, S.S.: OC-FakeDect: Classifying deepfakes using one-class variational autoencoder. In: Proceedings of the IEEE/CVF Conference on Computer Vision and Pattern Recognition Workshops, pp. 656–657 (2020)

20. Montserrat, D.M., et al.: Deepfakes detection with automatic face weighting. In: Proceedings of the IEEE/CVF Conference on Computer Vision and Pattern Recognition Workshops, pp. 668–669 (2020)

21. Wang, Y., Dantcheva, A.: A video is worth more than 1000 lies. Comparing 3DCNN approaches for detecting deepfakes. In: 2020 15th IEEE International Conference on Automatic Face and Gesture Recognition (FG 2020), pp. 515–519. IEEE (2020)

22. Yang, X., Li, Y., Lyu, S.: Exposing deep fakes using inconsistent head poses. In: ICASSP 2019–2019 IEEE International Conference on Acoustics, Speech and Signal Processing (ICASSP), pp. 8261–8265. IEEE (2019)

23. Baltrušaitis, T., Robinson, P., Morency, L.-P.: Openface: an open source facial behavior analysis toolkit. In: 2016 IEEE Winter Conference on Applications of Computer Vision (WACV), pp. 1–10. IEEE (2016)

24. Fydanaki, A., Geradts, Z.: Evaluating OpenFace: an open-source automatic facial comparison algorithm for forensics. Forensic Sci. Res. **3**(3), 202–209 (2018)

25. He, K., Zhang, X., Ren, S., Sun, J.: Deep residual learning for image recognition. In: Proceedings of the IEEE Conference on Computer Vision and Pattern Recognition, pp. 770–778 (2016)

26. Simonyan, K., Zisserman, A.: Very deep convolutional networks for large-scale image recognition. arXiv preprint arXiv:1409.1556 (2014)
27. Albahli, S., Nawaz, M., Javed, A., Irtaza, A.: An improved faster-RCNN model for handwritten character recognition. Arab. J. Sci. Eng. 1–15 (2021)
28. Albahli, S., Nazir, T., Irtaza, A., Javed, A.: Recognition and detection of diabetic retinopathy using densenet-65 based faster-RCNN. Comput. Mater. Contin **67**, 1333–1351 (2021)
29. Albattah, W., Nawaz, M., Javed, A., Masood, M., Albahli, S.: A novel deep learning method for detection and classification of plant diseases. Complex Intell. Syst. **8**(1), 507–524 (2021). https://doi.org/10.1007/s40747-021-00536-1
30. Dolhansky, B., et al.: The DeepFake detection challenge dataset. arXiv preprint arXiv:2006.07397 (2020)

Deep Learning Based Speech Synthesis with Emotion Overlay

Abhijnya Bhat$^{(\boxtimes)}$ ⓘ, Sejal Priyaⓘ, Abhijit Sethiⓘ, Kedar U. Shetⓘ,
and Ramamoorthy Srinathⓘ

Department of Computer Science and Engineering, PES University,
Bangalore, Karnataka, India
abhijnya.bhat@gmail.com, srinath@pes.edu

Abstract. This paper proposes a Text-to-Speech System that synthesizes emotional speech. We have identified the lack of style diversity within audios produced for the same emotion, and have thus, come up with techniques of combining emotion embeddings with style embeddings in a novel weighted-controlled manner to produce speech with style varying based on target speakers and emotion depending on the emotion-category specified. We also provide different model variations which suggest different methods of injecting emotion embeddings into the TTS and various tactics of combining them with style embeddings. Tests conducted on our model variation which overlays emotion embeddings on the encoder outputs during inference and combines them with style embeddings in a 3:7 weight-ratio as per our novel approach gives a Mean Opinion Score of 3.612 which shows the more-than-satisfactory performance of our models in synthesizing style varying emotional speech.

Keywords: Emotional TTS · Modified GST-Tacotron · Style and Emotion Embeddings

1 Introduction

Natural language processing is a technologically advanced discipline that is constantly expanding. Due to its numerous applications, text-to-speech (TTS) synthesis is a widely recognised subdomain.

There has been a lot of advancements in the technologies being used for TTS. All these advancements however have one aspect in common: The audio that the models generates does not contain emotional expression. Numerous applications can benefit from the injection of emotion into the audio waveform, namely, audiobooks, sign language to speech conversion, etc. We identified this area as a research gap as it possesses massive scope for technological development.

Through our work, we intend to build Text-To-Speech Systems which are capable of synthesizing speech which convey emotions. We also focus on bringing style variations within the voices that are generated for a particular emotion.

A. Bhat, S. Priya, A. Sethi, K. U. Shet, R. Srinath—These authors contributed equally to this work.

M. Singh et al. (Eds.): ICACDS 2023, CCIS 1848, pp. 304–315, 2023.
https://doi.org/10.1007/978-3-031-37940-6_25

Thus, various models suggesting different ways of combining the style embeddings and the emotion embeddings are discussed. The style embeddings capture the speaking style of the target speakers while the emotion embeddings focus on injecting the emotions in the speech. All the functionalities are built on top of the GST-Tacotron model [1] which in turn uses the Tacotron 2 [2] as TTS model and the Wavenet model as vocoder [3].

2 Related Works

To synthesise speech, Tacotron creates an end-to-end model. Due to the fact that individual component defects do not compound, it performs better than a multi-stage model [4]. Tacotron 2 creates mel-spectrograms from the character embeddings of the input text using a sequence-to-sequence network [2]. It transitions the mel-scale spectrograms into time-domain spectrograms using the WaveNet model [3]. Concatenative TTS and parametric TTS can both be replaced with WaveNet, a fully convolutional neural network-based system.

The GST (Global Style Tokens)-Tacotron [1] introduces global style tokens, which provide the synthesised voice more expressivity. These embeddings were trained using Tacotron and lack explicit labels as well as prosodic labels.

Since the GST-Tacotron closely resembles the reference audio's generic speaking style, it is challenging to add user-defined emotions to the synthesised speech. [5] discusses how to use the ratio of the distance between intra-emotion cluster vectors and inter-emotion cluster vectors to find representative embedding vectors for each emotion category. It was discovered that the representative emotion embedding vectors obtained using this method differed slightly from those obtained by averaging all the emotion-related vectors. We utilise the more straightforward approach—also supported in [6]—of using the average weight vectors as the representative embedding vector for an emotion category in our research.

The use of a style tag dataset is proposed in [7]. The idea is to feed the intensity, speed, tone and emotion of the audio in the form of a tag to the model which works like a natural language explanation of the audio being fed. The paper also makes use of pre-trained language models to understand the semantic connection between the speech style and the text description embedding. The use of language models like BERT and GPT-3 for similar purposes is also seen in [8] and [9] respectively. [10] describes a technique for transferring emotion from an emotional speaker to the targeted neutral speech while keeping the goal intonation. In order to segregate sounds and provide emotion-independent and timbre-independent acoustic representations, two simultaneously trained neural network branches were used in the research. The work proposed in [11] takes phoneme level emotion information and integrates it with the intensity of the emotion in that phoneme's utterance to overcome the monotony in synthesized speech.

One of the evident problem existing in synthesizing emotional speech is being able to obtain style diversity among the audios synthesized for a particular emotion, that is, how to make two audios synthesized under the same emotion label to sound different. Thus, to tackle this problem, we propose the use of a weighted-controlled combination of style embeddings obtained from the audios of different target speakers and emotion embeddings obtained from audios of speakers conveying the same emotion. We make use of the Global Style Token Tacotron as our base TTS model and incorporate the injection of emotion embeddings in different ways which are discussed in detail in the following sections.

3 Methodology

3.1 Style and Emotion Embeddings Generation

Our foundation model was the GST-Tacotron [1], and features were added on top of that. The GST-Tacotron uses a style token layer with an attention module that has a bank of global style tokens, which are randomly initialised embeddings. In order to preserve the prosody of the reference audio, this is used to create the style embedding. The tacotron model and this style embedding are jointly trained before being sent to the text encoder.

We use a GST model to create style tokens for each emotion category from a set of audio sequences that relate to specific emotion categories in order to build the emotion-embeddings. The average of all the style vectors created for a given emotion category is then used to create a representative emotion vector for that emotion category.

At this stage, we have global emotion-embeddings along with local style embeddings for each data point. We use a combination of both these embeddings in order to preserve the user specific styles while injecting emotion into the output waveform.

3.2 The General Flow

In order to figure out the most efficient technique of combining the style and emotion embeddings to produce emotional synthesized speech, different variations of models were built, trained and tested which are discussed in detail in Subsect. 3.3. However, the general flow followed by most of the variations depicted is Fig. 1 and is as follows:

- To create style tokens for each audio sequence, pass a set of reference audio sequences associated with specific emotion categories through a GST-Tacotron model.
- These sequences are fed into a model that creates a vector that represents each type of emotion.
- Both the style embedding and the emotion embedding are input into the model where they are weighted to manage their contributions in order to preserve style variance. The TTS model is then given this.
- The model produces an output mel-scale spectrogram, which is then transformed by a vocoder into its matching time-domain waveform.

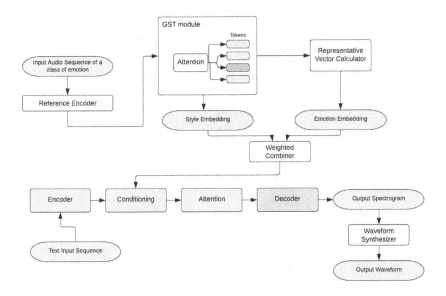

Fig. 1. Model architecture depicting the general flow. Reference audio sequences of specific emotions are used to generate style tokens, which are then used to produce representative vectors for the emotion categories. A weighted combiner is used to generate a convex combination of style and emotion embeddings.

3.3   The Different Model Variations

In order to figure out the most optimal way of overlaying the emotion embeddings, we built a number of different model variations as discussed below.

3.3.1   GST-Tacotron with Emotion Embeddings Added During Inference

All the audios belonging to a particular emotion were grouped and passed into a trained GST-Tacotron model. Thus embedding vectors belonging to every audio for an emotion category were obtained. The mean of all the output vectors acquired above was then calculated to produce a representative vector. Such representative vectors for each emotion category were found. In order to capture the speaking style of the target speaker, a reference audio is provided to the GST-Tacotron model during inference. This embedding is expanded and added to the encoder output of the GST-Tacotron model. Here we also expand and add the representative vector of the required emotion which had been calculated and stored. Thus, the input to the decoder has style and emotion embeddings injected in it and the corresponding mel-spectrogram is obtained which is then fed to the WaveNet vocoder for conversion to time-domain waveform, thereby producing the emotional synthetic speech. Figure 2(a) depicts the training process of a normal GST-Tacotron. Figure 2(b) depicts the inference process explained above.

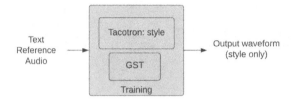

(a) Training process of GST-Tacotron

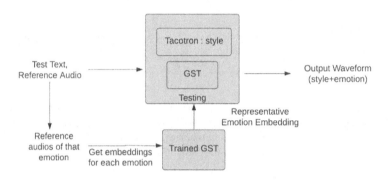

(b) Custom inference process for model variation 1

Fig. 2. GST-Tacotron with emotion embeddings added during inference. The emotion embeddings are obtained over the trained model and these are passed to the model during the testing phase.

3.3.2 GST Tacotron with Emotion Embeddings Added While Training

The emotion embeddings added during inference stage in model variation 1 were not involved during the training process of the GST-Tacotron. This model variation focuses on including the emotion embeddings during the training step so as to enable the model to train with the emotion embeddings. For this, the emotion embeddings that had been obtained in model variation 1 were added to the style embeddings during training to the encoder outputs. For example, during training, for an audio belonging to the emotion 'x', the previously stored emotion embeddings for the emotion 'x' are injected and concatenated with the style embeddings produced then during training. These are then added to the encoder outputs. The rest of the training process through the attention module, decoder module and the vocoder remains the same as before. Figure 3 represents the altered training process of the model variation 2.

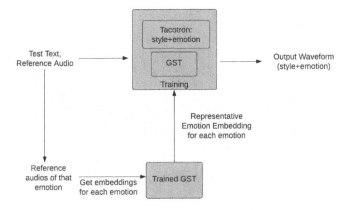

Fig. 3. GST Tacotron with emotion embeddings added while training. The emotion embeddings are obtained before hand, and these are injected into the model while training.

3.3.3   GST-Tacotron with Trainable Emotional Embeddings

In the model variation 2, the emotion embeddings being injected during training had been previously obtained and stored from another trained instance of the GST-Tacotron. These emotion embeddings were thus not changing with the training progression of the current GST-Tacotron instance. To enable this, the emotion embeddings had to be included in such a way that they were not constant through all the epochs and rather kept changing as per the model weights. So, in each epoch, the audio samples belonging to particular emotion categories were passed into the GST-Tacotron model, a representative vector was then calculated from these vectors and was added to the encoder outputs along with usual style vector. In this manner, the loss calculated included that of the emotion embeddings, which in turn helped the model to train.

Figure 4 represents the altered training process of the model variation 3.

3.3.4   GST-Tacotron with Manual Weighted Embeddings

In the original base GST-Tacotron model, the style embedding was only being added once during training. In the model variations 1,2 and 3 the embeddings were being added twice (style as well as emotion) and this was suspected to be the reason of introducing additional noise in the model. To solve this, a weighted combiner was employed which produce a convex combination of both the embeddings in order to make an illusion of adding just one embedding. The formula:

$$(\alpha * \text{StyleEmbedding}) + ((1 - \alpha) * \text{EmotionEmbedding})$$

was used to achieve the same. Here α and $(1 - \alpha)$ are the coefficients of the style and emotion embeddings respectively. Different values of α such 0.2, 0.3, 0.5, 0.7 were experimented with and the synthesized voice outputs for each were compared.

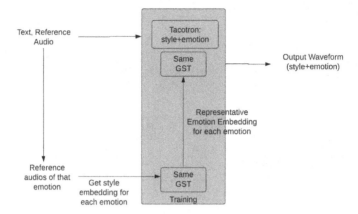

Fig. 4. GST-Tacotron with trainable emotional embeddings. The emotion embeddings are generated during the training phase to facilitate the learning of these embeddings.

3.3.5 GST-Tacotron with Automated Weighted Embeddings

The model variation 4 required manual specification of α values in the weighted combiner. This process was laborious and thus, an automated alternative was required.

For this, a sequence of linear layers followed by ReLU activation functions was used to first pass the emotion embeddings. The same was done for the style embeddings. The outputs of these blocks were then expanded to the size of the encoder outputs and added to them. In this manner, the manual weighing system was transformed into an automated one which would employ trainable weights to increase the accuracy and skip the manual intervention.

3.3.6 Using One Hot Encoded Vectors as Emotion Vectors

The GST-Tacotron model in the model variations up until now were obtaining gradients from two sources - from the style embeddings as well as from the emotion embeddings. To remove any possibility of uneven training due to this, the origin of the two embeddings were now separated so that the GST-Tacotron model would only train with respect to one of them. The previous method to obtain emotional embeddings was replaced by employing one hot encoding on the emotions to obtain 5 different emotion vectors. These vectors were then passed through a block of neural network consisting of linear layers and activation functions. They were expanded to the size of the encoder's outputs. This constituted the global emotion vectors. For the local prosody related vectors, the GST-Tacotron model was used, unchanged.

In this manner, different models were responsible for obtaining different hierarchical embeddings and thus, the training process was compartmentalized.

4   Experimental Details

The dataset chosen for the experimentation was the Emotional Speech Dataset (ESD) [12] made available by National University of Singapore (NUS) and Singapore University of Technology and Design (SUTD). The part of the dataset containing utterances spoken by 10 English speakers, both male and female was chosen. Each data sample consists of an audio recording and a text transcript that corresponds to the audio, both of which depict one of the five emotions listed: neutral, happy, angry, sad, or surprised.

We divided the dataset into training, testing, and validation sets after segregating and preprocessing it. For the training set, the global emotion embeddings were generated by passing the audio inputs present in each emotion category. The local style tokens for each data point was produced by iterating through the dataset.

The data samples were sent into the model variations in batches of 32 data points. An initial learning rate of 0.001 was maintained, which was further reduced using a learning rate scheduler as the training progressed. The model variations were trained for 200 epochs and the model was saved corresponding to the lowest loss on the validation file. This saved model checkpoint was then used to carry out tests on the test set. Help for implementation of the base GST-Tacotron was taken from [13] and the modifications and additions for all the different model variations were then done on top of the base model.

The initial model variations were trained on both male and female speaker audios together in a training session. With further investigation, it was found that the emotion vectors belonged to male voices and adding this to female outputs produced slightly distorted voices due to incompatible vectors. To solve this, the training for model variations 4 and 5 was done with two subsets of the dataset: male voices and female voices separately. This was achieved by filtering the dataset to separate the female occurrences from the male ones. In this manner these 2 model variations were trained on two different datasets. In this process the emotion embeddings obtained were gender specific, thereby preserving the gender-specific nuances.

5   Results and Discussion

5.1   MOS Test

The test findings were evaluated using the Mean Opinion Score (MOS) test. It is a quantitative indicator of how well-rounded an event or experience is perceived by people, in this case, the expressivity of the emotional audio. The most common rating scale is from 1 (poor) to 5 (outstanding).

A form with audio outputs for the 5 emotions was created and sent to individuals to obtain unbiased opinions. It was made sure that for the same text, synthesized audios belonging to all 5 emotions were sent. The audios were unlabelled and the listeners were requested to guess the emotions conveyed by the speech while also grading it on a scale of 1–5 based on how well the emotion was expressed and the quality of the audio.

The responses of over 50 individuals were collected and the average MOS score for each emotion as well as the total MOS was calculated. Table 1 contains the details of the MOS test conducted. The overall MOS was found to be 3.612.

Table 1. MOS Test Results

Emotion	Sad	Angry	Happy	Surprised	Neutral	Average
MOS	3.52	3.91	3.85	3.36	3.60	3.612

5.2   Emotion Embedding Plots

All the emotion embeddings obtained by the model were plotted in order to visualize them. We used the t-distributed stochastic neighbor embedding (t-SNE plot) [14], which deals with linearly non separable data. This plot is used to understand high dimensional data by projecting it into lower dimensional spaces. Figure 5 shows the t-sne plot obtained for the emotion embeddings obtained from the ESD dataset. It can be seen from the plots that the emotion embeddings belonging to the same emotion are highly clustered and 5 distinct clusters are formed corresponding to the 5 different emotions present in the dataset. This validates the method opted for production of the emotion embeddings.

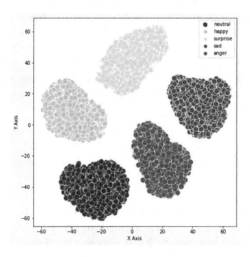

Fig. 5. Visualization of the emotion embeddings obtained from ESD Dataset

5.3 Discussions on the Various Model Variations

The following discussions are in reference with the various models discussed under Subsect. 3.3 of Sect. 3:

- In the model variation 1, adding emotion vectors only during testing led to the emotion vectors not being included during train time. Thus, they were not incorporated in the loss or in the back-propagation during training.
- In the model variation 2, the emotional vectors were generated from a separate GST module which had been previously trained, so they remained fixed during the entire training process and did not evolve with the increasing epochs.
- In the model variation 3, 4 and 5, incorporating the emotion vectors during training made them change each epoch and thus, the loss calculated included that of the emotion embeddings making the model train taking emotion vectors into perspective.
- In the model variation 5, the model was allowed to learn weights automatically by passing the style and emotion embeddings through a neural network block which helped in overcoming the manual weight specification burden in model variation 4.
- In the model variation 6, changing the source of production of the emotion embeddings by using one hot encoding vectors as base emotion vectors helped in compartmentalizing the production of the two embeddings, thereby reducing the confusion the model faced due to gradients coming from two sources.

Each of these variations had their own pros and cons, with some producing speech with more emotional expressivity, some capturing the speaker style better and the others producing better quality audio. Table 2 provides a quick summary covering the performance of the various model variations.

Table 2. Performance property Vs Model variations

Property	Model Variation
Better Quality Audio	Base GST model (adding emotion embeddings usually introduced noise)
Higher emotional expressivity	Variations 4,5 (with higher emotion embedding weightage)
Greater attention to speaker style	Variation 2 (with higher speaker embedding weightage)

Through experiments, it was also found out that training on single-speaker data produces better emotional audio output than the models trained on multi-speaker data.

6   Conclusion and Future Work

All the model variations were built focusing on finding the most optimal way of combining emotion embeddings with style embeddings so as to inject emotional expressivity in the speech that is synthesized by the Text-To-Speech system while also maintaining style diversity within the emotional voices produced depending on the local prosody properties of the target speaker. All of the model variations formed are capable of taking text as input and synthesizing an emotional speech based on the target speaker chosen and the emotion category selected.

More computational-resource expensive TTS models like FastSpeech2 can be used in place of the GST-Tacotron for the base of all the model variations in the future. This work can also be extended to produce audio for textual conversation between two or more people.

References

1. Wang, Y., et al.: Style tokens: unsuper-vised style modeling, control and transfer in end-to-end speech synthesis. In: International Conference on Machine Learning (2018)
2. Shen, J., et al.: Natural TTS synthesis by conditioning wavenet on mel spectrogram predictions. In: 2018 IEEE International Conference on Acoustics, Speech and Signal Processing (ICASSP), pp. 4779–4783 (2018). https://doi.org/10.1109/ICASSP.2018.8461368
3. Oord, A., et al.: Wavenet: A generative model for raw audio (2016)
4. Wang, Y., et al.: Tacotron: towards end-to-end speech synthesis. In: Interspeech (2017)
5. Um, S., Oh, S., Byun, K., Jang, I., Ahn, C.H., Kang, H.-G.: Emotional speech synthesis with rich and granularized control. In: ICASSP 2020–2020 IEEE International Conference on Acoustics, Speech and Signal Processing (ICASSP), pp. 7254–7258 (2019)
6. Kwon, O., Song, E., Kim, J.-M., Kang, H.-G.: Effective parameter estimation methods for an excitnet model in generative text-to-speech systems. ArXiv abs/1905.08486 (2019)
7. Shin, Y., Lee, Y., Jo, S., Hwang, Y., Kim, T.: Text-driven emotional style control and cross-speaker style transfer in neural TTS. ArXiv abs/207.06000 (2022)
8. Kim, M., Cheon, S.J., Choi, B.J., Kim, J.J., Kim, N.S.: Expressive text- to-speech using style tag. In: Interspeech (2021)
9. Yoon, H.-W., et al.: Language model-based emotion prediction methods for emotional speech synthesis system. ArXiv abs/206.15067 (2022)
10. Lei, Y., Yang, S., Zhu, X., Xie, L., Su, D.: Cross-speaker emotion transfer through information perturbation in emotional speech synthesis. IEEE Signal Process. Lett. **29**, 1948–1952 (2022). https://doi.org/10.1109/LSP.2022.3203888
11. Im, C.-B., Lee, S.-H., Kim, S.-B., Lee, S.-W.: EMOQ-TTS: emotion intensity quantization for ne-grained controllable emotional text-to-speech. In: ICASSP 2022–2022 IEEE International Conference on Acoustics, Speech and Signal Processing (ICASSP), pp. 6317–6321 (2022). https://doi.org/10.1109/ICASSP43922.2022.9747098

12. Zhou, K., Sisman, B., Liu, R., Li, H.: Seen and unseen emotional style transfer for voice conversion with a new emotional speech dataset. In: ICASSP 2021–2021 IEEE International Conference on Acoustics, Speech and Signal Processing (ICASSP), pp. 920–924 (2021). https://doi.org/10.1109/ICASSP39728.2021.9413391
13. Deng, C.: A PyTorch implementation of Style Tokens: Unsupervised Style Modeling, Control and Transfer in End-to-End Speech Synthesis. https://github.com/KinglittleQ/GST-Tacotron (2019)
14. van der Maaten, L., Hinton, G.: Visualizing data using t-SNE. J. Mach. Learn. Res. **9**, 2579–2605 (2008)

An Ensemble Deep Learning Algorithm to Predict PM2.5 Concentration Levels in Bengaluru's Atmosphere

Tushar Patil(ID), Lichingngamba Tensubam(ID), Nivedan Yakolli$^{(\boxtimes)}$(ID), and Divya Biligere Shivanna(ID)

Department of Computer Science and Engineering, M S Ramaiah University of Applied Sciences, Bengaluru 560058, India
nivedanyakolli361@gmail.com

Abstract. For the past few years, the air quality in Bengaluru has been hazardous. Particulate Matter 2.5 Air Quality Index (PM2.5 AQI) is the measure of air quality of the breathing air. Forecasting the PM2.5 content in the atmosphere is crucial to alert the inhabitants and also help governments in taking required actions. Deep learning architectures have proven to be efficient in predicting PM2.5. The objective of the research work intends to estimate PM2.5 in the next hour by the time series analysis of PM2.5 values in the Silk Board area of the Bengaluru region using the proposed ensemble time series algorithm. An experiment was conducted to evaluate the promising existing time series analysis models and an ensemble algorithm was designed considering best-performing models as base learners. The obtained Mean Absolute Error was 0.031 and the R2 score was 0.993.

Keywords: Deep Learning (DL) · Ensemble Learning · Gated Recurrent Unit (GRU) · Long Short-Term Memory (LSTM) · Particulate Matter 2.5 (PM2.5) · Time-series analysis

1 Introduction

The air quality in Bengaluru, Karnataka, India's silicon city, has been dangerous for many years. One of the major environmental issues facing humanity is air pollution, which is also a popular topic all around the world. Several studies paint a complete picture of Bengaluru's air quality conditions. Estimated city emissions for 2015, total 31,300 tonnes of $PM_{2.5}$, 56,900 tonnes of NO_x, 335,550 tonnes of CO, 83,500 tonnes of NMVOCs, 5300 tonnes of SO_2, and 67,100 tonnes of PM_{10}. Transportation is the major source of PM2.5 and PM10 emissions in Bengaluru, comprising 56% and 70% respectively. Industries (17.8%), open waste burning accounts for 11.0%, and home cooking, lighting (6.5%), and heating are the next largest sources of PM2.5 emissions [1]. The effective management of human wellness and the capacity of the government to make decisions for environmental monitoring greatly depends on the precise forecast of PM2.5 [2]. Air Quality Index (AQI) is used to set a range for the pollutant in their

© The Author(s), under exclusive license to Springer Nature Switzerland AG 2023
M. Singh et al. (Eds.): ICACDS 2023, CCIS 1848, pp. 316–327, 2023.
https://doi.org/10.1007/978-3-031-37940-6_26

measured unit, $\mu g/m^3$ for PM2.5, to a standard range in index form. Particulate Matter 2.5 (PM2.5) are fine molecules having a width of two and a half micrometers or even less than that. They will be seen as fine droplets that are present in the air. These fine particles are formed when a fuel source is burnt and smoke is generated. Smoke sources such as vehicle exhaust emissions from cars, trucks, buses, and bonfires including various industries. The molecules that make up PM include organic contaminants, acidic salts, metals, and biological components such as endotoxins, antigens, and pollen particles. When levels of fine PM in the air are excessive, it is a great risk not only to human health but also for animals. Asthma and other breathing-related issues have been identified in many people [3].

Hospitalizations for cardiorespiratory conditions including lung cancer, cardiovascular disease, and cerebrovascular disease have both increased in correlation with PM2.5 concentrations [4]. Because to the mobility of atmosphere contaminants, the PM2.5 contents has a significant spatiotemporal correlation. [5]. Another study that was conducted in 2018 showed that pregnant women who are exposed to a high level of pollutants are a higher possibility of delivering babies with low weight [6]. Deep learning (DL) has been extensively used to predict air quality, with particularly impressive results obtained using prediction models that take into account the geographical and temporal correlation of data [7]. Mean Absolute Error (MAE), R2 Score, and Root Mean Square Error (RMSE) metrics can be utilized to assess how well the prediction of these models performs [8].

The objective of the research work was to protect the locals from numerous health-related disorders by predicting PM2.5 one hour ahead. The research that is being suggested in this study takes into account Bengaluru's air pollution, which is a major problem throughout the winter.

1.1 Air Quality Monitoring in Bengaluru

Locations Description: Ten locations have been considered in Bengaluru city, for which good quality data was maintained by the website. Central Pollution Control Board (CPCB) and its subsidiary state board, Karnataka State Pollution Control Board (KSPCB), have set up 10 stations across the City for measuring the Air Quality of Bengaluru. Following are the locations 1. Silk Board, 2. Peenya, 3. Jayanagar 5th Block, 4. Hombegowda Nagar, 5. Hebbal, 6. City Railway Station, 7. Bapuji Nagar, 8. BWSSB Kadabesanahalli, 9. BTM Layout, 10. Sanegurava Halli.

2 Related Works

Unjin et al. proposed a technique to estimate the level of Beijing City's daily mean PM2.5 concentration for the following day by considering the spatiotemporal relation as it relates to the PM2.5 [2]. A study suggest an innovative method to train spread of PM2.5 prediction in a region using historical temporal feature data, PM2.5 database and meteorological database [9]. Another study aimed to examine how well multilayer perceptrons (MLP) and multiple linear regressions (MLR) predicted PM2.5 levels to forecast PM2.5 levels up to a week in advance [10]. Sangwon et.al suggested a prediction model that can react to airborne PM, a sign of poor air quality in South Korea. To estimate

PM concentrations, the model interpolates meteorological and air quality data before using a CNN [11]. Mehdi et.al applied Random forest (RF), DL, and ML approaches in the anaysis for the prediction of PM2.5, feature significance in the city area of Tehran [8]. Rui et.al suggested using a linear regression which used multivariate model to predict PM2.5 over short periods [12]. Jinghui et al. used hourly PM2.5 content by air quality numerical prediction system based on the Weather Research and Forecasting to create a model that predicts the everyday PM2.5 contents in Shanghai, China [13].

Ke Gu et al. considered entropy characteristics in the spatial and transform domains to construct naturalness statistics (NS) models utilizing a large collection of images that were taken in favorable weather circumstances with very low PM2.5 concentration [14]. Fei Xiao et.al, used weighted archival PM time series data, and some nearby surrounding sites were selected as the central site's neighbors of Beijing, Tianjin, and Hebei [5]. Weibiao et.al came up with a new model based on stacked auto-encoder (SAE) LSTM. First, wavelet transform (WT) was used to divide the PM2.5 time series into many low and high-frequency elements depending on various instances. Six research locations from China were used for the study [15]. PM2.5 was also studied at Central Taiwan [16]. From April to November 2018, Shrivallabha et.al gave a time series study of PM10 in Bengaluru. ARIMA (Auto-Regressive Integrated Moving Average) was employed for forecasting PM10 [3].

3 Data Acquisition

3.1 Data Acquisition

Information was acquired for PM2.5 concentrations on an hourly basis from the Central Control Room for Air Quality Management website. Here PM2.5 concentrations was mentioned in terms of AQI. The most popular way to track air pollution and air quality can be better understood with the help of the PM2.5 AQI table [17]. The relationship between them can be seen in Table 1.

Table 1. PM 2.5 Air Quality Index Table.

AQI	PM2.5 ($\mu g/m^3$)	Remark
0–50	0–12	Good
51–100	12.1–35.4	Satisfactory (Modulate)
101–150	35.5–55.5	Unhealthy for sensitive groups
151–200	55.5–150.4	Unhealthy
201–300	150.5–250.4	Very Unhealthy
301–400	>250.5	Hazardous

Data was obtained for 2 years (from 1st January 2019 to 31st December 2021) on hourly basis. The recommended PM2.5 concentration level given by the CPCB was 0–60 μg/m^3. Description of the features in the dataset were PM2.5, PM10, Nitric oxide, Nitrogen Dioxide, Nitrous Oxides, Ammonia Sulfur Dioxide, Carbon Monoxide, Ozone, Benzene, Toluene, Ethylbenzene, Meta-Xylene, Wind speed, Wind Direction, Temperature, Relative Humidity, Solar Radiation, Buoyancy Pressure, Vertical Wind Speed, Methane, Xylene or Dimethylbenzene, and Atmospheric Temperature. For the analysis, the values of PM2.5 were considered.

For the study of PM2.5 trend, data for the 'Silk Board' location was selected out of all the available area's data because this location acts as an intersection between the Hosur Road area and Outer Ring Road (all the location mentioned are present in Bangalore, India). It is also a gateway to major IT clusters such as Outer Ring Road and Electronic City locations. Due to this factor the amount of traffic it faces daily, traffic bottlenecks are high as compared to other locations in Bangalore. Also, the residential area is closer to Silk Board which gives a detailed insight into the amount of PM2.5 concentrations that the people are exposed daily. From a dataset perspective, the number of missing values and outliers in the data are less when compared with data from other locations.

3.2 Data Analysis and Pre-processing

Many times, the stations may malfunction or will be pulled down for maintenance. In such cases, it is not possible to record the concentration levels of PM2.5. Due to this, there were many missing values seen in the dataset. To overcome these challenges, the Interpolation and Normalization methods to pre-process the data were used.

Interpolation: The proposed model used Linear Interpolation for computing data gaps and it was observed that the PM2.5 concentration levels have recurring patterns. Equation 1 gives the formula for Interpolation.

$$y = y_1 + \frac{(x - x_1) \times (y_2 - y_1)}{x_2 - x_1} \tag{1}$$

where x is an index of the missing PM2.5 value, x_1, and y_1 are indexes and their PM2.5 value for a point, x_2, and y_2 are the indexes and their PM2.5 value for another point.

Normalization: The Min-Max normalization technique was used to normalize the given data. The concentration levels of PM2.5 was observed to be in the range of 1 to 334 for the collected data. Equation 2 was used to normalize the data.

$$y_{normalized} = \frac{y - y_{min}}{y_{max} - y_{min}} \tag{2}$$

where y is the cause variable, y_{min} is the cause variable's minimum value, and y_{max} is the cause variable's maximum value.

To see the nature of the data, Fig. 1 illustrates the visualization plots for a month's sample from the dataset in the region.

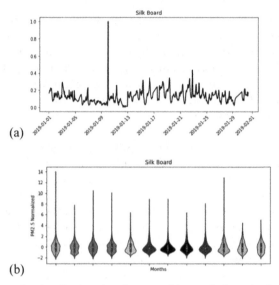

Fig. 1. (a) Illustrates the visualization plots for a month's sample from the dataset for Silk board region. (b) Violin plots for the same month and region

4   The Problem Definition

An ideal model should accurately and far in advance forecast the PM2.5 concentrations in Bengaluru's atmosphere, allowing for the implementation of protective measures to protect human life from its potentially lethal effects. The proposed model precisely dealt with the prediction of time-series analysis of PM2.5 concentrations in Bengaluru using a combination of deep learning algorithms. The proposed model used the Ensemble technique to reduce the rate of error, thereby bridging the gap between the ideal model and the current models. The proposed model is also aimed to deploy in a real-world scenario so that effective and preventive measures can be taken to tackle the concentration of PM2.5 around Bengaluru.

5   Various Existing Time Series Analysis Models

Recurrent Layers such as LSTM and Gated Recurrent Unit (GRU) were used while building the models. The LSTM is a network of recurrent neurons that makes use of feedback connections to learn the order of the data. A Cell Unit, an Input Gate, an Output Gate, and a Forget Gate are all available in LSTM. Similar to LSTM, GRU also learns from the sequence of data using gates such as Forget Gate and Input Gate. The only difference between LSTM and GRU comes in terms of parameters. There are fewer parameters in GRU as compared to LSTM as GRU does not have an Output Gate.

The presented technique's performance was evaluated by comparison with a number of other existing models. These models were taken into consideration:

Residual GRU: Residual GRU is in the family of the Gated Recurrent Unit neural network architecture that incorporates residual connections between the hidden layers of the

GRU model. **Residual LSTM:** Residual LSTM is a kind of LSTM neural network architecture that incorporates residual connections between the hidden layers of the LSTM model. A two-layer bidirectional LSTM is used for the prediction of the values. **Model using LSTM and Convolution 1D:** A three-layer architecture with 'Batch Normalisation' and 'Rectified linear unit' (relu) activation function was used in the convolution 1D layer, and the output was transmitted to the bidirectional LSTM for comparison in the study. **Residual model using LSTM and Convolution 1D:** The architecture is the same as that of the LSTM and convolution model but here the predicted result obtained is applied to the residual function thereby increasing its potential to model long-term dependencies. **Model using LSTM and Depth wise convolution 1D:** The combination of LSTM and depth wise 1D convolution allows the network to consider both long-term dependencies and local patterns in the data, potentially improving its performance. A three-layer architecture was used where 'Batch Normalization' and activation function 'relu' were used in a depth wise convolution layer and the resultant is passed to the bidirectional LSTM while building the model. **Residual Model using LSTM Layer and Depth Wise Convolution 1D layer:** The base model used here is an untrained LSTM and Depth wise Convolution1D model. The whole model is wrapped in a Residual layer. The layer performs summation on the output of the base model and input. **Model using LSTM Layer and Separable Convolution 1D layer:** The model consists of a Separable Convolution 1D layer followed by an LSTM layer. Batch-Normalization is applied to the output before it is passed to the LSTM layer. **Residual Model using LSTM and Separable Convolutional 1D layer:** The base model used here is an untrained LSTM and Separable Convolution 1D model. The base model is wrapped in a Residual Layer. **Residual Model using GRU and Separable Convolution 1D layer:** The base model used here is an untrained GRU and Separable Convolution 1D model. The base model is wrapped in a Residual Layer. **Residual Model using RNN and Separable Convolutional 1D layer:** The base model used here is an untrained RNN and Separable Convolution 1D model, previous Residual Model, the base model is wrapped in a Residual Layer.

6 The Proposed Model

Ensemble Modeling is the process of combining multiple pre-trained models that have better-optimized features of the dataset and have different perspectives in predictions. From the comparative study of models, it is evident that LSTM, GRU and RNN, when used with a Residual Wrapper, give better performance than any other models. For Ensemble learning, these three models were selected. The predictions of individual models were computed and the absolute difference was considered. For the two models having a lesser difference between the predictions, the average of their predictions was taken to get a final output. It was observed that, while using the proposed ensemble technique, the MAE of the presented model was comparatively lesser than the best model.

6.1 The Algorithm

The selected models were (a) the Residual model using LSTM and Separable Convolutional 1D Layer, (b) the Residual model using GRU and Separable Convolution 1D Layer, (c) the Residual model using RNN and Separable Convolution 1D Layer. The algorithm shown below was used in the proposed model.

Algorithm: Proposed Ensemble algorithm

1:   BM: list of best three models
2:   Preds ← initialize a list of predictions to 0
3:   D ← initialize a list of differences to 0
4:   X ← cause variable (Previous PM2.5 Time-Series Value)

5:   **FOR** each model in BM **DO**
6:       p ← model(X)
7:       Add p to Preds list
8:   END FOR

9:   **FOR** each s = 0 to 3 **DO**
10:     **FOR** each t = s+1 to 3 **DO**
11   d ← absolute(Preds [s] - Preds [t])
12:         Add d to D list
13:     **END FOR**
14: **END FOR**

15: initialize z:= 0
16: **IF** D [z] < D [z+1] and D [z] < D [z+2]
17:      **THEN** OUTPUT = (Preds [z] + Preds [z+1]) / 2
18: **ELSE**
19:      **IF** D [z+1] < D [z] and D [z+1] < D [z+2]
20:          **THEN** OUTPUT = (Preds [z] + Preds [z+2]) / 2
21: **ELSE** OUTPUT = (Preds [z+1] + Preds [z+2]) / 2
22: **PRINT** OUTPUT

In the algorithm, the *Best Models* (BM) consist of the selected best three models, and the difference in each model's prediction is stored in the *D* list. *Step 1* of the algorithm expects the *BM* list to have the object of the best model in descending order. *Step 2* and *Step 3* of the algorithm initialize the Prediction and Difference list. *Step 4* assigns the *X* variable with a cause variable (Previous PM2.5 Time-Series value). *Steps 5 to 8* append prediction *p* to *Preds* for each model in the *BM list* on given data. *Step 9 to Step 14* appends the absolute difference between two predictions in the *Preds* list. *Step 15 to Step 22* compares these differences to find the one that has less difference as compared to all the difference values *(D)*. After finding the difference value, the last PM2.5 prediction value was computed by taking the mean of the Prediction values that gave the least difference.

7 The Performance Analysis

The model was built using the Tensor Flow framework and trained on the data. To check the performance of the considered models, two metrics were considered. The metrics are compiled while building the model and they are as follows:

MAE: It is computed by averaging the absolute value of the dissimilarity between the true and projected values of PM2.5 on the entire sample. Equation 3 gives the formula to calculate the MAE.

$$MAE = \frac{1}{S}\sum_{j=1}^{S}\left|a_j - p_j\right| \tag{3}$$

where S is the sample count, a_j is the jth sample of the actual value for PM2.5 from the dataset and p_j is the jth sample of the estimated value for PM2.5 from the model.

R2 Score: R2 Score defines the proportion of variance from the previously known values of PM2.5 that can be explained from the predicted and true PM2.5 values. It gives a clear picture of how close the predicted value is to the true PM2.5 value. Equation 4 gives the formula to calculate the R2 Score.

$$R^2 = 1 - \frac{\sum_{j=1}^{S}(a_j - p_j)^2}{\sum_{j=1}^{S}\left(a_j - a_m\right)^2} \tag{4}$$

where S is the sample count, a_j is the jth sample of the true value for PM2.5 from the dataset, p_j is the jth sample of the estimated value for PM2.5 from the model, a_m is the mean of the true value for the given sample.

7.1 Results

The graph for actual vs. predicted values is shown in Fig. 2 and it can be observed that the proposed model almost predicted the 24-h data very close to the actual value. Thus, it can be said that the suggested model is the best for the prediction of PM2.5. The resultant R2 score bar plot and MAE plot are shown in Fig. 3 and 4 respectively. They demonstrated that the proposed ensemble model's performance is better since it results in substantially low MAE values and also it displays the highest R2-Score value when compared with any of the other models.

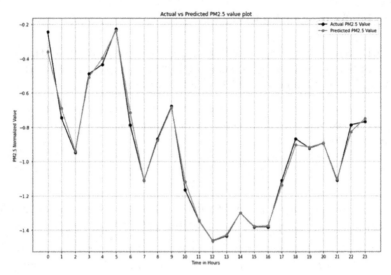

Fig. 2. Actual vs. Predicted PM2.5 value by the proposed model for 24-h data.

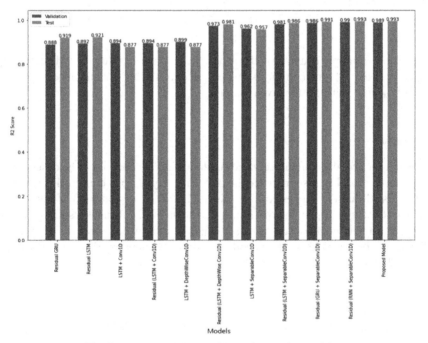

Fig. 3. R2 score of the considered time-series models.

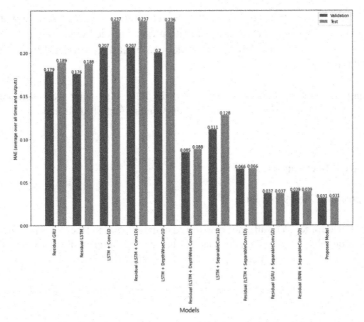

Fig. 4. Mean absolute error of the time-series models.

8 Discussion

The paper looked into Bengaluru's PM2.5 pollution particle estimation. The information was gathered from the website of the Central Pollution Control Board (CPCB). The Silk Board area was taken for analysis between January 1, 2019, and December 31, 2021. In the performance analysis section, in most of the time points, Fig. 2 showed that the actual and predicted values were nearly identical. In the proposed work, the model's performance was assessed primarily using two metrics: the first was MAE and the second metric was R2-score. Figure 3 and 4 demonstrated how well the suggested model predicted the PM2.5 particles and gives the model's performance. Among all the models, the proposed model's R2-Score is the highest value and MAE is lowest value. The suggested model was compared to other recent efforts under Sect. 5, and it was found that the presented ensemble model was the best-performing model. The technique can further be improved in the areas where it can predict multiple locations or it can be updated to predict a location given the dataset of some nearby location.

9 Conclusion

This work proposes an ensemble algorithm for time series analysis. The residual model using LSTM and Separable Convolutional 1D Layer, the Residual model using GRU and Separable Convolution 1D Layer, Residual model using RNN and Separable Convolution 1D Layer had less error, so these were chosen as base learners. The obtained Mean Absolute Error was 0.031 and the R2 score was 0.993, which is suitable for real-time.

The study has obtained very good results when compared with the existing works. The limitation of the study was, only a single location was considered and other regions were not completely explored. The proposed model can be further studied for various other mentioned locations.

Acknowledgments. We would like to acknowledge the different agencies of the Bengaluru region for providing the data under the unified portal of the Central Control Room and Central Pollution Control Board.

Statements and Declarations.

Ethical Approval. Not Applicable.

Consent to Participate. Yes, all authors approved of the content, and they all explicitly agreed to submit and acquire permission from the relevant authorities. Consent to Publish: Yes, all authors explicitly gave their consent to publish, and they all agreed with the content.

Publish. Yes, all authors explicitly gave their consent to publish, and they all agreed with the content.

Authors Contributions. The study's inception and design involved input from all authors. The following individuals prepared the materials, collected the data, and conducted the analysis: Tushar Patil, Lichingngamba Tensubam, Nivedan Yakolli, and Divya Biligere Shivanna. Nivedan Yakolli wrote the manuscript's initial draught, and all of the other authors offered feedback on earlier draughts. The final manuscript was read and approved by all writers.

Funding. The authors affirm that they did not receive any money, grants, or other kind of support for the creation of this publication.

Competing Interests. There are no material financial or non-financial interests to disclose for the authors.

Availability of Data and Materials. https://app.cpcbccr.com/ccr/#/caaqm-dashboard-all/caaqm-landing/data.

References

1. Guttikunda, S.K., et al.: Air quality, emissions, and source contributions analysis for the Greater Bengaluru region of India. Atmos. Pollut. Res. **10**, 941–953 (2019)
2. Pak, U., et al.: Deep learning-based PM2.5 prediction considering the spatiotem-poral correlations: a case study of Beijing, China. Sci. Total Environ. S0048-9697(19)33481-3(2019). https://doi.org/10.1016/j.scitotenv.2019.07.367
3. Shrivallabha, S., Nelavigi Kumaresh, P.: Time series analysis of atmospheric particulate matter of Bengaluru City. Int. J. Sci. Res. Math. Stat. Sci. **6**(5), 83–85 (2019)
4. Pérez, P., Trier, A., Reyes, J.: Prediction of PM2.5 concentra-tions several hours in advance using neural networks in Santiago, Chile. Atmos. Environ. **34**, 1189–1196 (2000)

5. Xiao, F., Mei, Y., Fan, H., Fan, G., Al-Qaness, M.A.A.: An improved deep learning model for predicting daily PM2.5 concentration. Sci. Rep. **10**, 20988 (2020). https://doi.org/10.1038/s41598-020-77757-w

6. Prafulla, S., et al.: Ambient and indoor air pollution in pregnancy and the risk of low birth weight and ensuing effects in infants (APPLE): a cohort study in Bangalore, South India. Wellcome Open Res. **3**, 133 (2020)

7. Erden, C.: Genetic algorithm-based hyperparameter optimization of deep learning models for PM2.5 time-series prediction. Int. J. Environ. Sci. Technol. **20**, 2959–2982 (2023)

8. Zamani Joharestani, M., Cao, C., Ni, X., Bashir, B., Talebiesfandarani, S.: PM2.5 prediction based on random forest, XGBoost, and deep learning using multisource remote sensing data. Atmosphere **10**, 373 (2019)

9. Yang, X., Zhang, Z.: An attention-based domain spatial-temporal meta-learning (ADST-ML) approach for PM2.5 concentration dynamics prediction. Urban Climate **47**, 101363 (2023). https://doi.org/10.1016/j.uclim.2022.101363. ISSN 2212-0955

10. Saiohai, J., Bualert, S., Thongyen, T., Duangmal, K., Choomanee, P.: Szymanski: statistical PM2.5 prediction in an urban area using vertical meteorological factors. Atmosphere **14**, 589 (2023)

11. Chae, S., Shin, J., Kwon, S., Lee, S., Kang, S., Lee, D.: PM10 and PM2.5 real-time prediction models using an interpolated convolutional neural network. Sci. Rep. **11**, 11952 (2021)

12. Zhao, R., Gu, X., Xue, B., Zhang, J., Ren, W.: Short period PM2.5 prediction based on multivariate linear regression model. PLoS One **13**(7), e0201011 (2018). https://doi.org/10.1371/journal.pone.0201011

13. Ma, J., Yu, Z., Qu, Y., Xu, J., Yu, C.: Application of the XGBoost machine learning method in PM2.5 prediction: a case study of Shanghai. Aerosol. Air Qual. Res. **20**, 128–138 (2020)

14. Ke, G., Qiao, J., Li, X.: Highly efficient picture-based prediction of PM2.5 concentration. IEEE Trans. Industr. Electron. **66**(4), 3176–3184 (2019)

15. Qiao, W., Tian, W., Tian, Y., Yang, Q., Wang, Y., Zhang, J.: The forecasting of PM2.5 using a hybrid model based on wavelet transform and an improved deep learning algorithm, vol. 7, p. 142815 (2019). https://doi.org/10.1109/ACCESS.2019.2944755

16. Chen, Y.-C., Li, D.-C.: Selection of key features for PM2.5 prediction using a wavelet model and RBF-LSTM. Appl. Intell. **51**(4), 2534–2555 (2020). https://doi.org/10.1007/s10489-020-02031-5

17. Du, X., Varde, A.: Mining PM2.5 and traffic conditions for air quality. In: IEEE 7th International Conference on Information and Communication Systems (ICICS), pp. 33–38 (2016)

Deep Learning Model Based on a Transformers Network for Sentiment Analysis Using NLP in Sports Worldwide

Luis Baca[1] (ORCID), Nátali Ardiles[2] (ORCID), Jose Cruz[2] (ORCID), Wilson Mamani[2(✉)],
and John Capcha[1] (ORCID)

[1] Universidad Cesar Vallejo, Lima, Peru
{lbacaw,jcapchaco}@uvcvirtual.edu.pe
[2] Universidad Nacional del Altiplano, Puno, Perú
{n.ardiles,josecruz}@unap.edu.pe, wilmamanimac@est.unap.edu.pe

Abstract. Currently, the amount of information processed through the Internet reaches 396 exabytes per month, almost triple what was generated in 2017, and the number of users has gone from 18 billion to almost 29 billion in 2022. This information, generally extracted from unstructured sources, must be processed to obtain data of interest. Users of social networks in 2022 reached a number of 4.7 billion, increasing by 5% compared to last year. Artificial Intelligence, specifically Natural Language Processing, offers an alternative to information processing. In this sense, the model based on Transformer DistilBERT carried out based on BERT allowed the analysis of sentiments of the tweets issued on the Twitter platform related to the Qatar 2022 World Cup. For this, the Google platform located in the cloud was used, as well as Vertex IA, Storage, and BigQuery services. Another resource used was the Tweeter API to collect the data and the CRISP-DM methodology. The results showed the relationship between the variables, with a normal distribution and few outliers. To improve the work in future research, it is proposed to eliminate the outliers and tune some hyperparameters of the model.

Keywords: Sentiment Analysis · Red Transformer · Sport · GCP

1 Introduction

According to Cisco, the volume of data transferred over the Internet currently reaches 396 exabytes per month, almost triple what was generated in 2017, likewise the growth of users since 2017 rose from 18 billion to almost 29 billion in the year 2022. All this traffic that has data that can be turned into valuable information must generally be extracted from unstructured sources such as PDFs, slides, graphics, and currently through social networks. For this, some type of processing is needed to obtain the data, process it, and obtain information of interest. Artificial intelligence with one of its branches "natural language processing" offers an alternative to information processing. Natural language processing (NLP) is a field that combines linguistics and artificial intelligence (AI) to enable computers to understand human or natural language [1]. Currently, the

M. Singh et al. (Eds.): ICACDS 2023, CCIS 1848, pp. 328–339, 2023.
https://doi.org/10.1007/978-3-031-37940-6_27

prominence of NLP has increased due to the large amount of unstructured text data that is generated every day. Some of the most common and popular NLP techniques include named entity recognition, sentiment analysis, machine translation, theme modeling, and text summarization [2–5], and [6]. As soccer is the most popular sport worldwide, it is important not only at a sporting level, as a game and fun, but also at a social level, since it unites social groups, clubs, or even nations. Likewise, it is one of the sports that generate the most money in all the regions where it is practiced, even in regions that previously did not consider it, such as the current organizer of the Qatar World Cup. So, the soccer world cup promotes cultural integration between nations.

2 Literature Review

This interaction can be analyzed through the so-called social networks, to extract information through natural language processing (NLP) networks. These large-scale pre-trained networks, generally contain millions of parameters and with high computational requirements, for example, the data to train the network can be 10 GB or more in size, which needs high-performance cluster processing to be processed. For training the deep neural network, therefore, technology transfer models are used for this purpose. Models that have been pre-trained can be tuned to perform well on different tasks while maintaining the flexibility of larger models. For example, [7] by using triple loss, achieved a 40% smaller Transformer network with similar performance to the original network, with a 60% reduction in execution time.

Internet usage is growing around the world, and it is being used as a profitable platform for sharing information. This is partly due to the rapid expansion of social media. There are many social media platforms that allow people to share their opinions about particular products, organizations, or situations. Attitudes and feelings are an essential part of evaluating an individual's behavior and can be further analyzed in relation to a specific entity, known as sentiment analysis or opinion mining. For example, [8] performs a sentiment analysis of the data obtained from Tweets, grouping them into two groups of techniques: The Naive Bayes, Maximum Entropy, and SVM, which reached an accuracy of 80%, while the second one based on ensembles and hybrids reached an accuracy of 80%. Classification accuracy of approximately 85%. In the context of the Coronavirus, [9] collected tweets over 20 days in March in four European states to analyze the outbreak of the new coronavirus. [10] used unsupervised machine learning techniques to analyze collected data on the coronavirus. [11] used a Naive Bayes classifier to analyze and classify the data extracted from Tweets also related to the coronavirus. [12] examined tweets posted in January 2020, emphasizing word patterns and sentiment recognition through the use of bigrams and trigrams. [13] applying the Latent Dirichlet Allocation technique found patterns of fear due to the scourge of the coronavirus. [14] designed an analysis of the COVID-19 outbreak to analyze sentiment using different machine-learning approaches.

With sentiment analysis, you can interpret the emotions or feelings of users and classify them into different categories. Generally, these data, in large volumes, are not structured, so pre-processing is needed to build the database from which the reactions to be classified will be taken. Also in the coronavirus environment, [15] and [16] used the tweets social network to categorize feelings by pre-processing the data.

By using sentiment analysis, we can interpret other people's emotions and feelings and classify them into different categories, which can help an organization better understand people's emotions and act accordingly. Sentiment analysis depends on the results you want to obtain; for example, the analysis of the text according to its polarity and emotions, the feedback on a particular feature, or the analysis of the text in different languages requires the detection of the respective language.

Therefore, the contributions of the research are:

- Use of a Transformers network for natural language processing.
- Analysis of reactions using a distilbert network for the analysis of sentiments in sport.

3   Methodology

In summary, this investigation consists of two parts. The first is the use of the DistilBERT network, which works internally with a Transformer model for natural language processing with its respective pre-processing. While the second is the analysis of emotions using a Cloud type architecture which uses Google Cloud services to manage the project workflow.

Figure 1 shows the architecture of a Transformers network, which is based on the use of a series of attention layers to process a text input. These attention layers allow the network to "look" at a sequence of words and discover useful patterns for the text classification task. These patterns are used to process the input and produce output that can be used for classification. The architecture also includes a classification layer to produce the final output. This classification layer takes the output from the attention layer and uses it to produce a prediction about the content of the input. The architecture of a Transformer's network also includes a memory system to allow the network to remember previously processed information. This allows the network to make more accurate predictions about the content of the post. The memory system also allows the network to learn from previously processed information, allowing it to improve its results as more data is received.

3.1   DistilBERT

It is a distilled version of BERT that retains the performance capabilities of BERT but uses only half the parameters, is faster, and is smaller. DistilBERT uses a technique called "distillation" in which looks a lot like Google's big neural network with a smaller one. Figure 2 shows a comparison of the growth in the number of parameters used by the different processing algorithms depending on the year in which they were created. It clearly shows that DistilBERT using fewer parameters for its training has performance comparable to other algorithms that use numerous parameters.

In Fig. 3 the flowchart of the DistilBERT model that has 6 transformer layers is shown since the base model contains 12 layers.

For example, DistilBERT compared to natural language recognition networks such as BERT contains a performance of 97%, contributing to a fast and efficient classification without losing the mother or main architecture that this type of Transformers model has.

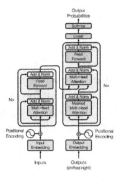

Fig. 1. Architecture of the Transformer model.

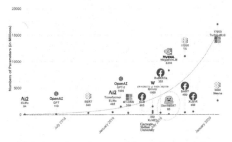

Fig. 2. Text processing models based on the year of creation and the number of parameters used for training [17].

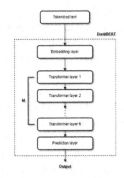

Fig. 3. Flowchart DistilBERT.

Fig. 4. GCP architecture.

To execute the DistilBERT model, the Google architecture called Google Cloud Platform (GCP) is depicted in Fig. 4.

The general architecture is described where the data extraction process consumed by the Tweets API will be carried out to later perform a pre-processing of the information obtained, which will then be classified into 6 categories, these mentioned steps will be carried out in the Vertex AI service; then the information is stored in Cloud Storage in a file with ".csv" format to later send it through an iteratively programmed query to the BigQuery service and finally, with the data obtained, perform the analysis of the data in the final dataset.

3.2 For the Analysis of Emotions

Interacting with GCP services is done by following the general flowchart shown in Fig. 5.

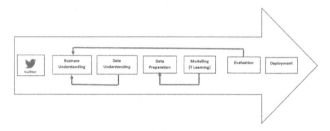

Fig. 5. General Flowchart for the classification of emotions.

Each of the stages is detailed below:

Business Understanding. At this stage, the focus is on defining the objectives and requirements of the project. In this case: analyzing the reactions using a DistilBERT network for the analysis of sentiments in sports, using a Transformers network for natural language processing. The data will be extracted from the tweet's social network.

Data Understanding. Data understanding involves the initial collection of data, in this case, the tweets related to the World Cup in Qatar. Here are the data to be classified is also described in a general way: tweets in any language. The way in which the data for the research was obtained was through its API, although it has limitations such as being able to make up to 300 requests for 15 min per application; 500,000 tweets per month per project; In this case, the collection was carried out every 20 min to carry out the analysis of repeated tweets and not be considered in the final result, using the two available platforms Vertex Ai and Storage.

Data Preparation. Also called pre-processing, it is the data preparation phase that covers all the steps necessary to create the final data set (the data that will be provided by the modeling tool). Preparation tasks include data selection, data cleansing, creating new variables, data integration, and data formatting. This process is described in Fig. 6.

First, headline tokenization is performed, where the headline text was split into individual words based on a given delimiter. Then, all the tokens are converted to their lemma or root form, this is to relate an inflected or derived word with its canonical or lemma form to reduce the number of words to be analyzed, ignoring the differences and

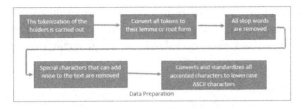

Fig. 6. Flowchart of the Pre-Processing stage.

combining all these variants in the same term. Subsequently, all the empty words are eliminated, that is, those tweets that are empty or only have tags that do not provide information. Then special characters that can add noise to the text are removed, such as tildes in the Spanish language. Finally, all accented characters are converted and standardized to lowercase ASCII characters for better processing.

Modelling. In this phase, the natural language processing model to be used for emotion recognition is applied, in this case, DistilBERT, which will run on the Google Cloud platform. The Process carried out is shown in the flowchart of Fig. 7. First, the pre-trained model is downloaded, in this case, DistilBERT, into Google's Vertex AI platform. Then the data is downloaded through the API provided by Tweets and is stored in the Storage of the Google platform. Data that has already been pre-processed is translated into English, as adapting individual machine learning models to each language can be redundant and means that the accuracy of the predictions varies from model to model. Likewise, the data is consistent in all languages thanks to a unified English machine-learning model. Most importantly, a large English dataset containing text data for training classified into their respective sentiments facilitates multilingual English translation and generalizes the learning and prediction methods of English sentiment ML models [17].

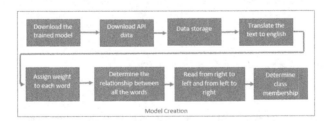

Fig. 7. Flowchart of the Model Creation stage.

Then the DistilBERT model was applied: assigning a weight to each word and storing this in a vector, to later determine the relationship between the words, again assigning a value to the determined matrix. Subsequently, the numerical vector is read in both directions: from left to right and from right to left to find the relationship between the analyzed words, once again storing the results in a numerical vector. Finally, based on the numerical values found, belonging to one of the classes of emotions to be classified is determined.

Evaluation. O In this case, the results obtained are evaluated, determining their usefulness for the established objective. At this stage, the model is established, and from a data analysis point of view, the model should be of high quality. This process is shown in Fig. 8.

Fig. 8. Flowchart of the Model Evaluation stage.

To do this, the results obtained are first stored, later compare with the previous data and then replace the data stored at the beginning with the new data: all this process is done to avoid having repeated tweets that do not influence the final classification in the categories of feelings to classify.

Deployment. Also called deployment, here the model was used in a production environment, being a live process (in continuous change) that was carried out on Google's Storage platform and the Vertex AI platform of the same company. This process is detailed in Fig. 9.

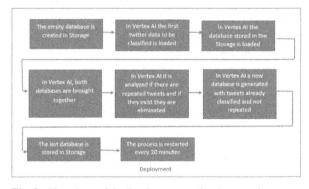

Fig. 9. Flowchart of the Deployment or implementation stage.

The empty database (only column names) is created in the Storage, and then Vertex AI processes the tweets to be classified and merges them with the database created in the Storage. We proceed to analyze if there are repeated twitters and if they exist, they are eliminated. The new database without repeated tweets is generated in Vertex AI. Finally, this database is transferred to Storage for storage. This process is repeated every 20 min, thus ensuring the non-existence of repeated tweets for sentiment analysis in the proposed system.

4   Results

This section presents the results obtained after applying the DistilBERT model to the data. The model is equipped with the database obtained from Tweets described in the Data Understanding section. The DistilBERT-based model was used after pre-processing the data to categorize them into 5 groups of emotions: Joy, Sadness, Anger, Fear, Love, and Surprise.

Table 1. Variables statisticians used.

	Friends_count	Listed_count	Statuses_count	Followers_count
Count	31999	31999	31999	31999
Mean	1792,189	292,0382	74699,26	112458,6
Std	15836,35	1725,077	175036,4	711890,7
Min	0	0	1	0
25%	119	0	2361	98
50%	448	2	14863	533
75%	1287	23	68443	3243,5
Max	651441	58134	7869669	22296952

It is worth noting that the output of the system is a label: it goes from the number 0 to 4, which represents the aforementioned categories. Table 1 shows the statistics of some variables used in the sentiment analysis.

Figure 10 shows the result of the sentiment analysis in the categories mentioned above.

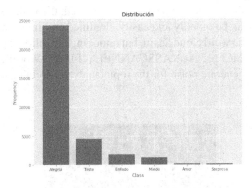

Fig. 10. Results of the sentiment analysis in the established categories.

From Fig. 10 it can be seen that the emotion expressed by the users of Tweets in order of quantity is: Joy, Sadness, Anger, Fear, Love, and finally surprised.

Figure 11 shows the number of tweets with respect to time. It is clearly noticeable that as the phases of the World Cup progressed, the interest increased, translated into a greater number of tweets created and analyzed.

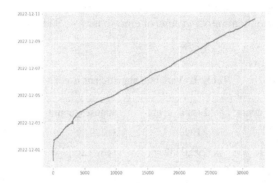

Fig. 11. Number of tweets analyzed as a function of time.

Fig. 12. Obtained word cloud.

Figure 12 shows the word cloud of the most found Tweets. Word clouds are a tool that speeds up the analysis of text-type data since through them, the most relevant words in the analyzed text can be quickly and easily identified and interpreted.

Figure 13 shows the word cloud again, but removing the following tags: 'Qatar2022', 'RT', 'Rt', 'FIFAWorldCup', 'ALKASSCANNEL', 'FIFA', 'World', 'Cup", 'Live" that are obvious and that generate noise for the topic analyzed and that do not contribute to sentiment analysis.

Fig. 13. Noiseless new word cloud.

To indicate the validity of the results, some correlations between the variables used for the classification are presented below. For example, for the variable "text-classification_num" in Fig. 14 the correlation with 7 other variables is shown.

	text-classification_num
text-classification_num	1.000000
statuses_count	0.056681
listed_count	0.027621
id	0.026912
favourites_count	0.023064
followers_count	0.017975
friends_count	-0.001634
retweet_count	-0.003300

Fig. 14. Correlation matrix with respect to the variable "text-classification_num".

Another way of expressing the correlation between the variables is the use of heat maps, which show the correlation between all the variables of the proposed system, which for the present investigation is shown in Fig. 15.

Another criterion to determine the validity of the results is that the variables used have a normal or close to-normal distribution so that the system is consistent. This is shown in Fig. 16 for the variables "Followers Count", and "Friends Count". All of them show an almost normal distribution, which is verified by the graph since the measures of central tendency: mean, and mode are at very close points.

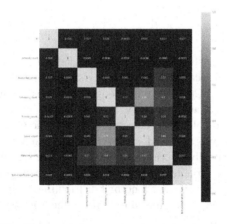

Fig. 15. Correlation matrix using color diagram.

Finally, one of the problems found in the classification is the outliers values that are abnormal and extreme observations in a statistical sample or time series of data that can potentially affect the estimation of its parameters. For the present investigation, we used a box plot, which is a graphic form of statistical presentation intended, to highlight aspects of the distribution of observations in one or more series of outliers data. This is shown in Fig. 17 for the variables "Followers Count", and "Friends Count". Although it is true

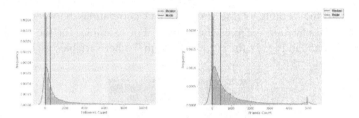

Fig. 16. Frequency distribution of quantitative variables.

that all the variables show outliers, these are not very representative of the classification. However, for future research, they could be eliminated to refine the model.

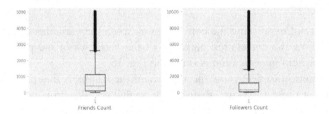

Fig. 17. BoxPlot for quantitative variables.

5   Conclusions

The model based on Transformer DistilBERT carried out based on BERT allows the analysis of sentiments based on the messages issued on the Twitter platform related to the Qatar 2022 World Cup. For this, Google platforms located in the cloud were used: Vertex IA, Storage, and BigQuery. In combination with the previous services, plus the Tweeter API to collect the data and following the CRISP-DM methodology, the analysis of feelings was achieved, classifying them into 5 categories: Joy, Sadness, Anger, Fear, Love, and finally surprise from the related tweets. To validate the results, the relationship between the variables used for the analysis of feelings was shown. Likewise, it was shown that the variables have a normal distribution, with few outliers. In future studies, the outliers could be eliminated to improve the accuracy of the model or tune the DistilBERT model used.

References

1. Krishnan, K., Rogers, S.P.: Social Data Analytics: Collaboration for the Enterprise. Morgan Kaufman, Walthman, MA, USA (2015)
2. Nadeau, D., Sekine, S.: A survey of named entity recognition and classification, Lingvistic Investigationes. Int. J. Linguist. Lang. Resour. **30**(1), 326 (2007). https://time.mk/trajkovski/thesis/li07.pdf

3. Liu, B.: Sentiment Analysis: Mining Opinions, Sentiments, and Emotions. Cambridge Univ. Press, New York, NY, USA (2015)
4. Wu, Y., et al.: Google's neural machine translation system: bridging the gap between human and machine translation (2016). arXiv:1609.08144. http://arxiv.org/abs/1609.08144
5. Blei, D.M.: Probabilistic topic models. Commun. ACM **55**(4), 77–84 (2012). https://doi.org/10.1145/2133806.2133826
6. Torres-Moreno, J.M. (ed.): Automatic Text Summarization. Wiley, Hoboken, NJ, USA (2014)
7. Vaswani, A., et al.: Attention is all you need. In: NIPS, 2017 (2017)
8. Alsaeedi, A., Khan, M.: A study on sentiment analysis techniques of tweets data. Int. J. Adv. Comput. Sci. Appl. (IJACSA) **10**(2) (2019). https://doi.org/10.14569/IJACSA.2019.0100248
9. Dubey, A.D.: Tweets sentiment analysis during COVID19 outbreak (2020). SSRN 3572023
10. Medford, R.J., Saleh, S.N., Sumarsono, A., Perl, T.M., Lehmann, C.U.: An infodemic: leveraging high-volume tweets data to understand public sentiment for the covid-19 outbreak. medRxiv (2020)
11. Alhajji, M., Al Khalifah, A., Aljubran, M., Alkhalifah, M.: Sentiment analysis of tweets in Saudi Arabia regarding governmental preventive measures to contain COVID-19 (2020)
12. Rajput, N.K., Grover, B.A., Rathi, V.K.: Word frequency and sentiment analysis of tweets messages during coronavirus pandemic (2020). Preprint arXiv: 2004. 03925
13. Prabhakar Kaila, D., Prasad, D.A., et al.: Informational flow on tweets-corona virus outbreak-topic modelling approach. Int. J. Adv. Res. Eng. Technol. (IJARET) **11**(3) (2020)
14. Muthusami, R., Bharathi, A., Saritha, K.: Covid-19 outbreak: tweet based analysis and visualization towards the influence of coronavirus in the world. Gedrag en Organisatie **33**(2) (2020)
15. Kaur, C., Sharma, A.: Tweets sentiment analysis on coronavirus using textblob. Technical report, EasyChair (2020)
16. Pastor, C.K.: Sentiment analysis of filipinos and effects of extreme community quarantine due to coronavirus (COVID-19) pandemic (2020). SSRN 3574385
17. Sanh, V.: Smaller, faster, cheaper, lighter: Introducing DistilBERT, a distilled version of BERT. [Internet], 28 August 2019. Disponible en https://medium.com/huggingface/distilbert-8cf338 0435b5
18. Jojoa, M., Eftekhar, P., Nowrouzi-Kia, B., et al.: Natural language processing analysis applied to COVID-19 open-text opinions using a DistilBERT model for sentiment categorization. AI Soc. (2022).https://doi.org/10.1007/s00146-022-01594-w

An Effective Framework for Sentiment Analysis Using RNN and LSTM-Based Deep Learning Approaches

Brajesh Kumar Shrivash(✉) ⓘ, Dinesh Kumar Verma ⓘ, and Prateek Pandey ⓘ

Jaypee University of Engineering and Technology, A-B Road, Raghogarh, Distt. - Guna, M.P., India

brajesh.kumar.shrivash@gmail.com, {dinesh.verma, prateek.pandey}@juet.ac.in

Abstract. Sentiment Analysis (SA) is an active research field in modern times by studying people's sentiments posted on different web platforms over the internet. To study these sentiments, different machine and deep learning techniques are in use. Deep learning is an efficient approach for SA. It gives more accurate results by learning text features from sentiments using a deep neural network. In this paper, we proposed a framework for sentiment classification;. The model was, trained, and tested on Recurrent Neural Network (RNN) and Long Short-Term Memory (LSTM) Deep Learning (DL) algorithms for SA. The performance of proposed framework was evaluated using accuracy, precision, recall, and F-1 score values. To calculate the performance, the data set was split in the ratio of 70:30, 80:20, and 90:10 for training and testing the model and observed that LSTM performs well with 88% accuracy and precision value, 87% recall value, and 87% F-1 measures.

Keywords: Sentiment analysis · LSTM, RNN · deep learning · classification

1 Introduction

Analysis of people's opinion/views in a variety of contexts is the subject of SA, which is often referred to as opinion mining. In today's time, large amount of sentiments data is being recorded in digital form. Sentiment's emergence and its rapid growth coincide with social media, user's reviews on different sites on internet like product or services reviews post, blogs, micro-blogs, forum discussions, Twitter, and social networks. To study the sentiments and find the insights from these to draw the conclusion helps in decision making to frame effective strategies. SA is an active study field of Natural Language Processing (NLP).

In addition to data mining and web mining, information retrieval and text mining are also included in SA. It has extended from computer science to other fields which include marketing, finance, political science, communications, health science etc. Sentiments are pervasive because they are crucial to virtually all facets of human behaviour. Our judgements and views about the world are heavily impacted by the opinions and

M. Singh et al. (Eds.): ICACDS 2023, CCIS 1848, pp. 340–350, 2023.
https://doi.org/10.1007/978-3-031-37940-6_28

evaluations of others. Now days, organisations are routinely consult with users before making a decision. Not just for individuals, but also for businesses to develop efficient business strategies.

Overall, the purpose of SA is to determine whether a text expresses a positive, negative, or neutral sentiment. Text data, including social media posts, product evaluations, customer feedback, and news articles, can be subjected to SA. There are different applications for SA, including market research, customer service, brand reputation management, product development, and political analysis. By analysing sentiment in text data, businesses can gain insights into customer opinions and preferences, enhance customer satisfaction, identify trends and patterns, and make decisions based on data.

It has been found from the previous studies that SA is the most acceptable way of understanding people's reactions, their needs, and societal challenges. Different methods of SA exist; Fine-grained SA (very positive, optimistic, neutral, negative, and extremely detrimental), aspect-based SA (positive and negative), emotion identification, and multilingual SA are a few examples of these types of analyses [1].

Users share their thoughts, including misspellings, slang, grammatical mistakes, and other things that make it hard to figure out how people feel. Users also do sentence framing in different ways, making it hard to read the text. There may be a lot of uncertainty about how keywords are defined. This means that words may have different meanings depending on how they are used and in what context they are used. Machine learning techniques that can be used to read people's emotions are essential. The advanced SA methods may help humans and machines work together to do SA more simply. It is possible to use SA to look at both structured and unstructured data.

Existing studies exhibit many machine learning approaches have been used to study how people feel [2, 3]. In the history of ML, DL is thought to be the next step. It is also called ANN (artificial neural network), and it is a way to connect algorithms that try to make the human brain work the same way as it does. It has made it possible to use machine learning in many different ways, like customer service automation and self-driving cars. The performance of deep-learning-based SA has been the subject of several researches in recent years. SA is an important research field as it analyses consumer feedback, online reviews, social media posts, and other textual data in order to gain insight of people's opinions and attitudes regarding various topics, products, or services. Here are some specific reasons why SA is crucial in the field of research:

1. **Understanding Customer Satisfaction:** One of the primary applications of SA is analysing consumer feedback and reviews to determine a product's level of satisfaction. By analysing these, researchers can identify the positive and negative aspects of a product or service and gain insight into consumer preferences and expectations.
2. **Improving Product Development:** By analysing customer's reviews, researchers can identify areas where product modifications or enhancements is necessary. This information can assist businesses in making informed decisions regarding product development and prioritising customer-focused features.
3. **Brand Management:** SA can also be used to monitor brand reputation and identify potential issues that may impact the company's image.

4. **Political Analysis:** By analyzing social media posts and other online content, researchers can gain insights into public opinion on political issues and candidate performance. This information can be used to develop more effective political campaigns and messaging strategies.

This paper explores SA of people during pandemic COVID 19 using baseline deep learning techniques. The baseline algorithms RNN and LSTM were used to classify the sentiments. The main contribution of this research work gives a direction to the researcher's, society, psychologists and people about the thinking of an individual during pandemic time. So, in future if such kind of situation or related situation occurs then this work can help to identify people's sentiments quickly and necessary actions can be taken on time to convince the people.

2 Literature Review

RNNs and LSTM networks are two of the most common DL models that are used for SA. Throughout this literature study, we are going to review different researches that have used RNNs and LSTMs to do SA.

In [1], the researchers proposed a novel LSTM–CNN based grid search model to analyses sentiment's data. The Convolutional Neural Networks (CNN), LSTM, N.N., K-NN, and CNN–LSTM were used to compute the evolutionary parameters on multiple datasets. The study's findings describe that hyper-parameter adjustment leads to an improvement in model performance, and the suggested model beats all previous baseline methods with an accuracy of better than 96%.

In [2], the researchers proposed LSTM approach to SA. In this study, emotional intelligence and attention mechanisms were incorporated into the LSTM network, which enhanced the performance. **In** [3], the researchers discovered that the number of positive tweets was higher in both circumstances, indicating that individuals had a more favourable attitude regarding the pandemic during the lockdown.

In [4], the findings of this analysis were based on a survey conducted for SA using ML/DL models. The study revealed that the performance measures of deep learning techniques were better than machine learning models like SVM. **In** [5], the paper demonstrated the use of CNN and GRU (Gated Recurrent Networks) in conjunction with each other for SA. GRU has also learned about long-term dependencies, which are in addition to the local features provided by CNN; the proposed model makes use of global features generated by CNN.

In [6], this paper explores aspect-based SA. The experimental analysis has been carried out on several SemEval datasets to validate the model. The suggested model beat baseline models in accuracy and Macro-F1, with improvements of 2.14% and 1.33%, respectively, above baseline models. **In** [7], using a BERT-based method; the model outperforms. The jargon converted into plaintext and tweets data were classified using BERT. The concept can be used in a wide range of languages.

In [8], a model was proposed based on a bidirectional CNN-RNN deep model with attention-based analysis. Both previous and future scenarios have been included in the investigation. Three Twitter datasets were taken into account for the study. **In** [9], aspect-based SA performed using a NNN. To evaluate the model's performance, it was trained on different data sets, among other things. The accuracy and F-1 score of the model were measured, among other things.

In [10], for SA, a hybrid model that incorporates DL techniques was deployed. Word2vec and BERT algorithms were used on SVM, CNN, and LSTM in order to conduct performance evaluations. One model was compared to four other hybrid models that were built. Research shows that hybrid models are more reliable than any single model when predicting sentiment polarity and it discovered that the SVM technique is superior to employing an individual model. **In** [11], this paper presents neutrosophic sets for analysis using seven membership functions. The findings reveal that the multi-refined neutrosophic groups are influential in judging the sentiments represented in texts.

Convolutional and recurrent neural networks (CNN and RNN) exhibit relatively good overall accuracy when [12] investigated the performance of the RNN approach and found good results. **In** [13], this paper identified the knowledge gaps in the estimation of software maintainability metrics. The suggested LSTM algorithm performs better than the competition. It is also clear that, in contrast to our LSTM algorithms, the majority of other works' machine learning calculations fail to mention which metrics affect a product's feasibility.

In [14], to assess travel sentiments, Pham employed several layers of knowledge representation on the service parameters, to represent customer reviews and extract more potent sentiment features; a unique layered architecture was presented. Pham; determined the sentiment scores and levels of significance of product features. A different strategy suggested by [15] builds an LSTM model based on emotion identification by combining sentiment and semantic information.

In [16], the researchers talked about different types of deep learning architectures and how they could be used for SA. The study demonstrated that DL techniques had shown the best results for many different tasks when it comes to analyzing people's emotions. Wang's used hybrid approach was used to improve a preliminary recommendation list created by combining collaborative filtering and content-based approaches improved by using SA of reviews of movies.

In [17], two sentiment datasets—IMDB and Stanford Sentiment Treebank were used to verify the proposed model. The findings demonstrated that deep learning models' inability to handle brief texts can be solved using CNN and RNN models.

In [18], a cloud recommender system employed deep learning for SA on a food dataset. The tests reveal that RNN-based Deep-learning SA enhances behaviour by increasing accuracy, which improves user suggestions and helps decision making.

Tweets concerning diabetes were subjected to an ontology-based, aspect-level SA technique by [19]. A suggested method makes use of ontology to efficiently identify features of diabetes in tweets. The findings demonstrate that the "N-gram around" approach produced the best outcomes, with precision, recall, and measure 81.93%, 81.13%, and 81.24%, respectively.

In [20], although the majority of DL techniques and current trends were not included, the paper reveals a thorough understanding of current trends as well as novel techniques for SA that employ DL techniques including classical attention mechanisms, cognitive attention-based models, DRL models, and GANs. It has been determined that deep learning methods work effectively.

In [21], SA was performed using LSTMs on a Chinese microblogging website. They preprocessed the text data and trained on a large dataset of user-generated content using LSTM. The model detected positive, negative, and neutral sentiment with high precision.

In a study [22], a hierarchical RNN architecture was devised for SA. The model was trained on a dataset of product evaluations and outperformed other SA methods in terms of performance. In another study [23], they compared the performance of different DL models, such as RNNs and LSTMs, for SA using a dataset of movies reviews; they discovered that LSTM models performed more accurately than other models.

In [24], a model that incorporates RNNs and CNNs for sentiment analysis was proposed. Using a dataset of customer reviews, they discovered that their model outperformed other deep learning models in terms of accuracy.

In [25], they proposed LSTM model based on attention for sentiment analysis. The model was trained using a dataset of product evaluations and outperformed all other deep learning models.

In [26], RNNs were used to develop a multi-task learning framework for SA using product reviews dataset; they discovered that their model accurately predicted sentiment and product aspect categories.

In [27], a dual-channel LSTM model was proposed for SA. The model detects the sentiment of social media messages with high accuracy using both text and image data.

Literature Review Summary

From the literature reviews; we found that DL has proven to be an effective technique for SA. This technique has its ability to autonomously extract and learn complex features from raw data without requiring manual feature engineering. This is especially useful for SA, which requires the processing of large set of text data to identify subjective opinions, attitudes, and emotions. Large datasets of labeled text can be used to train DL models, such as RNNs and CNNs, to automatically recognize patterns and relationships between words and phrases that indicate positive, negative, or neutral sentiments. These models can also learn to take into consideration contextual information and subtle linguistic nuances, such as sarcasm, irony, and figurative language, which can affect the tone of a sentence.

In addition, DL models can be trained on structured or unstructured data, allowing for flexibility in the quantity and type of data. This enables the training of SA models on a wide variety of domains and languages with high precision and robustness.

Overall, DL has become a popular and effective method for SA due to its ability to autonomously learn and extract complex features from text data, its flexibility in managing different types and quantities of data, and its ability to generalise well to new and untested data.

3   Research Methodology

This section provides an overview of the research approach that was used. The study primarily comprises of data collecting, data preprocessing, feature extraction, and data splitting into training and testing sets with the ratios of 70:30, 80:20, and 90:10, respectively, for each set of data. RNN and LSTM were used to develop a model for categorization of sentiment, which was trained, tested, and validated. The accuracy, precision, recall, and F-1 score were measured to validate models. The following is an outline of overall methodology:

3.1. Dataset – The dataset of COVID 19 sentiments was considered for the conduction of this study. The data set was collected through the Kaggle site and processed. The data set was explored to determine the relationship between data variables, the structure of the dataset, the presence of outliers, and the data value distribution. After transformation, the performance was measured to different machine and deep learning algorithms.

3.2. Data Preprocessing is a significant approach to prepare the dataset before analyse it. We often started by removing web links, numbers, punctuations, stopwords and convert text to lowercase which used in parsing. After this, we got a list of clean words. In line with this, we did tokenize a text which entails separating characters into tokens and deleting punctuation and stopwords at the same time after this stemming and lemmatization performed.

3.3. Feature Extraction - following Tokenization, we perform Term Frequency-Inverse Document Frequency (Tfidf) Vectorizer for feature extraction. TFIDF is a method of transforming textual data into numeric form, and Vectorization is the process of turning words into numbers.

The relative term frequency can be calculated for each term within each document as

$$TF(t, d) = \frac{\text{number of the times term(t) appears in document(d)}}{\text{total number of terms in document(d)}}$$

Inverse Document Frequency quantifies how significant the word is to distinguish each document. We can derive this using the below-mentioned formula –

$$IDF(t, D) = \log\left(\frac{\text{total no of document(D)}}{\text{number of documents with the term(t) in it}}\right)$$

Overall, TFIDF is calculated by:

$$X_{i,w_j} = \frac{1 + log(t_i, w_j)}{1 + log(\sum_i^N t_i, w_j)} * log(\frac{N}{\sum_{w_j} t_i, w_j})$$

Source: [19]

Where t_i, w_j is the frequency of word w_j appears in document i.

3.4. Applying Deep Learning Algorithms – after feature extraction, we did text classification using DL classification models, and the performance on different parameters was evaluated. The details have been described in Fig. 1.

3.5. Proposed Model – the following figure has described the proposed model of the work.

Evaluating the model - the model was evaluated using 70:30, 80:20, and 90:10 ratio's for training and test data using accuracy, precision, recall, and the F-1 score measures.

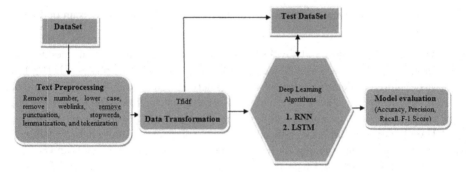

Fig. 1. A proposed framework based on RNN and LSTM deep learning approaches.

4 Results and Discussion

The performance evaluation of RNN and LSTM techniques were measured uisng accuracy, precision, recall and F-1 score parameters using 70:30, 80:20, and 90:10 ratios for training and test data. The details are described in Table 1, Table 2, and Table 3, respectively.

Table 1. Performance evaluation on Run 1 of the model with 70% training and 30% test data.

Model	Accuracy	Precision	Recall	F-1 Score
RNN	81%	82%	80%	81%
LSTM	**86%**	**88%**	**85%**	**86%**

From Table 1, results reveal that RNN performed with 81% accuracy, 82% precision, 80% recall, and 81% F-1 score while highest accuracy **86%**, precision, **88%**, recall **85%**, **and** F-1 score values **86% were recorded for the LSTM classifier.** The performance evolution for the LSTM classification was recorded better than RNN. Hence, it is stated that the performance of LSTM technique exhibits greater than RNN.

The graphical representation of the outcome shows that the performance evaluation of LSTM on different evaluation parameters.

Table 2 shows the outcomes of the second run of the model with 80% training and 20% test data. It has been observed that the performance measures for the LSTM classification were recorded better than RNN. The highest accuracy, **87.3%,** the precision **of 89%,** recall of **86%, and** F-1 score **of 87% were recorded for the LSTM classifier.** Hence, it is stated that the performance evaluation for the LSTM technique for second run again exhibits better.

Figure 2 demonstrated the graphical representation of the outcome shows with 80% training and 20% test dataset. It has been observed that the performance measures of LSTM were higher. The accuracy rate on Run2 of the model has increased compared to Run 1.

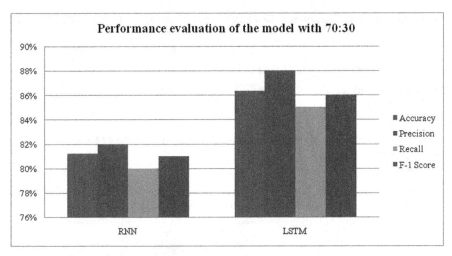

Fig. 2. Graphical representation of performance measures on run 1 of the model.

Table 2. Performance evaluation on Run 2 of the model with 80% training and 20% test data.

Models	Accuracy	Precision	Recall	F-1 Score
RNN	80.90%	82%	79%	81%
LSTM	**87.30%**	**89%**	**86%**	**87%**

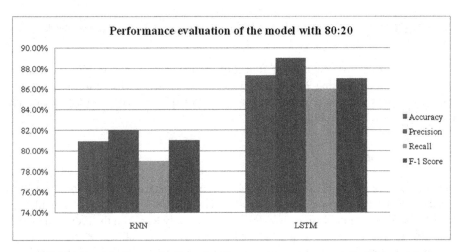

Fig. 3. Graphical representation of performance measures on run 2 of the model.

Table 3 demonstrated the third run of the model with 90% training and 10% test data. The results show that the **LSTM** outperformed with the highest accuracy, **88%**, the precision **of 88%**, recall of **87%, and** F-1 score **of 87%**.

Table 3. Performance evaluation on Run 3 of the model with 90% training and 10% test data.

Models	Accuracy	Precision	Recall	F-1 Score
RNN	82%	84%	80%	82%
LSTM	**88%**	**88%**	**87%**	**87%**

Hence, it is stated that the performance evaluation for the LSTM technique exhibits greater than RNN (Fig. 4).

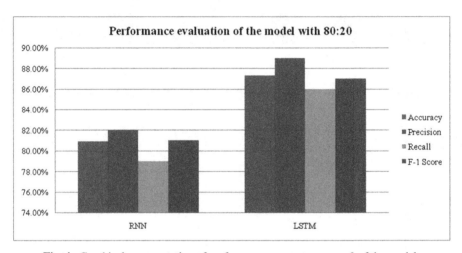

Fig. 4. Graphical representation of performance measures on run 3 of the model.

The graphical representation of Run 3 of the model is shown in Fig. 3. Figure 3 showed the outcomes using 90% training and 10% test dataset. The performance evaluation of LSTM with accuracy, precision, recall, and F-1 score. The accuracy rate of the model was higher in Run3 than Run 2.

5   Conclusion

The outcomes of the study reveal that when we ran the model with 70% testing and 30% test data; the LSTM performed well with 86% accuracy, 88% precision, 85% recall, and 86% F-1 score in comparison with RNN. When we apply Run 2 on the proposed model with 80% testing and 20% test data; again, LSTM performed well with 87.3% accuracy, 89% precision, 86% recall, and 87% F-1 score in comparison with RNN. The second run of the model observed that the performance measures on mentioned parameters were increased. During Run 3 of the proposed model with 90% testing and 10% test data, again LSTM performed well with 88% accuracy, 88% precision, 87% recall, and 87% F-1 score. It has been recorded that on Run 3 of the model, the performance measures on mentioned parameters were increased.

The proposed model result indicates that the model's performance was increased during the second and third run, respectively. Hence, the proposed model outperforms. The future work aligns with the parameter tuning on LSTM technique to validate the results.

References

1. Priyadarshini, I., Cotton, C.: A novel LSTM–CNN–grid search-based deep neural network for sentiment analysis. J. Supercomput. **77**(12), 13911–13932 (2021). https://doi.org/10.1007/s11227-021-03838-w
2. Huang, F., Li, X., Yuan, C., Zhang, S., Zhang, J., Qiao, S.: Attention-emotion-enhanced convolutional LSTM for sentiment analysis. IEEE Trans. Neural Netw. Learn. Syst. (2021)
3. Priyadarshini, I., Mohanty, P., Kumar, R., Sharma, R., Puri, V., Singh, P.K.: A study on the sentiments and psychology of Twitter users during the COVID-19 lockdown period. Multimed. Tools Appl. 1–23 (2021)
4. Shrivash, B.K., Verma, D.K., Pandey, P.: An analysis on machine learning approaches for sentiment analysis. In: Somani, A.K., Mundra, A., Doss, R., Bhattacharya, S. (eds.) Smart Systems: Innovations in Computing. SIST, vol. 235, pp. 499–513. Springer, Singapore (2022). https://doi.org/10.1007/978-981-16-2877-1_46
5. Zhao, N., Gao, H., Wen, X., Li, H.: Combination of convolutional neural network and gated recurrent unit for aspect-based sentiment analysis. IEEE Access **9**, 15561–15569 (2021)
6. Lu, Q., Zhu, Z., Zhang, G., Kang, S., Liu, P.: Aspect-gated graph convolutional networks for aspect-based sentiment analysis. Appl. Intell. **51**(7), 4408–4419 (2021). https://doi.org/10.1007/s10489-020-02095-3
7. Pota, M., Ventura, M., Catelli, R., Esposito, M.: An effective BERT-based pipeline for Twitter sentiment analysis: a case study in Italian. Sensors **21**(1), 133 (2021)
8. Basiri, M.E., Nemati, S., Abdar, M., Cambria, E., Acharya, U.R.: ABCDE: an attention-based bidirectional CNN-RNN deep model for sentiment analysis. Futur. Gener. Comput. Syst. **115**, 279–294 (2021)
9. Srividya, K., Sowjanya, A.M.: NA-DLSTM–A neural attention-based model for context-aware Aspect-based sentiment analysis. Materials Today: Proceedings (2021)
10. Dang, N.C., Moreno-García, M.N., De la Prieta, F.: Sentiment analysis based on deep learning: a comparative study. Electronics **9**(3), 483 (2020)
11. Kandasamy, I., Vasantha, W.B., Obbineni, J.M., Smarandache, F.: Sentiment analysis of Tweets using refined neutrosophic sets. Comput. Ind. **115**, 103180 (2020)
12. Alharbi, A.S.M., de Doncker, E.: Twitter sentiment analysis with a deep neural network: an enhanced approach using user behavioural information. Cognit. Syst. Res. **54**, 50–61 (2019)
13. Pham, D.H., Le, A.C.: Learning multiple layers of knowledge representation for aspect-based sentiment analysis. Data Knowl. Eng. **114**, 26–39 (2018)
14. Gupta, U., Chatterjee, A., Srikanth, R., Agrawal, P.: A sentiment-and-semantics-based approach for emotion detection in textual conversations (2017). arXiv preprint arXiv:1707.06996
15. Zhang, L., Wang, S., Liu, B.: Deep learning for sentiment analysis: a survey. Wiley Interdiscip. Rev. Data Min. Knowl. Discov. **8**(4), e1253 (2018)
16. Hassan, A., Mahmood, A.: Deep learning approach for sentiment analysis of short texts. In: 2017 3rd International Conference on Control, Automation and Robotics (ICCAR), pp. 705–710. IEEE (2017)
17. Preethi, G., Krishna, P.V., Obaidat, M.S., Saritha, V., Yenduri, S.: Application of deep learning to sentiment analysis for recommender system on the cloud. In: 2017 International Conference on Computer, Information and Telecommunication Systems (CITS), pp. 93–97. IEEE (2017)

18. Salas-Zárate, M.D.P., Medina-Moreira, J., Lagos-Ortiz, K., Luna-Aveiga, H., Rodriguez-Garcia, M.A., Valencia-Garcia, R.: Sentiment analysis on tweets about diabetes: an aspect-level approach. Comput. Math. Methods Med. (2017)
19. Ain, Q.T., et al.: Sentiment analysis using deep learning techniques: a review. Int. J. Adv. Comput. Sci. Appl. **8**(6) (2017)
20. Rojas-Barahona, L.M.: Deep learning for sentiment analysis. Lang. Linguist. Compass **10**(12), 701–719 (2016)
21. Zhou, B., et al.: Learning deep features for discriminative localization. In: Proceedings of the IEEE Conference on Computer Vision and Pattern Recognition, pp. 2921–2929 (2016)
22. Tang, D., et al.: Sentiment embeddings with applications to sentiment analysis. IEEE Trans. Knowl. Data Eng. **28**(2), 496–509 (2015)
23. Severyn, A., Moschitti, A.: Twitter sentiment analysis with deep convolutional neural networks. In: Proceedings of the 38th International ACM SIGIR Conference on Research and Development in Information Retrieval, pp. 959–962 (2015)
24. Li, X., et al.: Exploiting BERT for end-to-end aspect-based sentiment analysis (2019). arXiv preprint arXiv:1910.00883
25. Yang, M., et al.: Attention based LSTM for target dependent sentiment classification. In: Proceedings of the AAAI Conference on Artificial Intelligence (2017)
26. Ma, X., Zhou, C., Yang, X., Huang, Y., Zhu, X.: Modeling sentences with LSTM for emotion detection in textual conversations. In: Proceedings of the 56th Annual Meeting of the Association for Computational Linguistics (ACL), pp. 1426–1436 (2018)
27. Wang, K., et al.: Relational graph attention network for aspect-based sentiment analysis (2020). arXiv preprint arXiv:2004.12362

Spatial Domain Method for Image Analysis: A Grey-Level Computation Approach

Kumari Deepika[1]([⊠]) ⓘ, Deepika Punj[2] ⓘ, and Jyoti[2] ⓘ

[1] Symbiosis Institute of Computer Studies and Research, Symbiosis International (Deemed University), Atur Centre, Model Colony, Pune 411016, India
kumari.deepika@sicsr.ac.in

[2] J C Bose University of Science and Technology, 6, Mathura Rd, Sector 6, Faridabad, Haryana 121006, India

Abstract. This paper discusses image enhancement, which involves adjusting captured images to improve their visual quality and suitability for display or analysis using techniques like filtering, histogram equalization, and contrast adjustment. The spatial domain method is used for manipulating image grey levels, and the study analyses the transformed images using BRIQUE, NIQE, and PIQE quality metric descriptors. The study shows that the type of image used for processing depends on the application, and binary images have the highest PIQE score. Power law transformation or Gamma-corrected images can reduce or enhance contrast, depending on the gamma's value. The transformed images have different scores based on the quality metric used. The study also finds that the sum of the global threshold values of an input image and its inverse is 1, and the scores for binary images using both thresholds are similar. Finally, the study shows that BRISQUE and PIQE scores are higher for gamma and more significant than 1; NIQE scores are similar for both gamma values.

Keywords: Gray Level Transformations · No reference Quality Metrics · Image Enhancement

1 Introduction

Image is termed as the function of spatial coordinates that returns the amplitude or intensity of that image at that pixel. There are four categories of images. They are as follows: (a) Gray Image- In a gray image, pixel value stores the information related to its intensity. There is no color information held in it. The range for Gray Scale is from 0 to 255(0 shows the least intensity indicating black, and 255 shows the most intensity indicating with white color. A grayscale image is also known as a monochrome image, as there is only one color in it. (b) Color Image- The color image consists of all three channels- red, Blue, and Green; in other words, it is called 3-a band monochrome image. Each band needs 8 bits, so 24 bits are required to store the information of all three planes. It is represented in vector form or stored in a 3-dimensional array, and the first two dimensions indicate the location, and the third dimension indicates the color information.

M. Singh et al. (Eds.): ICACDS 2023, CCIS 1848, pp. 351–366, 2023.
https://doi.org/10.1007/978-3-031-37940-6_29

(c) Index Image- An indexed image is a type of image representation where each pixel's value is indexed into a static color map or palette rather than representing actual color values. It uses a limited set of colors to represent the image, usually 256 or less, making it efficient for low-color graphics and animations. The color map or palette is a table that lists the actual RGB color values for each index, and the pixel values in the image correspond to the indices in the color map. This image representation helps reduce the image's file size and improve graphics performance. (d) Binary Image- A binary image is an image representation where each pixel is stored as a single bit with a value of either 0 or 1, indicating black or white, respectively. Binary images are used in image processing and computer vision for object recognition, edge detection, and thresholding tasks. They are also used for image compression, as the simple 2-color representation significantly reduces the amount of data needed to store the image compared to grayscale or color images. Binary images are typically represented as 2D arrays, with the elements of the array representing the pixel values. Figure 1 provides a visual representation of the different image categories.

Fig. 1. Types of images

Image processing involves transforming an image to extract meaningful information from it. The input is an image, and the output can be either an improved version of the image or features extracted from the image. The steps involved in the image processing area. (a) Image acquisition involves capturing an image using a camera or scanner. (b) Image enhancement involves improving the quality of the image by adjusting its brightness, contrast, and other attributes. (c) Image restoration involves removing noise and restoring the image to its original form. (d) Image segmentation divides the image into segments to separate objects or regions of interest. (e) Image representation involves converting the image into a form that can be used for further processing, such as a matrix. (f) Image analysis involves extracting features from the image and performing various operations on the image to extract meaningful information. (g) Image interpretation involves making sense of the information obtained from the image and using it to make decisions or draw conclusions. (h) Image output involves displaying the processed image or presenting the extracted information in a helpful format. Figure 2 illustrates the various stages involved in image processing.

Image Acquisition is made by capturing images using a hardware device sensor. Image may contain noise because of several factors like heat generated while capturing, the transmission of the image from sender to receiver, errors due to synchronization, ISO factors, memory cell failure, etc. It causes distortion and degrades the quality of the image. Due to this, it needs to be filtered out to improve its quality. Several types

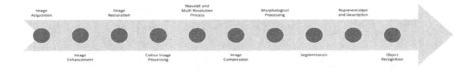

Fig. 2. Steps involved in image processing

of noise are present in an image, such as gamma noise, add noise, exponential noise, uniform noise, gaussian noise, periodic noise, salt, and pepper noise/impulse noise/shot noise/spike noise, etc. Noise can be removed by applying filters [10]. The filter is the process of removing noise from an image by preserving the attributes of the image through the filters. Table 1 summarizes the type, characteristics, and filter mechanisms applied to filter the noise [14].

Table 1. Types of noise and the corresponding filters applied to remove it.

Noise	Characteristics	Filters
Salt and Pepper Noise	Randomly scattered white or black or both pixels over the image	Mean Filtering, Median Filtering Gaussian Filtering
Gaussian	Addition of random values to an image	Mean Filtering/Convolution Median Filtering, Gaussian Filtering
Speckle Noise	Multiplication of random values to an image	Mean Filtering, Median Filtering
Uniform Noise	Gray Level values of the noise are evenly distributed	The filter is chosen based on the Nature of the task performed by the filter, filter behavior, and type of the data

A linear filter in image processing is a mathematical operation that modifies the pixel values of an image by a linear combination of their neighboring pixel values. The result is a filtered image with specific enhanced or suppressed image features. Examples of linear filters include mean filtering, Gaussian filtering, and edge detection filters. A non-linear filter in image processing is a mathematical operation that modifies the pixel values of an image using a non-linear function. Unlike linear filters, non-linear filters do not perform a simple linear combination of neighboring pixels but instead use a more complex function to modify the image pixels. Non-linear filters are often used to preserve or enhance certain image features that are not easily captured by linear filters. Examples of non-linear filters include median filtering, morphological filtering, and color correction filters. Figure 3 illustrates the various types of filters categorized.

However, the work to be done here will cover only software-related implementations, so the first step is to start from the pre-processing phase. Pre-processing is applied to the intensity image to get an intensity image. The image is said to be intense as these

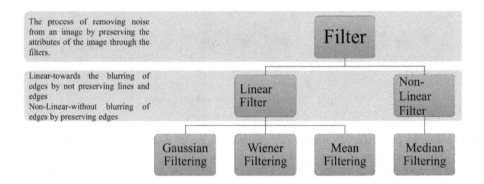

Fig. 3. Categories of Filter

images specify the information related to the intensity. Different techniques are involved in this phase to be applied to the intensity image to compute the pixel. It is depicted in the below figure. The mind map of Image processing is depicted in Fig. 4.

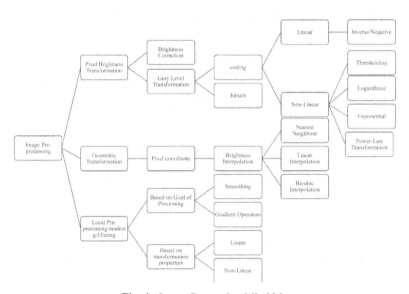

Fig. 4. Image Processing Mind Map

Pixel Brightness Transformation: The transformation applied to modify the pixels' values that depend on pixel properties. There are two classes of Pixel Brightness Transformations. These are Brightness corrections and Gray Level Transformations. Brightness corrections- If degradation is systematic, then this approach is applied. It generally occurs due to an error in capturing the device. In this approach, the position is also considered a parameter for modifying the brightness of an image represented in

Eq. (1) and Eq. (2).

$$f(i, j) = e(i, j) \, g(i, j) \tag{1}$$

$$g(i, j) = f(i, j)/e(i, j) = cf(i, j)/fc(i, j) \tag{2}$$

where, g (i, j)- input image

e(i, j)- error coefficient

f(i, j)-output image

Gray Scale Transformations- In this approach, the position is not considered a parameter for modifying the brightness of an image. Linear Gray level- Image Negative or Inverse Transformation-this transformation is used to reverse the image. Non-Linear Gray Level-Thresholding-This transformation converts the image into a binary image (black and white). Various non-linear gray-level techniques can be employed to adjust the contrast of brighter regions, such as logarithmic, exponential, and power law transformations. These methods can either reduce or enhance contrast and are effective in enhancing the visual quality of images. Logarithmic transformation reduces the contrast of brighter regions by lower the gray levels. In contrast, exponential transformations enhance the contrast of brighter regions, i.e., compressing low-contrast edges while expanding high-contrast edges. Hence, exponential transformations are not recommended. Non-Linear Gray Level-Logarithmic Transformation-lower gray values. Equation (3) represents logarithmic transformation.

$$s = c * \log(1 + r) \tag{3}$$

Non-Linear Gray Level-Exponential Transformation is represented in Eq. (4).

$$\text{I output } (i, \ j) = e^{(\text{I input } (i, \ j)} = c[(1 + \text{alpha}) \, \text{input}(\text{I}, \ j) - 1] \tag{4}$$

In the above Eq. (4), subtraction of -1 is done not to generate output if the input is not present. Non-Linear Gray-Level-Power Law Transformation or Gamma correction-When any source of the physical device generates light; its intensity is not always a linear function of the signal applied. A sensor generates a signal at the moment of capturing the picture. The actual intensity generated on the surface of the display is as approximate as the applied voltage raised to the power(energy) of 2.5. The calculated numerical value of the exponent of this energy function is represented as gamma. If gamma < 1, then logarithmic transformation, else exponential transformation.

2   Study of Related Work

Image Enhancement involves improving the visual quality of an image. Some standard techniques include: "(i) Brightness and contrast adjustment-Adjusting the brightness and contrast to improve the visibility of features in an image. (ii) Noise reduction-Removing random noise from an image to improve its quality. (iii) Sharpening-Enhancing the edges in an image to make it appear more precise and detailed. (iv) Colour correction-Adjusting the colors in an image to make them appear more natural or to change the mood of the image. (v) Histogram equalization-Adjusting the distribution of intensities in an image to improve its global contrast. (vi) Dehazing-Removing the haze or mist from an image to make it more transparent. (vii) Image smoothing-Blurring the image to remove small-scale details and reduce noise. (viii) Image scaling- Decreasing or decreasing an image's size while preserving its essential features.

In [3], the ROI Extraction process involves applying a Gaussian Low Pass Filter to smooth out image noise, computing the Gradient of the image, applying a K-means clustering algorithm to create a binary image, and eroding and dilating it. The Euclidean Distance is then calculated to find the peak location, and the centroid position is checked by rotating the image. The circle with the maximum radius is selected, and the image is normalized into rectangular strips using bilinear interpolation. Discriminative edge features are extracted using a gradient-based encoding method and a LOP-based encoding scheme to transform the image.

In [4], the binary image is created using Otsu's thresholding, and the Hand boundary is identified to extract ROI. The Reference Point is established at the bottom center of the wrist, and Peaks, Valleys, and local extrema are identified. ROI Enhancement techniques include Histogram Equalization, Adaptive Histogram Equalization, and Wiener Filtering. In ROI Transformation, the Texture Code Matrix encodes the image using a robust transformation scheme, generating ordered code patterns that describe the grayscale values' spatial arrangements in each subregion of the Palm ROI.

In [10], the ROI Extraction process involves binarizing the input image and applying a Border Tracing Algorithm to identify the hand boundary. The distance between contour points and the mid-point of the wrist is calculated, and four local minimum locations are identified from the distance distribution, corresponding to finger-web locations in the palm image. Two finger web points, F1 and F3, are used to locate the key points, C1 and C2, which lie in the middle of line segments F1p1 and F3P2. The palm image is then rotated by an angle Θ, calculated using C1 and C2 coordinates. Finally, a Gaussian low pass filter smooths and improves image quality.

In [5], a gray Scale Image is converted to a Binary Image using a Low Pass Filter. The boundary of the holes H1 and H2 are detected, and the tangents t1 and t2 are calculated. The tangents are aligned to determine the Y-axis of the palmprint coordinate system. Finally, the system's origin is established by drawing a line perpendicular to the Y-axis through the midpoint of T1 and T2.

In [2], the ROI Extraction process involves creating a Binary Image using Global thresholding, identifying the Hand Contour using a Contour Following Algorithm, and locating the Fingertips and Valleys. A Square ROI is created based on the distance between valley points and their corresponding fingertips. In the Geometry Normalization stage, the Karhunen - Loeve (K L) Transformation is applied to the feature space (eigenspace).

In [2], the palmprint recognition process includes several stages. First, the image is binarized using the Otsu Algorithm and rotated to the correct orientation. The ROI Extraction stage involves identifying the hand contour using valley points P1 and P2 as reference points. In the Normalization stage, a linear interpolation algorithm is applied.

The image is then denoised using Wavelet Thresholding and enhanced with Histogram Equalization.

In [8, 11], the palmprint recognition process involves creating a Square ROI with a resolution of 64*64 and enhancing the image using Contrast Limited Adaptive Histogram Equalization (CLAHE). A 2D Gaussian low pass filter is then used to remove the palm print. In Image Segmentation, Dynamic Thresholding is followed by morphological opening and closing using a circular structuring element with a radius of 3. The Image Skeletonization stage involves applying an Isolated Line Removal Algorithm and a Short Branch Pruning Algorithm to remove isolated lines and short branches.

In [6], image processing for license plate recognition involves several stages. Noise Reduction is performed first to improve image quality, followed by Binarization using Local Adaptive Thresholding with Otsu's method. The Canny operator is used for Edge Detection, which involves convolving the input image with a 2D-Gaussian filter, applying the gradient operator, and calculating the gradient magnitude and orientation. Then, the nonmaximal suppression algorithm is applied, followed by hysteresis thresholding using Otsu's algorithm with two thresholds. In the ROI Extraction stage, the license plate region is identified, and finally, Normalization is performed to adjust image size and improve recognition accuracy.

In [7], the hand recognition process involves identifying the ROI by performing Hand localization and palm localization. Normalization is then done to standardize the size and shape of the ROI. Hand Segmentation uses morphological operations, and the hand extremities are localized by contour extraction using the Freeman algorithm. The Euclidean distance between the considered point and its two neighbors among the reference points is then minimized to refine the hand extremities' locations.

In [1], the input image is converted into a Grayscale image and then a Binary Image using a thresholding technique. PCA is performed on the binary image to extract the Eigenvectors of the hand shapes that capture dominant variations of the hand shapes in the dataset. Next, the morphological opening is applied to identify the ROI of the hand by removing small objects and noise from the image and isolating the hand region.

In [12], the palmprint recognition process involves segmentation, localization, and feature extraction. The image is first thresholded using Otsu's algorithm to extract the foreground region. The palmprint region is then localized using template matching or feature extraction, and the ROI is extracted. The ROI can be enhanced using local histogram equalization or a Gabor filter to improve image quality for feature extraction. These steps are critical for accurate palmprint recognition.

In [9], a Gaussian smoothing low pass filter is applied to the palmprint image, followed by conversion into a binary image. Boundaries of the gaps between the fingers are determined using a boundary tracking algorithm, and the tangent of the two gaps is calculated. The Y-axis of the palmprint coordinate system is determined by the line passing through the midpoint of two points, and the origin of the coordinate system is determined by finding a line perpendicular to the Y-axis. A sub-image of fixed size is then extracted based on the coordinate system. The extracted ROI can be further enhanced using the Log-Gabor filter, a derivative of the standard Gabor filter with a Gaussian frequency response in the logarithmic frequency scale.

Table 2. Techniques-their approaches and challenges

Technique	Approach	Challenges
Brightness and contrast adjustment [8, 12]	Improves visibility of features in the image	Loss of information
Noise reduction [6, 9]	Improves image quality by removing random noise	Loss of detail or blurring
Sharpening [6, 9]	Enhances edges in the image, making it appear more apparent and detailed	increase noise or artifacts
Color correction [14]	Adjusts colors to make them appear more natural	oversaturation or loss of contrast
Histogram equalization [2, 4, 8, 12, 13]	Adjusts distribution of intensities to improve global contrast	Over-enhancement of noise
Dehazing [8, 12]	Removes haze or mist from the image, making it more transparent	Need for more details
Image smoothing [3, 5]	Removes small-scale details and reduces noise	Loss of important information
Image scaling [8, 12]	Changes the size of the image while preserving essential features	Loss of information or distortion
ROI extraction using Gaussian Low Pass Filter and K-means clustering [3]	Accurately extracts the ROI using a robust method	computationally intensive
ROI extraction using Otsu's thresholding and texture code matrix [4]	Accurately extracts the ROI using a robust method	not work well with images that have poor texture
ROI extraction using border tracing algorithm [11]	Accurately extracts the ROI using a robust method	It does not work well with images that have poor edge definition
ROI extraction using low pass filter and hole detection [5]	Accurately extracts the ROI using a robust method	It does not work well with images that have poor contrast
ROI extraction using contour following algorithm and K L transformation [2]	Accurately extracts the ROI using a robust method	Computationally intensive
Palmprint recognition using Otsu's algorithm and wavelet thresholding [2]	Accurately recognizes palmprints with denoising and enhancement	Computational intensive
Palmprint recognition using CLAHE and image segmentation [2]	Accurately recognizes palmprints with noise reduction and segmentation	Loss of information

(continued)

Table 2. (*continued*)

Technique	Approach	Challenges
Hand recognition using the Freeman algorithm and Euclidean distance [7]	Accurately recognizes hand extremities with contour extraction	Computationally intensive

A comprehensive summary of the literature review is presented in Table 2, which examines the strengths and weaknesses of different image enhancement techniques and methods used for extracting the region of interest (ROI) in palmprint recognition.

3 Results and Findings

An image is a collection of signals, and the term Image Quality refers to the degree of accuracy that any imaging system can use to capture, store, process, compress, transmit, and display the image in the form of signals. An image contains many attributes, some of which are visually significant, and a weighted combination of them is referred to as image quality. There are two aspects to define image quality; the first focuses on signal processing characteristics in various imaging systems, whereas the second focuses on visual or perceptual assessment to make a pleasant view of an image for the viewers. Image quality is measured in terms of quality metrics that give objective scores to determine image quality. There are two commonly used algorithms – 1. Full reference algorithm, and 2. No reference algorithm. In the Full reference algorithm, the input image is compared against a new reference image having no distortion. In contrast, in the No reference algorithm, statistical features of the input image are compared against a list of features derived from an image database.

The graphs in Figs. 5 through 9 demonstrate the effects of different image transformations. Each figure corresponds to a specific transformation, and the graph within each figure shows the resulting changes in image quality or characteristics. These transformations may involve changes in contrast, brightness, or other visual features, and the graphs allow for represent transformations that impact the image.

An inverse image in MATLAB 2018a is created by subtracting each pixel value from the maximum pixel value to create an inverted version of the original image. The graph of an inverse image shows the input pixel values of the original image on the x-axis and the output pixel values of the transformed image on the y-axis. An inverse transformation enhances the edges and boundaries of objects in an image and can create artistic effects or highlight specific features. However, it should be used carefully to avoid losing detail and image artifacts (Fig. 6).

```
input_image=imread('C:\Users\Deepika\Desktop\001_1.JPG');

inverse_image=255-input_image;
```

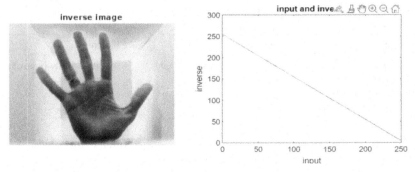

Fig. 5. Inverse Image and Graph of Inverse Image (MATLAB 2018a)

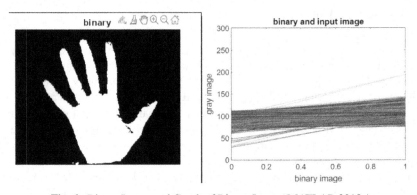

Fig. 6. Binary Image and Graph of Binary Image (MATLAB 2018a)

A binary image is a simplified representation of each pixel with only two possible values: black or white. MATLAB 2018a can be represented graphically with a binary matrix where 0 represents black and 1 represents white. Binary images are widely used in image processing and computer vision applications such as object detection and segmentation. They identify the presence or absence of objects or features in an image by thresholding the input image and generating a binary image where white pixels represent the objects or features. Binary images help simplify complex images and perform image processing tasks, but they may not capture all the details of the original image.

$$output\_image = \begin{cases} 0 & input\_image < threshold \\ 255 & otherwise \end{cases}$$

input_image=imread('C:\Users\Deepika\Desktop\001_1.JPG');

gray_image=rgb2gray(input_image);

threshold_input=graythresh(gray_image);

threshold_input=0.415686274509804

binary_image=imbinarize(gray_image,threshold_input);

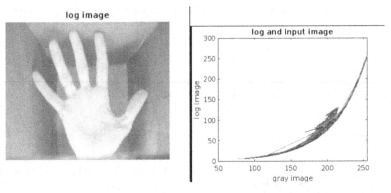

Fig. 7. Log Image and Graph of Log Image (MATLAB 2018a)

The log transformation in MATLAB 2018a is a non-linear image transformation that alters the brightness and contrast of an image, resulting in a broader range of tones in the darker regions of the image compared to the brighter regions. The transformation function has a steeper curve for lower input pixel values and a flatter curve for higher input pixel values, represented in Fig. 7. This enhances the darker regions of the image, making it useful for enhancing contrast in images with predominantly dark areas or reducing the effect of noise in an image. However, it should be used cautiously to avoid losing detail and image artifacts.

```
input_image=imread('C:\Users\Deepika\Desktop\001_1.JPG');

double_image=double(input_image);

c=1;

log_image=c*log(double_image+1);

T = 255/(c * log(256));

T=45.9859044283357

output_log_image=uint8(T*log_image);
```

Gamma correction with a value less than 1 (G < 1) makes an image appear brighter than the original image as it enhances the brighter regions of the image more than the darker regions. The graph of a gamma-corrected image has the x-axis representing the input pixel values of the original image and the y-axis representing the output pixel values of the gamma-corrected image represented in Fig. 8. The gamma function is applied to each input pixel value to obtain the corresponding output pixel value. When a gamma correction value of G < 1 is used, the gamma function has a flatter curve in the lower input pixel values and a steeper curve in the higher input pixel values. This results in the brighter regions of the image is enhanced more than the darker regions, leading to an overall brighter image. Gamma correction with G < 1 can enhance the contrast in images with predominantly bright areas, such as those in bright daylight or high-key lighting. It can also create a dreamy or ethereal effect in an image. However, overuse of

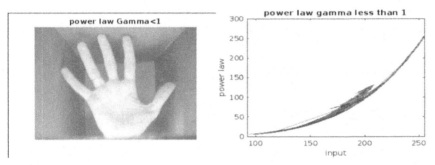

Fig. 8. Power Law or Gamma corrected Image and Graph of Power Law, or Gamma corrected Image(G < 1)(MATLAB 2018a)

gamma correction with G < 1 can lead to loss of detail and image artifacts, so it should be used carefully.

```
input_image=imread('C:\Users\Deepika\Desktop\001_1.JPG');

double_image=double(input_image);

c=1

G=0.25

output_power_law=c*(double_image.^G);

t_power_law=(255/(c*(255.^G)));

t_power_law=63.8124082979258

output_power_law_transformation=uint8(output_power_law*t_power_law);
```

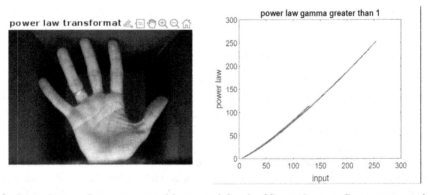

Fig. 9. Power Law or Gamma corrected Image and Graph of Power Law, or Gamma corrected Image(G > 1)(MATLAB 2018a)

Gamma correction is a non-linear transformation that changes the brightness and contrast of an image. When G > 1, the resulting image appears darker because the gamma correction enhances the darker regions more than, the brighter regions. The graph's x-axis represents the input pixel values, and the y-axis represents the output pixel values after gamma correction, represented in Fig. 9. A gamma correction value of G > 1 has a steep curve in lower input pixel values and a flatter curve in higher input pixel values, which enhances darker regions more. Gamma correction with G > 1 helps enhance contrast in dark areas or create a dramatic effect but should be used carefully to avoid loss of detail or artifacts (Table 3).

```
input_image=imread('C:\Users\Deepika\Desktop\001_1.JPG');

double_image=double(input_image);

c=1

G1=1.25

output_power_law1=c*(double_image.^G);

t_power_law1=(255/(c*(255.^G)));

t_power_law1=0.250244738423238

output_power_law_transformation1=uint8(output_power_law1*t_power_law1);
```

Table 3. Image with their Scores (BRISQUE, NIQE, PIQE)

	BRISQUE	NIQE	PIQE
Input Image	16.8546	4.6646	6.1436
Gray Image	16.8546	4.6646	6.1436
Inverse Image	16.8546	4.6646	6.1436
Binary image using input threshold	44.4229	12.5936	100
Binary image using inverse threshold	44.4281	11.8464	100
Log Image	19.5780	4.6301	14.7243
Power law Image (Gamma < 1)	27.3728	4.6368	10.3084
Power law Image (Gamma > 1)	16.610	4.7626	5.8497

The table shows the results of different image processing techniques on three quality assessment metrics: BRISQUE, NIQE, and PIQE. The "Input Image" row shows the scores for the original image. The scores for all three metrics are relatively low, indicating the image is of good quality. The "Gray Image" row shows the scores for a grayscale version of the image. The scores are the same as for the input image, which is expected since the grayscale image is a simple transformation of the original. The "Inverse Image" row shows the scores for an inverted version of the original image. Again, the scores are the same as for the input image, which suggests that the image inversion did not affect the perceived quality. The "Binary Image using input threshold" and "Binary Image using inverse threshold" rows show the scores for binary images created using different thresholding techniques. The scores are higher than the original image, which implies that the quality of the image has decreased due to the thresholding. The "Log Image" row shows the scores for an image transformed using a logarithmic transformation. The BRISQUE and PIQE scores have increased compared to the original image, while the NIQE score has decreased. The perceived quality of the image may have increased due to the transformation. The "Power law Image (Gamma < 1)" row shows the scores for an image transformed using a power law transformation with a gamma value less than 1. The BRISQUE and PIQE scores have increased compared to the original image, while the NIQE score has decreased. The perceived quality of the image may have increased due to the transformation. The "Power law Image (Gamma > 1)" row shows the scores for an image transformed using a power law transformation with a gamma value greater than 1. The BRISQUE and PIQE scores have decreased compared to the original image, while the NIQE score has increased. It observes that the perceived quality of the image decreased due to the transformation. Figure 10 presents a graph illustrating the image scores for a given quality assessment metric.

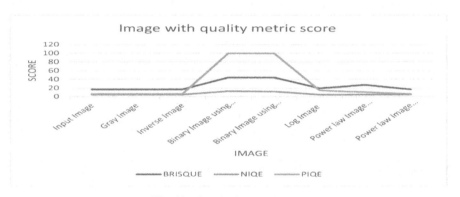

Fig. 10. Graph of Image Scores

4 Conclusion

RGB image, Gray Image, and Inverse image have the same score. The type of image to be considered for processing depends on the application where it will be used. Binary image has the highest PIQE score, approximately 100%. If the value of gamma is less than 1, then Power law transformation or gamma corrected image would be similar to log transformation, i.e., reduces the contrast of the brighter regions by lowering the gray values applying the log function. If the value of gamma is more significant than one, then power law transformation or gamma corrected image would be similar to exponential transformation, i.e., enhances the contrast of the brighter regions by applying the exponential function to the input image. It is also observed here that the log-transformed and power law-transformed images (Gamma less than 1) have identical NIQE scores. However, there is a slight variation in the BRIQUE and PIQE scores. The sum of the global threshold value of the input image and that of the inverse image is 1. The three scores (BRIQUE, NIQE, PIQE) of binary images computed using the threshold value of input and inverse images are similar. Comparing the scores of power law transformation in which gamma is less than one and gamma is more significant than 1, BRISQUE and PIQE scores are more in gamma than 1. However, NIQE scores are similar in both gamma values.

5 Future Work

There are several potential avenues for future work in the field of image processing and image quality metrics that include further exploration of the impact of different image processing techniques on metrics beyond BRISQUE, NIQE, and PIQE, as well as the development of new metrics, specifically designed for certain types of images or transformations. Additionally, conducting user studies to evaluate the subjective impact of different techniques on image quality, investigating the impact of combining multiple techniques, and developing new techniques for specific applications such as medical imaging or low-light conditions are also recommended. Further explore the impact of different file formats on image quality, particularly about compression techniques such as JPEG.

Acknowledgment. The authors like to acknowledge IIT Delhi for providing their database.

References

1. Abeysundera, H.P., Eskil, M.T.: Palmprint verification using SIFT majority voting. In: Gelenbe, E., Lent, R., Sakellari, G. (eds.) Computer and Information Sciences II, pp. 291–297. Springer, London (2011). https://doi.org/10.1007/978-1-4471-2155-8_37
2. Arunachalamand, M., Amuthan, K.: Finger knuckle print recognition using MMDA with fuzzy vault. Int. Arab J. Inf. Technol. 17(4), 554–561 (2020)
3. Jaswal, G., Poonia, R.C.: Selection of optimized features for fusion of palm print and finger knuckle-based person authentication. Expert Syst. 38(1), e12523 (2021)
4. Jaswal, G., Nigam, A., Kaul, A., Nath, R., Singh, A.K.: Bring your own hand: how a single sensor is bringing multiple biometrics together. Soft Comput. 23(19), 9121–9139 (2018). https://doi.org/10.1007/s00500-018-03709-2
5. Lu, G.M., Adams, K.: Online palmprint identification system for civil applications. J. Comput. Sci. Technol. 20(1), 70–76 (2005)
6. Palma, D., Montessoro, P.L., Giordano, G., Blanchini, F.: Biometric palmprint verification: a dynamical system approach. IEEE Trans. Syst. Man Cybern. Syst. 49(12), 2676–2687 (2017)
7. Poinsot, A., Yang, F., Brost, V.: Palmprint and face score level fusion: hardware implementation of a contactless small sample biometric system. Opt. Eng. 50(2), 027002–027002 (2011)
8. Rehman, A., Harouni, M., Omidiravesh, M., Fati, S.M., Bahaj, S.A.: Finger vein authentication based on wavelet scattering networks. Comput. Mater. Contin. 72(2), 3369–3383 (2022)
9. Taleb-Ahmed, A.: Efficient palmprint biometric identification systems using deep learning and feature selection methods. Neural Comput. Appl. 34(14), 12119–12141 (2022)

10. Tran, N.C., Wang, J.H., Vu, T.H., Tai, T.C., Wang, J.C.: Anti-aliasing convolution neural network of finger vein recognition for virtual reality (VR) human–robot equipment of metaverse. J. Supercomput. **79**(3), 2767–2782 (2023)
11. Wang, J., He, Y., Zhu, J., Gao, X., Cui, Y.: Palm vein for efficient person recognition based on 2D Gabor filter. In: Biometric and Surveillance Technology for Human and Activity Identification X, vol. 8712, pp. 117–125. SPIE, May 2013
12. Wang, R., Wang, G., Chen, Z., Zeng, Z., Wang, Y.: A palm vein identification system based on Gabor wavelet features. Neural Comput. Appl. **24**(1), 161–168 (2013). https://doi.org/10.1007/s00521-013-1514-8
13. Yang, F., Ma, B., Xia Wang, Q., Yao, D., Fang, C., Zhao, S.: Information fusion of biometrics based-on fingerprint, hand-geometry and palmprint. In: 2007 IEEE Workshop on Automatic Identification Advanced Technologies, pp. 247–252. IEEE, June 2007
14. Zhao, L., Zhu, Q.: Edge detail enhancement algorithm for high-dynamic range images. J. Intell. Syst. **31**(1), 193–206 (2022)
15. IITD Palmprint Database. http://web.iitd.ac.in/~ajaykr/Database_Palm.htm

Crop Yield Prediction for Smart Agriculture with Climatic Parameters Using Random Forest

Ghassan Faisal[1,2] , S. Sreelakshmi[2(✉)] , and Vinod Chandra S. S.[2]

[1] Directorate of Agriculture in Wasit Governorate, Kut, Iraq
[2] Machine Intelligence Research Lab, University of Kerala, Thiruvananthapuram, India
{sreelakshmis,vinod}@keralauniversity.ac.in

Abstract. It is estimated that more than 50% of the world's population depends on agriculture, making it a pillar of the global economy. Variations in weather, climate, and other environmental factors are now a significant threat to the continued success of agriculture. Machine learning offers data-driven decision support capabilities for crop yield prediction that can help with decisions about which crops to produce and what to do when those crops are in the growing season. Even though there are machine learning-powered approaches reported in the literature, there are still a lot of avenues where the potential of machine learning can be exploited to its full potential. The proposed approach incorporates climatic elements to build models for crop yield prediction in the context of smart farming. The experiments conducted using random forest, one of the robust prediction models, reported an accuracy of 95.9%, comparable to state-of-the-art approaches. This study further strengthens the need for incorporating climate parameters to build better models for smart agriculture applications.

Keywords: Climate change · Crop-yield prediction · Machine learning · Random forest · Smart agriculture

1 Introduction

The agriculture industry will be more directly impacted by anthropogenic climate change than many industries because it relies on the weather. The type and magnitude of these effects are influenced by the evolution of the climate system and by the relationship between agricultural production and weather [1]. Crop yield forecasting is a crucial yet difficult issue that must be solved for sustainable intensification and effective use of natural resources. Many participants in the agri-food chain, such as farmers, agronomists, commodities merchants, and policymakers, find value in crop production estimates. The many crop-specific features, environmental factors, and management choices that affect crop output make it challenging to develop an accurate and explicable model [2,12].

Machine learning, a subset of Artificial Intelligence (AI) that focuses on learning, is a useful method that can estimate yields more accurately utilizing a variety of characteristics. Machine learning (ML) may extract knowledge from datasets by finding patterns, correlations, and associations. The models must be trained using datasets with results modeled based on prior knowledge. Multiple features are used to build the predictive model, and as a result, the parameters of the models are established using historical data during the training phase. A portion of the previous data from the training phase is used for performance evaluation during the testing phase [13].

Machine learning is a crucial decision-support tool for predicting agricultural yields, enabling choices about which crops to cultivate and what to do while they are in the growing season. Research on agricultural production prediction has been supported by the application of many machine learning algorithms and which offer a viable method for enhancing crop output projections using a data-driven or empirical modeling approach to uncover useful patterns and connections from input data [3]. An approximation function that connects characteristics or predictors to labels, such as crop yield, is produced by machine learning algorithms. Machine learning can use the results of other methods as features, much like statistical models. Additionally, these have several noteworthy advantages, including the ability to model non-linear relationships between various data sources, a tendency for performance to increase with the availability of more training data, and the ability to adapt to noisy data by using regularization techniques to reduce variance and generalization errors. Machine learning, therefore, has the potential to combine the advantages of various techniques, such as crop growth models and remote sensing, with data-driven modeling to make reliable crop yield predictions.

Higher agricultural crop output is the primary goal of crop yield prediction, and this goal is pursued by employing multiple tested models. Due to its effectiveness in many fields, including forecasting, fault detection, pattern identification, etc., machine learning is now utilized globally. When there is a loss due to unfavorable conditions, the machine learning algorithms also increase crop yield production rate. Regardless of the distracting environment, machine learning algorithms are used for the crop selection approach to reduce yield production losses. In this study, the authors employed Random Forest, one of the most powerful machine learning models, to forecast crop yields. In numerous sectors, the usefulness of RF has been demonstrated. For example, Su et al. [4] used it when studying dam displacement prediction, and Behrens et al. [5] used it for inflation prediction. Additionally, RF aggregates the predictions from numerous decision trees, scores the significance of each predictor variable, and, compared to other models, clearly outperforms them in handling high-dimensional and nonlinear situations.

2   Related Work

The usefulness of data mining methods for forecasting agricultural yields based on climatic input variables. The developed website is user-friendly, and all additional grains and areas selected for the analysis should have a reliability of prediction over 75%, suggesting a higher predictive performance Chinese website was created to forecast agricultural production using information from that region. Every user employs the crop option [6]. By merging many architectures for their distinct benefits, this work offers a viable route for deep learning techniques. This would greatly reduce the computational complexity and help to provide an accurate and reliable forecast of crop suggestions. It is discovered that the suggested ACO-IDCNN-LSTM recommender model is successful in selecting an appropriate crop. Future research will concentrate on auto-encoder-based deep learning mechanisms in order to improve outcomes and investigate the use of the suggested strategy on agricultural datasets [7]. Though the algorithms like deep-Q learning and an artificial neural network predict With comparable results, the suggested method beats all other algorithms thanks to transparency, little parameter adjustment, quick execution, and reduced over-fitting. Implementing strategies and procedures in a way that is as clear and understandable as feasible is the main goal of the suggested effort [8].

The results of this study indicate that the country's maize production is not only determined by the climatic variables of rainfall, soil moisture, and minimum and maximum temperatures. Non Climatic factors, such as better seeds, fertile soils, and agricultural techniques, among others, may also be to blame for the country's variation in maize output [9]. When comparing the outcomes of simulation models to ML approaches, ML techniques outperform simulation models. We also came to the conclusion that, provided sufficient input parameters are available, simulation models may be the most appropriate for estimating the results of those highly expensive and time-consuming tests. The genetics governing yield and waterlogging should be enhanced to adapt to climate change. Under future temperatures, genotypes with the best genetic characteristics and useful management choices might reduce the yield penalty by up to 15%. Our findings give plant breeders and agronomists information based on model-aided crop ideotypes to help them decide which traits are crucial for crop yields under challenging future climate change conditions [9]. It is advised that regression analysis be used with all model variables for the next AquaCrop calibration investigations. By keeping more thorough records of crop and field experimentation, the outcomes can be further enhanced [10]. Current and future maize production systems will be negatively impacted by climate change, but these effects can be reduced by using management and adaptation techniques like changing the planting dates (30 January in spring rather than mid-February), increasing nitrogen fertilizer by 20%, using 90% fertigation irrigation, and creating heat- and drought-tolerant hybrids [11].

Table 1. Description of crop yield prediction features.

Feature	Description
Nitrogen	For plant color and growth
Phosphorus	Reaches down the roots and helps produce blooms
Potassium	Promote all-around well-being
Temperature	influences the majority of plant processes, including germination, blooming, respiration, transpiration, and photosynthesis
Humidity	Is important to make photosynthesis possible
pH value	The pH range of 5.5 to 6.5 is ideal for plant growth since nutrients are readily available
Rainfall	establishes a crop's rate of growth from seed, as well as its maturity for harvest.

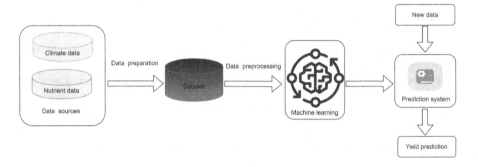

Fig. 1. Architecture of the proposed methodology.

3   Proposed Methodology

The machine learning method for agricultural yield prediction is described in detail in this section. During the first stage, we collected the raw data from the Kaggle website, which is freely downloadable, in our research. The dataset includes information on temperature, humidity, pH, rainfall, and nutrient variables that affect crop yield, such as nitrogen (N), phosphorus (P), and potassium (K). Brief descriptions of these features are listed in Table 1. For this investigation, a total of 22 crops were used. The second stage of the proposed methodology involves pre-processing the data to normalize it and eliminate the problems with class imbalance. Additionally, data is preprocessed to remove inaccurate and noisy unformatted data. The prediction model is then given the processed input. The performance measures of various machine learning models are compared, and it is discovered that random forests are better suited for predicting crop yields. Additionally, it is discovered from the literature that a significant amount of data is continuously produced with the development of digital technology in the field of agriculture. As a result, agriculture data is now a part of the big data universe. In this research, a method for analyzing crop yield prediction in a big data context is proposed. Figure 1 displays the block diagram of the Random Forest (RF) model for studying crop yield prediction. A combination model made up of numerous regression trees is called Random Forest Regression (RFR). A regression tree is described as a flowchart-like structure that begins with a sequence of terminal nodes and repeatedly divides the input dataset into increasingly homogeneous subsets at each node. The means of the response variables are collected within the terminal node into which the observation fell while the input variables are passed down the tree, a regression tree can be trained using training data to provide predictions for fresh observations. With replacement from the original data set, each regression tree in an RFR is built using a subset of training samples that have been separately chosen. Only a tiny sample of variables is randomly chosen to determine the split for each node per tree. With this approach, the model becomes more resilient and the variety across trees is increased to prevent over-fitting. By averaging the values over all trees, the final RFR predictor is created.

4   Result and Discussion

The performance analysis and explanation of the proposed model are provided in this section. A total number of seven features are used in this study including nutrient and environmental. Figure 2 illustrates the correlation between these different features. Figure 3 and 4 reports the distribution of the crop yield predicting features using box and distribution plots. The crop yield prediction random forest model performed remarkably well, with an accuracy of 0.95 according to the data. The effectiveness of random forest algorithms is displayed in Figs. 5 and 7. Here, we evaluate how well agricultural yield predictions made using machine learning models perform. The Linear regression (LR) algorithm scored 0.89, and the Decision Tree (DT) obtained 0.92 for the accuracy score. The Support Vector regression (SVR)scores were 0.90. The LR algorithm scored the lowest accuracy score out of all the baselines chosen, followed by SVR. The suggested RF model has attained an accuracy score of 0.95 which shows that the RF outperforms all the other baselines in this context for crop yield prediction. In Fig. 6, we display the results of the performance comparison of RF with baseline models. The outcomes unequivocally demonstrate that the random forest method outperforms others. Different measures are discussed below for validating the prediction accuracy of classifiers. The average squared error of the regression is computed using the Mean Squared Error (MSE). In other words, it calculates the square of the difference between the actual value and the predicted value before averaging them. The calculation formula of MSE is shown in (1), Yi represents the actual value and Yi bar denotes the anticipated value for an ith data and 'i' ranges from 1 to N. A smaller value of MSE suggests that the classifier's accuracy is good. The MSE value for a perfect classifier would be 0.

$$MSE = 1/N \sum_{i=1}^{N} (Yi - \bar{Y}i)_i^2 \qquad (1)$$

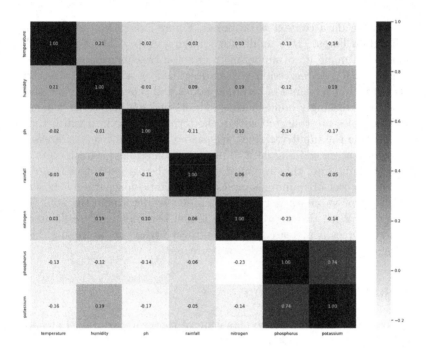

Fig. 2. The Pearson correlation heatmap showing the correlation between different features influencing crop yield prediction; darker colors show high correlation (close to 1), and lighter colors show low correlation (close to 0).

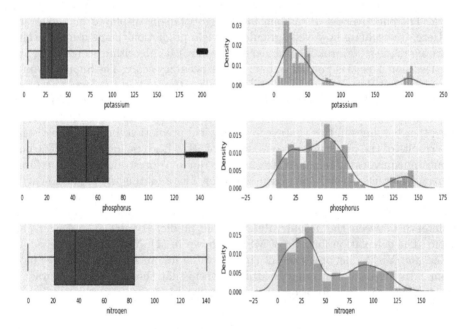

Fig. 3. Distribution of different nutrient parameters which affect crop yield prediction.

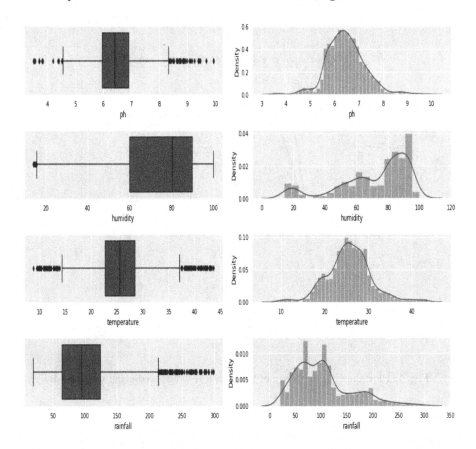

Fig. 4. Distribution of different climatic parameters affecting crop yield prediction.

The square root of MSE is called Root Mean Squared Error (RMSE). It is the prediction error standard deviation. First, the difference between the model output value and the actual target value is determined. The root of the mean value is then calculated once this difference has been squared and averaged over all data points. RMSE is expressed as in (2)

$$RMSE = \sqrt{1/N \sum_{i=1}^{N}(Yi - \bar{Y}i)2} \tag{2}$$

The average of the absolute differences between the target values and the forecasted values is used to calculate mean absolute error (MAE). It is written as in (3)

$$MAE = 1/N \sum_{i=1}^{N} \left| Yi - \bar{Y}i \right| \tag{3}$$

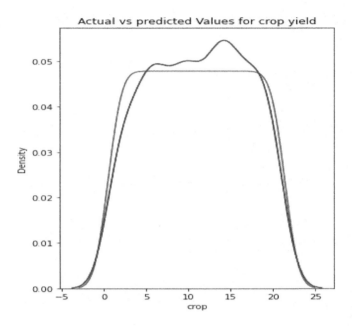

Fig. 5. Plot showing accuracy of random forest for predicting crop yields

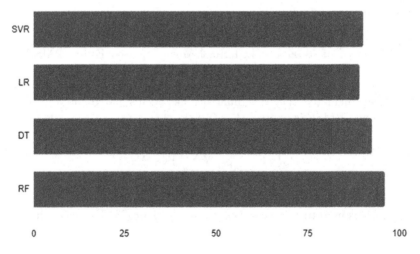

Fig. 6. Accuracy comparison of random forest with baseline models.

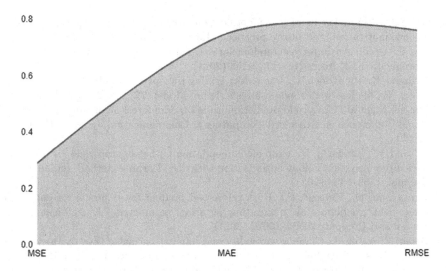

Fig. 7. Plot shows the performance of random forest.

5 Conclusion

Crop yield is affected by different soil properties, an unstable climate, pest infestations, and persistent crop illnesses. The ability of machine learning to enhance agriculture by combining various parameters to forecast productivity is stunning. By incorporating the random forest algorithm, we have proposed a reliable method for agricultural yield prediction in this work. This technology can be used to accurately predict crop yield based on both nutrient and environmental parameters. The suggested method can produce significantly more accurate predictions than its deep learning and standard machine learning competitors. Regarding the future study, we will assess it on more datasets gathered from various farmers in various places in order to further validate the recommended model's general efficacy, by preserving more thorough records of crop and field experimentation, the results could be further optimized.

Acknowledgements. The researchers and employees at the Machine Intelligence Research Lab at the Department of Computer Science, University of Kerala, are grateful to the authors for giving all the resources and assistance needed to conduct such a large-scale study.

References

1. Crane-Droesch, A.: Machine learning methods for crop yield prediction and climate change impact assessment in agriculture. Environ. Res. Lett. **13**(11), 114003 (2018)
2. Paudel, D., et al.: Machine learning for large-scale crop yield forecasting. Agric. Syst. **187**, 103016 (2021)

3. SS, V.C., Shaji, E.: Landslide identification using machine learning techniques: review, motivation, and future prospects. Earth Sci. Inf. **15**, 1–28 (2022)
4. Su, Y., et al.: An improved random forest model for the prediction of dam displacement. IEEE Access **9**, 9142–9153 (2021)
5. Behrens, C., Pierdzioch, C., Risse, M.: Testing the optimality of inflation forecasts under flexible loss with random forests. Econ. Model. **72**, 270–277 (2018)
6. Suresh, N., et al.: Crop yield prediction using random forest algorithm. In: International Conference on Advanced Computing & Communication Systems (ICACCS) (2021)
7. Mythili, K., Rangaraj, R.: Crop recommendation for better crop yield for precision agriculture using ant colony optimization with deep learning method. Indian J. Sci. Technol. 1583–6258 (2021)
8. Elavarasan, D., Vincent, P.D.R: A reinforced random forest model for enhanced crop yield prediction by integrating agrarian parameters. J. Ambient Intell. Humanized Comput. 10009–10022 (2021)
9. Atiah, W.A., et al.: Climate variability and impacts on maize (Zea mays) yield in Ghana, West Africa. Sci. Weather Inf. Tech. (SWIFT) Africa **148**, 185–198 (2021)
10. Yan, H., et al.: Crop traits enabling yield gains under more frequent extreme climatic events. Sci. Total Environ. **808**, 152–170 (2022)
11. Batool, D., et al.: A hybrid approach to tea crop yield prediction using simulation models and machine learning. Plants **11**, 1925 (2022)
12. Yasin, M., et al.: Climate change impact uncertainty assessment and adaptations for sustainable maize production using multi-crop and climate models. Environ. Sci. Pollut. Res. 18967–18988 (2022)
13. Van Klompenburg, T., Kassahun, A., Catal, C.: Crop yield prediction using machine learning: a systematic literature review. Comput. Electron. Agric. **177**, 105709 (2020)

Implementation of XGBoost Regression for Calories Burnt Prediction Using R

Vijay Gaikwad$^{(\boxtimes)}$ ⓘ, Moreshwar Khodke ⓘ, Sujay Shahare ⓘ, Pranav Terkar ⓘ, and Rohit Talmale ⓘ

Vishwakarma Institute of Technology, Pune, India
{vijay.gaikwad,moreshwar.khodke,sujay.shahare20,pranav.terkar20,
rohit.talmale20}@vit.edu

Abstract. Nowadays, people are curious about the exercises they do in order to maintain a balanced weight. The proposed paper aims to conduct a comparison of machine learning algorithms to predict the calories burn during a workout or any other exercises in R. To implement this, different machine learning models, including Decision Tree, Simple Linear Regression, and Multiple Linear Regression, Random Forest algorithm, Support Vector Machine (SVM), K-Nearest Neighbor, and XGBoost regressor are taken into consideration. There have been significant efforts to identify research publications that employed many supervised machine learning algorithms for prediction. Additionally, we compared eight research articles based on how accurate their chosen algorithms were, highlighting the best algorithm from them. At last, prediction of calories burnt during some exercise is carried out using the mentioned algorithms and a comparison is made using validation methods. In comparison to other models the XGBoost method, with a Root Mean Square Error (RMSE) score of 0.2965203, has provided the most accurate predictions.

Keywords: Machine learning · calories prediction · XGBoost · Support Vector Regressor · Random Forest · KNN Regression

1 Introduction

It is a known fact that the heartbeat and temperature of human body increase outstandingly while doing a workout or any other exercise. Some important factors to consider while measuring total energy burned are weight gain, weight loss, etc. If a person wishes to reduce their weight, they are required to burn more calories than they take in [1]. It is important to note that a lot many factors significantly affect how much calories are burned by a person each day. Out of them, some are person's height, age, gender, weight, heart rate, body temperature, duration of exercise, and so on [1, 2].

The proposed paper has deep roots in machine learning (ML) algorithms. Speaking of which, ML is a subset of artificial intelligence that incorporates certain algorithms to gain insight from present data and predict future outcomes based on that data. Broadly speaking, ML let us to feed specific algorithms with gigantic amount of data which in turn

M. Singh et al. (Eds.): ICACDS 2023, CCIS 1848, pp. 377–390, 2023.
https://doi.org/10.1007/978-3-031-37940-6_31

can be used to make certain predictions or classifications. Thus, the paper aims to predict the calories burnt during any exercise using machine learning algorithms using R. The proposed paper employs different machine learning models specifically, Random Forest (RF), Support Vector Machine (SVM), Multiple Linear Regression (MLR), and Simple Linear Regression (SLR), K-Nearest Neighbor, and XGBoost regressor. The prediction of calories burnt during some exercise is predicted using the mentioned algorithms and a comparison is made.

2 Literature Review

Many researchers are working on regression analysis and have worked on different techniques to obtain various results. Historically, the German mathematician Carl Friedrich Gauss was a dominant figure in this area. Together with Andrey Markov, he published a version of the Gauss–Markov theorem, which acts as the foundation of regression analysis [3].

By contrasting several machine learning methods, we aim to compare the available literature. Based on that, we examined the final eight papers that were chosen for comparison using the algorithms, what has been predicted and the accuracy of algorithms is also mentioned. To avoid selection bias, we extracted articles in which more than one algorithm were taken into consideration. In [4], a hybrid decision support system is proposed for early detection of heart disease using machine learning algorithms. The system combines feature selection and pre-processing techniques, which was then tested on the Cleveland heart disease dataset, and classifiers including naive Bayes, random forest, Support Vector Machine (SVM), logistic regression, and AdaBoost. With the random forest classifier, the system achieved the highest accuracy of 86.6%.

In [5], the authors identified different machine learning classifiers with the highest accuracy for heart disease prediction. The authors applied several supervised learning algorithms, including Multilayer perceptron (MLP), LR, DT, KNN, RF, and AdaboostM1 (ABM1). The study found that KNN, DT, and RF provided maximal accuracy, sensitivity, and specificity, with the RF method achieving 100% accuracy, sensitivity, and specificity as such. The use of machine learning techniques for predicting heart disease is also mentioned in [6], where nine different algorithms are used. In terms of accuracy and other evaluating metrics, the Random Forest performed the best with an accuracy of 95.60%.

In [7], the use of the Extreme Gradient Boosting (XGB) algorithm in credit evaluation based on big data is examined. The empirical study using open data from the Lending Club Platform in the USA shows that XGB has advantages in both classification performance feature selection.

In [8], a comparative analysis is conducted on two datasets using MATLAB platform. The algorithms that were considered for analysis are neural networks, support vector machines, and k-nearest neighbor. Among these, SVM predicted more accurately with an accuracy of 99.38%. In [9], a dataset of 1474 identified patients containing metabolomic data is considered for evaluating the likelihood of coronary artery disease. The algorithms that were taken into consideration include Lasso Regression (also known as L1 regression) and Random Forest classifier. From these two models, L1 regression performed more accurately with an accuracy value of 0.767.

In [10], the author discusses the results of classifying web pages using machine learning algorithms. Three algorithms—Artificial Neural Networks (ANN), AdaBoost, and Random Forests (RF)—are utilized for categorization. AdaBoost and Artificial Neural Network are less accurate than Random Forest (RF), which has a higher accuracy. And is more accurate at classifying than ANN & AdaBoost classifier. The Random Forest (RF) is quicker, typically completes in a few minutes, and is simpler to train.

In [11], a hybrid algorithm is proposed for short-term load forecasting which incorporates similar days selection, empirical mode decomposition (EMD), and long short-term memory (LSTM) networks. According to the research, the SD-EMD-LSTM method has proven to be a reliable technique for predicting electricity usage, making it a viable option for short-term load forecasting. In [12], artificial neural network (ANN) and logistic regression (L1) are compared in order to predict Cardiovascular Autonomic (CA) dysfunction in a dataset consisting of 2,092 individuals. Out of these, the ANN predicted accurately with an accuracy of 0.714.

In [13], performance of three decision tree algorithms is evaluated using data mining techniques. The algorithms are implemented on an educational dataset to predict a student's achievement in an exam. To predict how well students would perform on the final test, all algorithms are applied to the data from their internal assessments. The algorithms used are ID3, C4.5 and CART respectively. Additionally, the ID3 and CART algorithms displayed a tolerable degree of accuracy.

The utilization of data mining techniques to predict breast cancer survivability was conducted in [14]. It is done to forecast breast cancer survival using data mining techniques. To create prediction models, logistic regression is combined with data mining techniques like Artificial Neural Networks and Decision Trees. Accuracy, sensitivity, and specificity are three performance metrics that are used to assess the usefulness and efficiency of the algorithms. With an accuracy of 93.6%, the decision tree performed the best, according to the findings.

A thorough investigation of Table 1 reveals the algorithms that were taken into consideration with their performance as such. From this, we notice that most of the articles considered diverse algorithms for implementation. As a matter of fact, most of the studies aimed at predicting the likelihood of one disease.

As seen earlier, prediction of calories is carried out using seven regression algorithms including XGBoost regression, which is not the case with reviewed articles. Also, the whole paper was carried out using R programming language [15] which only a handful of reviewed articles have done. With insightful data visualization, we can infer as to how the data is distributed as well. Also, validation of proposed algorithms is accomplished via three methods namely, R-Squared error (RSE), Root Mean Square Error (RMSE) and Mean Absolute Error (MAE) which is not found in the reviewed articles.

Table 1. Comparison of referred articles with types of data and performance evaluation.

Reference	Prediction	Algorithms compared	Type of data	Data points	Performance of best one(s)	Best one (s)
Rani et al. (2021) [4]	Heart disease	SVM, Naïve Bayes, L1, RF, AdaBoost	Heart data	303	Accuracy (86.60%), Sensitivity (84.14%), Specificity (89.02%), Precision (88.46%), F-Measure (86.25%)	RF
Ali et al. (2021) [5]	Heart disease	MLP, KNN, DT, RF, LR, ABM1	Heart data	1025	Accuracy (100%), Sensitivity (1.000), Specificity (1.000)	RF
Katarya et al. (2021) [6]	Heart disease	RF, FCM, SVM, LDA, L1	Health related data	N/A	Accuracy (95.60%), RMSE (0.0439), MAE (0.0440), Precision (0.5528), Recall (0.9768)	RF
Sethi et al. (2017) [8]	Wine quality prediction	KNN, SVM, CNN	Wine quality data	6500	Accuracy (99.38%)	SVM
Forssen et al. (2017) [9]	Coronary artery disease	L1, RF	Metabolomic data	1474	Accuracy (0.767), AUC (0.765), Sensitivity (0.949), Specificity (0.339), PPV (0.760), NPV (0.750)	L1

(*continued*)

Table 1. (*continued*)

Reference	Prediction	Algorithms compared	Type of data	Data points	Performance of best one(s)	Best one (s)
Ansam (2017) [10]	Webpage classification	ANN, RF, AdaBoost	Health data from webpage	269	Precision (92.94%), Recall (82.24%), F-Measure (97.26%)	RF
Tang et al. (2013) [12]	Cardiovascular Autonomic Dysfunction prediction	ANN, L1	CA dysfunction data	2092	Sensitivity (0.777), Specificity (0.704), PPV (0.373), NPV (0.932), Accuracy (0.714)	ANN
Delen et al. (2005) [14]	Breast cancer prediction	ANN, DT	Breast cancer data	433272	Accuracy (93.6%), Sensitivity (0.9602), Specificity (0.9066)	DT

3 Methodology

3.1 Description of Dataset

The dataset which is being used is named fmendes-DAT263x-demos which is available in Kaggle repository available at https://www.kaggle.com/datasets/fmendes/fmendesda t263xdemos. It mainly contains two datasets with 15,000 data-points. The attributes with further description of input variables are provided in Table 2.

3.2 Data Preprocessing

In the considered dataset, there were no null or missing values for handling but some variables did have inconsistent values which would affect the prediction output and may also cause other problems. Then the p-values of the predictor variables were calculated and compared to check their significance in the prediction. The User_Id attribute had very low p-values and hence, was not considered further in the prediction models.

3.3 Splitting of Data into Training and Testing Data

The given dataset is split into training and testing data for the model, with the former being used to train the prediction model and the latter to evaluate the model's effectiveness. In

Table 2. Description of input variables of considered dataset.

Sr. No.	Observation	Description	Values
1	Gender	Gender of subject	Male/Female
2	Age	Age (years)	Continuous
3	Height	Height (cm)	Continuous
4	Weight	Weight (kg)	Continuous
5	Duration	Duration of exercise in minutes	Continuous
6	Hart_Rate	Pulse per minutes	Continuous
7	Body_Temp	Temperature of body after exercise in $^{\circ}$C	Continuous

this experiment, the dataset is split in half, with 30% used for testing and 70% used for training.

3.4 Data Visualization

For better understanding the distribution of the data and correlation between the target and predictor variables, inbuilt plotting methods in R were used as the data visualization tool. The study used the ggplot2 library in R to create data visualizations. Histograms and scatterplots were used to show the distributions and relationships between the variables. The 'hist()' function was used to plot histograms for the variables 'calories', 'heart rate', and 'duration', while the 'plot()' function was used to plot scatterplots. Figure 1 depicts the distribution of data as such.

3.5 Algorithms Used

Simple Linear Regression (SLR). It is the simplest form of statistical method which is used to predict the results based on the value of a single independent variables also known as input variable. Also, the variable that we wish to predict is called as predictor or dependent variable. Equation of SLR is given by

$$y = b_0 + b_1x \tag{1}$$

where, y is the dependent variable, x is the independent variable, b_0 is the y-intercept of the line, the point where the line crosses the y-axis, and b_1 is the slope of the line, indicating the rate of change of y with respect to x.

Multiple Linear Regression (MLR). Like SLR, multiple linear regression is also a statistical method which is used in prediction of results based on the value of multiple input variables. Like SLR, the algorithm is also a linear model that finds the best fit line for the given data points of the training set. Equation of MLR is given by

$$y = b_0 + b_1x_1 + b_2x_2 + \ldots + b_nx_n \tag{2}$$

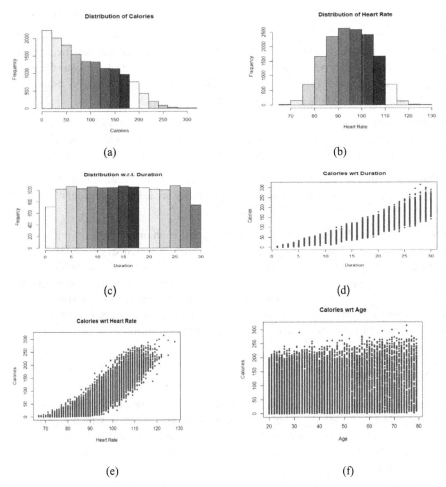

Fig. 1. Data visualization using inbuilt plotting methods in R: (a) distribution of "Calories" vs frequency, (b) distribution of input variable "Heart_Rate" vs frequency, (c) distribution of input variable "Duration" vs frequency, (d) distribution of input variable "Duration" with respect to "Calories", (e) distribution of input variable "Heart_Rate" with respect to predictor "Calories" and (f) distribution of input variable "Age" with respect to predictor "Calories."

where y is the dependent variable, $x_1, x_2,..., x_n$ are the independent variables, b_0 is the y-intercept of the line, the point where the line crosses the y-axis, and $b_1, b_2,..., b_n$ are the coefficients for each independent variable, indicating the effect of each variable on the dependent variable.

Decision Tree (DT). Decision trees develop a tree-like structure when building regression or classification models. It separates a dataset into more manageable parts. The accompanying DT is developing gradually at the same time. In fact, the root node in a

DT is the highest decision node that correlates to the best predictor. Both numerical and numerical data can be operated by DTs.

Random Forest (RF). It is an algorithm which is based on tree-based classification. RF works very well with regression as well as classification problems. Clearly as the name suggests, it selects random samples from the training dataset and generates n-numbers of decision trees considering only a subset of attributes.

Support Vector Regression (SVR). As the name suggests, a supervised learning approach that predicts discrete values is support vector regression which classifies cases using a separator. Support Vector Regression utilizes the same principle as the SVM.

K-Nearest Neighbor (KNN) Regression. KNN regression algorithm is used for classifying alike cases. It generates outputs only when they are asked to. Hence, it is known as a lazy learner since there is no learning cycle.

Extreme Gradient Boosting (XGBoost). XGBoost is an ensemble learning method. It works on a specific technique known as boosting, in which trees are constructed in succession, with each tree aiming to reduce the mistakes of the one before it. The residual errors are updated when each tree learns from its forerunner. Thus, subsequently generated tree learns from an updated version of the residuals in the sequence. One drawback is that having a large number of trees might cause data to be overfit. Hence, it is required to deliberately choose the boosting stopping criterion.

3.6   Validation Methods

R-Squared Method. In statistics, the coefficient of determination, which is designated as R-squared (R2), is the proportion of variance of an input variable, which is described by one or more input a regression model's variables. It elucidates the region in which the variation of one variable contributes to the variance of the second variable. The formula for R-squared is

$$R^2 = 1 - \frac{SS_{res}}{SS_{tot}} \tag{3}$$

where SS_{res} is the residual sum of squares, and SS_{tot} is the total sum of squares (the sum of the squared differences between the actual values and the mean of the actual values).

Root Mean Square Error (RMSE). It is one method of determining how inaccurately a model predicts the data. It is the square root of the mean of the squares of all the errors, to put it another way. The standard deviation () of the distribution of errors can be approximated well by the RMSE, which is why it is employed. The formula for RMSE is

$$RMSE = \sqrt{\frac{1}{n} \sum_{i=1}^{n} (y - \hat{y})^2} \tag{4}$$

where n is the number of observations, Σ is the sum over all observations, $y^{\hat{}}$ is the value predicted by the model for a given observation, and y is the true value for that observation.

Mean Absolute Error (MAE). It is also another standard which is used to calculate the accuracy for continuous variables. It is the mean of the absolute values of the individual prediction errors of the whole test set. The formula for MAE is

$$MAE = \frac{1}{n} \sum_{i-1}^{n} |y - \hat{y}| \tag{5}$$

where n is the number of observations, Σ is the sum over all observations, $y^{\hat{}}$ is the value predicted by the model for a given observation, and y is the true value for that observation.

4 Result and Discussions

The study so far has been focusing on the predictive analysis of various algorithms on the fmendes-DAT263x-demos which is available in Kaggle repository. The accuracy for some algorithms is greater while some models give and less for other models. The SLR model considers the duration (Duration) and heart rate (Heart_Rate) which is a significant variable of the above explained data-frame. Therefore, The SLR model is trained for those two variables separately and further prediction is done. Figure 2 gives the linear model interpreting the correlation between the dependent variable "calories" and the independent variables "Duration" and "Heart_Rate."

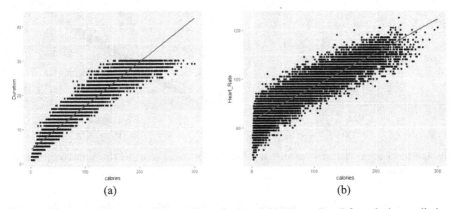

(a)   (b)

Fig. 2. Linear model by comparing (a) "Duration" and (b) "Heart_Rate" for calories prediction.

The output of the Simple Linear Regression gives the predicted values which in turn are compared with the actual existing values. SLR provides very less accuracy where we get R2 = 0.9146842 for the first linear model which makes it less significant for prediction.

The Multiple Linear Regression considers all the significant features that are selected through comparison of the p-values. Similar to SLR, the accuracy of MLR is quite low as compared to the other algorithms with the R2 value as 0.9676041 which is greater than SLR and Decision Tree. Figure 3 shows the contrast between the actual and the predicted values obtained from Multiple Linear Regression with the help of linear model.

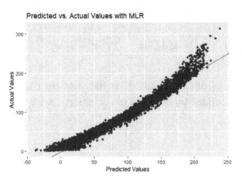

Fig. 3. Actual vs predicted values for MLR.

Another algorithm that was being employed to carry out the prediction was Decision Tree regression algorithm. Figure 4(a) shows the correlation between the actual and the predicted values obtained from DT regression.

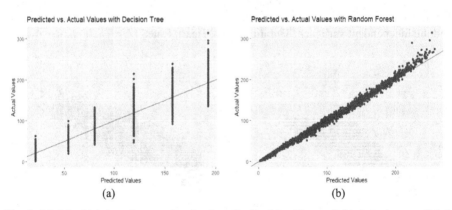

Fig. 4. Model of (a) Actual vs predicted values for Decision Tree and (b) Actual vs predicted values for Random Forest.

The output of the DT regression gives the predicted values which in turn are compared with the actual existing values. DT provides very less accuracy where we get an RMSE value of 19.2256527 for the model which makes it less significant for prediction as such.

Random Forest also gives an accurate model where the predictions in the output are having values closer to the actual values and the RMSE value is equal to 4.1162736, which is pretty good, compared to other algorithms discussed above. The comparison for the actual vs the predicted values for the RF model can be seen in Fig. 4(b).

Support Vector Regression (SVR) considered 162 support vectors mapped in the hyperplane. Using the best hyperplane, it gives a comparatively accurate model with the RMSE value equal to 3.1897168. Figure 5(a) shows the comparison between the actual and predicted values obtained through predictive analysis with Support Vector Regression.

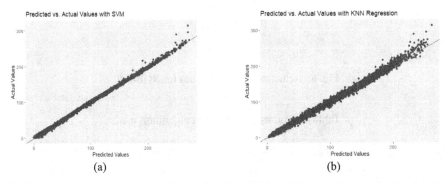

(a)   (b)

Fig. 5. Model of (a) Actual vs predicted values for SVM and (b) Actual vs predicted values for KNN regression.

The KNN regressor resamples the results across tuning parameters and gives output as shown in Table 3.

Table 3. Evaluation of hyperparameter k on KNN regressor using RMSE, R-Squared, and MAE.

Hyperparameter "k"	RMSE value	R-Squared (R^2) value	MAE value
5	6.5571	0.9891	4.8353
7	6.2628	0.9901	4.6004
9	6.0924	0.9907	4.4572

These results lead to selection of value for k as 9 due to the smallest Root Mean Square value for k = 9. Figure 5(b) shows the predicted vs actual plot for KNN regression model. Xtreme Gradient Boosting (XGBoost) model so far gives the most accurate prediction for the considered dataset with the RMSE value of 0.2965503. The iterative model comprises 1000 iterations and has maximum depth of 2. The output predictions and the actual price values can be seen plotted in Fig. 6.

The accuracy of the algorithms used is done with the help of the R-Square, values for Mean Absolute Error (MAE) and Root Mean Square Error (RMSE). Table 4 demonstrates the accuracy of the algorithms used.

The graphical representation of performance evaluation is represented in Fig. 7. It is clear that the XGBoost algorithm has made the most accurate predictions as compared to other models with a RMSE of 0.2965503.

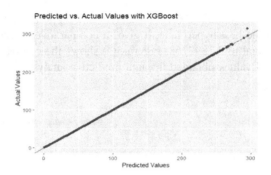

Fig. 6. Actual vs predicted values for XGBoost.

Table 4. Evaluation metrics for algorithms used.

Sr. No.	Algorithm	R-Squared	RMSE	MAE
1	SLR	0.9146	18.7988	13.7419
2	MLR	0.9676	11.4160	8.4042
3	DT	0.9042	19.2256	14.5097
4	RF	0.9956	4.1163	2.4192
5	SVM	0.9973	3.1897	2.5312
6	KNN	0.9973	3.1897	NA
7	XGB	0.9999	0.2965	NA

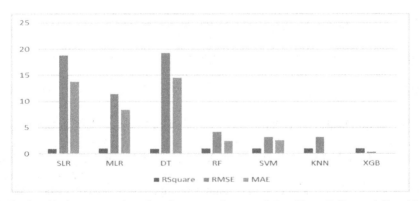

Fig. 7. Graphical representation of performance of proposed algorithms: R-Squared, Root Mean Square (RMS) and Mean Absolute Error (MAE).

5 Conclusion

This paper has reviewed, analyzed and examined most of the current research on note-worthy attributes and algorithms of the calories burned prediction system, used to predict calories burned during some exercise. This paper also studies the attributes and algorithms applied by the previous researchers to predict the calories and also made some amends to overcome their flaws. The paper also shows the everlasting potential of the SVM, Random Forest and XGBoost algorithms while predicting the burned calories during exercise. As mentioned earlier, the dataset which was being used is named fmendes-DAT263x-demos which is available in Kaggle repository. The prediction of calories burnt during some exercise is predicted using the mentioned algorithms and a comparison is made where it was found that XGBoost predicted calories most accurately. This indicates that the algorithm can be preferred for any regression analysis concerned with value prediction.

Acknowledgments. We would like to thank Dr R. Jalnekar, Honorable Director of the Institute for providing us with the opportunity to work on this paper.

References

1. Cox, C.: Role of physical activity for weight loss and weight maintenance. Diab. Spectr.: Publ. Am. Diab. Assoc. **30**(3), 157 (2017)
2. Jaiswal, A.: A study on the intake and expenditure of calories among the manufacturing workers. Hum. Biol. Rev. **1**, 1–10 (2012)
3. Shaffer, J.P.: The Gauss-Markov theorem and random regressors. Am. Stat. **45**, 269–273 (1991). https://doi.org/10.2307/2684451
4. Rani, P., Kumar, R., Ahmed, N.M.O.S., Jain, A.: A decision support system for heart disease prediction based upon machine learning. J. Reliable Intell. Environ. **7**(3), 263–275 (2021). https://doi.org/10.1007/s40860-021-00133-6
5. Ali, M.M., Paul, B.K., Ahmed, K., Bui, F.M., Quinn, J.M.W., Moni, M.A.: Heart disease prediction using supervised machine learning algorithms: performance analysis and comparison. Comput. Biol. Med. **136**, 104672 (2021). ISSN 0010–4825. https://doi.org/10.1016/j.compbiomed.2021.104672
6. Katarya, R., Meena, S.K.: Machine learning techniques for heart disease prediction: a comparative study and analysis. Heal. Technol. **11**(1), 87–97 (2020). https://doi.org/10.1007/s12553-020-00505-7
7. Li, H., Cao, Y., Li, S., Zhao, J., Sun, Y.: XGBoost model and its application to personal credit evaluation. IEEE Intell. Syst. **35**(3), 52–61 (2020). https://doi.org/10.1109/MIS.2020.2972533
8. Sethi, K., Gupta, A., Gupta, G., Jaiswal, V.: Comparative analysis of machine learning algorithms on different datasets. In: International Conference on Innovations in Computing (ICIC), pp. 87–91 (2017)
9. Forssen, H., et al.: Evaluation of machine learning methods to predict coronary artery disease using metabolomic data. In: Stud Health Technol Inform, vol. 235, pp. 111–115 (2017). PMID: 28423765
10. AbdulHussien, A.A.: Comparison of machine learning algorithms to classify web pages. Int. J. Adv. Comput. Sci. Appl. **8**(11) (2017). https://doi.org/10.14569/IJACSA.2017.081127

11. Zheng, H., Yuan, J., Chen, L.: Short-term load forecasting using EMD-LSTM neural networks with a Xgboost algorithm for feature importance evaluation. Energies **10**, 1168 (2017). https://doi.org/10.3390/en10081168

12. Tang, Z.H., Liu, J., Zeng, F., Li, Z., Yu, X., Zhou, L.: Comparison of prediction model for cardiovascular autonomic dysfunction using artificial neural network and logistic regression analysis. PLoS ONE **8**(8), e70571 (2013). https://doi.org/10.1371/journal.pone.0070571

13. Priyam, A., Gupta, R.K., Rathee, A., Srivastava, S.K.: Comparative analysis of decision tree classification algorithms. Int. J. Curr. Eng. Technol. **3**(2), 334–337 (2013)

14. Delen, D., Walker, G., Kadam, A.: Predicting breast cancer survivability: a comparison of three data mining methods. Artif. Intell. Med. **34**(2), 113–127 (2005). https://doi.org/10.1016/j.artmed.2004.07.002

15. James, G., Witten, D., Hastie, T., Tibshirani, R.: An Introduction to Statistical Learning: with Applications in R. Springer, New York (2013). https://doi.org/10.1007/978-1-4614-7138-7

Text Data Augmentation Using Generative Adversarial Networks, Back Translation and EDA

Premanand Ghadekar, Manomay Jamble⑩, Aditya Jaybhay⑩, Bhavesh Jagtap$^{(\boxtimes)}$ ⑩,
Aniruddha Joshi⑩, and Harshwardhan More⑩

Vishwakarma Institute of Technology, Pune, India
{premanand.ghadekar,manomay.jamble20,aditya.jaybhay20,
bhavesh.jagtap20,aniruddha.joshi20,harshwardhan.more20}@vit.edu

Abstract. Data or information augmentation techniques have been explored in NLP to create greater textual information for training. But, the overall performance advantage of present strategies is regularly marginal. Text augmentation can generate additional variations of the original text and improve the generalization ability of a machine learning model that processes natural language text data. The paper represents the performance of Generative Adversarial Networks for the overall performance in text classification. The results show that the Generative models give the best overall performance advantage over the EDA or back translation accuracy. The proposed models process text augmentation using GANs compared to methods like Easy Data Augmentation and back translation. The goal of EDA is to generate new, semantically similar sentences from an existing sentence, to increase the size and diversity of a dataset for training natural language processing models. Back Translation is an improved method with increased accuracy. Generative Adversarial Networks (GANs) generate new, synthetic data that is similar to a given training dataset using neural network architecture. In the context of text augmentation, GANs can be used to generate new, realistic text similar to a given text dataset.

Keywords: Text augmentation · NLP · GANs · Generative models · neural networks

1 Introduction

Text augmentation is a technique used to artificially increase the size of a dataset of text by generating new, synthetic text that is similar to the original text in various ways, depending on the specific technique used. The main purpose of text augmentation is to enhance the performance of natural language processing (NLP) models by providing them with more diverse and varied training data.

There are a variety of text augmentation techniques that can be used, including.

Synonym replacement: This technique involves replacing certain words in the text with synonyms to create new, semantically equivalent sentences.

M. Singh et al. (Eds.): ICACDS 2023, CCIS 1848, pp. 391–401, 2023.
https://doi.org/10.1007/978-3-031-37940-6_32

Random insertion: This technique involves inserting random words or phrases into the text to create new, semantically similar sentences.

Random deletion: This technique involves deleting random words or phrases from the text to create new, semantically similar sentences.

Text rewriting: This technique involves rephrasing or rewording the text to create new, semantically similar sentences.

Text generation: This technique involves using machine learning models, such as generative adversarial networks (GANs) or language models, to generate new, synthetic text that is similar to the original text.

Text augmentation can be an effective way to improve the performance of NLP models, particularly when working with small or limited datasets. However, it is important to use text augmentation techniques carefully, as generating synthetic text that is not representative of the original data can lead to models that are biased or perform poorly on real-world tasks.

2 Literature Review

[1] The paper describes that NLP is at an early stage. Another option is to apply Data augmentation compared to computer vision. They highlighted the various promising ideas and the crucial differences. But it hasn't been tested in NLP yet. Data augmentation is facilitated by a model that describes the tools, like regularization, controllers, offline and online augmentation pipelines, and consistency.

[2] As the paper states, because of the challenges posed by the discrete nature of language data, this area remains quite unexplored. The paper focuses on the survey and research. As a result, no implementation is carried out. It motivates data augmentation for NLP. Next, the paper states the features used in NLP applications and tasks.

[3] In the proposed system, sentiment analysis was implemented and compared to the three cases. They also conducted an error analysis. EDA i.e., exploratory data analysis of the key techniques that they have used and which ported their process. The datasets that they have used are from SST (Stanford Sentiment Treebank) and CrisisLex. NLP is the only deep learning concept applied by them.

[4] In this paper, the author describes and focuses mostly on an open-source tool that can aid users in creating automated test data for numerous programs under test. The model's tool works by clustering input data from a corpus folder and generating generative models for each of the clusters. The models have a recurrent NN (Neural Network) structure and are trained and sampled in parallel using Tensorflow.

[5] This paper proves that data augmentation mostly uses Generative Adversarial Networks. Low-quality generated images and unstable training is two drawbacks. To address those issues, this research was presented. To boost the ability to produce details and produce dependent scenes and objects, distance components in the image are connected.

[6] Random insertion, synonym replacement, random swap, and deletion are effective, straightforward procedures that boost EDA. The paper demonstrates that EDA enhances performance for both convolutional and recurrent neural networks on five

text categorization tasks. EDA demonstrates robust results for comparatively smaller datasets.

[7] Deep convolutional neural networks have outperformed various tasks associated with computer vision. However, these networks are dependent on huge amounts of data to avoid overfitting. Unfortunately, many real-life applications, such as medical image analysis, do not have access to extensive data. This paper explains Data Augmentation, which provides aid to the problem of limited data.

[8] This research paper basically analyzes and compares methods in image classification, traditional image transformations such as rotating, cropping, zooming, histogram-based approaches, and style transfer, as well as typical examples of data augmentation. Based on image style transfer, a new data augmentation method is presented. The method creates new image data of high-quality resolution that combines the features of an original image with others. To improve the efficiency of the training process, pre-training is done with the help of newly created images.

3 Problem Statement

In many natural language processing tasks, the performance of machine learning models depends on the learning and quality of the training data. However, collecting and annotating large amounts of data can be time-consuming and costly and may not be feasible for all applications. This challenges researchers and practitioners seeking to build accurate and effective models for tasks such as text classification, language translation, and language generation.

One potential solution to this problem is to use data augmentation techniques, which involve generating synthetic data that is similar to real data to increase the data. However, traditional augmentation techniques such as adding noise or applying random transformations may not be effective for text data, as they can easily result in nonsensical or meaningless sentences. This raises the question of whether there are alternative approaches for augmenting text data that are more effective and better suited to natural language processing tasks.

The proposed system aims to explore different text augmentation techniques, including synonym replacement, random deletion, back translation and GANs, evaluate each technique's effectiveness on the tasks and dataset, and determine which techniques are most effective at improving the model's performance.

4 Algorithm

The steps required to perform random deletion as a text augmentation technique are:

1. Select a text sample to work with. This can be a single sentence or a longer piece of text.
2. Divide the text into smaller units, such as individual words or phrases.
3. Select a random unit from the text and delete it.
4. Repeat the process a specified number of times, either by selecting additional random units to delete or by generating multiple versions of the text with different units removed.

5. Use the resulting modified text samples as additional data for training a machine learning model (Fig. 1).

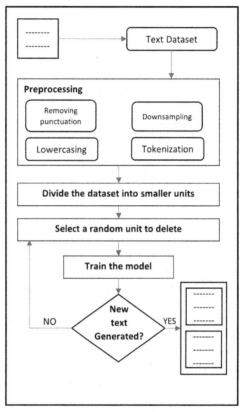

Fig. 1. Dataset is given as an input which was well augmented before. A flowchart gives the complete workflow and results are getting as per the workflow.

It is important to note that random deletion may not always be the most effective text augmentation technique, as it can significantly alter the meaning of the original text.

To perform text augmentation using back translation, the following steps were followed:

1. Select a text sample to work with. This can be a single sentence or a longer piece of text.
2. Identify the languages that you want to translate the text between. You will need access to machine translation systems for these languages.
3. Translate the text to any different language using a machine translation system.
4. Translate the text back to the original language using a different machine translation system.
5. Use the resulting translated text as an augmented version of the original text.

6. Repeat the process a specified number of times, either by generating multiple versions of the text with different translations or by incorporating the translated text into the model's training process through techniques such as data sampling or augmentation on the fly.

It is important to note that the quality of the augmented data generated using back translation will depend on the accuracy of the machine translation systems used.

To assisting conduct an experiment on text augmentation using GANs and transformers, these general steps were followed:

1. Collect and preprocess a dataset of text that you want to use for training the GAN. This may involve cleaning and normalizing the text, and possibly also generating additional data by performing tasks such as text summarization or translation. Cleaning text includes removing stopwords, punctuations and other characters which will not contribute to the performance of the model while normalizing text means reducing the randomness and bringing the text in a standard format.
2. Train a transformer-based GAN on the dataset. This may involve implementing the GAN in a programming language such as Python, using a deep learning framework such as TensorFlow, and adjusting various hyperparameters such as the learning rate and batch size.
3. Evaluate the performance of the GAN on the task of generating synthetic text. This may involve using various metrics such as perplexity or BLEU score to measure the quality of the synthetic text and comparing it to the real text in the dataset.
4. Iterate on the design of the GAN and the training process as needed, adjusting the model architecture and hyperparameters as needed to improve the performance of the GAN.
5. Generate additional synthetic text data for text augmentation using the trained GAN. This may involve using the GAN to generate a large number of synthetic text samples and using these samples to augment the original dataset of text.
6. Evaluate the impact of the synthetic text on downstream tasks such as language modelling or text classification, and compare the results to those obtained using only the real text data (Fig. 2).

It is important to keep in mind that the process of experimenting with GANs for text augmentation can be quite complex and time-consuming, and may require significant expertise in deep learning and natural language processing. It may also be necessary to try several different approaches and configurations before finding a GAN architecture and training process that produces good results.

The loss function in the GAN model used in the project, with the concept of binary-cross entropy loss can be written as:

$$L(y', d) = [d * \log y' + (1 - d) * \log(1 - y')]. \tag{1}$$

where, d = original data

y' = reconstructed data.

While training the discriminator, the data coming from $p_{\text{data}}(x)$ is $d = 1$ (real data) and $y' = K(i)$.

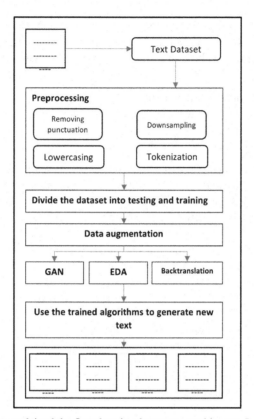

Fig. 2. The figure has explained the flowchart in a better executable way. It describes the end-to-end flow of the process.

Substituting this in the above loss function the output is,

$$L(K(i), 1) = \log(K(i)) \tag{2}$$

And for data coming from the generator, the label is $d = 0$ (fake data) and $\acute{y} = D(G(z))$. So, in this case,

$$L(K(G(z)), 0) = \log(1 - K(G(z))) \tag{3}$$

Now, the discriminator tries to correctly distinguish between synthesized and original datasets. So Eqs. (1) and (2) need to be maximized and we get the final loss function for the discriminator as,

$$L^{(K)} = \max\left[\log(K(i)) + \log(1 - K(G(z)))\right] \tag{4}$$

Here, the generator is in the opposite role as that of the discriminator. So, the equations need to be minimized and the loss function is,

$$L^{(G)} = \min\left[\log(K(i)) + \log(1 - K(G(z)))\right] \tag{5}$$

Combining Eqs. (4) and (5) and write as,

$$L^{(G)} = \min \max \left[\log(D(i)) + \log(1 - K(G(z))) \right] \qquad (6)$$

The expectation of the above equation considering the entire dataset will be given as,

$$\min - \max V(K,G) = \min \max \left(E_{X \sim Pdata(X)} \left[\log K(i) \right] + E_{z \sim Pz(z)} \left[\log(1 - K(G(z))) \right] \right)$$
$$\qquad (7)$$

5 Existing System

The existing systems contain methods for text augmentation like random deletion, and random swapping, and use neural networks. These methods are simpler to implement and require fewer computing resources. However, they may not produce as high-quality or diverse examples as GANs, and may not be as effective at improving the performance of a machine-learning model.

In summary, GANs can be a powerful tool for augmenting text data, but they may not always be the best choice depending on the specific needs and resources of a given system. It is important to consider the trade-offs between the various approaches and choose the one that is most appropriate for the task at hand.

6 Proposed System

The proposed system is a text generation model that takes the input of a series of texts and returns modified text, retaining its context. In the case of a binary text classification problem, a case of an imbalanced dataset can come up where the count for several positive data points is way larger than that of negative data points. One option is to delete or remove the excess data but this can lead to data loss. The other option is to feed the negative data into the text generation model which will return a large amount of negative data remembering the context which you can then use in your data modelling process.

The proposed system is a GAN and hence it naturally consists of two networks: A generator and a Discriminator.

Below is the representation of these two networks (Fig. 3):

The generator network consists of 3 layers. The first layer is an embedding layer which takes tokenized sequences as input. The second layer is an LSTM layer. LSTM stands for Long Short Term Memory and this layer helps us preserve more information by passing a portion of initial values till the end of the network (Fig. 4).

The generator is trained to create synthetic data that is indistinguishable from the real data, while the discriminator is trained to properly discern between real and synthetic data in a zero-sum game framework. Both the discriminator and the generator are trained using a mix of actual and artificial data.

The discriminator becomes better at telling the difference between genuine and fake data as the training goes on, while the generator gets better at producing synthetic data that is comparable to actual data. As a consequence, a generator and a discriminator can successfully generate artificial data that is very identical to the original data and tell the difference between the two.

```
Model: "sequential_15"

Layer (type)                 Output Shape              Param #
=================================================================
embedding_10 (Embedding)     (None, None, 128)         1280000

lstm_10 (LSTM)               (None, 128)               131584

dense_14 (Dense)             (None, 1000)              129000

=================================================================
Total params: 1,540,584
Trainable params: 1,540,584
Non-trainable params: 0
```

Fig. 3. The figure gives an idea about the layers, algorithms, and denseness of layers used to develop it.

```
Model: "sequential_16"

Layer (type)                 Output Shape              Param #
=================================================================
embedding_11 (Embedding)     (None, None, 128)         128000

lstm_11 (LSTM)               (None, 128)               131584

dense_15 (Dense)             (None, 1)                 129

=================================================================
Total params: 259,713
Trainable params: 259,713
Non-trainable params: 0
```

Fig. 4. The figure gives an idea about the layers, algorithms, and denseness of layers used to develop it.

7 Scope of the Project

This system gives an end-to-end solution to generate text data using various Machine learning and deep learning techniques. The performance of two algorithms with the best performance, random change and random deletion, has been analyzed using accuracy data from convolutional neural networks (Deep Learning). GAN is the uniqueness in terms of the techniques that the paper is implemented with. This will lead to innovations in the respective domain in terms of techniques. So, the model now can generate the text data by previously generated data. It can lead to avoiding problems such as Overfitting, and randomness in any such Machine learning models.

8 Limitations

One potential drawback of random deletion is that it can alter the meaning of the original text, particularly if a large number of units are deleted. This can result in modified text samples that are significantly different from the original and may not convey the same meaning. It is important to consider the potential impact of random deletion on the meaning of the text when using this technique for text augmentation.

One potential drawback of back translation is that it can alter the meaning of the original text, particularly if the translations are not accurate. This can result in modified text samples that are significantly different from the original and may not convey the same meaning. It is important to consider the potential impact of backtranslation on the meaning of the text when using this technique for text augmentation.

This can make it difficult to achieve good performance with GANs, especially for tasks with large and diverse datasets. GANs are prone to mode collapse, where the generator produces only a limited set of outputs rather than a diverse range of outputs. This can lead to repetitive or nonsensical generated text, which may not be suitable for many applications. Evaluating the quality of the generated text with GANs can be difficult, as there is no clear metric for measuring the realism or diversity of the output.

9 Results and Discussion

The paper was successfully able to improve the performance of the model to a large extent after augmenting the dataset using this technique. Evaluating a dataset consisting of course reviews which were highly imbalanced consisting of a large number of positive samples but very less negative ones. In the first method, the algorithm simply resampled the data using downsampling which caused a very large data loss. The following is the performance of that model (Fig. 5):

	precision	recall	f1-score	support
-1	0.90	0.96	0.93	9487
0	0.91	0.90	0.90	14793
1	0.96	0.94	0.95	24822
accuracy			0.93	49102
macro avg	0.92	0.93	0.93	49102
weighted avg	0.93	0.93	0.93	49102

Fig. 5. Based on the evaluation matrix, precision, recall, and f1-score, support such parameters are drawn out and it is saying a lot about the model evaluation.

In this model, a count vectorizer was used followed by a random forest ensemble to predict the results.

	precision	recall	f1-score	support
-1.0	0.85	0.99	0.91	6109
0.0	0.83	0.98	0.90	9134
1.0	1.00	0.99	1.00	299298
accuracy			0.99	314541
macro avg	0.89	0.99	0.94	314541
weighted avg	0.99	0.99	0.99	314541

Fig. 6. Based on the evaluation matrix, precision, recall, f1-score, and support, such parameters are drawn out and it is saying a lot about the **GAN** model evaluation.

After implementing GANs and increasing the data of the class with very fewer data, a very high increase in performance was witnessed when evaluated using the same machine-learning process as above (Fig. 6).

The results thus show that Generative Adversarial Networks provided a better performance when used as a data augmentation technique as compared to methods like back-translation, easy data augmentation and other traditional methods.

10 Conclusion

In conclusion, this research has shown that using GANs to augment text data can significantly improve the performance of a classifier. By training a GAN to generate synthetic data that is similar to the real data, the model was able to effectively increase the size of the training dataset and improve the generalization ability of the classifier. The results suggest that GANs may be a useful tool for text augmentation in a variety of applications, and further research is needed to fully explore the potential of this approach. Overall, this work highlights the potential of GANs for improving the performance of machine learning models in natural language processing tasks.

11 Future Scope

There are several directions in which this research can be extended in the future. One potential avenue for future work is to explore the use of GANs for text augmentation in other natural languages processing tasks, such as machine translation, text classification, and language modelling. Additionally, it would be interesting to study the effect of different GAN architectures and training procedures on the performance of the augmented classifier, as well as the potential for combining GANs with other data augmentation techniques.

Another interesting direction for future research is to investigate the potential for using GANs to augment data in other domains beyond natural language processing, such as image or audio data. It is also worth exploring the potential for using GANs in a semi-supervised learning setting, where a small amount of labelled data is augmented using synthetic data generated by a GAN.

Overall, the use of GANs for text augmentation is a promising area of research with the potential to significantly improve the performance of natural language processing systems. Further study is needed to fully understand the capabilities and limitations of this approach, and to identify the best practices for using GANs in this context.

References

1. Shorten, C., Khoshgoftaar, T.M., Furht, B.: Text data Augmentation for deep learning. J. Big Data **8**, 1–34 (2021)
2. Feng, S.Y., et al.:A Survey of Data Augmentation Approaches for NLP (2021)
3. Bayer, M., Kaufhold, M.-A., Buchhold, B., Keller, M., Dallmeyer, J., Reuter, C.: Data augmentation in natural language processing: a novel text generation approach for long and short text classifiers. Int. J. Mach. Learn. Cybern. **14**, 1–16 (2022). https://doi.org/10.1007/s13042-022-01553-3
4. Paduraru, C., Melemciuc, M.C., Paduraru, M.: Automatic test data generation for a given set of applications using recurrent neural networks (2019)
5. Luo, Y., Zhang, J.: Data augmentation based on generative adversarial network with mixed attention mechanism. In: 2020 IEEE International Conference on Big Data (Big Data), pp. 778–787. IEEE (2020)
6. Min, S., Eom, S.H.: Easy Data Augmentation Techniques for Boosting Performance on Text Classification Tasks. arXiv preprint arXiv:1809.08047 (2018)
7. Cubuk, E.D., Zoph, B., Mane, D., Vasudevan, V., Le, Q.V., Shlens, J.: Understanding data augmentation for classification: when to warp? arXiv preprint arXiv:1902.09665 (2019)
8. Mikołajczyk, A., Grochowski, M.: Data augmentation for improving deep learning in an image classification problem. Int. Interdisc. PhD Workshop (IIPhDW) **2018**, 117–122 (2018)
9. Wang, Lu., Feng, Y., Hong, Yu., He, R. (eds.): NLPCC 2021. LNCS (LNAI), vol. 13029. Springer, Cham (2021). https://doi.org/10.1007/978-3-030-88483-3
10. Félicité, K., Lemaître, C., Hérault, J.: Data augmentation with text generation for sentiment analysis. In: Proceedings of the 2021 International Conference on Artificial Intelligence and Natural Language (AINL 2021) (2021)
11. Wei, Y., Zou, C., He, X.: A survey on data augmentation for natural language processing. arXiv preprint arXiv:1901.11196 (2019)
12. Fadaee, M., McCallum, A., Canny, J.: Data augmentation for neural machine translation. arXiv preprint arXiv:1709.04615 (2017)
13. Zhang, Y., LeCun, Y.: Text understanding from scratch. arXiv preprint arXiv:1502.01710 (2015)
14. Sennrich, R., Haddow, B., Birch, A.: Improving neural machine translation models with monolingual data. arXiv preprint arXiv:1511.06709 (2016)
15. Edunov, S., Auli, M., Grangier, D.: Understanding back-translation at scale. arXiv preprint arXiv:1808.09381 (2018)
16. Klementiev, A., Roth, D.: Inducing syntactic structure for unstructured text. In: Proceedings of the 2012 Joint Conference on Empirical Methods in Natural Language Processing and Computational Natural Language Learning, pp. 691–701. Association for Computational Linguistics (2012)

Machine Learning-Based Temperature Monitoring and Prediction

Sonam Kumari Bharti$^{(\boxtimes)}$ ⓘ, Priyadarshi Anand ⓘ, and Shradha Kishore ⓘ

Birla Institute of Technology, Mesra, Patna, India
{btech15111.19,btech15125.19,skishore}@bitmesra.ac.in

Abstract. Critical monitoring of ambient temperature at all times is necessary for all areas of pharmaceutical storage inventory where a sudden temperature change can damage the medicine efficiency as sensitive medicines can only be stable when deviations from a required condition are minimal; otherwise, unexpected temperature changes might alter the physical-chemical and pharmacological properties of the medication, and potentially endangering the patient's health. The temperature of a storage site is determined and data is produced from an LM35 temperature sensor, an analogue linear temperature sensor which has been interfaced to the Wi-Fi ESP8266 microcontroller. The generated data from the system is pushed to the cloud and can be fetched from any remote place where the internet is available, and by performing data analysis, the prediction of the temperature for the next instance of time is obtained. In this paper, the machine learning algorithms used for temperature prediction are linear regression and polynomial regression. To find the correct degree so it does not under-fit and over-fit with points and predict the temperature for the next instance of time.

Keywords: LM35 Temperature Sensor · ESP8266 Wi-Fi Microcontroller Module · Internet of Things · Machine Learning · Linear Regression · Polynomial regression · Mean Absolute Error

1 Introduction

The complete ecosystem of IoT (Internet of things) and ML (machine learning) [1] finds applications in several areas. This system can be implemented in pharmaceutical industries where monitoring of temperature is very crucial in production as well as storage. It can also be implemented in blood banks where continuous temperature monitoring is required.

In India, over 6 lakh liter of blood is discarded every year due to poor handling and unattended storage facilities in blood banks. Pharmaceutical companies face huge losses due to poor temperature monitoring of their products. Poor environmental monitoring not only affects pharmaceutical and medical industries but also leads to valuable data loss and machine damage in data centers in server rooms across the country. The problem lies in the traditional data lockers and chart recorders which are present in such facilities and it does not offer real-time visibility and alert facilities. So, this traditional system can be replaced with the IoT-based monitoring and alert system where there is real-time visibility and it is cost-effective and scalable. The two main objectives of this paper are:

1. To design a temperature monitoring model for industrial purpose.
2. Predicting the future temperature by finding the most accurate model.

The methodology used in obtaining the objectives of this paper by using the ESP8266 Wi-Fi module and LM35 temperature sensor to monitor and send temperature data to the cloud for analysis and visualization. Machine learning algorithms, including linear and polynomial regression, are used to predict future temperatures and the one with the least mean absolute error determines the most accurate model. Various graphs and charts illustrate the best prediction model for temperature, closely matching actual observed temperatures.

2 Literature Review

Over the past few decades, after the introduction of the internet, the Internet of Things (IoT) comes into existence that allows the system to monitor and control other connected devices using sensors, electronic circuits, and programming, (by sending alerts or automatically) remotely, creating an environment in which physical object can communicate with each other to improve the efficiency and accuracy for achieving financial and environmental benefits [2]. Temperature Sensors can be used to measure the temperature at the pharmaceutical storage and manufacturing unit. There are some common models used for temperature monitoring of surroundings such as DHT11 [3], Ultrasonic sensor HC-SR04, and LM35 model [4]. R A Atmoko et al. 2017, have discussed how the requirement for high-quality data is quickly rising, and how data monitoring and collection systems in real-time are necessary to achieve system accuracy. MQTT is one of the data communication protocols used in the Internet of Things [5]. With the fast growth of technology in the information era, data has become available and accessible everywhere in today's world, since it is gathered in massive amounts in many forms and from broad range of sources in an unstructured or semi-structured manner [6]. The temperature of the high-voltage transmission line is monitored in the past study [7]. The data is collected via Zigbee and LoRa temperature monitoring equipment. The convolutional layer recovers the first features by convolution, and additional convolutional layers extract numerous features from data.

3 Hardware Description

The major components used are – The Wi-Fi microcontroller module (ESP8266), LM35 (temperature sensor), power source, and USB for supplying power to the Wi-Fi microcontroller module (ESP8266).

3.1 Wi-Fi Microcontroller Module (ESP8266)

The Wi-Fi based microcontroller module (ESP8266) is connected with the temperature sensor (LM35) and the data (temperature values at different instants) collected by temperature sensor is constantly sent via Wi-Fi module to cloud and on cloud this dataset is analyzed and visualized with the help of certain code or instruction given to the microcontroller.

Fig. 1. Pin diagram for ESP8266 Wi-Fi based microcontroller module.

The Fig. 1 depicts the pin diagram of the Wi-Fi module. The ESP8266 module has 1 analog pin(A0), 5 digital I/O (Input and Output) pins, transmitting (Tx) and receiving pins (Rx), 3.3 V/5 V power pins, and a ground pin. UART is used for data communication, allowing for serial or parallel communication between devices. The ESP8266 is connected to an LM35 temperature sensor, with temperature data sent via Wi-Fi to the cloud for analysis and visualization.

3.2 Temperature Sensor (LM35)

The LM35 sensor detects ambient temperature and produces an analog voltage, which is converted by the Wi-Fi-enabled microcontroller module to a 10-bit digital number from 0 to 1023. This data is transmitted to the cloud for visualization. The analog value is transformed to actual temperature values using the formula: (analog value * 100)/1023. The digital data is then plotted for visual representation.

Fig. 2. LM-35 (Temperature sensor).

The Fig. 2 depicts the circuit diagram for the LM-35 temperature sensor. The three pins mainly indicate Supply Voltage (V_s), Output (V_{out}), and a ground pin. This sensor is interfaced over a Wi-Fi-based microcontroller module for collection of temperature readings. The range of the LM-35 sensor is from -55 to $155\ ^\circ$C with an accuracy of ± 1 $^\circ$C.

3.3 Hardware Connection

The Vs supply voltage pin of the LM35 has been connected to the 5 v of the Wi-Fi microcontroller module (ESP8266) and the Vout, the output pin of the LM35 sensor has been connected to the A0 (Analog input pin) of the Wi-Fi module whereas the GND, the ground pin of the LM35 sensor has been connected to the ground pin of the Wi-Fi module.

4 Machine Learning

Machine learning [8] is a part of artificial intelligence that focuses on using data and algorithms to simulate how humans learn. Without being explicitly programmed to do so, the algorithm based on the ML builds the model from a set of examples used to fit the parameters, known as training data, to make predictions or decisions. Machine learning is primarily divided into three types based on the techniques and modes of learning, which are: Supervised Machine Learning- learn from the given set of input data and output, Unsupervised Machine Learning- learn through the input data without knowing output and, Reinforcement Learning- learn based on obtaining the maximum reward.

With the help of the labeled dataset and supervised machine learning, we can precisely identify the classes of objects. These algorithms are useful for making predictions [9] about the results based on past performance.

One of the uses of machine learning is predictive modeling, in which the algorithm uses statistics to forecast [10] outcomes. Predictive modeling can be used even though the event being predicted is typically in the future. To predict behavior and events, predictive analytics is used. Predictive analytics [11], in contrast to other BI (Business Insights) technologies, is forward-looking and makes predictions based on the past. Prediction analysis techniques include:

1) linear regression
2) polynomial regression

4.1 Linear Regression

The linear regression [12] is used to determine the relationship between dependent output and independent input variables. Relationships are modeled in such a way using linear functions as a predictor so that the linear graph covers less distance from each point and gives the least deviated predicted value.

Linear regression is a well-researched and widely employed in practice in the past decade. This is because linearly connected models are more easily fitted than non-linearly related models, and the statistical data of the obtained desired value are simpler to evaluate. A data visualization that depicts the linear characteristics between an independent and a dependent variable is known as linear regression. It is used to demonstrate how the actual data are close to the obtained ones and cause the dispersion of the data visually.

By using data visualization, we can find the equation without handling the dataset and observing its slope and intercept, but behind the data visualization, the mathematical model is built. These mathematical models have a given set of data namely, $\{y_i, x_{i_1}, \ldots, x_{ip}\}_{i=1}^n$ of n sampled units is presented. A disturbance term, often known as an error variable, is an unobserved random variable that inserts "noise" into the linear relationship between the dependent variable and the regressors. As a result, the model adopts the following shape:

$$y_i = \beta_0 + \beta_1 x_{i_1} + \ldots + \beta_p x_i p + \varepsilon_i = x_i^T \beta + \varepsilon_i, \quad i = 1, \cdots, n \tag{1}$$

where T indicates the transpose and $x_i^T \beta$ is the inner product between vector x_i and β.

The equations are written in matrix form as show below:

$$y = X\beta + \varepsilon \tag{2}$$

where,

$$y = \begin{bmatrix} y_1 \\ y_2 \\ \vdots \\ y_n \end{bmatrix} \tag{3}$$

$$X = \begin{bmatrix} x_1^T \\ x_2^T \\ \vdots \\ x_n^T \end{bmatrix} = \begin{bmatrix} 1 & x_{11} & \cdots & x_{1p} \\ 1 & x_{21} & \cdots & x_{2p} \\ \vdots & \vdots & \ddots & \vdots \\ 1 & x_{n1} & \cdots & x_{np} \end{bmatrix} \tag{4}$$

$$\beta = \begin{bmatrix} \beta_0 \\ \beta_1 \\ \beta_2 \\ \vdots \\ \beta_p \end{bmatrix} \tag{5}$$

$$\varepsilon = \begin{bmatrix} \varepsilon_1 \\ \varepsilon_2 \\ \vdots \\ \varepsilon_n \end{bmatrix}. \tag{6}$$

y is the vector of the observed value y_i, where $i = 1, \cdots r \cdots, n$, which is known independent variable.

X is the matrix of row vector x_i which are known as independent variables.

β is a (p + 1) dimensional parameter vector, where β_0 is the intercept term. Its constituents are referred to as effects or regression coefficients (although the latter term is sometimes reserved for the estimated effects).

ε is an error vector values consist of ε_i. This is also known as the noise, or disturbance component.

4.2 Polynomial Regression

Regression analysis in which the equation between the independent variable x and the dependent variable y is established as an nth-degree of polynomial in x is known as polynomial regression [13].

For complex set of data (temperature), machine learning algorithm polynomial regression can be used for better accuracy and results of prediction [14–16]. Data set is taken into consideration and we try to deduce a polynomial equation which best mimics the dataset. A polynomial equation is in the form of y = f(x), where f(x) is the function of sum of different parts of x. To find the polynomial which best mimics the dataset, we need to tweak the constants until the equation matches the dataset. As a result, polynomial regression is regarded as a subset of multiple linear regression. The polynomial regression model:

$$y = \beta_0 + \beta_1 x_i + \beta_2 x_i^2 + \ldots + \beta_k x_i^k + \varepsilon_i (i = 1, 2, 3, \cdots r \cdots, n) \tag{7}$$

The equation expressed in matrix form as a design matrix x, a response vector $\overrightarrow{y}$, a parameter vector β, and a random error vector $\overrightarrow{\varepsilon}$. The i-th row of x and y will include the x and y values for the i-th data sample. The model is then expressed as a series of linear equations:

$$\begin{bmatrix} y_1 \\ y_2 \\ y_3 \\ \vdots \\ y_n \end{bmatrix} = \begin{bmatrix} 1 & x_1 & x_1^2 & \cdots & x_1^k \\ 1 & x_2 & x_2^2 & \cdots & x_2^k \\ 1 & x_3 & x_3^2 & \cdots & x_3^k \\ \vdots & \vdots & \vdots & \ddots & \vdots \\ 1 & x_n & x_n^2 & \cdots & x_n^k \end{bmatrix} \begin{bmatrix} \beta_0 \\ \beta_1 \\ \beta_2 \\ \vdots \\ \beta_k \end{bmatrix} + \begin{bmatrix} \varepsilon_1 \\ \varepsilon_2 \\ \varepsilon_3 \\ \vdots \\ \varepsilon_n \end{bmatrix} \tag{8}$$

which is written in pure matrix notation as

$$\overrightarrow{y} = X \overrightarrow{\beta} + \overrightarrow{\varepsilon} \tag{9}$$

The estimated polynomial regression coefficients vector

$$\overrightarrow{\beta} = \left(x^T x\right)^{-1} x^T \overrightarrow{y} \tag{10}$$

Because X is a Vandermonde matrix, where the invertibility condition is necessary if all x_i values are different, assuming m is smaller than n, which is required condition for the matrix to be invertible. This is the sole least-squares solution.

5 Experimental Layout

5.1 Data Collection

Data collection is the first and essential step for using a prediction algorithm which is generated out by the hardware kit for temperature monitoring in our case, which offers a set of data in a predetermined amount of time in a closed room. The data generated is pushed to the cloud using a set of python code.

Table 1. Temperature dataset recorded from 7:00 AM to 7:00 PM.

Time	Temperature (in °C)				
	Day-1	Day-2	Day-3	Day-4	Day-5
7:00 AM	14.66	13.66	13.32	12.42	10.04
8:00 AM	14.78	13.51	13.49	12.6	10.1
9:00 AM	15.08	13.58	13.95	12.86	10.92
10:00 AM	15.83	13.96	14.57	13.31	12.13
11:00 AM	16.82	14.74	14.82	14.05	13.69
12:00 PM	17.33	15.18	14.92	14.7	14.97
1:00 PM	17.75	15.8	14.96	15.28	15.88
2:00 PM	17.51	16.23	15.3	15.52	16.84
3:00 PM	17.6	15.97	15.37	15.19	17.5
4:00 PM	16.92	15.61	15.11	14.62	16.59
5:00 PM	16.16	14.93	14.74	13.37	14.93
6:00 PM	15.48	14.67	14.61	12.66	13.26
7:00 PM	15.2	14.58	14.55	12.18	13.36

The dataset as shown in Table 1 used in this project is recorded between the hours of 7 AM and 7 PM. Eliminating the temperature at 7 PM and using the data from 7 AM to 6 PM, the temperature value is predicted to be at 7 PM and this forecast is compared to the actual value to determine how accurately the machine learning model predicted the temperature [17–19].

5.2 Data Pre-processing

This step involves converting non-integer and non-float data types in the project. Excel or Python can be used for data pre-processing and analysis. Excel can manually convert data to numbers and then export as a CSV file. CSV files use a specific format to store table-structured data. The CSV file can then be opened in a Python interpreter for further processing.

5.3 Importing Libraries and Temperature Reading File

Python's import function makes it easier to refer to code—specifically, functions and objects—written in another file. Additionally, it is used to import Python libraries and packages that are used in our code but have been installed using Pip (the Python package manager). For this paper, libraries such as pandas and numpy are used in data analysis whereas seaborn and matplotlib are used for plotting the graph. Importing the temperature readings file in the python interpreter by the function called "read_csv". Now the data is ready to use for further analysis.

5.4 Modelling and Finding the Best Model

Linear and polynomial regression algorithm models are executed in order to depict the future temperature values. By comparing anticipated and measured values, the accuracy of the predicted values is determined. Mean absolute error is obtained in both the cases to find the most suitable algorithm for temperature prediction. As a result, confirming that the predicted value is correct and taking preventative steps are required when the temperature rises to unlivable levels.

Figure 3 shows the above flow processes involved in prediction analysis of temperature.

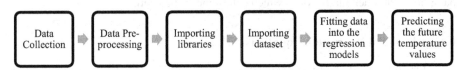

Fig. 3. Process flow for temperature prediction.

6 Result Analysis

The predicted linear and polynomial regression equations and their graphs are obtained respectively and a comparison between the two model is made to find the best model for temperature prediction.

6.1 Linear Regression Equation and Graph.

From the linear regression model of machine learning, y intercept and coefficients of equation are obtained.

y-intercept: 14.33
slope: 0.15965035

$$y = mx + c \qquad (11)$$

where, m and c are slope and y intercept respectively of the linear equation.

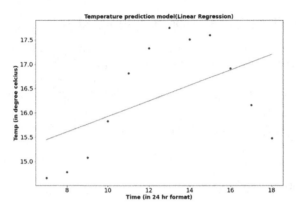

Fig. 4. Temperature prediction using linear regression model.

By substituting the value of slope and y-intercept in the Eq. (11), we get the final equation of linear graph as:

$$y = 0.16x + 14.33 \tag{12}$$

In Fig. 4, the predicted linear graph is the best fit line as it is closest to all the points but it does not touch many points which determines it behavior.

6.2 Polynomial Regression Equation and Graph

From the polynomial regression model of machine learning, y intercept and coefficients of polynomial is obtained.

y_intercept = 64.52168706761573

polynomial coefficients = [0.00000000e + 00, −1.83658365e + 01, 2.36467184e + 00, −1.24300456e−01, 2.28911713e−03].

From the above we get the value of coefficients as:

coef = 0; coef1 = −18.36; coef2 = 2.36; coef3 = −0.12; coef4 = 0.0022.

The polynomial equation is given by

$$Y = y\_intercept + coef + coef1 * x + coef2 * x * *2 + coef3 * x * *3 + coef4 * x * *4 \tag{13}$$

By substituting the value of coefficient and y-intercept in the Eq. (13), we get the final equation of polynomial graph as:

$$Y = 64.52 - 18.36x^1 + 2.36x^2 - 0.12x^3 + 0.0022x^4 \tag{14}$$

In Fig. 5, the predicted graph is in the form of polynomial and touches almost every point and the polynomial graph is not under-fit as well as over-fit. By using these graphs, the next predicted values are found for different days.

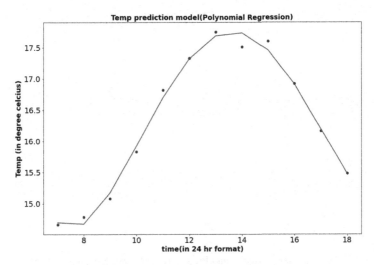

Fig. 5. Temperature prediction using polynomial regression model.

6.3 Comparison Between Linear Regression Model and Polynomial Regression Model

The comparison table as shown in Table 2 was obtained using two different regression models, the linear regression model, and the polynomial regression model [17–19]. The data chart includes the actual temperature (in °C), linear Regression Prediction Temperature (in °C), polynomial Regression Prediction Temperature (in °C) for deg = 4 polynomial, their absolute difference and their respective MAE values. The MAE shows that the actual value and observed polynomial predicted value are closer than the linear predicted value. The graph is also obtained using Table 2 as shown in Fig. 6.

Table 2. Comparison table

Day	Actual Temperature at 7:00 PM (in °C)	Linear Regression Prediction Temperature (in °C)	Polynomial Regression Prediction Temperature (in °C) for deg = 4	MAE for Linear Regression (in °C)	MAE for Polynomial Regression (in °C)
1	15.2	17.36	14.96	2.056	0.478
2	14.58	16.03	14.08		
3	14.55	15.48	14.24		
4	12.18	14.7	11.2		
5	13.36	16.58	13		

The Fig. 6 shows the actual, linear predicted, and polynomial predicted values of temperature on different days. From the Fig. 6, it is notable that the polynomial predicted

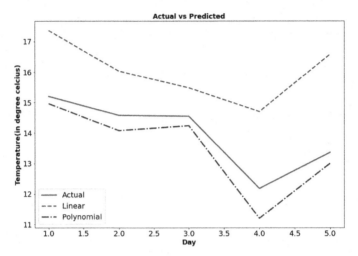

Fig. 6. Actual and Predicted temperature graph

temperature values are closer to the actual temperature value than the linear predicted temperature values [21].

7 Conclusion

From the result analysis, it can be concluded that polynomial regression is the suitable algorithm for temperature prediction as it is more accurate, has minimal complexity and, provides the lowest mean absolute error (MAE) than linear regression. Temperature is the most complex data set hence linear regression does not depict future temperature values accurately. Linear regression gives the mean absolute error of 2.056 °C whereas polynomial regression has 0.478 °C (refer Table 2), which states that the polynomial is much closer to the actual value. In the Fig. 6, the polynomial predicted line is closer to the actual value line as comparison to the linear predicted line graph.

This paper concludes that the polynomial regression is the better suited for non-linear nature graphs like temperature and it includes that while using the polynomial regression model, the correct degree of polynomial must be used so that, the curve does not over-fit and under-fit which gives the incorrect temperature value.

References

1. Maulud, D., Abdulazeez, A.M.: A review on linear regression comprehensive in machine learning. J. Appl. Sci. Technol. Trends 1(4), 140–147 (2020). https://doi.org/10.38094/jas tt1457
2. Akkaş, M.A., Sokullu, R.: An IoT-based greenhouse monitoring system with Micaz motes. Procedia Comput. Sci. 113, 603–608 (2017). https://doi.org/10.1016/j.procs.2017.08.300
3. Srivastava, D., Kesarwani, A., Dubey, S.: Measurement of temperature and humidity by using Arduino tool and DHT11. Int. Res. J. Eng. Technol. (IRJET) 05, 876–878 (2018)

4. Kanase, P., Gaikwad, S.: Smart hospitals using internet of things (IoT). Int. Res. J. Eng. Technol. (IRJET) **03**, 1735–1737 (2016)
5. Atmoko, R.A., Riantini, R., Hasin, M.K.: IoT real time data acquisition using MQTT protocol. In: International Conference on Physical Instrumentation and Advanced Materials. Citation: Atmoko, R.A., et al., 2017 IOP Conf. Series: Journal of Physics: Conf. Series 853, 012003 (2017)
6. Subahi, A.F.: Edge-based IoT medical record system: requirements, recommendations, and conceptual design. IEEE Access **7**, 94150–94159 (2019)
7. Subahi, A.F., Bouazza, K.E.: An intelligent IoT-based system design for controlling and monitoring greenhouse temperature. IEEE Access **8**, 125488–125500 (2020)
8. Chen, W., Yang, Q., Gao, S., Zhang, T., Han, H.: Temperature monitoring and prediction under different transmission modes. Comput. Electr. Eng. **92**, 107140 (2021). https://doi.org/10.1016/j.compeleceng.2021.107140
9. Shalev-Shwartz, S., Ben-David, S.: Understanding machine learning: from theory to algorithms. Cambridge Univ. Press **1**, 19–21 (2014)
10. Hwang, J.-R., Chen, S.-M., Lee, C.-H.: Handling forecasting problems using fuzzy time series. Fuzzy Sets Syst. **100**(2), 217–228 (1998)
11. Rani, R.U., Rao, T.K.R.K.: An enhanced support vector regression model for weather forecasting. IOSR J. Comput. Eng. (IOSR-JCE) **12**, 21–24 (2013)
12. Sharapov, R.V.: Using linear regression for weather prediction. In: Wave Electronics and its Application in Information and Telecommunication Systems (WECONF), St. Petersburg, Russian Federation, pp. 1–4 (2022). https://doi.org/10.1109/WECONF55058.2022.9803493
13. Menon, S.P., Bharadwaj, R., Shetty, P., Sanu, P., Nagendra, S.: Prediction of temperature using linear regression. In: International Conference on Electrical, Electronics, Communication, Computer, and Optimization Techniques (ICEECCOT), pp. 1–6 (2017). https://doi.org/10.1109/ICEECCOT.2017.8284588
14. Radhika, Y., Shashi, M.: Atmospheric temperature prediction using support vector machines. Int. J. Comput. Theory Eng. **1**(1), 55–58 (2009). https://doi.org/10.7763/IJCTE.2009.V1.9
15. Oyebola, B., Toluwani, O.: LM35 based digital room temperature meter: a simple demonstration. Equatorial J. Comput. Theor. Sci. **2**(1), 6–15 (2017)
16. Fisher, G., Daly, J.C., Recksiek, C.W., Friedland, K.D.: A programmable temperature monitoring device for tagging small fish: a prototype chip development. IEEE Trans. Very Large Scale Integr. (Vlsi) Syst. **5**(4), 401–407 (1997). https://doi.org/10.1109/92.645066
17. Wang, X., Yan, K.: Fault detection and diagnosis of HVAC system based on federated learning. In: 2022 IEEE Intl Conf on Dependable, Autonomic and Secure Computing, Intl Conf on Pervasive Intelligence and Computing, Intl Conf on Cloud and Big Data Computing, Intl Conf on Cyber Science and Technology Congress (DASC/PiCom/CBDCom/CyberSciTech), Falerna, Italy, pp. 1–8 (2022)
18. Wu, J., Liu, C., Cui, W., Zhang, Y.: Personalized collaborative filtering recommendation algorithm based on linear regression. In: 2019 IEEE International Conference on Power Data Science (ICPDS), Taizhou, China, pp. 139–142 (2019). https://doi.org/10.1109/ICPDS47662.2019.9017166
19. Murphy, K.P.: Machine Learning: A Probabilistic Perspective, pp. 219–265. MIT Press, Cambridge (2012)
20. Bargarai, F., Abdulazeez, A., Tiryaki, V., Zeebaree, D.: Management of wireless communication systems using artificial intelligence-based software defined radio. Int. J. Interact. Mob. Technol. (iJIM) **14**(13) (2020) https://doi.org/10.3991/ijim.v14i13.14211

Pathrank Algorithm: Ranking Proteins in *Mycobacterium Tuberculosis* and Human PPI Weighted Bipartite Graph Network

Merina Dhara[✉][iD], Veeky Baths[iD], and Aiswarya Subramanian[iD]

BITS-Pilani K K Birla Goa Campus, Sancoale, Goa 403726, India
merina.dhara.paul@gmail.com, veeky@goa.bits-pilani.ac.in,
20150585g@alumni.bits-pilani.ac.in

Abstract. Graph theory and Computational methods have been applied to analyze host(human) pathogen(tuberculosis) PPI network. Our goal is to find most influential tuberculosis proteins and also their interactions with the host(human) proteins through predicting host pathogen protein interaction pathways. Interactions pathways of the predicted proteins will give insight about their involvement with host metabolic pathways and thereby impart better insight about the effect of the proteins on the network. This PPI network is reconstructed as bipartite graph. Later this graph has been converted to weighted bipartite graph where pagerank values and node degree have been assigned to all the vertices as weightage separately. We have ranked the protein paths using a new algorithm Pathrank and presented the most influential paths in the network. Finally we find influential proteins within these path and analyze their drugability. Predicted proteins are the family members of acyl-CoA dehydrogenase enzyme which play a crucial role in cholesterol catabolism in *M. tuberculosis*. Therefore these proteins may need further study as possible potential drug target.

Keywords: Bipartite Graph · Pagerank · Protein Protein interaction (PPI)Network · *Mycobacterium tuberculosis* (M*tb*)

1 Introduction

The application of graph theory and algorithm together in biological network analysis is very much effective. Biological networks are of different type such as metabolic and biochemical network, gene regulator network(GRN), signal transduction network, brain functional network, protein-protein interaction network (PPI). Among them, PPI network is very much important to understand various biological processes in details as proteins are involved with cellular process vigorously. In our present study, we have used human and tuberculosis PPI network, also called host-pathogen protein-protein interaction network, for analysis with

M. Singh et al. (Eds.): ICACDS 2023, CCIS 1848, pp. 414–425, 2023.
https://doi.org/10.1007/978-3-031-37940-6_34

the help of graph theoretic and computational approach. Tuberculosis is caused by *Mycobacterium tuberculosis*, a gram negative, slow growing, aerobic, and acid fast bacterium. Tuberculosis is an airborne disease. According to WHO report, about one third population of the world are infected with tuberculosis disease actively or latently [WHO, Global tuberculosis report,2017] [1]. Hence, to prevent this disease we need to look for new methods to find out potential drug target. Therefore, we are studying tuberculosis human PPI network to find out most influential proteins. In finding most influential proteins, centrality analysis such as degree centrality [2], closeness centrality [3], betweenness centrality [4], eigenvector centrality [5], subgraph centrality [6], are very much instrumental and have already been well explored. In other words these centrality measures give idea about the position of the certain node in the given network that is how the node is maximally connected to other nodes in the network or how it is positioned between two nodes (vertices) hub. But in our present study, we are looking for the paths having most influential vertices with the help of pagerank values. Google search uses Pagerank algorithm to rank the web pages [8]. Recently pagerank has many applications in network analysis and Pagerank may be used to rank protein in a given networks. We have created a bipartite graph with our PPI network having one set of vertices as tuberculosis proteins and another set of vertices as human proteins. Interactions between these two sets of vertices that is interaction between tuberculosis and human proteins have been considered as edges. Interaction between same species has been ignored as our graph is bipartite. Here we are looking for paths which have vertices with maximum weightage. Therefore, first we have calculated the pagerank and node degree for all the proteins. Then these values have been assigned as weightage for every vertices in two sets of experiment respectively. Now as a result we are having two weighted bipartite graphs. We have searched for the all possible paths in the entire graph or network and then ranked the paths with score. Paths having vertices with maximum values will have high score. This method gives us the sets of high ranked paths which we are looking for. Then we find the vertices present in all high ranked paths with higher values. These vertices not only have high values but also present in high ranked paths and thus involved in a series of biological reactions. Thereby these vertices (proteins) may need further study for possible drugability.

2   Literature Survey

Many research works have been conducted to identify possible drug targets by analyzing disease networks applying graph theoretic principle to explore network topological properties along with computational approach. In particular, various centralities methods have been applied widely to identify potential drug targets, as these methods are very useful to locate the influential proteins within a vast network. Many researcher have applied these methods to analyze and identify potential drug targets within M*tb* network. For instance, betweennes, eigenvector, degree and closeness centrality methods have been utilized by Melak and

Gakkhar to identify drug targets in M*tb* network [9]. They have also analyzed their network using graph theoretic principles. In a similar fashion, Mazandu et al. [10] also applied network topological properties, mainly various centrality methods to analyze the functional network of the M*tb*. They integrated different genomic data to construct this functional interacting network. Recently, identification of key proteins in M*tb* human interactions have been conducted in the literature [11] . Alongside with finding hub nodes in the interactions networks applying various centralities methods, they have also identified motifs which are important in disease progressing. Raman et al. [12] provided a path score methods using edge centrality on the identified shortest paths from M*tb* PPI networks to indicate the appearance of drug resistance. In an another approach to find the drug targets from the M*tb* network, M.P. Raman et al. [13] created Molecular Interaction Map (MIP) with the possible drug targets mined from the databases. They have used molecular complex detection (MCODE) algorithm [14] to find the cluster of highly interacting nodes within the MIP networks. Further different centrality methods have been applied to narrow down the final target nodes. From the above survey, it is very much evident that the one of the main theory adopted by many researcher to find hub nodes as potential drug targets by analyzing M*tb* network is application of different centralities along with computational approach. These findings motivated us to find not only the hub proteins but also the pathways involved in the host pathogen network to gain further insight of the possible effects of the influential protein on the interaction network.

3   Methodology

We have described the methodology of the present work in step wise manner. At the beginning step, we have collected human(host) and tuberculosis(pathogen) protein-protein interaction network [7]. Next step includes reconstructing the PPI network into a bipartite graph. Pagerank value has been applied to give weightage to all the proteins in the entire network. Thus we convert our graph into a weighted bipartite graph. This weighted bipartite graph has been analyzed to find most influential proteins. In the final step we search for the paths which have been predicted with the proposed Pathrank algorithm on the basis of interactions between host and pathogen proteins in the network. Then we ranked these predicted paths with score value. In next three subsections we portray the methodology in details. Below the outline of the work flow has been given in the Fig. 1.

3.1   M*tb* and Human PPI Network

Protein protein interaction plays a crucial part in biological activities. PPI network are based on the interactions between same species or cross species. Therefore analyzing this network gives insight about many cellular aspect of a given species and also interaction with other in case of cross species. There are many

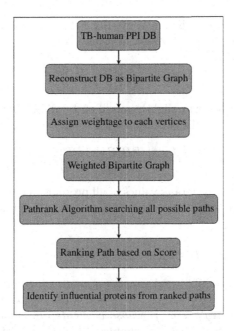

Fig. 1. Flow diagram of the proposed method.

Protein protein interaction plays a crucial part in biological activities. PPI network are based on the interactions between same species or cross species. Therefore analyzing this network gives insight about many cellular aspect of a given species and also interaction with other in case of cross species.

There are many databases which house a huge amount of information about both human and tuberculosis proteins such as Uniport [15], HPRD [16], MINT [17], InAct [18], DIP [19], BioGRID [20], BIND [21] etc.

In our study we are analyzing host(human) pathogen (Mtb) interaction (HPI) network. This HPI network has been collected from [7] to predict interaction between tuberculosis and human proteins.

3.2 Preliminaries

In this section first, we introduce pagerank algorithm along with some basic mathematical description of graph theory which have been applied in our present study.

Pagerank Algorithm : Pagerank algorithm [8] is applied for Google search. It is very much effective in finding most relevant (most searched) page in web. The same idea is reciprocated in biological net- work analysis for finding influential proteins in the network. In calculation of pagerank, it is stated that an imaginary user who is clicking on random pages will stop clicking at some point of time. In this regard damping factor has been introduced to measure the probability of continuation of clicking the pages by user. The damping factor for Pagerank algorithm is set at 0.85. The formula of pagerank given by Page is following:

$$PR(A) = 1 - d + d(\frac{PR(B)}{L(B)} + \frac{PR(C)}{L(C)} + \frac{PR(D)}{L(D)}) \tag{1}$$

Later this algorithm was slightly modified by dividing $1 - d$ with N (total number of documents in the collection). Therefore the formula becomes:

$$PR(A) = \frac{1-d}{N} + d(\frac{PR(B)}{L(B)} + \frac{PR(C)}{L(C)} + \frac{PR(D)}{L(D)}) \tag{2}$$

We have calculated pagerank values of all proteins in our human tuberculosis PPI database. Later these values have been used as Weightage to our bipartite graph.

Definitions From Graph Theory :

Definition 1 (Graph [22]). *A graph (or undirected graph) G consists of a tuple (V, E), where $V = \{v_1, v_2, \ldots, v_n\}$ is a finite set, known as vertices and another set $E = \{e_1, e_2, \ldots, e_m\}$, whose elements are known as edges so that every edge e_k is associated with an unordered pair (v_i, v_j) of vertices.*

Definition 2 (Path [22]). *A path consists of alternating finite sequence of edges and vertices so that both the terminal vertices are distinct and no vertex appears twice.*

Remark 1. We generally denote a path P by $P = < v_{i_1}, v_{i_2}, \ldots, v_{i_k} >$ (that is, there is an directed edge starting from v_{i_1} to v_{i_2} and then v_{i_2} to v_{i_3} and so on).

Definition 3 (Pathlength [22]). *Pathlength is the total number of edges in a path.*

Definition 4 (Degree [22]). *The degree of a vertex v_i (denoted as $d(v_i)$) is represented by the total edges associated with v_i.*

Definition 5 (Bipartite Graph [23]). *Bipartite graph is defined as a graph $G = (V, E)$ where V is divided in two sets V_1 and V_2, where $u_1, u_2 \in E$ indicates either $u_1 \in V_1$ and $u_2 \in V_2$ or vice versa.*

Definition 6 (Weighted Graph). *In weighted graph, every vertex is assigned some particular weightage and the weights are real numbers.*

Remark 2. Here weights are nothing but pagerank values.

3.3   Weighted Bipartite Graph Paths to Find Most Influential Proteins

Now we describe about the methods to find influential paths along with influential proteins. First we compute the weight of all the possible paths by assigning individual weightage to all the vertices of the host pathogen interaction network. Next we define the score to rank the paths having most influential proteins with maximum weightage. To do so, we now define the weight of path i.e. cumulative weights all the vertices of a path and then, maximum score of path which basically reflects the scoring methods to find out the most influential paths containing the proteins with higher weightage. These two scoring methods are the main backbone of our proposed Pathrank algorithm.

Definition 7 (Weight-of-path). *For a weighted graph $G = (V, E)$ with $V = \{v_1, \ldots, v_Q\}$ such that weight of $v_i \geq$ weight of v_j for $i \leq j$ and $\#V = Q$, where # V is cardinality of the set V.*
Then For a path $P = < v_{i_1}, v_{i_2}, \ldots, v_{i_k} >$ of the graph G,

$$weight\text{-}of\text{-}path \; P = \sum w_{i_j} \tag{3}$$

where the sum is over all the vertices of the path P and w_{i_j} is the pagerank of the jth vertex v_{i_j} (of the path P).

Definition 8 (max-score-of-path). *For a vertex v_i, we define $N(v_i) = Q - i + 1$ for $i = 1, 2, ..., n$. Now for a path $P = < v_{i_1}, v_{i_2}, \ldots, v_{i_k} >$, we define*

$$max\text{-}score\text{-}of\text{-}path \; P = \#\left\{ v_ij | N(v_ij) \geq \frac{Q}{2} \right\} \tag{4}$$

Remark 3. Max score can range from 0 to k, where k is the number of vertices of the path.

Bipartite graph can give useful insight about interactions between two sets of vertices. The left set of vertices is composed of tuberculosis proteins while right set of vertices indicates human proteins. The edges between these two sets of vertices are the interaction between tuberculosis and human proteins. The interaction between tuberculosis proteins that is within left vertices and interaction between human proteins that is within in right vertices have not considered. After constructing the network in a bipartite graph, we are searching for the paths having vertices with maximum pagerank values. Though pagerank values helps in predicting the important vertices, we are here searching for the paths where these high ranked proteins belong. Hence we are also able to predict the interaction of pathogen proteins with the host proteins. That way the bipartite graph becomes weighted bipartite graph. Therefore these paths will also be able to predict a series of interaction within proteins in a path. This bipartite graph is undirected one. Our algorithm is scripted in python to find desired paths with maximum pagerank values. Afterwards the vertices in the paths and also the vertices repeating in most of the important paths have been considered to be influential.

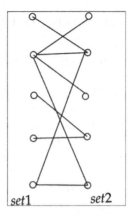

set1   set2

Fig. 2. Schematic representation of the host pathogen bipartite graph. From the Fig. 2 one can observe that there are two sets of vertices. Set1 represents the TB proteins and Set2 represents human proteins. This is a bipartite graph therefore there is no edge within set1 or within set2 rather edges are between set1 and set2. These edges represent the interaction between the host proteins and the pathogen proteins.

3.4 The Pathrank Algorithm

Here, We state the Pathrank algorithm implemented in python. This algorithm searches for all the paths in the entire network and ranks them with the score. The working principles of this algorithm have been described in five major steps below:

Step 1: The Standard Pagerank Algorithm has been applied on the graph to find the most influential vertex (proteins) in the entire graph. The pagerank value is calculated using the power method for finding pagerank of a graph.

Step 2: We modify our bipartite graph, by giving each of the vertex, a weight equal to that of it's page rank value and thereby our graph becomes weighted bipartite graph.

Step 3: All the paths in the weighted graph have been searched (This has been done using Depth First Search approach).

Step 4: For each path P, we find the weight-of-path and max-score-of-path for P (see Definition 7.).

Step 5: After the max-score-of-path is calculated for all paths in the graph, the value is normalized by dividing each max-score-of-path by the sum of all max-score-of-path of all paths in the graph.

Finally for a path P,

$$\text{path-score of } P = x \times \left(\frac{\text{max-score of } P}{\sum_{P_1} \text{max-score of } P_1} \right) + (100 - x) \times (\text{weight-of-path})$$

(5)

where the sum is over all possible paths in G.

Remark 4. During final scoring we have given the maximum weightage to the max-score of P over the weight- of-path. Therefore, in case of normalization, we have set x value as 75 and multiplied max-score of P with 75 while weight- of-path with 25. Here, the main and foremost objective is to identify the important proteins. Therefore, we have set x as 75 as by definition max-score of P ensures the inclusion of the highest ranked proteins in the final scoring or ranking of path P.

4 Experimentation

4.1 Dataset

The PPI dataset of tuberculosis and human proteins has been collected from the literature [7]. These dataset contains tuberculosis and human proteins and their interactions. Afterwards, we have reconstructed these dataset into a weighted bipartite graph network.

4.2 Experiment

The raw dataset has been pre-processed and transformed in .csv format to run the experiment. The experimental analyses in this study are carried out using ubuntu 16.04 LTS system with proposed Pathrank algorithm in python2.

First we have computed the Pathrank algorithm using pagerank value as weightage to the entire network in search of influential proteins along with the interaction pathways. Next We have replicated the same experiment taking node degree values as weightage. We present the result in the following section.

5 Result

We have used Pathrank algorithms to retrieved the most influential pathways from the host (human) pathogen(tuberculosis) interaction (HPI) database. Afterwards we have ranked all the paths with above said scoring methods. We have used both pagerank and node degree as weightage for scoring the pathways. Finally, we present two sets of predicted paths with highest score listed below:

A. Set one of predicted pathway with pagerank as weightage:

Pathrank score $=$ (76.72)

I. P63427 _ Mtb $\rightarrow$ P45954_ HUMAN $\rightarrow$ O06164 _ Mtb $\rightarrow$ Q6JQN1 _ HUMAN $\rightarrow$
P95208_ Mtb $\rightarrow$ Q92947 _ HUMAN $\rightarrow$ P95280 _ Mtb $\rightarrow$ Q709F0 _ HUMAN.

B. Set two of predicted pathway with node degree as weightage:

Pathrank score = (19.25)

I. P63427 _ M*tb* → P45954 _ HUMAN → O06164 _ M*tb* → Q6JQN1_ HUMAN →

O33229 _ M*tb* → P28330 _ HUMAN → P96844 _ M*tb* → Q92947_ HUMAN →

P95208 _ M*tb* → Q9UKU7 _ HUMAN → O86319 _ M*tb* → P16219_ HUMAN →

O53815 _ M*tb* → Q15067 _ HUMAN.

Here we have articulated two sets of pathways. These predicted pathways may give insight about the proteins which involved in metabolic pathways. Pagerank values give an idea of a vertex with higher connections or degrees only. But in Pathrank method, we are able to find the most important proteins within paths. These paths have been created on the basis of interactions between vertices (proteins) of host and pathogen. In biological systems, all the protein interaction pathways are consists of series of interacting proteins. Therefore we are able to predict the pathways first and then look for influential proteins within that pathway. This method predicts the desired proteins in a different perspective within paths as these paths are predicting a series of interrelated proteins between both host and pathogen proteins. Hence we will have better insight about the impact of the targeted protein in the host pathogen PPI network. Therefore it is very much important to study the drugability of those predicted proteins.

From two sets of pathways, the influential tuberculosis (pathogen) proteins and their interactions with human (host) proteins have been articulated in the following two tables with their respective UniportKB IDs, which mainly predicted the cholesterol catabolism pathway as retrieved from the database (Tables 1, 2 and 3).

Table 1. Most influential tuberculosis proteins from high ranked predicted paths from set one with pagerank as weightage.

UniportKB ID M*tb* Proteins	Protein	Gene	Organism (M*tb* strain)
P63427	Acyl-CoA dehydrogenase fadE25	fadE25	(H37Rv/ ATCC 25618)
O06164	Acyl-CoA dehydrogenase	MT2575	(Strain CDC/ 1551)
P95208	Acyl-CoA dehydrogenase FadE7	fadE7	(H37Rv/ ATCC 25618)
P95280	acyl-CoA dehydrogenase FadE17	fadE17	(H37Rv/ ATCC 25618)

Table 2. Most influential tuberculosis proteins from high ranked predicted paths from set two with node degree as weightage.

UniportKB ID Mtb Proteins	Protein	Gene	Organism (Mtb strain)
P63427	Acyl-CoA dehydrogenase fadE25	fadE25	(H37Rv/ATCC 25618)
O06164	Acyl-CoA dehydrogenase	MT2575	(H37Rv/ ATCC 25618)
O33229	Acyl-CoA dehydrogenase FadE20	fadE20	(H37Rv/ ATCC 25618)
P96844	Acyl-CoA dehydrogenase	MT3667	(Strain CDC 1551)
P95208	Acyl-CoA dehydrogenase FadE7	fadE7	(H37Rv/ ATCC 25618)
O86319	Acyl-CoA dehydrogenase	fadE13	(Strain CDC 1551)
O53815	Acyl-CoA dehydrogenase	MT0776	(Strain CDC 1551)

Table 3. Influential pathogen (tuberculosis) proteins interacting with host(human) proteins.

UniportKB ID (Mtb Proteins)	Interaction with host proteins (Human)
P63427	P45954 (Short/branched chain specific acyl-CoA dehydrogenase)
O06164	Q6JQN1 (Acyl-CoA dehydrogenase family member 10)
P95208	Q6JQN1 (Acyl-CoA dehydrogenase family member 10)
	Q92947 (Glutaryl-CoA dehydrogenase)
	Q9UKU7 (Isobutyryl-CoA dehydrogenase)
P95280	Q92947 (Glutaryl-CoA dehydrogenase)
	Q709F0 (Acyl-CoA dehydrogenase family member 11)
O33229	Q6JQN1 (Acyl-CoA dehydrogenase family member 10)
	P28330 (Long-chain specific acyl-CoA dehydrogenase)
P96844	P28330 (Long-chain specific acyl-CoA dehydrogenase)
	Q92947 (Glutaryl-CoA dehydrogenase)
O86319	Q9UKU7 (Isobutyryl-CoA dehydrogenase)
	P16219 (Short-chain specific acyl-CoA dehydrogenase)
O53815	P16219 (Short-chain specific acyl-CoA dehydrogenase)
	Q15067 (acyl-coenzyme A oxidase 1)

6   Conclusion

In this article, Pathrank algorithm has been proposed to find influential proteins within biological network. Here we have transformed tuberculosis and human PPI network into a weighted bipartite graph by assigning pagerank values and node degree as weightage to all the vertices. Pathrank algorithm looks for the paths having vertices with maximum weightage and ranks them accordingly. These paths consist of the influential vertices we are looking for. There are many existing methods such as different centrality measures which find influential proteins based on the connectivity of that proteins with all other proteins in the entire network. In our proposed method, one can be able to find not only the influential pathogen proteins but also the host proteins interacting with that pathogen proteins. It is possible because we are looking for the paths in the host pathogen PPI network. In the present study we have predicted many paths based on the interaction. In result section we have listed top two sets of paths. Acyl-CoA dehydrogenase fadE25, acyl-CoA dehydrogenase, acyl-CoA dehydrogenase FadE7, and acyl-CoA dehydrogenase FadE17 are most influential pathogen(tuberculosis) proteins from the predicted path with pagerank as weightage. These proteins are the family members of acyl-CoA dehydro/genase enzyme. This enzyme plays a crucial role in cholesterol catabolism in M*tb*. If catabolism of cholesterol is inhibited then it accumulates within the cell. This accumulation is toxic to cell thereby arresting the cell growth eventually leading to cell death. Therefore these proteins become an attractive potential drug target [24]. We have also predicted paths with node degree as weightage and got all the proteins related to acyl-CoA dehydrogenase pathway. This supports the efficiency of our proposed Pathrank algorithm.

In future we will try to strengthen our network by integrating multi-omic databases, so that we will be able to predict more and more interactions between host and pathogen proteins. We will also construct directed bipartite graph for Pathrank algorithm and assigning different weightages like molecular Gene Ontology (GO) terms together to analyze the effect of the predicted proteins along with interactions pathways within host pathogen PPI network for a better understanding.

Competing Interests: The authors have affirmed that no competing interests exist.

Acknowledgement. Authors are thankful to the anonymous reviewers for their valuable suggestions in improving the article.

References

1. World Health Organization. Global Tuberculosis Report 2013. https://apps.who.int/iris/handle/10665/91355. Accessed 4 July 2017
2. Jeong, H., Mason, S.P., Barabási, A.L., Oltvai, Z.N.: Lethality and centrality in protein networks. Nature **411**(6833), 41–42 (2001)
3. Wuchty, S., Stadler, P.F.: Centers of complex networks. J. Theor. Biol. **223**(1), 45–53 (2003)
4. Newman, M.E.J.: A measure of betweenness centrality based on random walks. Soc. Netw. **27**(1), 39–54 (2005)
5. Bonacich, P.: Power and centrality: a family of measures. Am. J. Sociol. **92**(5), 1170–1182 (1987)
6. Estrada, E., Rodriguez-Velázquez, J.A.: Subgraph centrality in complex networks. Phys. Rev. E Stat. Nonlinear Soft Matter Phys. **71**(5), 056103 (2005)
7. Zhou, H., et al.: Stringent DDI-based prediction of H. sapiens-M. tuberculosis H37Rv protein-protein interactions. BMC Syst. Biol. **7**(6), S6 (2013)
8. Brin, S., Page, L.: The anatomy of a large-scale hypertextual web search engine. Comput. Netw. ISDN Syst. **30**, 107–117 (1998)

9. Melak, T., Gakkhar, S.: Comparative genome and network centrality analysis to identify drug targets of mycobacterium tuberculosis H37Rv. Biomed. Res. Int. **2015**, 212061 (2015)

10. Mazandu, G.K., Mulder, N.J.: Generation and analysis of large-scale data-driven mycobacterium tuberculosis functional networks for drug target identification. Adv. Bioinform. **2011**, 1–14 (2011)

11. Verma, R.N., Malik, M.Z., Singh, G.P., et al.: Identification of key proteins in host-pathogen interactions between mycobacterium tuberculosis and homo sapiens: a systematic network theoretical approach. Healthc. Anal. **2**, 100052 (2022)

12. Raman, K., Chandra, N.: Mycobacterium tuberculosis analysis unravels potential pathways to drug resistance. BMC Microbiol. **8** (2008)

13. Raman, M.P., Singh, S., Devi, P.R., Velmurugan, D.: Uncovering potential drug targets for tuberculosis using protein networks. Bioinformation **8**(9), 403–406 (2012)

14. Bader, G.D., Hogue, C.W.: An automated method for finding molecular complexes in large protein interaction networks. BMC Bioinform. **4** (2003)

15. Boutet, E., et al.: UniProtKB/Swiss-Prot, the manually annotated section of the UniProt KnowledgeBase: how to use the entry view. Methods Mol. Biol. **1374**, 23–54 (2016)

16. Mishra, G.R., Suresh, M., Kumaran, K., Kannabiran, N., Suresh, S., et al.: Human protein reference database - 2006 update. Nucleic Acids Res. **34**, D411–D414 (2006)

17. Zanzoni, A., Montecchi, P.L., Quondam, M., Ausiello, G., Helme, C.M., Cesareni, G.: MINT: a molecular INTeraction database. FEBS Lett. **513**(1), 135–140 (2002)

18. Aranda, B., Achuthan, P., Alam-Faruque, Y., Armean, I., Bridge, A., et al.: The IntAct molecular interaction database in. Nucleic Acids Res. **38**, 525–531 (2010)

19. Xenarios, I., Rice, D.W., Salwinski, L., Baron, M.K., Marcotte, E.M., Eisenberg, D.: DIP: the database of interacting proteins. Nucleic Acids Res. **28**(1), 289–291 (2000)

20. Stark, C., Breitkreutz, B.J., Reguly, T., Boucher, L., Breitkreutz, A., Tyers, M.: BioGRID: a general repository for interaction datasets. Nucleic Acids Res. **34**, 535–539 (2006)

21. Bader, G.D., Betel, D., Hogue, C.W.: BIND: the biomolecular interaction network database. Nucleic Acids Res. **31**(1), 248–250 (2003)

22. Deo, N.: Graph Theory with Applications to Engineering and Computer Science. Prentice-Hall of India Private Limited, New Delhi (2004)

23. Pavlopoulos, G.A., Secrier, M., Moschopoulos, C.N., et al.: Using graph theory to analyze biological networks. BioData Min. **4** (2011)

24. Ouellet, H., Johnston, J.B., Montellano, P.R.: Cholesterol catabolism as a therapeutic target in Mycobacterium tuberculosis. Trends Microbiol. **19**(11), 530–539 (2011)

Circ RNA Based Classification of SARS CoV-2, SARS CoV-1 and MERS-CoV Using Machine Learning

M. Vinayak$^{(\boxtimes)}$ ⓘ, Harishchander Anandaramⓘ, S. Sachin Kumarⓘ, and K. P. Somanⓘ

Centre for Computational Engineering and Networking (CEN), Amrita School of Engineering, Coimbatore, Amrita Vishwa Vidyapeetham, Coimbatore, India
cb.en.p2dsc21029@cb.students.amrita.edu, {a_harishchander, s_sachinkumar}@cb.amrita.edu, kp_soman@amrita.edu

Abstract. The SARS-CoV-2 virus has demonstrated its ability to adapt and spread in various environments, making it a challenging target for identification and prediction. While current studies in the field concentrates on utilization of transcriptome sequence classification to identify the virus, circular RNAs (circRNAs) have shown potential as a diagnostic marker for viral diseases. These single-stranded, covalently closed RNA molecules possess unique features such as RNA binding capacity and expression regulation, making it a promising source for potential biomarkers to create a new classification model. In this study, we propose a circRNA-based classification model utilizing the dna2vec algorithm to extract distributed representations of variable-length k-mers, combined with classical machine learning algorithms. The results demonstrate superior performance of the model, with Random Forest classifier achieving an accuracy of 99.99%, highlighting the efficacy of circRNA-based classification for SARS-CoV-2 identification and the potential of circRNAs as diagnostic markers for viral diseases.

Keywords: circRNAs · dna2vec · Machine learning · SARS CoV-2 · SARS CoV-1 · MERS-CoV

1 Introduction

The virus SARS-CoV-2, which originated in Wuhan, China, has had a significant impact on the global population, spreading throughout the world and leading to millions of infections. In situations like this, it is important to determine whether the epidemic is caused by a previously known virus or a new one. This helps scientists decide on the most effective strategies for identifying, controlling, and mitigating the potential consequences of novel viruses like SARS-CoV-2. Proper classification of these viruses is crucial for taking the appropriate actions to address the outbreak and protect public health.

Correctly identifying SARS-CoV-2 can be challenging because it is genetically similar to other viruses in the Coronaviridae family. This can lead to false positives in

M. Singh et al. (Eds.): ICACDS 2023, CCIS 1848, pp. 426–439, 2023.
https://doi.org/10.1007/978-3-031-37940-6_35

detection as other viruses in the family may also be present. Additionally, people with SARS-CoV-2 may experience symptoms that are similar to those of other respiratory viral infections [1]. It is important to accurately identify and characterize SARS-CoV-2 in order to effectively diagnose patients and manage the spread of the virus.

Circular RNAs (circRNAs) are RNA molecules that are covalently closed and single-stranded, and can be detected across a broad spectrum of organisms, ranging from viruses to mammals. Considerable progress has been made in comprehending the processes involved in the biogenesis, regulation, localization, degradation, and modification of circRNAs. These molecules serve various biological functions, such as serving as transcriptional regulators, microRNA (miRNA) sponges and protein templates. Recent evidence has also shown that some circRNAs can function as protein decoys, scaffolds, and recruiters. Aberrant circRNA expression, induced by viral infections, has the potential to function as a novel diagnostic biomarker for detecting these infections. Identifying particular circRNAs in cells, tissues, or body fluids could lead to the diagnosis of viral infections. In addition, circRNAs play a crucial role in regulating the host immune response and viral replication [2].

Circular RNAs are receiving more attention for their possible use in healthcare. CircRNAs are stable and can be detected in various bodily fluids and tissues, which makes them potentially useful for identifying and predicting the course of various viral diseases [3]. There is evidence that vcircRNAs may be more effective than current RNA diagnostic markers. For instance, EBERs are often used to diagnose Epstein-Barr virus (EBV) infections, but research indicates that circRPMS1 can be detected in certain PTLDs with a high number of EBV genomes where EBER cannot, which suggests that EBER tests may produce false negatives [3].

To address the issue of identifying SARS CoV-2 from other Coronaviruses, a novel method is proposed here in which circular RNA sequences are given as inputs to train various machine learning models so that it can predict unknown viral sequences.

2 Related Works

There has been a significant effort within computational biology to find ways to represent genome sequences in a way that is meaningful and useful. Traditional methods, such as BLAST and BWA, rely on aligning viral genes and are effective at finding sequence similarities [4], but can be computationally intensive and assume that genes are homologous, which is not always the case [5]. To address these limitations, several alignment-free approaches have been developed for predicting DNA-protein binding [6]. One such approach is DeepFam, which uses a feedforward convolutional neural network and does not require the alignment of sequences. It has been shown to be more accurate and faster than methods that do require alignment, as well as those that do not. Another alignment-free tool is the MLDSP-GUI, which is designed to address issues with aligning DNA sequences [7].

Various techniques, including probabilistic, distributive, and similarity-based approaches, are used to represent biosequences [8]. Initial methods used numerical mapping, but later methods included encoding bases as binary numbers [9], one-hot encoding, and the bag-of-words approach [10]. Advances in natural language processing have led to the development of word embedding methods, such as Word2Vec, ProtVec, seq2vec, and dna2vec [11–13]. To capture global contextual information, methods such as LSTM, RNNs, and BERT have been used [14, 15].

Recent studies have employed AI and DL approaches to classify SARS-CoV-2 genome sequences, with notable examples including the development of a CNN and Bi-LSTM-based algorithm [16] achieving 99.95% classification accuracy and detection of regulatory motifs with 99.76% accuracy, and the COVID-DeepPredictor [17], a deep learning tool capable of identifying unknown pathogens with 100% accuracy on validation data and 99.51–99.94% accuracy on test datasets. Furthermore, a study of sequence embeddings utilizing GloVe and FastText n-gram representation [18, 19] showed that FastText-based embeddings produced the most significant sequences, as demonstrated by higher classification accuracy and clearer visualization.

3 Methodology

Gene sequences can be represented numerically using sequence embeddings, which capture the meaning and relationships between sequences more effectively than other methods. In this approach, sequences are divided into smaller units called "tokens," similar to the way words are processed in natural language processing (NLP). In this work, these tokens, known as "k-mers," are mapped to a vector space using dna2vec algorithm. One major benefit of using sequence embedding representations is that they can capture similarities between sequences while also providing a fixed-length representation for all sequences. These vectors are then used for classification using different machine learning algorithms and are evaluated.

3.1 Dna2vec

Vector encoding of k-mers using one-hot vectors can suffer from the curse of dimensionality. Dna2vec is a library that addresses this issue by training distributed representations of variable-length k-mers. It is based on the word2vec model, which uses a shallow neural network, and is able to consistently represent variable-length k-mers. Studies [13] indicate that the arithmetic of dna2vec vectors is similar to nucleotide concatenation. Moreover, the Needleman-Wunsch algorithm's similarity score for two k-mers is related to the cosine distance between the respective dna2vec vectors.

3.2 Dataset Description

Experiments in this study were conducted using the dataset collected from the VirusCircBase database [20], which is a comprehensive repository of circular RNA molecules produced by viruses. This database was created using computational tools such as CIRI2, which analysed viral infection-related RNA-seq data from the National Center for Biotechnology Information's (NCBI) Short Read Archive (SRA). The most recent version of the VirusCircBase database includes 46,440 viral circRNAs from 26 different viruses.

In this study, the predicted circRNA datasets [21] of MERS-CoV, SARS-CoV-1, and SARS-CoV-2 were analysed. Using data mining of RNA-sequencing data associated with viral infections, 28754, 720, and 3437 circRNAs were discovered for MERS-CoV, SARS-CoV-1, and SARS-CoV-2. Identification of these viral circRNAs was carried out using three de novo techniques, namely circRNA_finder, find_circ, and CIRI2. In comparison to SARS-CoV-1/2, MERS-CoV exhibited a higher capacity to encode circRNAs in all parts of its genome. It is important to note that viral circRNAs exhibits low expression levels and are primarily expressed during late stages of infection.

The data was obtained as FASTA files, containing information like accession ID, sequence, and comments for each sequence. These FASTA files were converted to CSV format, resulting in a total of 28,754 circRNA sequences for MERS-CoV, 3437 circRNA sequences for SARS-CoV-2, and 720 circRNA sequences for SARS-CoV-1. Different k-mers were generated from the dataset to determine the ideal "k" value and embedding dimensions for effectively representing the k-mers in experimental settings (Table 1).

Table 1. Dataset overview

circRNA	Number of sequences	Minimum sequence length	Maximum sequence length
SARS-CoV-1	720	78	29690
SARS-CoV-2	3437	79	29800
MERS-CoV	28754	65	29951

3.3 Proposed Experimental Design

In this study, our focus was on circRNA sequences of MERS-CoV, SARS-CoV-1, and SARS-CoV-2. We conducted a comprehensive analysis of the sequence embeddings and performed multi-class classification on the resulting embeddings. The sequences were divided into k-mers, with lengths of 4, 5, and 6. The generated k-mers were used to generate sequence embeddings using the dna2vec trained model with various window sizes and embedding dimensions. To assess the effectiveness of the sequences, different machine learning algorithms were employed to classify the obtained sequence embeddings. The proposed architecture's block diagram is illustrated in Fig. 1.

Data Collection   Pre-processing   Dna2vec   Classification   Evaluation

Fig. 1. Block diagram for proposed architecture

4 Experimental Setup

4.1 Sequence Embeddings

To classify sequences effectively, the initial stage involves acquiring meaningful embeddings for the gene sequences. In this work, the dna2vec model was used to obtain sequence embeddings. Experiments were conducted with different values of "k" to determine the optimal value for k-mer representation. Each k-mer was embedded using the dna2vec model with different parameters, including vector dimensions of 100 and 200, and window lengths of 3, 4, and 5. Additionally, the dna2vec pretrained model was used to generate sequence embeddings.

4.2 Sequence Classification

Machine learning algorithms, namely K-Nearest Neighbor, Naive Bayes (Gaussian), SVM Classifier (Linear, Polynomial), and Random Forest with 10-fold cross-validation, were employed to classify the acquired sequence embeddings. In the case of K-Nearest Neighbor algorithm, the optimal number of neighbors was determined to be 4 via Grid-SearchCV. As for the Random Forest algorithm, the Gini Index was utilized as the splitting criterion, and the number of estimators was fixed at 100. Moreover, a value of 1 was used for the parameter C in all SVM classifiers. The results for classification task are discussed in the Tables 2, 3, 4 and 5.

Table 2. 10-Fold cross validated accuracy scores of different algorithms for k-mer 4

Parameters	KNN	Random Forest	Logistic Regression	Naive Bayes	SVM Linear	SVM Polynomial
100dim, window-3	0.9142	0.9433	0.8759	0.6763	0.8730	0.9076
100dim, window-4	0.9185	0.9437	0.8737	0.6877	0.8744	0.9100
100dim, window-5	0.9184	0.9442	0.8733	0.7076	0.8744	0.9126
200dim, window-3	0.9110	0.9434	0.8689	0.6500	0.8630	0.8958
200dim, window-4	0.9201	0.9454	0.8678	0.6591	0.8670	0.9032
200dim, window-5	0.9208	0.9437	0.8706	0.6650	0.8687	0.9015
DNA2Vec Pretrained	0.8958	0.9382	0.8760	0.6326	0.8727	0.9099

(*continued*)

Table 2. (*continued*)

Parameters	KNN	Random Forest	Logistic Regression	Naive Bayes	SVM Linear	SVM Polynomial
SMOTE						
100dim, window-3	0.9488	0.9955	0.7909	0.6242	0.8303	0.9414
100dim, window-4	0.9509	0.9970	0.7791	0.6440	0.8147	0.9448
100dim, window-5	0.9499	0.9966	0.7914	0.6371	0.8276	0.9442
200dim, window-3	0.9515	0.9973	0.8166	0.6329	0.8465	0.9368
200dim, window-4	0.9553	0.9975	0.8080	0.6324	0.8453	0.9431
200dim, window-5	0.9554	0.9967	0.8158	0.6381	0.8469	0.9447
DNA2Vec Pretrained	0.9442	0.9956	0.7699	0.6300	0.7997	0.9370
RANDOM OVERSAMPLING						
100dim, window-3	0.9556	0.9977	0.7762	0.5982	0.8144	0.9348
100dim, window-4	0.9571	0.9988	0.7652	0.6058	0.7975	0.9359
100dim, window-5	0.9568	0.9983	0.7766	0.6110	0.8130	0.9368
200dim, window-3	0.9577	0.9989	0.8034	0.6062	0.8317	0.9311
200dim, window-4	0.9610	0.9983	0.7944	0.6054	0.8312	0.9402
200dim, window-5	0.9614	0.9985	0.8034	0.6081	0.8343	0.9418
DNA2Vec Pretrained	0.9507	0.9974	0.7571	0.6034	0.7860	0.9294

Table 3. 10-Fold cross validated accuracy scores of different algorithms for k-mer 5

Parameters	KNN	Random Forest	Logistic Regression	Naive Bayes	SVM Linear	SVM Polynomial
100dim, window-3	0.9187	0.9411	0.8772	0.6995	0.8821	0.9262
100dim, window-4	0.9218	0.9412	0.8746	0.6968	0.8788	0.9155
100dim, window-5	0.9252	0.9418	0.8622	0.7116	0.8701	0.9101
200dim, window-3	0.9270	0.9399	0.8567	0.6729	0.8524	0.8998
200dim, window-4	0.9380	0.9410	0.8606	0.6803	0.8603	0.9018
200dim, window-5	0.9463	0.9378	0.8537	0.6974	0.8493	0.8944
DNA2Vec Pretrained	0.9118	0.9382	0.8748	0.6477	0.8755	0.9088
SMOTE						
100dim, window-3	0.9533	0.9974	0.8173	0.6630	0.8460	0.9544
100dim, window-4	0.9562	0.9972	0.8086	0.6618	0.8478	0.9562
100dim, window-5	0.9568	0.9972	0.8154	0.6878	0.8514	0.9508
200dim, window-3	0.9621	0.9988	0.8396	0.6679	0.8681	0.9455
200dim, window-4	0.9677	0.9984	0.8338	0.6791	0.8667	0.9537
200dim, window-5	0.9697	0.9987	0.8412	0.7018	0.8677	0.9510
DNA2Vec Pretrained	0.9529	0.9964	0.7815	0.6480	0.8048	0.9559
RANDOM OVERSAMPLING						
100dim, window-3	0.9592	0.9994	0.8049	0.6129	0.8351	0.9510

(*continued*)

Table 3. (*continued*)

Parameters	KNN	Random Forest	Logistic Regression	Naive Bayes	SVM Linear	SVM Polynomial
100dim, window-4	0.9616	0.9990	0.7964	0.6164	0.8317	0.9537
100dim, window-5	0.9628	0.9991	0.8037	0.6232	0.8370	0.9474
200dim, window-3	0.9671	0.9997	0.8301	0.6155	0.8604	0.9432
200dim, window-4	0.9715	0.9994	0.8245	0.6144	0.8587	0.9514
200dim, window-5	0.9743	0.9993	0.8338	0.6262	0.8574	0.9505
DNA2Vec Pretrained	0.9580	0.9980	0.7676	0.6092	0.7904	0.9543

5 Results

The experiments were conducted using Zephyrus g14 laptop with AMD Ryzen 7 4800HS processor and 16 GB RAM. The 10-Fold cross validated accuracy scores of k-mer 4,5 and 6 with different dimensions and window size are given in the Tables 2, 3 and 4 with results before and after using SMOTE and random oversampling technique.

It was observed that accuracy scores increased with increasing k-mer size and vector dimension. Random forest was the best performing algorithm with k-mer 6, vector dimension 200 and window size 3 after random oversampling technique. It obtained an accuracy score of 0.99 with precision 0.99, recall 0.99 and F1-score 0.99. This suggests that the model can generalize effectively to new data, without being overfitted to the training data. The best parameter combination of different algorithms is given in Table 5. Confusion matrices for the same is shown in Fig. 2.

The t-Distributed Stochastic Neighbor Embedding (t-SNE) is an unsupervised approach that visualizes high-dimensional data in lower dimensions. For our experiments, we utilized t-SNE plots with a perplexity value of 30. The t-SNE plots for dna2vec embeddings, shown in Fig. 3, demonstrate a clear differentiation between the various virus classes, indicating that the dna2vec embeddings were effective in capturing the unique features of each virus.

6 Discussion

The emergence of SARS-CoV-2 has resulted in a global pandemic, highlighting the need for accurate and efficient diagnostic tools. In this study, we explored the use of circRNA-based classification for SARS-CoV-2 from other Coronaviruses.

Table 4. 10-Fold cross validated accuracy scores of different algorithms for k-mer 6

Parameters	KNN	Random Forest	Logistic Regression	Naive Bayes	SVM Linear	SVM Polynomial
100dim, window-3	0.9283	0.9393	0.9005	0.7186	0.9073	0.9419
100dim, window-4	0.9275	0.9406	0.9020	0.7222	0.9070	0.9261
100dim, window-5	0.9299	0.9403	0.8839	0.7339	0.8927	0.9252
200dim, window-3	0.9386	0.9406	0.8840	0.6945	0.8917	0.9236
200dim, window-4	0.9532	0.9377	0.8965	0.7050	0.9014	0.9249
200dim, window-5	0.9635	0.9376	0.8866	0.7254	0.9047	0.9200
DNA2Vec Pretrained	0.9252	0.9397	0.8718	0.6604	0.8692	0.9092
SMOTE						
100dim, window-3	0.9561	0.9988	0.8865	0.7114	0.9177	0.9744
100dim, window-4	0.9578	0.9987	0.8785	0.7014	0.9147	0.9725
100dim, window-5	0.9595	0.9979	0.8760	0.7111	0.9178	0.9680
200dim, window-3	0.9671	0.9994	0.9012	0.6960	0.9364	0.9702
200dim, window-4	0.9729	0.9991	0.8973	0.7097	0.9424	0.9737
200dim, window-5	0.9772	0.9993	0.8998	0.7162	0.9455	0.9745
DNA2Vec Pretrained	0.9573	0.9975	0.7878	0.6583	0.8137	0.9620
RANDOM OVERSAMPLING						
100dim, window-3	0.9617	0.9995	0.8762	0.6372	0.9049	0.9708
100dim, window-4	0.9633	0.9997	0.8689	0.6254	0.9024	0.9709
100dim, window-5	0.9646	0.9993	0.8697	0.6402	0.9072	0.9671

(*continued*)

Table 4. (*continued*)

Parameters	KNN	Random Forest	Logistic Regression	Naive Bayes	SVM Linear	SVM Polynomial
200dim, window-3	0.9705	0.9999	0.8938	0.6198	0.9287	0.9692
200dim, window-4	0.9762	0.9996	0.8931	0.6254	0.9352	0.9728
200dim, window-5	0.9809	0.9998	0.8958	0.6354	0.9383	0.9743
DNA2Vec Pretrained	0.9621	0.9987	0.7740	0.6105	0.7969	0.9614

Table 5. Precision, Recall and F1 Scores of different algorithms for k-mer 6

Algorithm	Dimension	Window	Oversampling	Precision	Recall	F1 Score
KNN	200	5	Random	0.9819	0.9809	0.9807
Random Forest	200	3	Random	0.9999	0.9999	0.9999
Logistic Regression	200	3	SMOTE	0.9110	0.9012	0.8984
Naive Bayes	200	5	SMOTE	0.7196	0.7162	0.7157
SVM Linear	200	5	SMOTE	0.9498	0.9455	0.9443
SVM Polynomial	200	5	SMOTE	0.9760	0.9745	0.9742

Our results demonstrate the effectiveness of the circRNA-based classification model utilizing the dna2vec algorithm to extract distributed representations of variable-length k-mers, combined with classical machine learning algorithms. The superior performance of our model, with a Random Forest classifier achieving an accuracy of 99.99%, highlights the efficacy of circRNA-based classification for SARS-CoV-2 identification and the potential of circRNAs as a diagnostic marker for viral diseases, for which further research is to be done.

In addition, we found that increasing the k-mer size and vector dimension led to improved accuracy scores. This may be attributed to the fact that larger k-mer sizes and vector dimensions capture more sequence information. The confusion matrices in Fig. 2 provide a detailed breakdown of the model's predictions for each class, allowing for a deeper understanding of its performance. Furthermore, the t-SNE plots for dna2vec embeddings shown in Fig. 3, provided insight into the effectiveness and distinguishability of the sequence embeddings. The clear differentiation between the various sequence classes observed in t-SNE plots validates the effectiveness of circRNA-based classification model.

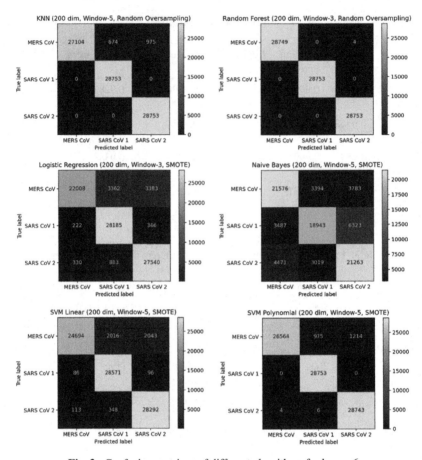

Fig. 2. Confusion matrices of different algorithms for k-mer 6

One limitation of our study is the relatively small sample size used for training and testing the model. Although we utilized various techniques such as random oversampling and SMOTE to address class imbalance, larger datasets with a diverse range of viral strains would be necessary to fully validate the accuracy and generalizability of circRNA-based classification model.

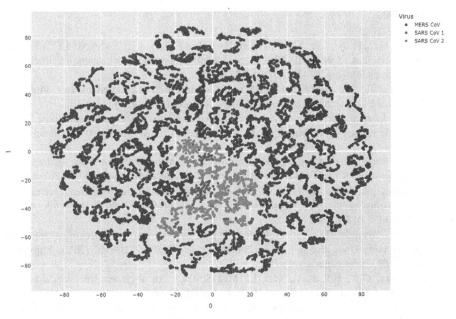

Fig. 3. t-SNE plot for dna2vec embeddings

7   Conclusion

This study explored the potential of using circular RNA sequences and the dna2vec algorithm to classify SARS-CoV-2, SARS-CoV-1, and MERS-CoV. We demonstrated that the use of circular RNA sequences offers unique advantages, including stability and potential as biomarkers. Results showed that as the k-mer size and vector dimension increased, the accuracy scores improved, and the t-SNE visualizations for the dna2vec embeddings demonstrated a clear differentiation between the various sequence classes.

We achieved the best results using the Random Forest algorithm with a k-mer size of 6, a vector dimension of 200, and a window size of 3, along with random oversampling. This combination achieved an accuracy score of 0.99, as well as precision, recall, and F1-score of 0.99. These findings suggest that the proposed method has high potential for use in clinical settings for the rapid and accurate identification of viral diseases, particularly in the case of emerging viruses such as SARS-CoV-2.

However, further validation studies are necessary to establish the generalizability of the proposed method for other viral diseases and to assess its performance in real-world clinical settings. Additionally, future studies could explore the use of other machine learning algorithms or combine multiple classification models to improve accuracy further. Overall, this study provides a promising foundation for further research on the use of circular RNAs and dna2vec embeddings for viral disease classification.

References

1. Metsky, H.C., Freije, C.A., Kosoko-Thoroddsen, T.-S.F., Sabeti, P.C., Myhrvold, C.: CRISPR-based surveillance for COVID-19 using genomically-comprehensive machine learning design (2020). https://doi.org/10.1101/2020.02.26.967026
2. Xie, H., et al.: The role of circular RNAs in viral infection and related diseases. Virus Res. **291**, 198205 (2021). https://doi.org/10.1016/j.virusres.2020.198205
3. Avilala, J., et al.: Role of virally encoded circular RNAs in the pathogenicity of human oncogenic viruses. Front. Microbiol. **12**, 657036 (2021). https://doi.org/10.3389/fmicb.2021.657036
4. Li, H., Durbin, R.: Fast and accurate short read alignment with burrows-wheeler transform. Bioinformatics **25**(14), 1754–1760 (2009). https://doi.org/10.1093/bioinformatics/btp324
5. Zielezinski, A., Vinga, S., Almeida, J., Karlowski, W.M.: Alignment-free sequence comparison: benefits, applications, and tools. Genome Biol. **18**(1), 1–17 (2017). https://doi.org/10.1186/s13059-017-1319-7
6. Zeng, H., Edwards, M.D., Liu, G., Gifford, D.K.: Convolutional neural network architectures for predicting DNA-protein binding. Bioinformatics **32**(12), i121–i127 (2016). https://doi.org/10.1093/bioinformatics/btw255
7. Randhawa, G.S., Hill, K.A., Kari, L.: MLDSP-GUI: An alignment-free standalone tool with an interactive graphical user interface for DNA sequence comparison and analysis. Bioinformatics **36**(7), 2258–2259 (2020). https://doi.org/10.1093/bioinformatics/btz918
8. Mikolov, T., Corrado, G., Chen, K., Dean, J.: Efficient estimation of word representations in vector space. In: Proceedings of the International Conference on Learning Representations (ICLR 2013), pp. 1–12 (2013)
9. Kwan, H., Arniker, S.: Numerical representation of DNA sequences, pp. 307–310 (2009). https://doi.org/10.1109/EIT.2009.5189632
10. Rizzo, R., Fiannaca, A., La Rosa, M., Urso, A.: A deep learning approach to DNA sequence classification. In: Angelini, C., Rancoita, P.M.V., Rovetta, S. (eds.) CIBB 2015. LNCS, vol. 9874, pp. 129–140. Springer, Cham (2016). https://doi.org/10.1007/978-3-319-44332-4_10
11. Asgari, E., Mofrad, M.: Continuous distributed representation of biological sequences for deep proteomics and genomics. PLoS ONE **10**, e0141287 (2015). https://doi.org/10.1371/journal.pone.0141287
12. Kimothi, D., et al.: Distributed representations for biological sequence analysis (2016). ArXiv abs/1608.05949
13. Ng, P.: dna2vec: consistent vector representations of variable-length k-mers (2017) arXiv preprint. arXiv:1701.06279
14. Lopez-Rincon, A., Tonda, A., Mendoza-Maldonado, L., et al.: Classification and specific primer design for accurate detection of SARS-CoV-2 using deep learning. Sci. Rep. **11**, 947 (2021). https://doi.org/10.1038/s41598-020-80363-5
15. Zhang, J., Chen, Q., Liu, B.: DeepDRBP-2L: a new genome annotation predictor for identifying DNA binding proteins and RNA binding proteins using convolutional neural network and long short-term memory. IEEE/ACM Trans. Comput. Biol. Bioinf. **18**, 1454–1463 (2019). https://doi.org/10.1109/TCBB.2019.2952338
16. Whata, A., Chimedza, C.: Deep learning for SARS COV-2 genome sequences. IEEE Access **9**, 59597–59611 (2021). https://doi.org/10.1109/ACCESS.2021.3073728
17. Saha, I., Ghosh, N., Maity, D., Seal, A., Plewczynski, D.: COVID-DeepPredictor: recurrent neural network to predict SARS-CoV-2 and other pathogenic viruses. Front. Genet. **12**, 569120 (2021). https://doi.org/10.3389/fgene.2021.569120
18. Ganesan, S., Sachin Kumar, S., Soman, K.P.: Biological sequence embedding based classification for MERS and SARS. In: Singh, M., Tyagi, V., Gupta, P.K., Flusser, J., Ören, T.,

Sonawane, V.R. (eds.) ICACDS 2021. CCIS, vol. 1440, pp. 475–487. Springer, Cham (2021). https://doi.org/10.1007/978-3-030-81462-5_43

19. Ganesan, S., Kumar, S.S., Soman, K.P. Deep Learning Based NLP Embedding Approach for Biosequence Classification. In: Chbeir, R., Manolopoulos, Y., Prasath, R. (eds) Mining Intelligence and Knowledge Exploration. MIKE 2021. Lecture Notes in Computer Science, vol. 13119, pp. 161–173. Springer, Cham (2022). https://doi.org/10.1007/978-3-031-21517-9_16

20. Cai, Z., et al.: VirusCircBase: a database of virus circular RNAs. Brief. Bioinform. **22**(2), 2182–2190 (2021). https://doi.org/10.1093/bib/bbaa052

21. Cai, Z., et al.: Identification and characterization of circRNAs encoded by MERS-CoV, SARS-CoV-1 and SARS-CoV-2. Brief. Bioinform. **22**(2), 1297–1308 (2021). https://doi.org/10.1093/bib/bbaa334

Ensemble Approach to Classify Spam SMS from Bengali Text

Abdullah Al Maruf[1] , Abdullah Al Numan[2] , Md. Mahmudul Haque[1] ,
Tasmia Tahmida Jidney[3], and Zeyar Aung[4(✉)]

[1] Bangladesh University of Business and Technology (BUBT), Dhaka, Bangladesh
[2] Bangladesh University of Engineering and Technology (BUET), Dhaka, Bangladesh
[3] Ahsanullah University of Science and Technology, Dhaka, Bangladesh
[4] Khalifa University, Abu Dhabi, United Arab Emirates
zeyar.aung@ku.ac.ae

Abstract. The Short Message Service (SMS) is a popular communication tool, but it has some security weaknesses, such as the influx of spam messages from cyber criminals. While several studies have been conducted on filtering and categorizing spam messages in various languages, including English, limited research has been done on detecting spam in Bengali (endonym Bangla) text. This study aims to fill this gap by classifying Bengali SMS messages as either spam or ham (legitimate messages). To accomplish this, the study used machine learning algorithms, including support vector machine (SVM) with a linear kernel and decision tree (DT), logistic regression (LR), and random forest (RF) with various parameters, as baseline models. Ensemble approaches, such as bagging, boosting, and stacking, were then used to enhance the performance of the models. The results show that the ensemble approach successfully identified spam messages in Bengali text, with XGBoost producing the most favorable outcome. The contribution of this study lies in its focus on Bengali text and the demonstration of the ensemble method's performance on a small dataset. The tool developed in this study can provide a secure and efficient SMS service to customers by reducing the burden of spam messages and improving the overall user experience. Additionally, the tool can be marketed as a value-added service for customers who are concerned about the security of their personal and financial information. Overall, this study highlights the importance of machine learning algorithms, specifically ensemble methods, in detecting spam messages in Bengali text and provides a valuable contribution to the field of SMS security.

Keywords: Bengali Text · Machine Learning · Ensemble Method · SPAM SMS · Ensemble · Classification

© The Author(s), under exclusive license to Springer Nature Switzerland AG 2023
M. Singh et al. (Eds.): ICACDS 2023, CCIS 1848, pp. 440–453, 2023.
https://doi.org/10.1007/978-3-031-37940-6_36

1   Introduction

Globally, a short messaging service (SMS) is one of the most popular and afford-able telecommunication service packages [20]. It is the most convenient mode of communication for the young generation. Nowadays, SMS marketing is an effec-tive way for businesses to reach their customers, as it allows for quick and direct communication. However, business companies must ensure they have obtained proper consent from their customers before sending any marketing messages via SMS. This is mainly because mobile marketing remains intrusive to the personal freedom of the subscribers [8]. Because SMS allows for immediate, direct engage-ment with clients, it can be an effective tool to inform or advertise. More or less, everyone in every country prefers to trust SMS over any online-based messaging service [7].

Unwanted messages can be sent electronically under the term "spam". While SMS messages are conveyed through the mobile network, spam e-mails are sent over the internet [11]. Additionally, spam SMS poses security risks since they may contain links that take consumers to malicious websites [7]. SMS market-ing's popularity can be a target for cybercriminals, who may use it to send spam or unsolicited messages to the receivers to get personal information or access dangerous websites. This will force them to divulge their private information to an unreliable source. Spammers may also use this information for targeted phishing attacks or to sell to third parties. The trust in SMS service is advan-tageous for spammers [5]. Additionally, it is quite unpleasant for the recipients. Which intentionally raises the alarm about personal information [17].

"The spam text" problem can be solved by employing more complex SMS fil-tering techniques, such as machine learning algorithms, to recognize and prevent spam texts automatically. This can help to limit the number of spam messages that reach the customer's inbox, making it easier for them to identify relevant information. That kind of classifier can be produced via a machine learning-based algorithm [2]. It is crucial to note that SMS filtering techniques, including machine learning algorithms, are not flawless. There is always the risk of false positives (genuine messages being labeled as spam) or false negatives (illegal mes-sages being marked as spam). Therefore, a new generation of intelligent security systems is being launched for data filtering [4]. It is challenging to filter datasets word-by-word in another language. Outdated models cannot control the escalat-ing number. Therefore, new algorithms may be efficient and require less work. So a mechanism for SMS filtering and spam SMS detection is created by this work.

However, it is important to note that developing a system capable of accu-rately identifying SPAM SMS written in Bengali (endonym Bangla) may be difficult. There may not be a standardized classification system in place, and the task may necessitate a significant amount of data and resources. Furthermore, the system would need to be updated and modified regularly to keep up with the changing nature of spam messages. There are several works to detect SPAM SMS from English text. But, Bengali text SMS currently lacks a consistent clas-sification system, making distinguishing between SPAM messages challenging. It will take a lot of data, resources, and knowledge in natural language processing (NLP) and machine learning to develop a system recognizing spam SMS. According to Google, Bengali is the seventh most widely-used language in the world. Many people use this language. However, some work has been done based on the Bengali language, and the outcomes of that work are not efficient. So, SPAM SMS filtering from Bengali text imposes some challenges. Due to this, we propose a system that can differentiate between spam and ham SMS from Bengali text using ML and NLP techniques. For classifying the SPAM SMS, we manually collected the dataset.

2   Related Work

Several initiatives have been completed in this sector, and this chapter showcases some of them. By examining these initiatives, we can identify their limitations and pinpoint areas where additional efforts are required. In the subsequent sections, we will explore the relevant completed works and assess any shortcomings or gaps that were encountered.

The analysis of message filtering was conducted in a study [12], where the authors employed dual filtering, K-Nearest Neighbor (KNN) classification, and rough set techniques. They proposed the concept of dual-filtering communication and used the primary classification techniques of Naive Bayes (NB), Support vector machine (SVM), and KNN. To achieve high accuracy and efficiency in their new method, the KNN algorithm re-filters communications previously classified as spam. In [23], researchers conducted a study describing the Fake-base-station Spam Ecosystem in China. They provided a comprehensive overview of the FBS spam environment by collecting three months' worth of actual FBS detection results. Over this period, they gathered FBS messages from spammers in China and acquired supplementary datasets to perform an in-depth analysis of the FBS ecosystem. The study found that contacts were evenly distributed across campaigns, with over 96% of campaigns involving a single connection for all messages. The effectiveness of several evolutionary learning classifiers for filtering mobile spam (SMS) was analyzed in a study [16] using a real-world dataset by researchers. The aim was to identify the best classifier for spam detection from a large pool of evolutionary and machine-learning classifiers. The study found that the supervised classifier system (UCS) produced a detection rate of more than 89% and a false alarm rate of 0% when used with the specified features. A trust-based management system for controlling SMS spam was developed and applied, as presented by the authors in a study [9]. They provided the prototype concept and implementation of the system, which demonstrated excellent accuracy, efficiency, and resilience for SMS messages on mobile devices. The study showed the positive commercial potential of the GTM system and its importance in addressing the growing issue of undesirable traffic lights (UTC) in communication networks.

SMS Classification: SMS classification is the process of sorting documents into predefined classes based on their content. The researchers in [11] use Word2Vec feature extraction to analyze SMS classification. In another research by Zhang et al. [22], an empirical comparison of English and Chinese spam text classification was made using publicly available Internet datasets. Information Gain (IG) IG-KNN has the highest accuracy and stability of classification among the four classifiers examined; however, it performs poorly when classifying Chinese.

Spam SMS Detection: Spam SMS detection is the process of identifying unwanted SMS and filtering them out of the inbox. Research is being done on filtering external mobile SMS spam by Androulidakis [5], and Merugu [17]. A weighting technique based on the propensity of words to indicate specific target classes (HAM or SPAM) has been employed to categorize new SMSs as SPAM or HAM in a study [17]. The False Positive (FP) to False Negative (FN) ratio is widely used in epidemiological studies as it plays a vital role in evaluating the effectiveness of various security software and systems.

Dual Filtering Message: Dual filtering of messages refers to using two different techniques to filter and process messages to improve the accuracy and reliability of that system. Another study [12] mentioned using dual filtering, KNN classification, and rough set techniques for analyzing message filtering. The concept of dual-filtering communications is suggested in the study. To achieve high classification accuracy and speed in the new method, the KNN algorithm re-filters these communications previously classified as spam.

Language-based Spam SMS Detection: Due to the variation in speech architecture across different languages, it is essential to develop a new algorithm for detecting SMS spam for each language. Turkish spam sms was detected by Onur (Spam SMS Detection for the Turkish Language with Deep Text Analysis and Deep Learning Methods) compared to Chinese and English spam text by Liumei [23]. Theodorus et al. created a dataset of Indonesian SMS used in their investigation [21]. In addition, they did categorize SMS into three categories: spam, ham, and promotional texts. The training was done in two clumps utilizing different dataset proportions. The classifiers were evaluated using a 10-fold cross-validation metric. The findings indicate that Extreme Gradient Boosting (XGBoost) 94.52%, Multinomial Logistic Regression (MLR) 94.57%, SVM 94.38%, and Random Forest (RF) 94.62% are some of the top models for a multiclass SMS classification.

Fake-base-station Spam Ecosystem: A description of "The Fake-base-station" is given by Chinese scholars [23]. This paper presented the first large-scale depiction of the FBS spam ecosystem by collecting three-month certifiable FBS discovery results. They managed FBS messages sent by genuine spammers in China over three months and gathered a large number of crucial secondary datasets to conduct an in-depth analysis of the FBS environment. Contacts are distributed more evenly throughout campaigns, with more than 96% of campaigns utilizing a single connection for all communications.

The literature reviewed above highlights the existing gaps in the field, particularly the significant challenge of detecting SPAM SMS from Bengali text. Examining these works further reveals the limitations and gaps in the existing research work.

3   Methodology

The proper steps lead to the highest accuracy of a model. So, we followed some steps to get our results accurately and effectively. We decided to break it down into several stages. Data collection, labeling, preprocessing, feature extraction, model training, and outcome analysis are the stages. Figure 1 depicts the work-flow for the suggested approach.

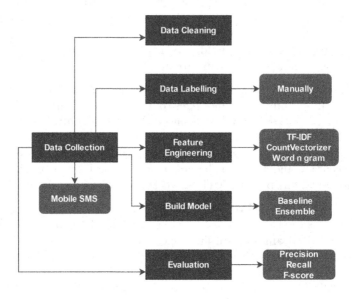

Fig. 1. Workflow of the proposed system.

3.1   Dataset Preprocessing and Labeling

The raw text data must undergo some preparation before being sent to the classifier [3]. Firstly, we clean the data. In the raw dataset, there can be numbers, emojis, hashtags, URLs, etc., which can impact the outcome of the system. So, data cleaning is a necessary step for this research. In these steps, we perform various activities. The actions performed in the data cleaning part are discussed below:

The Python library was used to remove the number, digits, emoji, and URLs. However, we do not use hashtags in our situation. We also employ additional standard text preparation techniques, such as punctuation removal, white space removal, accent mark removal, stop word removal, etc. Unlike the English lan-guage, Bengali does not have a designated stop word. This is dependent on a particular task. We utilized the Python library and regex to remove unnecessary words and punctuation from the text after carefully identifying all stop words

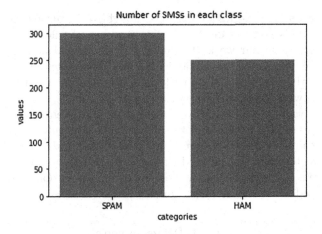

Fig. 2. Number of SMSs in two classes of dataset.

that did not include objectionable language. We collected the SMS from each of our phones. The SMS is slated to launch between 2020 and 2022. Only SMS written in Bengali is taken into consideration. We manually labeled every message in our dataset after it was assembled. We classified them into two groups (Spam, Ham). The number of each label is shown in Fig. 2. The expert in this field, who has experienced more than five years and has depth knowledge of the telecommunication sector, was the supervisor of data labeling. Some samples in the Bengali SMS dataset that we used are demonstrated in Table 1.

3.2 Feature Extraction

In the context of text classification, each model requires the input of a numerical vector that represents a text document. The process of converting the words in a text document into fixed-length floating-point or binary vectors is referred to as feature extraction or vectorization. Our research focuses on discussing the techniques used for feature extraction, as outlined below.

TF-IDF Vectorizer: The TF-IDF technique calculates the weight of a word by considering both its frequency (TF) within a document and the number of documents in which it appears (IDF). This allows for each word's weight to be effectively determined as the IDF takes into account the presence of a term across multiple documents [15].

Count Vectorizer: The bag-of-words approach is used by the count vectorizer to avoid textual structures. To extract information, just word counts are used. Each document will be converted to vector format. The input to the vector is a count of how many times each particular word appears in the material [10]. The n-grams and count vectorizer combinations approach works best in our research.

Word N-grams: The N-grams feature is thought to be quite effective for text classification. This consists of the next n words or characters. We employed the word N-grams in our work. To find the optimal model, we examined the performance of unigrams, bigrams, and trigrams. To achieve the overall scenario, we integrated the N-grams feature with other features [6]. We used Scikit-learn to experiment with features [19].

The count vectorizer, TF-IDF vectorizer, and word N-grams approach mentioned above were used for feature extraction. The best results for our model are obtained by combining word N-grams and TF-IDF vectorizer. TF-IDF converts text into a vector that can be used. It increases a system's precision. The N-grams feature is also regarded as the best for text classification. N-gram features (2, 3) were used to achieve the best possible result.

Table 1. Dataset example.

Example	English Translation	Data Label
	Dial for 25 GB internet at 150 Taka.	HAM
	Good news, you have won Grameenphone lottery selection. Now you enter the link.	SPAM
	Dear customer, you can get 20 taka discount if you buy Rin 1 kg powder! Buy today from your specific Chuck Bazaar; Visit the Love Bazaar booth located in the market to avail the offer. Offer valid from 11th January to 21st January.	HAM
1000 BDT:wa.me/ 8801917379501	Part Time Job Salary 1000 BDT1/day Base work :wa.me/8801917379501	SPAM

3.3   Classification Model

Data collecting and preprocessing are the initial part of our research. After that, to fit the data into the model, data were converted into vector formats by applying many feature extraction techniques. Finally, the model was built up. In our research, the models that are involved are shown here.

Baseline Models: We first used the baseline models to categorize the SPAM and HAM SMS. To categorize the SMS, we used support vector machine (SVM) with a linear kernel. Then, we applied decision tree (DT), logistic regression (LR), and random forest (RF) with various parameters.

When using the baseline model, we applied several parameters. Table 2 shows the parameters used for different baseline classifiers and their values.

The baseline algorithm produced unsatisfactory results. As a result, we choose to use some of the other models. Due to the ensemble approach's ability

Table 2. Baseline algorithms details (parameters).

Algorithm	Parameters	Values
DT	criterion, max_depth, max_features	entropy, 5, 10
RF	n_estimators, criterion, max_depth	100, entropy, 3
LR	penalty, class_weight, random_state, max_iter	l1, balanced, none, 200

to mix two or more models, we used it. Recently, Al Maruf et al. [1] used an ensemble model to classify the text, and then the accuracy was increased. We used bagging, boosting, and stacking models to detect SPAM and HAM SMSs.

Bagging: It stands for Bootstrapped Aggregation, is a machine-learning ensemble technique that involves aggregating the results of multiple base estimators to produce a more robust and accurate final prediction [13]. We applied different base estimator models, but When the base estimator is a LR model, and the n_estimators value is set to 100, it performs best. In this setup, 100 instances of the LR model are trained on different randomly sampled subsets (with replacement) of the original training data. Each of these 100 models makes predictions on a new data point independently. The final prediction is obtained by taking a majority vote (in the case of a classification problem) or by averaging the outputs (in the case of a regression problem). By aggregating the predictions of multiple models, the Bagging Ensemble Model can reduce the variance of the predictions and produce a more accurate final prediction compared to a single Logistic Regression model. Moreover, since each base estimator is trained on a different subset of the training data, the Bagging Ensemble Model has the ability to capture different aspects of the underlying data distribution, further improving the accuracy of the final prediction. In this case, it appears that a Bagging Ensemble Model with 100 LR base estimators performs better compared to using a different value for n_estimators. Bagging output depends on several parameters. We used different classifiers as base estimators in our work to get the best result. Firstly, we created the object of each baseline algorithm we used. Then each object was used as a base estimator. Finally, others parameters values were applied differently to get the best result. Finally, LR performs best as a base estimator. Other parameters used in our works and their values are (n_estimator=200, max_samples=0.8). When the n_estimator value was increased, the performance of the Bagging classifier was improved.

Boosting: The intention was used to bagging algorithm to increase the accuracy. But we found some training errors in our model. For this reason, To decrease training errors boosting model was used. It transforms a collection of weak learners into strong learners [14]. Boosting is an iterative technique where each iteration focuses on correctly classifying the data that was misclassified in the previous iteration. The final prediction is a weighted average of all the predictions made by the individual models. Boosting algorithms, such as Adaboost and XGBoost, are specifically designed to address the issue of overfitting.

Adaboost (Adaptive Boosting) assigns higher weights to the instances that were misclassified in previous iterations and train the base model on these instances to produce a better prediction. This helps to focus the model's attention on the more complicated samples, resulting in a more robust model.

XGBoost (eXtreme Gradient Boosting) is an optimized implementation of gradient boosting. It uses a tree-based model as the base model and trains the model using gradient descent to minimize a loss function. XGBoost also includes several techniques to prevent overfitting, such as regularization and early stopping, which help to control the complexity of the model and prevent it from overfitting to the training data. In summary, as the bagging model is facing overfitting issues, we applied boosting techniques like Adaboost or XGBoost to overcome these issues and improve the model's performance. Table 3 displays the Adaboosting and XGboosting parameters.

Table 3. Boosting ensemble parameters.

Adaboosting	XGboosting
base_estimator (RF)	base_estimator (DT)
n_estimators (350)	n_estimators (500)
max_samples (0.5)	max_samples (0.8)
–	learning_rate (0.2)

Stacking: It is a type of ensemble learning technique in machine learning that combines the predictions of multiple models to produce a more accurate prediction. It works by training multiple base models on the same training data and using their predictions as input features to train a higher-level model, known as the meta-model. The final prediction is made by the meta-model [18].

The main idea behind stacking is to leverage the strengths of different models and to produce a more robust and accurate model by combining their predictions. The base models can be of different types, such as DT, SVM, etc. Stacking can be performed in two ways: (i) level-0 models: Each base model makes its own independent prediction. These predictions are then combined with the meta-model to produce the final prediction, and (ii) level-1 models: The outputs of the base models are used as input features to train another model, which makes the final prediction. This second-level model is referred to as a meta-model. In our research, we used different algorithms as a base model and meta models. Finally, the base algorithms that performed well in our research when we used DT, RF, and KNN, and the meta classifier is LR.

4 Result and Discussions

As mentioned in the section above, we used bagging, boosting, and stacking approaches. First, we compute the accuracy of each model. The feature extraction techniques count vectorizer, word N-grams, and TF-IDF was utilized, with the combination of TF-IDF and word N-grams producing the best results.

We used a train-test split for model selection. From the dataset, 80% were randomly selected (with stratification) for training and the rest 20% of the data for testing.

The baseline models' performances are shown in Table 4. The training accuracy score indicates the accuracy of the model on the data it was trained on. For instance, the decision tree algorithm has a training accuracy of 76.54%, meaning that it correctly predicts the output for 76.54% of the training data. The testing accuracy score is the accuracy of the model on new, unseen data. For example, the decision tree algorithm has a testing accuracy of 55.98%, meaning that it correctly predicts the output for 55.98% of the new data. The training accuracy is generally higher than the testing accuracy because the model may overfit the training data, memorizing the training data instead of learning the underlying patterns. This can result in a lower accuracy on new, unseen data. In this case, Logistic Regression has the highest testing accuracy of 60.97%. This means that it has the best accuracy in predicting the output for new, unseen data among the four algorithms. However, it's important to note that accuracy is not the only measure of model performance. Other metrics such as precision, recall, and F1-score may also be necessary depending on the problem you are trying to solve. Typically, the result is not up to the mark.

Table 4. Accuracy of baseline models.

Algorithm	Training accuracy	Testing Accuracy
DT	76.54%	55.98%
RF	72.96%	54.97%
LR	77.54%	60.97%
SVM	73.54%	62.65%

To achieve better results, we applied the ensemble method and calculated all of the evaluation metrics. Table 5 shows the testing accuracy and training accuracy scores for four different ensemble models: Bagging, Adaboosting, XGboosting, and Stacking. Bagging has a testing accuracy of 75.89% and a training accuracy of 99.74%. This indicates that the model is overfitting to the training data, as the training accuracy is significantly higher than the testing accuracy. Adaboosting has a testing accuracy of 79.76% and a training accuracy of 92.36%. This is a better result compared to Bagging, as the difference between the training and testing accuracy scores is smaller. Adaboosting is another ensemble method that trains multiple base models, but it assigns different weights to the training samples based on their prediction errors. XGboosting has a testing accuracy of 83.87% and a training accuracy of 90.32%. This is a good result, as the testing accuracy is high, and the difference between the training and testing accuracy is not too extensive. XGboosting is an implementation of gradient boosting, an ensemble method that sequentially trains multiple base models, each model attempting to correct the errors of the previous model. Stacking has a testing accuracy of 82.65% and a training accuracy of 94.76%. This is a good result, similar to XGboosting. Stacking is an ensemble method that trains multiple base models and uses their predictions as input features to train a higher-level model that makes the final prediction. Overall, XGboosting and Stacking have the highest testing accuracy among the four models. The bagging, Boosting and stacking ensemble model performs better than the baseline model.

Table 5. Accuracy of ensemble models.

Model	Training Accuracy	Testing Accuracy
Bagging	99.74%	75.89%
Adaboosting	92.36%	79.76%
XGboosting	90.32%	83.87%
Stacking	94.76%	82.65%

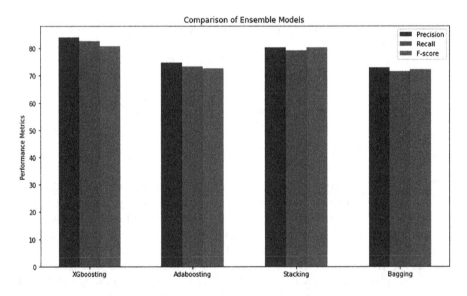

Fig. 3. Precision, recall and F-score of ensemble models.

All of the ensemble algorithms produced better results and among them Xboosting gave us the best result (some selected results shown in Table 6). Figure 3 represents the performance metrics (Precision, Recall, and F-score) of 4 different ensemble models. Comparing the metrics, XGboosting has the highest precision (83.90), followed by Adaboosting (74.65), Stacking (80.32), and Bagging (72.78). XGboosting also has the highest recall (82.55), followed by Stacking (79.22), Adaboosting (73.24), and Bagging (71.5). However, when it comes to F-score, XGboosting has the lowest score (80.78), followed by Bagging (72.11), Adaboosting (72.76), and Stacking (80.32).

Table 6. True labels and predicted labels using XGboosting.

Data	English Translation	Actual Label	Predicted Label
PROMO: DRL17 App:t.ly/sopi	Bike Car PROMO throughout March: DRL17 App:t.ly/sopi Hotline 09611991177.	SPAM	SPAM
https://wa.me/ 639279003599	Earn 2500 daily with your phone: https://wa.me/639279003599.	SPAM	SPAM
	Dial *141*100# to get emergency balance up to Tk 100	HAM	SPAM
SMS, :wa.me/ 8801917379501	Recharge today Rs.212 5 GB, 200 min and 50 SMS, validity: 30 days	HAM	HAM

In our research, we observed that if the dataset is not small, it may not be necessary to apply boosting models. Boosting models are typically used when the dataset is small (as in our case) and the goal is to improve the performance of a weak learning algorithm. If a large enough dataset is available, other machine learning algorithms, such as deep learning models, could be used to classify spam SMS from Bengali text. Researchers could explore other techniques to improve the performance of the model, such as data augmentation, feature engineering, or transfer learning. However, if the dataset is still not big enough and re-boosting is not possible, the performance of the classification model may be impacted.

5 Conclusion

This study aimed to classify and filter spam SMS in the Bengali language and has important implications for both business and cyber security. The use of bagging, boosting, and stacking ensemble models, along with preprocessing techniques, NLP modules, and the extraction of features using word N-grams, count vectorizer, and TF-IDF, resulted in high performance in detecting spam SMS. The combination of word N-grams and count vectorizer provided the best results.

The boosting ensemble performed well in the task of detecting spam SMS in Bengali, and there is potential for further improvement through the use of deep learning models with a larger dataset. From a business perspective, the ability to accurately detect spam SMS can help organizations protect their customers from malicious activities and maintain the integrity of their communication channels. This, in turn, can lead to improved customer trust and brand reputation. From a cyber security perspective, the classification of spam SMS has the potential to prevent phishing scams, prevent the spread of malware and other harmful content, and protect sensitive personal and financial information. Our research will also have an impact on Bengali NLP, and future efforts will focus on the classification of multi-class concerns such as spam, ads, and ham SMS in Bengali text. The results of this study and the provided dataset have the potential to make a significant impact in both the business and cyber security domains.

References

1. Al Maruf, A., Ziyad, Z.M., Haque, M.M., Khanam, F.: Emotion detection from text and sentiment analysis of Ukraine Russia war using machine learning technique. Int. J. Adv. Comput. Sci. Appl. **13**(12) (2022)
2. Al-Talib, G.A., Hassan, H.S.: A study on analysis of SMS classification using TF-IDF weighting. Int. J. Comput. Netw. Commun. Secur. **1**(5), 189–194 (2013)
3. Alasadi, S.A., Bhaya, W.S.: Review of data preprocessing techniques in data mining. J. Eng. Appl. Sci. **12**(16), 4102–4107 (2017)
4. Alshahrani, A.: Intelligent security schema for SMS spam message based on machine learning algorithms. IJIM **15**(16), 53 (2021)
5. Androulidakis, I., Vlachos, V., Papanikolaou, A.: Fimess: filtering mobile external SMS spam. In: Proceedings of the 6th Balkan Conference in Informatics, pp. 221–227 (2013)
6. Azmin, S., Dhar, K.: Emotion detection from Bangla text corpus using Naive Bayes classifier. In: 2019 4th International Conference on Electrical Information and Communication Technology (EICT), pp. 1–5. IEEE (2019)
7. Ballı, S., Karasoy, O.: Development of content-based SMS classification application by using Word2Vec-based feature extraction. IET Softw. **13**(4), 295–304 (2019)
8. Beatrix Cleff, E.: Privacy issues in mobile advertising. Int. Rev. Law Comput. Technol. **21**(3), 225–236 (2007)
9. Chen, L., Yan, Z., Zhang, W., Kantola, R.: Implementation of an SMS spam control system based on trust management. In: 2013 IEEE International Conference on Green Computing and Communications and IEEE Internet of Things and IEEE Cyber, Physical and Social Computing, pp. 887–894. IEEE (2013)
10. Dadhich, A., Thankachan, B.: Sentiment analysis of amazon product reviews using hybrid rule-based approach. In: Somani, A.K., Mundra, A., Doss, R., Bhattacharya, S. (eds.) Smart Systems: Innovations in Computing. SIST, vol. 235, pp. 173–193. Springer, Singapore (2022). https://doi.org/10.1007/978-981-16-2877-1_17
11. Delany, S.J., Buckley, M., Greene, D.: SMS spam filtering: methods and data. Expert Syst. Appl. **39**(10), 9899–9908 (2012)
12. Duan, L., Li, A., Huang, L.: A new spam short message classification. In: 2009 First International Workshop on Education Technology and Computer Science, vol. 2, pp. 168–171. IEEE (2009)

13. Gaikwad, D., Thool, R.C.: Intrusion detection system using bagging ensemble method of machine learning. In: 2015 International Conference on Computing Communication Control and Automation, pp. 291–295. IEEE (2015)
14. González, S., García, S., Del Ser, J., Rokach, L., Herrera, F.: A practical tutorial on bagging and boosting based ensembles for machine learning: Algorithms, software tools, performance study, practical perspectives and opportunities. Inf. Fusion **64**, 205–237 (2020)
15. Hakim, A.A., Erwin, A., Eng, K.I., Galinium, M., Muliady, W.: Automated document classification for news article in Bahasa Indonesia based on term frequency inverse document frequency (TF-IDF) approach. In: 2014 6th International Conference on Information Technology and Electrical Engineering (ICITEE), pp. 1–4. IEEE (2014)
16. Junaid, M.B., Farooq, M.: Using evolutionary learning classifiers to do MobileSpam (SMS) filtering. In: Proceedings of the 13th Annual Conference on Genetic and Evolutionary Computation, pp. 1795–1802 (2011)
17. Merugu, S., Reddy, M.C.S., Goyal, E., Piplani, L.: Text message classification using supervised machine learning algorithms. In: Kumar, A., Mozar, S. (eds.) ICCCE 2018. LNEE, vol. 500, pp. 141–150. Springer, Singapore (2019). https://doi.org/10.1007/978-981-13-0212-1_15
18. Pavlyshenko, B.: Using stacking approaches for machine learning models. In: 2018 IEEE Second International Conference on Data Stream Mining & Processing (DSMP), pp. 255–258 (2018). https://doi.org/10.1109/DSMP.2018.8478522
19. Pedregosa, F., et al.: Scikit-learn: machine learning in Python. J. Mach. Learn. Res. **12**, 2825–2830 (2011)
20. Shafi'I, M.A., et al.: A review on mobile SMS spam filtering techniques. IEEE Access **5**, 15650–15666 (2017)
21. Theodorus, A., Prasetyo, T.K., Hartono, R., Suhartono, D.: Short message service (SMS) spam filtering using machine learning in Bahasa Indonesia. In: 2021 3rd East Indonesia Conference on Computer and Information Technology (EIConCIT), pp. 199–203. IEEE (2021)
22. Zhang, L., Ma, J., Wang, Y.: Content based spam text classification: an empirical comparison between English and Chinese. In: 2013 5th International Conference on Intelligent Networking and Collaborative Systems, pp. 69–76. IEEE (2013)
23. Zhang, Y., et al.: Lies in the air: characterizing fake-base-station spam ecosystem in China. In: Proceedings of the 2020 ACM SIGSAC Conference on Computer and Communications Security, pp. 521–534 (2020)

A Scientific Study for Breast Cancer Detection Using Various Machine Learning Algorithms

Prashant Soni[1]([✉]) [iD], Sanjeev Kumar[1] [iD], and Dilip Kumar[2] [iD]

[1] Department of Computer Science Engineering, United University, Prayagraj, U.P 211012, India
prashant85soni@gmail.com
[2] Department of Computer Science and Engineering, United College of Enginnering and Research, Prayagraj, U.P 211001, India
dilipkmr322@gmail.com

Abstract. A noncommunicable chronic illness, cancer impacts people of all ages, races, nationalities, and socioeconomic statuses indistinctly. Having this diagnosis confirmed illness is usually concerning for the patient. Because it is extremely difficult to predict. A significant spot in the ranking is taken by breast cancer, particularly in women. However, there is a very good possibility of a cure if it is discovered quickly. The cancer of the breast is among the most common type of cancer in women. Mammography also is one of the most successful methods for the early diagnosis and detection of breast cancer lowers the death rate. Mammograms, which are digital images of the breast, are used to identify the beginning signs of breast cancer. These radiographic images improve diagnostic precision while decreasing human error in cyst detection and diagnosis time. This study provides a scholarly analysis of machine learning techniques for cancer in breast detection and classification. In this research, we investigate how different machine-learning approaches affect the automation of breast imaging categorization. This study gathers examples that show how the machine learning approach is used to address various issues that have been identified through various analytical scientific investigations. This work aims to gather and compare the various screening techniques, classifiers, and their efficacy. Along with details about the kind of dataset used, the author has stated a Research Gap analysis. Numerous authors have used various machine learning algorithms on this common dataset, but the outcomes of the algorithms vary. The author has discussed these various algorithms and analyzed the findings in tabulation format.

Keywords: Breast cancer detection · Naïve Bayes classifier · Wisconsin breast cancer original · Support Vector Machine · Random Forest

1 Introduction

The most serious disease that kills females, rising over 2.1 million women on a sporadic basis, is breast cancer. According to WHO estimates, 627,000 women will a breast cancer death in 2018. Its evolution stalled, and nearly 15% of all malignant growth deaths among women were caused by it [1]. We require early measures to identify and categorize the tumors in order to lower the number of deaths caused by cancer of the breast. Breast cancer can be classified as benign (not harmful) versus abnormal (malignant). Benign tumors are relatively slow-growing, do not invade nearby tissues, and do not disseminate to other bodily regions [2]. Breast cancer is frequently found after a woman discovers a tumor or during screenings before symptoms show. Most breast lumps and the biggest mass visible on a mammogram are regarded as mild (not cancerous). If cancer is found, tissue is usually taken from a needle biopsy (fine needle or larger heart needle) for microscopic analysis and less frequently from a surgical biopsy [3]. The review of prior diagnostic data and the gathering of useful information from the data are the main goals for the prompt and precise identification of this illness. The procedure will be aided by machine learning methods and medical imaging. The earlier diagnosis increases the likelihood of effective therapy, which enhances the illness prognosis. Women's breast tumors that develop into breast cancer are to blame [4]. The possibility of detection can be explored using one's own or one's family's medical history, physical exams, mammograms, ultrasound scans, biopsy, etc. An accurate diagnosis can lower the patient's chance of dying. A significant issue is that IA-classified tumors are so small—less than 4 mm—making it challenging for medical professionals to recognize them on CT scan pictures. Since IA tumors have a roughly 50/50 chance of being benign or malignant, a biopsy is necessary, but it can be very invasive and impractical [5]. Re-scanning is typically advised every 6–12 weeks to check for any symptoms of growth, which can more than double in size.

1.1 Doctor's Challenges During Diagnosis

1) Professionals in medical are faced a main challenge when dealing with classification is that, size of tumours which are less than the size of 4 mm are very difficult to diagnose via images of MRI and CT scan.
2) The process of biopsy needs to be conducted very invasive and infeasible to do as IA tumours have around There are 50% chance to be identify begin and malignant tumours.
3) The final result can be declared by general medical advice after scanning in 6–12 weeks. If growth rate of cell is twice in count.
4) Next important challenge faced by Doctor's is analysing CT scans is analysis fatigue

1.2 Breast Cancer

Cancer of the breast is a common disease major unavoidable health issue. Cancer is formed by formed by uncontrolled growth of cells. Pre-Mature Diagnosis of cancer tissues can be useful. Tumors are of two types Benign the harmless one and other is Malignant tumors, they are dangerous and cells together form as a lump and spreads throughout the body if not diagnosed on time. **Breast** cancer has been very typical disease recorded every year for the women. Breast cancer is discovered by the group of people belong from Ancient Greek and Egypt. The condition most frequently affecting women is breast cancer, which can affect up to 2 lakhs of them. Due to high death rate of breast cancer this disease is second in position for cause of death. If breast cancer detected in early stage, then doctors are able to provide best diagnosed with cancer. This review focuses on the involving the most accurate outcome for determining the stages of a tumour utilizing several machine learning techniques. As a result, this is very effective for the detecting the cancerous cell and of its.

1.3 Deep Learning

An End-to-end training approach of Deep learning is used for cancer detection. Amazing and remarkable accurate result can be recorded in recent year by applying Deep Learning to the oncology sector. Deep learning delivers extremely important outcomes in the field of cancer of the breast. The accuracy in the field of cancer that has been achieved rivals by radiologists and is suitable for implementation for the clinical tool. +

2 Datasets

For image processing, the original database "Wisconsin Breast Cancer" is utilized. The Utilised is the Wisconsin Diagnostic Breast Cancer (WDBC) dataset for the test several machine learning breast cancer detection algorithms. The algorithms utilised in this common dataset range widely. Based on a breast mass fine needle aspirate (FNA) digital image, features are calculated. It refers to the characteristics of the picture's discernible cell nuclei. Additionally, the UW CS ftp server provides access to this database's contents.

Following is the transfer protocol server. File transfer protocol: https://archive.ics. uci.edu/ml/datasets/Breast+Cancer+Wisconsin+%28Diagnostic%29 ftp ftp.cs.wisc.edu cd math-prog/cpo-dataset/machine-learn/WDBC.

Above mention link is also used to found the Machine learning repository.
Information related to Attribute are as follow.

1) First column is "ID number"
2) Diagnosis for type of cell (M = malignant, B = benign).

Features in number of counts is 10 real-valued of each cell.

a) The radius is the distance in meters between the mean center and every point on the outermost edge.
b) Gray- scale values-based texture of standard deviation.
c) Radius length as smoothness
d) area of cells
e) cell's perimeter.
f) The equation for tightness is perimeter2 / region-1.0.
g) The term concave areas are referred to as 1 m wide.
h) small portions of contour used as concave points
i) Facial features
j) coastline approximation as Infinity dimension

The three greatest values for the average, standard deviation, and worst the provided images are computed, yielding 30 characteristics. These are an example: -

1) Mean radius is third field.

2) Radius is in thirteen field.

3) Worst Radius is twenty-three field.

Significant digits four in count are used to recorded all feature values. Four significant figures are added to each feature value. The dataset has no missing attributes, and 357 benign and 212 malignant cell pictures have been categorized

3 Comparative Analysis

The online version of the volume will be available in LNCS Online. Members of institutes subscribing to the Lecture Notes in Computer Science series have access to all the pdfs of all the online publications. Non-subscribers can only read as far as the abstracts. If they try to go beyond this point, they are automatically asked, whether they would like to order the pdf, and are given instructions as to how to do so.

3.1 "Utilizing Data Mining Tools to Predict and Analyze Breast Cancer" [2]

Because of being simple, clear and fast Naïve Bayes classifier is used. It considers mutually independent attribute and Bayesian logistic used for problems of two class values. Simple CART is used for prediction and J48 provides a decision node in a tree to determine the class's anticipated value. WEKA data mining tool is used for classification and accuracy as shown in Fig. 1.

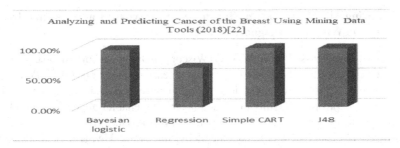

Fig. 1. Analyzing and Predicting Cancer of the Breast Using Mining Data Tools. [2]

The preprocessing of images is done using UCI repository of Original Wisconsin breast cancer dataset (WBCO) for cancer classification and prediction. For analysis of preprocessed images Bayes Classifier (Bayesian Logistic Regression, Naïve Bayes), Decision tree are used [6–8].

The accuracy of the algorithm is Naïve Bayes with 95.265%, Bayesian Logistic regression with 65.432%, Simple CART 98.1349%, and J48 with 97.274% accuracy.

3.2 "Evaluating Several Machine-Learning techniques for Cancer of breast Detection and Diagnosis" [9]

For image preparation, the Original Wisconsin Breast Cancer Dataset is utilized. SVM (Support Vector Machine), a machine learning method, is employed in the identification of cancer. Key patterns are found in support vector using SVM algorithm which further separates them making liner functions look like recursive BN (Bayesian Network) method based on RF (Random Forest). In this method, Selection of random data set one with replacement and other without replacement is done. After that, the data divided is utilised to detect breast cancer using ML Algorithm [10–13].

The algorithm's accuracy is 97%, and the values for precision and recall fall under the ROC (Receiver Operating Characteristic) as shown in Fig. 2.

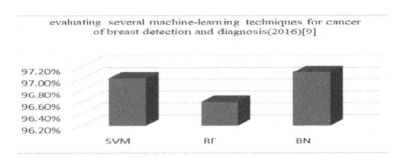

Fig. 2. Evaluating several machine-learning techniques for cancer of breast detection and diagnosis (2016). [9]

3.3 "Utilizing Machine Learning Techniques to Discover Cancer of the Breast Using the Wisconsin Diagnostic Dataset" [14]

The image is preprocessed using WDBC (Wisconsin Diagnostic breast cancer) dataset. Cancer detection is carried out using the machine learning technique GRU-SVM (Gated Recurrent Unit). Classification was done using linear regression by applying threshold. Multilayer perceptron, consisting of hidden layers that enables approximation of function, for optimization using nearest neighbor. SoftMax produces probability distribution for classes. SVM (Support vector machine) determine optimal hyperplane for separation of two classes in dataset by using binary classification. Thus, cancer tumors are identified [9, 15, 16].

GRU-SVM has 90.68% accuracy, Linear regression has 92.89% accuracy Multilayer perceptron has 96.92% accuracy, SoftMax has 97.36% accuracy, SVM has 97.7% accuracy. Training Accuracy of nearest neighbor is not recorded because it does not require training as shown in Fig. 3.

3.4 "A Comparison of Breast Tumor Detection Techniques Using Machine Learning" [17]

Data mining techniques have been used for cancer early detection. Here, the image is preprocessed using the Wisconsin breast cancer study and dataset, and the data is then corrected by deleting any missing values. For classification 3, the Naive Bayes method is employed to calculate the possibility that each occurrence belongs to a particular

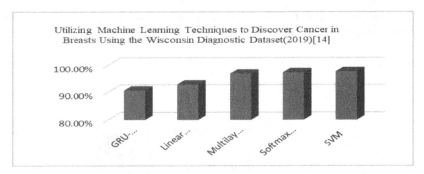

Fig. 3. Utilizing Machine Learning Techniques to Discover Cancer in Breasts Using the Wisconsin Diagnostic Dataset (2019). [14]

class. J48 algorithm use information entropy to decompose each data attribute into smaller datasets for verification of entropy difference. By replacing all the missing value globally, SMO (Minimal order optimization) converts nominal attributes to binary [18]. After applying preprocessing techniques, the accuracy of the report and result for J48 75.52% for breast cancer data and SMO 96.99% of WBC dataset has increased. Hence SMO is better than J48 as shown in Fig. 4.

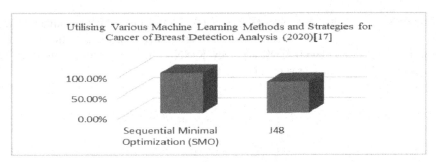

Fig. 4. Utilising Various Machine Learning Methods and Strategies for Cancer of Breast Detection Analysis (2020). [17]

3.5 "Using a New Genetic Algorithm-Based Feature Selection Method, an Efficient System for Diagnosing Cancer of the Lung" [19]

Breast cancer is detected through a hybrid genetic algorithm. Data preprocessing is done using the UCI machine learning repository's Wisconsin Breast Cancer dataset. Combination of GA (Genetic Algorithm) and MI (Mutual Information) is used for hybrid feature selection, they are good indicator of correlation between feature and class names. For the separation of two distinct classes, SVM (Support Vector Machine) is applied using hyperplane and KNN which considers distance between different node for identification of breast cancer tumor whether it is benign or malignant [20].

For SVM classifier AUC is 96.69 and correct rate is 98.44, for KNN classifier AUC is 96.78 and correct rate is 98.65 as shown in Fig. 5.

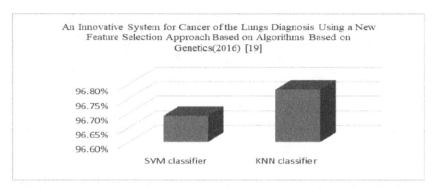

Fig. 5. An Innovative System for Cancer of the Lungs Diagnosis Using a New Feature Selection Approach Based on Algorithms Based on Genetics (2016). [19]

3.6 "Testing the Efficiency of ML Techniques for Wisconsin Breast Carcinoma Classification" [21]

With the Wisconsin Breast Cancer Diagnosis (WBCD) dataset, three machine learning algorithms—SVM, Decision tree, and K-NN—are utilized for the detection of breast cancer. Result are SVM with 97.9% accuracy, K-NN with 96.7% accuracy, and decision tree with 93.7% accuracy as shown in Fig. 6.

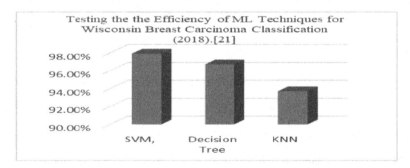

Fig. 6. The Efficiency of ML Techniques for Wisconsin Breast Carcinoma Classification (2018). [21]

3.7 "Pattern Recognition Using Fuzzy c-Means are Used in an Examination of the Disease for Cancer of Breast" [22]

Pattern recognition model and FCM classifier is used with Fuzzy c means algorithm. Results are **100% true +,0% false +** and **87% true -, 13% false –** as in Fig. 7.

3.8 "An Innovative Method for Cancer of Breast Diagnosis Using DM Techniques" [23]

Classification is a data mining technique applied to detect the cancer of breast. Classification method applied in WEKA are SMO (Sequential Minimal Optimization), IBK (K-NN classifier), BF (Best First) trees. Acquired result are BF tree has 95.46%, IBK has 95.90%, SMO has 96.19% accuracy as shown in Fig. 8.

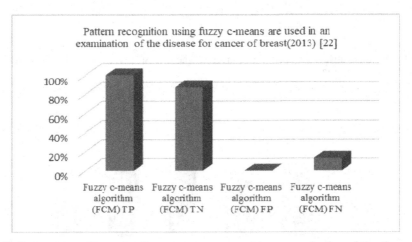

Fig. 7. Pattern recognition using fuzzy c-means are used in an examination of the disease for cancer of breast (2013). [22]

3.9 "Machine Learning Techniques for Cancer of the Breast Prediction: A Comparison" [24]

Utilizing BCCD, various ML classification *Techniques* are used to detect cancer of breast, WBCD dataset. DT, RF, SVM, NN, LR, models are applied for different datasets (BCCD). Results are DT has 68.6%, SVM has 71.4%, RF has 74.3%, LR has 65.7%, NN has 60.0% and for WBCD and accuracy of DT has 96.1%, SVM has 95.1%, RF has 96.1%, LR has 93.7%, NN has 95.6% as shown in Fig. 9 (Table 1).

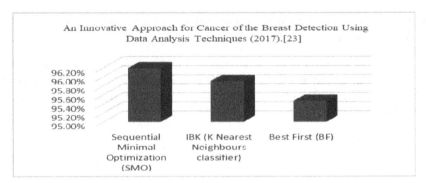

Fig. 8. An Innovative Approach for Cancer of the Breast Detection Using Data Analysis Techniques (2017). [23]

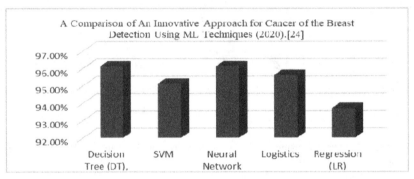

Fig. 9. A Comparison of An Innovative Approach for Cancer of the Breast Detection Using ML Techniques (2018). [24]

3.10 Table of Comparative Analysis

Table 1. Comparative Analysis.

S.NO	Year	Algorithm Used	Result	Drawback and Future Scope
1	2018	NB (Naïve Bayes), BL (Bayesian logistic), Regression, Simple CART, and J48 [2, 6–8]	95.26%, 65.42%, 98.13%, 97.27%	Applied Automatic Region Growing (AGR) for segmentation

(continued)

Table 1. (*continued*)

S.NO	Year	Algorithm Used	Result	Drawback and Future Scope
2	2016	SVM, RF and BN [9–13]	97.0%, 96.6%, 97.1%	The method can be checked for other data base with large images and accuracy can be increased using suitable algorithm
3	2019	GRU-SVM (Gated Recurrent Unit- Support Vector Machine), Linear Regression, Multilayer, SoftMax Regression, and SVM [14–16]	90.68%, 92.89%, 96.92%, 97.36%, 97.7%	Filtering technique is not applied to remove noise
4	2020	SMO(Sequential Minimal Optimization), J48 [17, 18]	96.99%, 75.52%	Statical parametric approach is used to improve the accuracy
5	2018	Genetic Algorithm, SVM, KNN [19, 20]	96.69%, 96.78%	Tuning the best machine learning classifiers and by applying best segmentation techniques accuracy of the system con be increased
6	2022	SVM, Decision Tree, KNN [21]	97.9%, 96.7%, 93.7%	Performance of the method can be increased using the best feature extraction technique and suitable machine algorithm
7	2013	Fuzzy c-means algorithm (FCM) [22]	TP-100%, TN-87%, FP-0%, FN-13%	The method is ideal for breast cancer detection
8	2017	SMO (Sequential Minimal Optimization), IBK (K Nearest Neighbours classifier), Best First (BF) [23]	96.19%, 95.90%, 95.46%	Efficiency of the system can be increased using efficient filtering technique, the ideal segmentation, feature extraction, and classification procedure
9	2018	Decision Tree (DT), RF, SVM, Neural Network (NN), Logistics, Regression (LR) [24]	96.1%, 95.1%, 96.1%, 95.6%, 93.7%	Cancer of the breast can be detected among with its Several stages like I, II, III and IV stage of cancer

4   Conclusion

The Detection of Breast Cancer is a daunting task because every algorithm gives different results so survey of all possible combination of algorithms with their accuracy and computation time mentioned in above papers. Dataset used in different papers for different technique are also mentioned. This work can be applied to further review of breast cancer screenings can be assessed using all possible methods. This will help in further implementation of prediction algorithm for detection cancer of breast. A roughly equivalent study will help in choosing classification algorithm which has higher accuracy.

References

1. Das, A., Mohanty, M.N., Mallick, P.K., Tiwari, P., Muhammad, K., Zhu, H.: Breast cancer detection using an ensemble deep learning method. Biomed. Signal Process. Control. **70**, 103009 (2021). https://doi.org/10.1016/J.BSPC.2021.103009
2. Singh, S.N., Thakral, S.: Using data mining tools for breast cancer prediction and analysis. In: 2018 4th International Conference on Computer Communication Autom. ICCCA 2018. (2018). https://doi.org/10.1109/CCAA.2018.8777713
3. Breast Cancer Occurrence 3 Breast Cancer Risk Factors 12 What Is the American Cancer Society Doing about Breast Cancer? 26 Sources of Statistics 30 References 32
4. Younis, Y.S., et al.: Early diagnosis of breast cancer using image processing techniques. J. Nanomater. **2022**, 1–6 (2022). https://doi.org/10.1155/2022/2641239
5. Houssein, E.H., Emam, M.M., Ali, A.A., Suganthan, P.N.: Deep and machine learning techniques for medical imaging-based breast cancer: a comprehensive review. Expert Syst. Appl. **167**, 114161 (2021). https://doi.org/10.1016/j.eswa.2020.114161
6. (8) (PDF) Knowledge based analysis of various statistical tools in detecting breast cancer, https://www.researchgate.net/publication/265797360_Knowledge_based_analysis_of_various_statistical_tools_in_detecting_breast_cancer. Accessed 04 Apr 2023
7. Kalmegh, S.: Analysis of WEKA Data Mining Algorithm REPTree, Simple Cart and Random Tree for Classification of Indian News (2015)
8. Liou, D.-M., Chang, W.-P.: Applying data mining for the analysis of breast cancer data. In: Fernández-Llatas, C., García-Gómez, J.M. (eds.) Data Mining in Clinical Medicine. MMB, vol. 1246, pp. 175–189. Springer, New York (2015). https://doi.org/10.1007/978-1-4939-1985-7_12
9. Bazazeh, D., Shubair, R.: Comparative study of machine learning algorithms for breast cancer detection and diagnosis. International Conference on Electronic Devices, System Applications (2017). https://doi.org/10.1109/ICEDSA.2016.7818560
10. Kourou, K., Exarchos, T.P., Exarchos, K.P., Karamouzis, M.V., Fotiadis, D.I.: Machine learning applications in cancer prognosis and prediction. Comput. Struct. Biotechnol. J. **13**, 8–17 (2015). https://doi.org/10.1016/J.CSBJ.2014.11.005
11. Berk, R.A.: Statistical Learning from a Regression Perspective (2020). https://doi.org/10.1007/978-3-030-40189-4
12. Powers, D.M.W., Ailab: Evaluation: from precision, recall and F-measure to ROC, informedness, markedness and correlation (2020)
13. Carvalho, V.R., Moraes, M.F.D., Braga, A.P., Mendes, E.M.A.M.: Evaluating five different adaptive decomposition methods for EEG signal seizure detection and classification. Biomed. Signal Process. Control. **62**, 102073 (2020). https://doi.org/10.1016/J.BSPC.2020.102073
14. Cruz, J.A., Wishart, D.S.: Applications of machine learning in cancer prediction and prognosis. Cancer Inform. **2**, 59 (2006). https://doi.org/10.1177/117693510600200030

15. Xiao, Y., Wu, J., Lin, Z., Zhao, X.: Breast cancer diagnosis using an unsupervised feature extraction algorithm based on deep learning. In: Chinese Control Conference CCC. 2018-July, pp. 9428–9433 (2018). https://doi.org/10.23919/CHICC.2018.8483140

16. Islam, M.M., Iqbal, H., Haque, M.R., Hasan, M.K.: Prediction of breast cancer using support vector machine and K-Nearest neighbors. In: 5th IEEE Region 10 Humanitarian Technology Conference 2017, pp. 226–229. R10-HTC 2017. 2018-January (2018). https://doi.org/10.1109/R10-HTC.2017.8288944

17. Mohammed, S.A., Darrab, S., Noaman, S.A., Saake, G.: Analysis of breast cancer detection using different machine learning techniques. Commun. Comput. Inf. Sci. 1234 CCIS, 108–117 (2020). https://doi.org/10.1007/978-981-15-7205-0_10/TABLES/7

18. Witten, I.H., Frank, E., Hall, M.A.: Data Mining: Practical Machine Learning Tools and Techniques, Third Edition. Data Min. Pract. Mach. Learn. Tools Tech. Third Ed. 1–629 (2011). https://doi.org/10.1016/C2009-0-19715-5

19. Lu, C., Zhu, Z., Gu, X.: An intelligent system for lung cancer diagnosis using a new genetic algorithm based feature selection method. J. Med. Syst. 38(9), 1–9 (2014). https://doi.org/10.1007/s10916-014-0097-y

20. Pradhan, K.S., Chawla, P., Tiwari, R.: HRDEL: High ranking deep ensemble learning-based lung cancer diagnosis model. Expert Syst. Appl. 213, 118956 (2023). https://doi.org/10.1016/J.ESWA.2022.118956

21. Obaid, O.I., Mohammed, M., Ghani, M.K.A., Mostafa, S., Taha, F.Y., AL-Dhief: Evaluating the performance of machine learning techniques in the classification of wisconsin breast cancer (2018)

22. Muhic, I.: Fuzzy Analysis of Breast Cancer Disease using Fuzzy c-means and Pattern Recognition. Southeast Eur. J. Soft Comput. 2 (2013). https://doi.org/10.21533/scjournal.v2i1.45

23. Chaurasia, V., Pal, S.: A novel approach for breast cancer detection using data mining techniques (2017). https://papers.ssrn.com/abstract=2994932

24. Thomas, T., Pradhan, N., Dhaka, V.S.: Comparative analysis to predict breast cancer using machine learning algorithms: a survey. In: Proceedings of 5th International Conference Inventive Computation Technologies ICICT 2020, pp. 192–196 (2020). https://doi.org/10.1109/ICICT48043.2020.9112464

YOLO Based Segmentation and CNN Based Classification Framework for Epithelial and Pus Cell Detection

V. Shwetha[1]($\boxtimes$)(iD), Keerthana Prasad[2](iD), Chiranjay Mukhopadhyay[3](iD), and Barnini banerjee[3](iD)

[1] Electrical and Electronics Department, Manipal Institute of Technology, Manipal Academy of Higher Education, Manipal 576104, Karnataka, India
shwetha.v@manipal.edu

[2] Manipal School of Information Sciences, Manipal Academy of Higher Education, Manipal 576104, Karnataka, India
Keerthana.Prasad@manipal.edu

[3] Department of Microbiology, Kasturba Medical College Manipal, Manipal Academy of Higher Education, Manipal 576104, Karnataka, India
chiranjay.m@manipal.edu, barnini.banerjee@manipal.edu

Abstract. Identifying the cells, such as pus and epithelial cells, from microscopic images is one of the important steps in medical diagnostics. Microscopic examination by hand is labor-intensive and unreliable. Therefore, it is helpful to have an automated approach for classifying these cells to enable quick and accurate diagnosis. Creating a model for automated cell identification is challenging because of the numerous variable parameters such as various stains and magnifications and cell overlapping. This paper offers a robust object detection model that detects the pus and epithelial cells images obtained from the microscopic analysis of direct samples of Gram-stained patient samples such as pus and sputum. This paper also presents a novel classifier that addresses the overlapping issues present in the cells. The proposed methodology offers an mAP of 0.87 and a classification accuracy of 94.5%.

Keywords: YOLO · CNN · epithelial cell · pus cell · Object detection model · Machine Learning · Classification · Augmentation

1 Introduction

Epithelial and pus cells establish the main line of protection against contaminating microorganisms [1,2], which forms the first line of defense against pathogenic organisms. These cells can be found in the specimen of urine, saliva, blood and sputum samples [3,4]. Epithelial and pus cells are mainly present in sputum smear samples. In which pus cells are obtained from Lower Respiratory Tract (LRT) specimens. In contrast, epithelial cells are obtained from the Upper Respiratory Tract (URT) saliva [5,6]. Counting the cell manually is labor intensive, tedious, and subjective, often prone to lab technician errors [7,8] due to additional stain, background debris and cell overlapping. With advancements in image Processing and CNN techniques, numerous segmentation and classification methods have been used to identify and classify these cell images [9–11], However, detection of pus cells and epithelial cells from sputum smear images using CNN is reported by significant less number of literature. The current study focused on detecting pus and epithelial cells from sputum smear images.

1. This paper presents a segmentation of Region of interest (ROIs) such as pus cells and Epithelial cells using YOLO based object detection model
2. The proposed detection model compared with the other YOLO-based object detection model, in which YOLOV5 performed better in comparison with YOLOV4 and YOLOV3 object detection models
3. This paper presents a CNN-based classification model to address the overlapping and non-overlapping issues in the cells.

2 Related Work

Many Machine learning and CNN-based approaches are used to detect the cells from the microscopic images. Many detection algorithms were proposed to detect the pus and epithelial cells from the microscopic images of urine samples [12–14]. The detection of cells WBC and epithelial cell were reported in work [15, 16] for blood sample images. Chan et al. [17] proposed a lobe counting (LC) method was presented to count the number of nucleus lobes in the pus cells from blood smear images. Sputum smear detection was reported by work [5, 18] used Machine Learning approaches to detect pus cells and epithelial cells. Azman et al. [19, 20] used Machine learning approaches. Steps such as contrast enhancement, segmentation process, removing noise, and k-means clustering were used to identify the cell. Out of 200 images, 193 images of Epithelial cells were detected with a detection accuracy of 96%. Crossman et al. [21] used 100 sputum smear images. 1600 image patches were created from the database. Computer vision approaches such as Histogram Oriented Gradient (HOG) with Support Vector Machine (SVM) and a Single covariance approach with kernel SVM (SC-KSVM) were used and proposed a multiple covariance approach with KSVM (MC-KSVM) were used to detect epithelial and leukocytes. The proposed system achieved a detection percentage of 86%. Carvajal et al. [22] Focused on finding the specific areas using low-magnification images to the existence of leukocytes. Eight slide images with 25x magnification images were created and used. Zhang et al. work proposed a HOG+SVM-based model to detect the pus cell from sputum smear images. The proposed model resulted in an accuracy of 86.8% [23]. Handzlik et al. [24] used an image-based CNN approach. This involves identifying and quantifying NETs formation based on live imaging and a CNN-trained model. The overall detection accuracy achieved is 70%. Detection of cells in sputum smear images was reported by a few works of literature in the past [19, 21, 22]. CNN-based algorithm suffered with low accuracy due to few images [23–25]. Detection model YOLO was used for segmenting the [26, 27] vehicle images. Further, it is further used for segmenting the cancer tumor from mammogram images [28], Lung CT images [29,30], and cardio ultrasound images [31–33]. The present study focused on the application of object detection in segmenting the pus cell and epithelial cells from the sputum smear images.

3   Methodology

The proposed method is used to segment and classify the pus and epithelial cell from the given images. Pre-processing steps include dividing the images into four images since the database image size is 2048 × 1533 pixels. After pre-processing, the individual size of each image is around 512 × 384. After the resize of the operation, using the YOLOV5-based object detection model, cells are identified. Using bounding boxes, ROIs are extracted from the background. Further, the

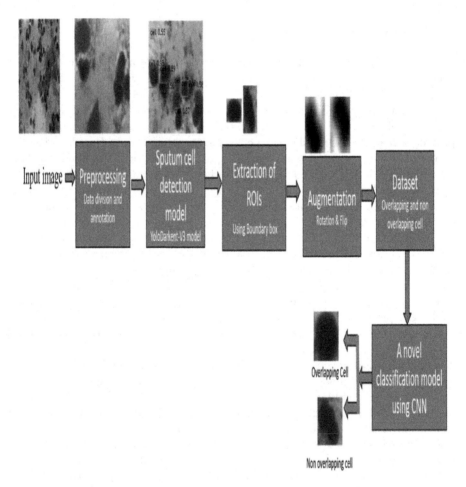

Fig. 1. Overall proposed methodology for the detection of pus and epithelial cell

CNN classifier is used to classify the overlapping and non-overlapping cells. Using the classifier to separate the cell into overlapped and non-overlapped ones is essential. Non-overlapping indicates an image comprising a single cell that might be a pus or epithelial cell, with no formation of clusters. The overall proposed methodology is shown in Fig. 1.

3.1   Dataset

Sputum cell is the combination of pus cell and epithelial cell obtained from the sputum of the throat and lung The proposed method is used in detecting pus cell and epithelial cells of images UQSNP_GCellDetect Dataset [21,22]. The sample images are shown in Fig. 2.

3.2   ROI Segmentation Using YOLO Detection Model

You Only Look Once (YOLO) is the real-time object detection model used to detect vehicles, parking signals, and people in groups and individuals. In this work, an attempt is made for microscopic image implementation. The methodology such as pre-processing steps includes dividing the images into four images since the database image size is 2048×1533 pixels. After pre-processing individual size of each image is around 512×384. Then images were annotated using the LabelImg tool. Overall, 600 images were annotated using LabelImg software to implement the object detection model. YOLO v5 is a convolution neural network. It consists of 24 convolutional layers followed by 2 fully connected layers. Each layer has its importance, and the layers are separated by their functionality.

3.3   ROI Extraction

After the object detection, ROIs are extracted by pointing to the coordinates of bounding boxes, which are created after the detection of ROIs from the YOLO detection models.

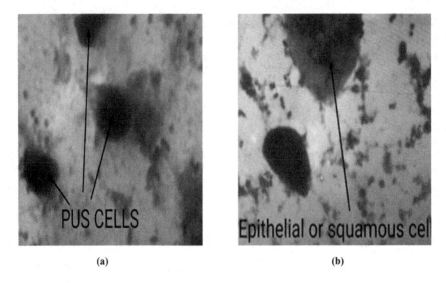

(a)   (b)

Fig. 2. a) and b) are the sample images used in the study

3.4   Data Augmentation

Data augmentation is a technique that can be used to artificially expand the size of the training set by creating modified data from existing data. It is good to use data augmentation to avoid overfitting or the initial data set is small to train or even to get better performance from the model. Data augmentation was used to increase the accuracy. Overall classification accuracy increased from 82%-83% to 94%.

3.5   Classification Model to Address the Overlapping Issue of Cell

The proposed CNN model for cell detection is shown in Fig. 3. it consists of four fully connected convolution layers and three max pool layers with Rectified Linear Unit (ReLU) activation function in each layer. Weights are initialized using the normal distribution. Batch normalization is added in the second layer, and dropouts are considered at the fifth layer, which helps in fast convergence to reduce the overfitting problem. Segmented ROIs are resized into 32 × 32 pixels with 32 batch sizes.

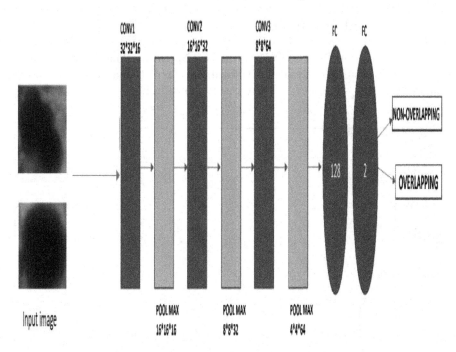

Fig. 3. Proposed CNN network for cell l overlapping and non-overlapping classification

3.6   Implementation Details

The dataset required to train a detector with YOLOv5. It contains two parameters namely images and labels. We implemented the model on the Collab notebook. Steps such as cloning and building the Darknet are followed, the next step is the model trained for 600 sputum smear images with YOLOv5 pre-trained weights. The classification model is run for 50 epochs for 32 batch sizes. Adam optimizer with a learning rate of 0.001 is used. An early stopping mechanism is used for the custom CNN classifier, which is trained from scratch.

3.7   Performance Metrics

Utilizing performance measures Average Precision (AP) and mean Average Precision to find the best object detection model for cell detection is explored in this study, whereas The area between the precision-recall curve represents the AP, which is specific to a particular class and mAP stands for the mean average of the AP. Classification accuracy, Precision and Recall are used to evaluate the CNN classifiers.

4   Result and Discussion

Object detection model, YOLOV5 is run for 100 epochs. The loss curve for the training indicates that the model is trained, and prediction accuracy is reported at 90.82% for the dataset. The loss curve is shown in Fig. 4.

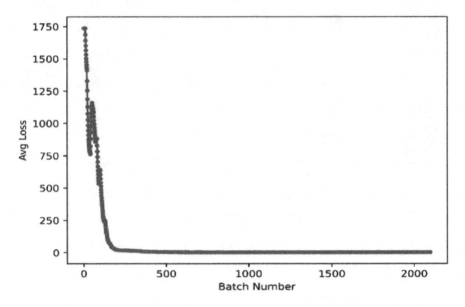

Fig. 4. Loss curve obtained for YOLOV5 object detection model training process

The result obtained for the prediction is shown in Fig. 5. The confidence value of each bounding box varies from 0.78 to 0.98 for the cell detection The proposed segmentation method is compared with other YOLO-based models such as YOLO v3 and YOLO v4. Table 1 details that the YOLO v5 model performed better than the other object detection model.

Table 1. Result obtained for various object detection model

Object detection model	Class	AP	mAP
YOLO v3	Pus cell	0.84	0.86
	Epithelial cell	0.87	
YOLO v4	Pus cell	0.83	0.84
	Epithelial cell	0.86	
YOLO v5	Pus cell	0.90	0.87
	Epithelial cell	0.85	

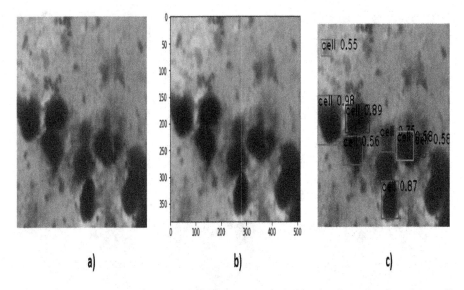

Fig. 5. a) Input images b) Normalized images and c) Obtained result after the predication

The prediction result obtained for overlapped cells is shown in Fig. 6. It indicates the model can detect the cell with a confidence value of 0.99. Still, it cannot identify them as overlapped or non-overlapped cells. This is resolved by using the CNN classifier model.

The proposed model is compared with VGG-16, EfficientNetB0 CNN classifier. All the models are run for 50 epochs. The models are trained for 3000 cell images. The performance metrics of the proposed classifier, along with other CNN classifiers, are detained in Table 2. it shows proposed CNN classifier performance is better in comparison with other classifiers.

Table 2. Classification comparison with CNN classifier

CNN classifier	Cell	Precision	Recall	Classification Accuracy
VGG-16	Overlapped Cell	82%	83%	84%
	Non-overlapped Cell	84%	84%	
EfficietNetB0	Overlapped Cell	78%	82%	80%
	Non-overlapped Cell	80%	78%	
Proposed CNN network	Overlapped Cell	94%	91%	94%
	Overlapped Cell	88%	96%	

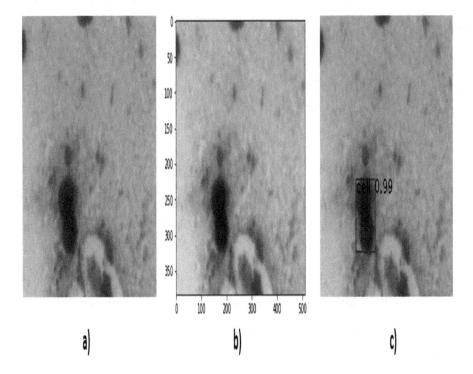

Fig. 6. a) Input image b) Normalized image and c) overlapped predicted image

5   Conclusion

This study proposed a methodology for detecting pus and epithelial cells in sputum smear images. The novelty of the proposed methodology lies in the application of an object detection model for segmenting regions of interest (ROIs) in the images. However, the segmentation model used in the study was unable to handle the issue of overlapping cells. To address this issue, propose a novel convolutional neural network (CNN) classifier to classify the segmented ROIs into either pus or epithelial cells. This classifier is likely trained on a dataset of labeled ROIs, which allows it to learn to differentiate between the two cell types. The application of the object detection model in segmenting ROIs is the novelty of the proposed methodology. Even though the segmentation model could not handle the overlapping issue in the cells. A novel CNN classifier is proposed, which differentiates the overlapped and non-overlapped cells. The proposed model is also compared with the other classifier and object detection models. YoloV5 is reported with an mAP of 0.87. CNN-based classier reported the highest accuracy of 94% in comparison with the VGG-16 and EffiecientB0 classifier models. The proposed method could be useful in diagnosing respiratory infections and other conditions that affect the respiratory tract. Detecting pus and epithelial cells from sputum smear images can be a useful diagnostic tool for various respiratory diseases, including tuberculosis, pneumonia, and bronchitis. The identification of these cells can provide important information about the type and severity of the infection. This approach is the automated image analysis technique to detect and quantify pus and epithelial cells in sputum smear images. This involves using computer algorithms to analyze the images and identify regions of interest that contain these cells. This can be faster and more efficient than manual examination and provide more objective and reproducible results. Detecting pus and epithelial cells from sputum smear images can be a valuable tool in diagnosing and managing respiratory diseases, and advances in automated image analysis techniques are likely to further improve the accuracy and efficiency of this approach. However, it is essential to note that the effectiveness of the method will depend on its accuracy and reliability, as well as its ability to be easily implemented in clinical settings. Further research and testing may also be necessary to determine the method's effectiveness and potential limitations. Overall, this study demonstrates a promising approach for detecting and classifying pus and epithelial cells in sputum smear images. However, the limitations of the segmentation model used in the study highlight the need for further research to address the issue of overlapping cells in medical imaging.

References

1. Prah, J.K., Amoah, S., Ocansey, D.W., Arthur, R., Walker, E., Obiri-Yeboah, D.: Evaluation of urinalysis parameters and antimicrobial susceptibility of uropathogens among out-patients at university of cape coast hospital. Ghana Med. J. **53**(1), 44–51 (2019)
2. Larsen, S.B., Cowley, C.J., Fuchs, E.: Epithelial cells: liaisons of immunity. Curr. Opin. Immunol. **62**, 45–53 (2020)
3. Dey, S., Saha, T., Narendrakumar, U.: Analysis of urine as indicators of specific body conditions. In: IOP Conference Series: Materials Science and Engineering, vol. 263, p. 022051. IOP Publishing (2017)
4. Moss, B., Smith, G.L., Gerin, J.L., Purcell, R.H.: Live recombinant vaccinia virus protects chimpanzees against hepatitis B. Nature **311**(5981), 67–69 (1984)
5. Hamid, R., Halim, N.A., Arshad, N.W., Naim, F., Jusof, M.F., Mohamed, Z.: Feature extraction of pus cells detection and counting in sputum slide images. In: 2013 Saudi International Electronics, Communications and Photonics Conference, pp. 1–6. IEEE (2013)
6. Lee, Y.J., et al.: Acceptability of sputum specimens for diagnosing pulmonary tuberculosis. J. Korean Med. Sci. **30**(6), 733–736 (2015)
7. Oei, R.W., et al.: Convolutional neural network for cell classification using microscope images of intracellular actin networks. PloS One **14**(3), e0213626 (2019)
8. Flight, R., Landini, G., Styles, I.B., Shelton, R.M., Milward, M.R., Cooper, P.R.: Automated noninvasive epithelial cell counting in phase contrast microscopy images with automated parameter selection. J. Microsc. **271**(3), 345–354 (2018)
9. Hung, J., et al.: Keras R-CNN: library for cell detection in biological images using deep neural networks. BMC Bioinform. **21**(1), 1–7 (2020)
10. Xue, Y., Ray, N.: Cell detection in microscopy images with deep convolutional neural network and compressed sensing. arXiv preprint arXiv:1708.03307 (2017)

11. Hanaa, A.-Z., Gamil, A.A.: Segmentation of epithelial human type 2 cell images for the indirect immune fluorescence based on modified quantum entropy. EURASIP J. Image Video Proc. **2021**(1), 1–19 (2021). https://doi.org/10.1186/s13640-021-00554-6

12. Goswami, D., Aggrawal, H.O., Gupta, R., Agarwal, V.: Urine microscopic image dataset. arXiv preprint arXiv:2111.10374 (2021)

13. Li, T., et al.: The image-based analysis and classification of urine sediments using a LeNet-5 neural network. Comput. Meth. Biomech. Biomed. Eng. Imaging Visual. **8**(1), 109–114 (2020)

14. Goswami, D., Aggrawal, H., Agarwal, V.: Cell detection and classification from urine sediment microscopic images

15. Jung, C., Abuhamad, M., Mohaisen, D., Han, K., Nyang, D.H.: WBC image classification and generative models based on convolutional neural network. BMC Med. Imaging **22**(1), 1–16 (2022)

16. Jung, C., Abuhamad, M., Alikhanov, J., Mohaisen, A., Han, K., Nyang, D.: W-Net: a CNN-based architecture for white blood cells image classification. arXiv preprint arXiv:1910.01091, 2019

17. Chan, Y.-K., Tsai, M.-H., Huang, D.-C., Zheng, Z.-H., Hung, K.-D.: Leukocyte nucleus segmentation and nucleus lobe counting. BMC Bioinform. **11**(1), 1–18 (2010)

18. Qiao, G., Zong, G., Sun, M., Wang, J.: Automatic neutrophil nucleus lobe counting based on graph representation of region skeleton. Cytometry A **81**(9), 734–742 (2012)

19. Azman, F.I., Ghazali, K.H., Mohamed, Z., Hamid, R.: Detection of sputum smear cell based on image processing analysis. ARPN J. Eng. Appl. Sci. **10**(21), 9880–9884 (2015)

20. Goździkiewicz, N., Zwolińska, D., Polak-Jonkisz, D.: The use of artificial intelligence algorithms in the diagnosis of urinary tract infections—a literature review. J. Clin. Med. **11**(10), 2734 (2022)

21. Crossman, M., Wiliem, A., Finucane, P., Jennings, A., Lovell, B.C.: A multiple covariance approach for cell detection of Gram-stained smears images. In: IEEE International Conference on Acoustics, Speech and Signal Processing (ICASSP), pp. 932–936. IEEE (2015). https://doi.org/10.1109/ICASSP.2015.7178106

22. Carvajal, J., et al.: An early experience toward developing computer aided diagnosis for gram-stained smears images. In: Proceedings of the IEEE Conference on Computer Vision and Pattern Recognition Workshops, pp. 62–68 (2017). https://doi.org/10.1109/CVPRW.2017.113

23. Zhang, T., et al.: SlideNet: fast and accurate slide quality assessment based on deep neural networks. In 2018 24th International Conference on Pattern Recognition (ICPR), pp. 2314–2319. IEEE (2018)

24. Manda-Handzlik, A., Fiok, K., Cieloch, A., Heropolitanska-Pliszka, E., Demkow, U.: Convolutional neural networks-based image analysis for the detection and quantification of neutrophil extracellular traps. Cells **9**(2), 508 (2020)

25. Pang, S., et al.: A novel YOLOv3-arch model for identifying cholelithiasis and classifying gallstones on CT images. PLoS ONE **14**(6), e0217647 (2019)

26. Bernhard, W., Rouiller, C.: Close topographical relationship between mitochondria and ergastoplasm of liver cells in a definite phase of cellular activity. J. Biophys. Biochem. Cytol. **2**(4), 73 (1956)

27. Ju, M., Luo, H., Wang, Z.: An improved YOLO V3 for small vehicles detection in aerial images. In: 2020 3rd International Conference on Algorithms, Computing and Artificial Intelligence, pp. 1–5 (2020)

28. Aly, G.H., Marey, M., El-Sayed, S.A., Tolba, M.F.: Yolo based breast masses detection and classification in full-field digital mammograms. Comput. Meth. Programs Biomed. **200**, 105823 (2021)
29. George, J., Skaria, S., Varun, V.V., et al.: Using YOLO based deep learning network for real time detection and localization of lung nodules from low dose CT scans. In: Medical Imaging 2018: Computer-Aided Diagnosis, vol. 10575, pp. 347–355. SPIE (2018)
30. Hwang, W.H., Jeong, C.H., Hwang, D.H., Jo, Y.C.: Automatic detection of arrhythmias using a YOLO-based network with long-duration ECG signals. Eng. Proc. **2**(1), 84 (2020)
31. Zhuang, Z., et al.: Cardiac VFM visualization and analysis based on YOLO deep learning model and modified 2D continuity equation. Comput. Med. Imaging Graph. **82**, 101732 (2020)
32. Wu, H., Wu, B., He, S., Liu, P.: Congenital heart defect recognition model based on YOLOV5. In: 2022 IEEE 16th International Conference on Anti-Counterfeiting, Security, and Identification (ASID), pp. 1–4. IEEE (2022)
33. Kuo, H.C., et al.: Detection of coronary lesions in kawasaki disease by scaled-YOLOv4 with HarDNet backbone. Front. Cardiovasc. Med. **9**, 1000374 (2022)

Employing a Novel Metaheuristic Algorithm to Optimize an LSTM Model: A Case Study of Stock Market Prediction

Amin Karimi Dastgerdi[ID] and Paolo Mercorelli[(✉)][ID]

Leuphana University of Lueneburg, 21335 Lueneburg, Germany
`paolo.mercorelli@leuphana.de`

Abstract. It has been long since researchers as well as investors and stakeholders who are actively pursuing financial markets are trying to analyze stock price movements and predict its trend more accurately. To minimize the forecasting risk and make the most profit, several methods have been used among which Deep Learning was at the center of attention in recent years. Deep learning techniques include analyses of historical data and recognizing patterns that can assist scientists to make a more precise prediction. This paper focuses on the application of a optimization approach called Neural Network Algorithm to optimize Long short-term Memory for the prediction of financial time series. The findings reveal that the utilization of an optimization technique such as Neural Network Algorithm to optimize Long Short-term Memory neural networks results in a notable improvement of 40%, 65%, 4%, and 85% in the MAPE, Theil U, R, and RMSE metrics, respectively. Consequently, this leads to even more accurate results and more precise predictions.

Keywords: Deep Learning · Time Series Forecasting · LSTM · Optimization · Metaheuristic Algorithms

1 Introduction

Extensive research has been conducted to predict future trends in financial markets for decision-making purposes, employing statistical and econometric approaches. Predicting financial market behavior is regarded as a highly challenging task due to its noisy and unstable nature. Numerous factors, such as economic conditions, long-term macro policies, short-term market expectations, and unexpected events, contribute to the complex dynamics of financial markets. The rapid growth of financial markets, coupled with the expansion of investor participation and advancements in sensor networks and communication technology, has resulted in an abundance of historical stock data. This abundance of data poses a challenge for statistical experts in extracting valuable information. Some researchers argue that the high noise in financial time series is what makes them difficult to predict [1]. They contend that applying statistical models to predict financial time series requires data preprocessing, which compromises the integrity and authenticity of the data. Other statistical researchers attribute the difficulty of financial market prediction to the nonlinearity and nonstationary nature of these markets. They assert that only higher-level models can accurately describe such nonlinear financial time series. Consequently, effectively analyzing massive stock data, accurately predicting changes in the stock market, and capturing the market's behavioral patterns have become the focal points of recent research. This is where deep neural networks come into play.

Although artificial neural networks have a simple architecture, they possess a large number of parameters, making them prone to issues such as overfitting complexity and gradient dispersion, resulting in lower prediction accuracy [2]. However, the efficiency of deep neural networks, which represent the most advanced architectures of neural networks, has recently been proven in handling prediction problems involving large datasets and complex nonlinear relationships. This has provided researchers with increased confidence in employing deep neural networks for pattern recognition and detecting financial market behaviors. The inception of employing neural networks for exchange markets prediction dates back to 1977 [3]. Deep neural networks are now capable of discovering complex patterns in data and extracting high-level abstractions. Recent developments in neural network structures and the introduction of dropout turned deep learning into a hot spot in predicting financial markets [3]. However, finding an optimal set of model hyperparameters that produces the most accurate results is still a huge challenge. In this paper, NNA was employed to optimize an LSTM model by finding the optimal number of LSTM units. The results show that employing a metaheuristic algorithm such as NNA to optimize LSTM neural networks leads to even more accurate results and precise predictions.

In line with this, the subsequent sections of this paper are structured as follows. The second section provides a description of the materials and methods employed in this study, followed by a literature review and an outline of the problem statement. The third section elucidates the model framework and the proposed methodology. Section four contains the data collection and description as well as pre-processing and further information regarding the dataset and implementation details. Section five contains a comparative analysis of results and section six concludes the paper with findings indicative of the results.

2 Literature Review

For a considerable period, financial market prediction has been a prominent subject of interest for researchers due to its vital role in both individual and national economies. To increase the accuracy in time series predictions, a host of supervised learning (Artificial Neural Networks [4]) and unsupervised learning (Generative Adversarial Networks [5]) techniques have been employed so far. Deep learning models commonly utilized in financial market prediction encompass various architectures such as Recurrent Neural Networks (RNN), Long Short-term Memory (LSTM), Bidirectional LSTM (BiLSTM), Convolutional Neural Networks (CNN), and Recurrent Convolutional Neural Networks (RCNN). In this section, related works in this area have been reviewed.

Initially, Recurrent Neural Networks (RNNs) were introduced to capture sequential patterns using internal loops. However, researchers soon discovered a major drawback of RNNs: the tendency to forget previous state information over time. This issue arises during the backpropagation process, commonly known as the problem of gradient vanishing or exploding. When the gradient is propagated backward through the activation function, it can become extremely small or large. To address this problem, Hochreiter et al. proposed a solution in 1997 by introducing memory cells and gates in place of neurons. These gate structures allow for selective retention or deletion of cell information, enabling the network to decide whether to remember or forget information for an extended duration. This innovation made RNNs more suitable for tasks requiring continuous prediction, as the memory cells and gates improved the model's ability to capture and retain relevant long-term dependencies. [6]. This upgraded version of RNN is referred to as Long Short-term Memory (LSTM) which was proposed by Hochreater and Schmidhuber [6] and improved and promoted by Gers et al. [7] and Grave [8]. LSTM can integrate long information and short information well and solve the problems of gradient dispersion and gradient explosion [9]. These networks have achieved considerable success on many issues and have been widely used especially in time series prediction.

Pang et al. [10] proposed a powerful adaptive online gradient learning algorithm based on LSTM to predict time series with outliers. Chandar [11] combines the LSTM depth network and Grey Wolf optimization algorithm to predict multistep time series. In their work, Huang et al. [12] focused on utilizing Long Short-term Memory (LSTM) as the primary model for stock prediction. They employed the Bayesian optimization method to dynamically select parameters and determine the optimal number of units within the LSTM architecture. The results of their study demonstrated a significant improvement in prediction accuracy, with a 25% enhancement compared to traditional LSTM models. Another study, conducted by Dastgerdi et al. [13], investigated the effectiveness of noise elimination techniques, specifically Wavelet Transform and Kalman Filter, in the context of stock market prediction. The findings of their research indicated that the application of various noise elimination techniques can be advantageous in improving the accuracy of stock market predictions.

While the main aim of artificial neural networks is mapping the input data to the target data and reducing the error between the predicted solutions and target solutions, in optimization problems, the purpose is to find a feasible optimal solution in the solution space using a defined strategy. Metaheuristic algorithms are widely used to discover near-optimal solutions for optimization problems with large search spaces. These algorithms are usually inspired by observing phenomena and rules seen in nature such as the Genetic Algorithm (GA), the Particle Swarm Optimization (PSO), the Harmony Search (HS), the Simulated Annealing (SA), and so on [14].

Hegazy et al. [15] in the year 2014 in a research employed particle swarm optimization and LS-SVM to predict stock prices. In this study, some technical indicators like a stochastic oscillator, MFI, EMA, MACD, and RSI were used. By combining Artificial Neural Networks and metaheuristics, Gocken et al. [16] in a research in 2016 employed a hybrid model in which a Genetic algorithm and Harmony search were used to facilitate the best selection of technical indicators for stock price prediction. In 2018 Chung et al. [17] employed GA to optimize LSTM to predict stock market price. Yang et al. [18] in the year 2019 employed a hybrid system using particle swarm optimization and Brain Storm Optimization to predict stock prices. Wang et al. [19] in the year 2020 used the model integrating genetic algorithm with high convergence and artificial neural network to predict the stock market.

In 2020 Hui Liu and Zhihao Long [20] presented an EWT-dpLSTM-PSO-ORELM hybrid framework to predict the stock market price. In this framework, to increase the stability they employed an empirical wavelet transform to turn the financial time series into a more reliable time series as a preprocessing task and they utilized the outlier robust extreme learning machine for the purpose of error correction and improving the prediction accuracy. They also used the Dropout strategy to prevent overfitting and the PSO algorithm for optimizing the LSTM model by selecting the best hyperparameters. The experimental outcomes of their study illustrated the effectiveness of their approach to predicting stock market prices with high accuracy. In a research study conducted in 2021 by Yangzi Zhao et al. [21], an attention-guided deep neural network algorithm for stock prediction was proposed. The algorithm utilized a combination of synthesized daily stock social media text emotion index and stock technology index as the data source. To predict the stock market, an LSTM model was employed. The stock emotion index was extracted by constructing a social text classification emotion model using BiLSTM with attention mechanism and glove word vector representation algorithm. Similarly, in 2021, Nan Jing et al. [22] proposed a hybrid deep learning approach for predicting stock prices. They utilized a CNN to classify the investors' hidden sentiments and an LSTM model to analyze the technical indicators of six key industries of the Shanghai Stock Exchange (SSE). The experimental outcome of their study illustrates that their prosed framework has achieved superior results and is more effective in predicting stock prices compared to the models without sentiment analysis. In 2022, Wang, J., et al. [23] Proposed a hybrid model which takes advantage of secondary decomposition for noise elimination and multi-factor analysis on an LSTM network to predict four major Asian stock markets. In this study, the authors supplemented the LSTM model with an attention layer to increase the weight of effective information. The outcomes of their study illustrated a 30% improvement in prediction accuracy. Ali, M., et al. [24] in the year 2023 proposed a framework based on a new version of empirical mode decomposition and LSTM to predict the KSE-100 index of the Pakistan Stock Exchange. The empirical mode decomposition was used to divide the stock data into multiple components and use the most correlated ones to feed the LSTM network. The predictions of this approach proved to be superior in comparison with SVM, Random Forest, and Decision Trees.

3   Research Methodology

In this study, daily price information from four different stock markets (Apple, Amazon, Microsoft, and Google) were used as inputs. After scaling and other preprocessing tasks, NNA was used to find the optimal number of units to train the model. In this study authors used NNA to optimize an LSTM model because the superiority of this algorithm over other metaheuristic algorithms has been approved [14]. To build an appropriate LSTM model for training, there isn't much concrete and general information that works well for all models and all purposes. As a result, machine learning engineers often rely on information related to previous studies and projects or adjust the required hyperparameters by trial and error. Since NNA is an unsupervised adaptive metaheuristic algorithm that doesn't require any information or initial fine-tuning of the parameters, it can be a suitable choice in this study. The NNA is a single-layer perceptron optimizer with a self-feedback mechanism that updates the solutions by learning from the environment. In this study, this process occurs in the model evaluation phase in which an LSTM model is evaluated by the NNA iteratively to find the optimal number of units. Once NNA finds the optimal unit numbers, a stacked LSTM neural network will be constructed based on the obtained unit numbers and will be trained using training data as input. Subsequently, the trained LSTM model will be utilized to make predictions on unseen data. The predictions obtained from the model will then be compared to the actual values. This comparative analysis serves as an evaluation of the effectiveness and efficiency of the proposed LSTM model. The flowchart of the proposed NNA-LSTM model is illustrated in Fig. 1.

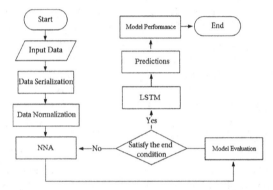

Fig. 1. Flowchart of the proposed NNA-LSTM model

3.1 Neural Network Algorithm

NNA is a new metaheuristic optimization algorithm, inspired by biological nervous systems and artificial neural networks (ANNs) which has been proposed by Sadollah et al. [14] for solving complex optimization problems. To generate new candidate solutions, NNA exploits the complicated configuration of the ANNs and their operators. The main advantage of NNA over other optimizers is its statistical superiority and fine-tuning of initial parameters effortlessly.

NNA begins with a population of pattern solutions ($X_{Pattern}$). As the ANNs mostly used for prediction purposes, it receives input data and target data and predicts the relationship between input and target data. Therefore, inspired by the ANNs, in the NNA, the best-obtained solution at each iteration is assumed as target data and the aim is to reduce the error (cost) among the target data and other predicated pattern solutions. Based on the defined concept, the NNA is developed for minimization problems (i.e., minimizing the error among the target and pattern solutions). It is worth pointing out that this target solution (X_{Target}, W_{Target}) is being updated at each iteration. The suggested strategy for the bias operator applied to new input solutions and updated weight matrix can be seen in Fig. 2.

```
For i in range (1, Npop):
  if rand ≤ β:
    #Bias for New Pattern Solution
    #No. of biased variables in population of new pattern solution
    Nb = Round (D × β)
    For j in range (1, Nb):
      XInput (i, Integer, rand [0, D]) = LB + (UB - LB) × rand
    #Bias for Updated Weight Matrix
    #No. of biased variables in updated weight matrix
    Nwb = Round (Npop × β)
    For j in range (1 to Nwb):
      WUpdated (j, Integer, rand [0, Npop]) = U (0,1)
```

Fig. 2. The suggested strategy for the bias operator in NNA [14].

Lower and upper bounds are the dimensions of domain spaces of LB and UB, and the β factor shows the extent of modification and alteration of the solution and determines the degree to which the solutions need to be modified. The value of β varies between 0 and 1 during the implementation of the algorithm. In the initial iterations of the algorithm, the β value is assumed to be higher because the exploration must be carried out. Then, the iteration rate is reduced to zero by a t-dependent linear or non-linear equation. The value of β in the final iterations of the algorithm should be reduced because the exploit is not to get away from the best solution obtained [14].

3.2 Long Short-Term Memory

The Long Short-term Memory (LSTM) architecture is specifically designed to address the challenges of vanishing or exploding gradients and learn from sequential data. With the inclusion of memory units, LSTMs are capable of effectively predicting time series data with time lags. These memory units can retain valuable time-related information for an indefinite period [25]. A memory cell in an LSTM comprises four key components: an input gate, a target gate, a forget gate, and a recurrent neuron. The input gate determines whether an incoming signal can impact the state of a memory cell, while the target gate determines whether the state of one memory cell can influence another. Moreover, the target gate also regulates the decision of whether to retain or forget the previous state [26]. The cell state, depicted as a horizontal line in Fig. 3, plays a crucial role in LSTM operations. Throughout the entire sequence, information flow within the cell state undergoes minor linear interactions without necessarily being modified. LSTMs modify the cell state information using gates, which are carefully regulated structures that facilitate the optional modification or passage of information.

First, the forget gate decides what information is going to be thrown away from the cell state by a sigmoid layer which looks at h_{t-1} and x_t and outputs a number between zero and one for each number in the cell state C_{t-1}.

$$f_t = \sigma(W_f[h_{t-1}.x_t]+b_f) \tag{1}$$

Then, LSTM decides what new information should be stored in the cell state. This step contains three parts. First, the input gate layer which is a sigmoid layer decides

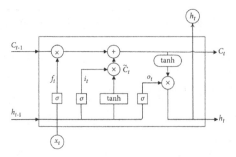

Fig. 3. Structure of an LSTM unit

which values should be updated. Next, a tanh layer creates a vector of new candidate values, $\tilde{C}_t$, that could be added to the state. Then, these two will be combined to create an update to the state [25].

$$i_t = \sigma(W_i[h_{t-1}.x_t] + b_i \tag{2}$$

$$\tilde{C}_t = \tanh(W_C[h_{t-1}.x_t] + b_C \tag{3}$$

In the third step, the old cell state, C_{t-1}, will be updated into the new cell state C_t. To forget the information that was decided to be forgotten, the old state will be multiplied by f_t. Then $i_t * \tilde{C}_t$ will be added to create the new candidate values, scaled by how much it has been decided to update each state value.

$$C_t = f_t * C_{t-1} + i_t * \tilde{C}_t \tag{4}$$

The output gate plays a crucial role in determining which outputs should be filtered or passed through. All of these gates utilize a sigmoidal nonlinearity function, allowing them to selectively decide which parts of the cell state should be included in the output. To ensure that the values are within a specific range, typically between -1 and 1, the cell state is passed through a hyperbolic tangent (tanh) function. The resulting values are then multiplied by the output of the sigmoid gate. This multiplication process ensures that only the parts of the cell state determined by the gates are outputted. Moreover, the modified cell state can serve as an additional input to other gating units, allowing for complex interactions within the LSTM architecture.

$$o_t = \sigma(W_o[h_{t-1}.x_t] + b_o) \tag{5}$$

4 Implementation

For this study, all the data utilized was sourced from "Yahoo Finance," a prominent media platform known for providing financial news, data, and analysis. Additionally, Yahoo Finance offers various online tools for personal finance management. The researchers employed the Yahoo Finance API in Python to collect the necessary data for their study. The data collected spanned a period of 10 years, encompassing all trading days between September 10, 2011, and September 10, 2021.

4.1 Project Environment

The experiment was conducted using the Python interpreter version 3.8 as the programming language. The network model structure was constructed using Keras, which is based on TensorFlow. To implement the proposed model, several Python packages were required, including NumPy, pandas, scikit-learn, matplotlib, and Plotly. All simulations were carried out on Google Colab, utilizing its Tesla T80 GPU for computational purposes.

4.2 Data Description

To increase the generalizability of our model and the achieved outcomes, the data from four different companies have been selected. This also allows us to hinder specific market situations affecting the credibility of the implemented model and providing a natural condition to assess its robustness and efficiency. The data includes the basic price information of Microsoft, Amazon, Apple, and Google for ten years from September 10, 2011, to September 10, 2021, which consists of 2517 trading date records. To convert this time-series data to a supervised learning problem, the data has been turned into 2417 samples of 100 historical time slots to predict one day ahead price. Furthermore, by using the data from four different markets with different conditions the stability of the results can be evaluated.

4.3 Feature Selection

Two sets of variables were utilized as input for the model. The first set, referred to as "basic information," consisted of the opening price, highest price, lowest price, last price, and volume of transactions. The second set of input variables comprised six commonly used technical indicators. These indicators were MACD (Moving Average Convergence Divergence), RSI (Relative Strength Index), WMA (Weighted Moving Average) and Ichimoku. These indicators are widely employed in technical analysis to gain insights into market trends and price movements. Following most of the machine learning projects, the obtained dataset is divided into three sets: training, validation, and test dataset with proportions of 0.8, 0.1, and 0.1 respectively. To compute the gradient in weight space with respect to the loss function and eventually train the LSTM model the backpropagation algorithm with a learning rate of 0.05 has been employed. The batch size and frequency of the training set have been set to 32 and 100 respectively. The dropout method with a rate of 0.3 for each LSTM layer has been used to prevent overfitting. The learning rate controls the model convergence which is a descending function of time. The architecture of the LSTM model which has been used in this study is illustrated in Fig. 4.

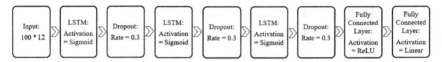

Fig. 4. LSTM model architecture

5 Results

While with a small amount of LSTM units the model, complexity is reasonable and our model can be trained in a short time, we will face underfitting which is a phenomenon that occurs when our model has not been trained well and cannot illustrate a good performance in predicting new data. Nevertheless, increasing the number of LSTM units directly influences the model fitting complexity and training time without any considerable performance improvements. Therefore, NNA can make an appropriate balance between poor performance and high complexity. The optimal number of LSTM units found by NNA varies for different financial markets and also for different periods in which the input data is selected.

Figure 4 displays the training and validation loss of LSTM models for various markets. It is evident from the different plots that as the number of training iterations increases, both the training loss and validation loss decrease. This trend indicates that the model's prediction error diminishes with additional training epochs. Eventually, the training loss and validation loss tend to converge, indicating a lower overall prediction error for the LSTM model.

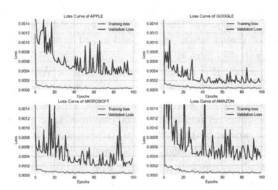

Fig. 5. Training loss and validation loss of the LSTM model for different companies

Figure 5 displays the training and validation loss of LSTM models for various markets and Fig. 6 represents the plot depicting the performance of NNA-LSTM for Amazon stock price versus actual values.

To evaluate the performance of the proposed deep LSTM models, four key error metrics were utilized: RMSE, MAPE, R, and Theil U. Table 1 presents the performance of the LSTM model with randomly set unit numbers (referred to as "LSTM") and the model with optimal unit numbers determined by the NNA algorithm (referred to as

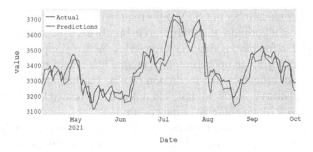

Fig. 6. Predictions of the NNA-LSTM model for Amazon price

"NNA-LSTM"). The table provides a comparison of the two models in terms of their performance based on the specified error metrics.

Table 1. Predictive accuracy is based on different metrics for different markets.

	Metric	Google	Microsoft	Amazon	Apple	Improvement
NNA-LSTM	MAPE	0.017	0.011	0.009	0.013	40%
LSTM		0.024	0.027	0.018	0.022	
NNA-LSTM	Theil U	0.010	0.006	0.006	0.008	65%
LSTM		0.025	0.022	0.027	0.023	
NNA-LSTM	R	0.932	0.963	0.889	0.943	4%
LSTM		0.888	0.915	0.827	0.904	
NNA-LSTM	RMSE	53.814	3.826	46.370	2.250	85%
LSTM		78.957	11.223	70.311	15.389	

It is clear from Table 1 that in all cases LSTM models which were constructed based on the NNA had better performance and lower error in predicting the actual value. That is, MAPE, Theil U, R, and RMSE metrics have shown levels of improvement of 40, 65, 4, and 85 percent respectively.

6   Conclusion

LSTMs have proved to be able to learn long-term dependencies of time-series data, especially longitudinal ones. However, their performance is affected by many hyperparameters such as time lag, unit numbers, initial learning rate, dropout rate, etc. Hence, a careless tuning of hyperparameters will lead to either overcomplexity of the model or poor results. In this regard, employing an appropriate algorithm that is capable of finding the optimal hyperparameters is of great importance. In order to enhance the accuracy of the LSTM neural network in predicting financial time series, this paper introduces a metaheuristic algorithm called NNA. The NNA algorithm is proposed as a means to determine the optimal number of LSTM units required for the prediction task. By leveraging metaheuristic algorithms, the study aims to improve the precision and reliability of the LSTM model in forecasting financial time series. With a glimpse of the final results, it is clear that employing an optimization technique like one of the metaheuristics can improve the model efficiency and prediction results. Furthermore, choosing the right number of LSTM units is a difficult task and it is done by researchers only through trial and error which is so demanding and time-consuming.

One of the effective factors in enhancing the performance of an LSTM model is to find the optimal hyperparameters. Although there are some methods available to search among the solutions and find the best, these methods only suggest an approximate of the best solution while using a metaheuristic algorithm such as NNA can help us to find the optimal solution faster and with lower computation cost. Future works can be evaluated and compared with the findings of the present study by using a combination of metaheuristic algorithms, using optimization methods to find other hyperparameters of LSTMs, using Japanese candlestick charts, ascending and descending patterns, and the combination of their results.

References

1. Gálvez, R.H., Gravano, A.: Assessing the usefulness of online message board mining in automatic stock prediction systems. J. Comput. Sci. **19**, 43–56 (2017)
2. Miao, M., Cai, W., Li, X.: Parameter estimation of gamma-gamma fading with generalized pointing errors in FSO systems. Wirel. Commun. Mob. Comput. **2021**, 1–21 (2021)
3. Göçken, M., Özçalıcı, M., Boru, A., Dosdoğru, A.T.: Stock price prediction using hybrid soft computing models incorporating parameter tuning and input variable selection. Neural Comput. Appl. **31**(2), 577–592 (2017). https://doi.org/10.1007/s00521-017-3089-2
4. Wanjawa, B.W., Muchemi, L.: ANN model to predict stock prices at stock exchange markets. arXiv preprint arXiv:1502.06434 (2014)
5. Zhang, K., et al.: Stock market prediction based on generative adversarial network. Procedia Comput. Sci. **147**, 400–406 (2019)
6. Hochreiter, S., Schmidhuber, J.: Long short-term memory. Neural Comput. **9**(8), 1735–1780 (1997)
7. Gers, F., Long short-term memory in recurrent neural networks. 2001, Verlag nicht ermittelbar
8. Graves, A.: Long Short-Term Memory. In: Supervised Sequence Labelling with Recurrent neural Networks, pp. 37–45. Springer, Heidelberg (2012)
9. Kong, H., et al.: A novel torque distribution strategy based on deep recurrent neural network for parallel hybrid electric vehicle. IEEE Access **7**, 65174–65185 (2019)
10. Pang, X., Zhou, Y., Wang, P., Lin, W., Chang, V.: An innovative neural network approach for stock market prediction. J. Supercomput. **76**(3), 2098–2118 (2018). https://doi.org/10.1007/s11227-017-2228-y
11. Kumar Chandar, S.: Grey Wolf optimization-Elman neural network model for stock price prediction. Soft. Comput. **25**(1), 649–658 (2020). https://doi.org/10.1007/s00500-020-05174-2
12. Huang, B., et al.: Stock prediction based on Bayesian-LSTM. in Proceedings of the 2018 10th International Conference on Machine Learning and Computing. 2018
13. Dastgerdi, A.K., Mercorelli, P.: Investigating the effect of noise elimination on LSTM models for financial markets prediction using Kalman filter and wavelet transform. WSEAS Trans. Bus. Econ. **19**, 432–441 (2022)

14. Sadollah, A., Sayyaadi, H., Yadav, A.: A dynamic metaheuristic optimization model inspired by biological nervous systems: neural network algorithm. Appl. Soft Comput. **71**, 747–782 (2018)
15. Hegazy, O., O.S, Soliman., Salam, M.A.: A machine learning model for stock market prediction. arXiv preprint arXiv:1402.7351 (2014)
16. Göçken, M., et al.: Integrating metaheuristics and artificial neural networks for improved stock price prediction. Expert Syst. Appl. **44**, 320–331 (2016)
17. Chung, H., Shin, K.-S.: Genetic algorithm-optimized long short-term memory network for stock market prediction. Sustainability **10**(10), 3765 (2018)
18. Yang, B., Zhang, W., Wang, H.: Stock market forecasting using restricted gene expression programming. Comput. Intell. Neurosci. **2019**, 1–14 (2019)
19. Wang, Y., Guo, Y.: Forecasting method of stock market volatility in time series data based on mixed model of ARIMA and XGBoost. China Commun. **17**(3), 205–221 (2020)
20. Liu, H., Long, Z.: An improved deep learning model for predicting stock market price time series. Digit. Sign. Process. **102**, 102741 (2020)
21. Zhao, Y.: A novel stock index intelligent prediction algorithm based on attention-guided deep neural network. Wirel. Commun. Mob. Comput. **2021**, 1–21 (2021)
22. Jing, N., Wu, Z., Wang, H.: A hybrid model integrating deep learning with investor sentiment analysis for stock price prediction. Expert Syst. Appl. **178**, 115019 (2021)
23. Wang, J., et al.: Asian stock markets closing index forecast based on secondary decomposition, multi-factor analysis and attention-based LSTM model. Eng. Appl. Artif. Intell. **113**, 104908 (2022)
24. Ali, M., et al.: Prediction of complex stock market data using an improved hybrid EMD-LSTM model. Appl. Sci. **13**(3), 1429 (2023)
25. How, D.N.T., Loo, C.K., Sahari, K.S.M.: Behavior recognition for humanoid robots using long short-term memory. Int. J. Adv. Rob. Syst. **13**(6), 1729881416663369 (2016)
26. Yu, W., Li, X., Gonzalez, J.: Fast training of deep LSTM networks. In: Lu, H., Tang, H., Wang, Z. (eds.) Advances in Neural Networks – ISNN 2019. ISNN 2019. Lecture Notes in Computer Science, vol. 11554, pp. 3–10. Springer, Cham (2019).https://doi.org/10.1007/978-3-030-22796-8_1

Driver Dozy Discernment Using Neural Networks with SVM Variants

Muskan Kamboj[1]([⊠]) [iD], Janaki Bhagya Sri[1] [iD], Tarusree Banik[1] [iD], Swastika Ojha[1] [iD], Karuna Kadian[1] [iD], and Vimal Dwivedi[2,3] [iD]

[1] Indira Gandhi Delhi Technical University for Women, Kashmere Gate, Delhi, India
muskankamboj001@gmail.com, karunakadian@igdtuw.ac.in
[2] University of Tartu, Tartu, Estonia
[3] Queen's University Belfast, Belfast, UK

Abstract. A driver's lack of concentration or distraction is one of the main reasons for causing road accidents. Thus, increasing the driver's awareness at the ideal moment will reduce the possibility of an accident of any kind. There were around 155 thousand accidents in India, and around 40 percent of accidents were caused by driver distraction, mainly due to driver drowsiness. Detecting drowsiness or fatigue prior to an accident will help reduce these accidents. There are several ways we may execute this. One of the easiest and most effective ways is through artificial intelligence and machine learning algorithms. We consider both physiological and behavioral categories, such as face movement and eye closure movements, to detect drowsiness. Further, training a particular model with different types of eye movements helps in detecting driver conditions. Driver drowsiness detection can be improved by continuously monitoring the driver via video, which helps in real-world applications, and by expanding the dataset through training, we get high accuracy and unrecognizable losses. Therefore, in this paper, we use the MRL dataset, which contains images from every angle and in every shade. To train the existing model with this dataset, we use image processing techniques and classification techniques to distinguish images of open and closed eyes on the basis of accuracy and loss function, a comparison of SVM (Support Vector Machine) and CNN (Convolutional Neural Network) models has been performed. As a result, CNN is considerably better than SVM and it is an effective technique for dozy detection.

Keywords: Driver Drowsiness · CNN · SVM · Image Processing · Greyscale Images · Eye Tracking

1 Introduction

A driver's lack of focus or attention is one of the major reasons for road accidents. Bus drivers on long-distance routes or overnight buses, as well as truck drivers who operate for extended periods of time (particularly at night), are more prone to this issue. Numerous people are hurt or killed in traffic accidents each year as a result of driver drowsiness. Therefore, due to its enormous practical relevance, the discernment of driver weariness

M. Singh et al. (Eds.): ICACDS 2023, CCIS 1848, pp. 490–501, 2023.
https://doi.org/10.1007/978-3-031-37940-6_40

and its portent is an important study area [10]. The definition of driver distraction includes four essential components. Firstly, a misallocation of recognition eyes off the road; Secondly, a diversion of attention towards a competing activity, inside or outside the vehicle, which may or may not be related to driving; Thirdly, a diversion of attention toward the competing activity may be required or encouraged by it; and lastly, there is an implicit or explicit, compensation or discipline for drawing attention away under the presumption that driving safety is negatively influenced.

Since actual on-road drowsiness acquisitions are typically dangerous, researchers frequently depend on simulated scenarios to create such systems. Researchers have created a number of techniques to identify driver sleepiness due to the dangers that weariness poses on the road, and each methodology has advantages and disadvantages of its own [11]. Despite the fact that driving while distracted impairs cognitive function, researchers have identified a number of distraction detection strategies, including visual, physical, and cognitive distractions proposed by Stancin et al. [14]. By holding their hands, eyes, or intellect (mechanical interruption, retinal hindrance, or cognitive diversion) aside from either the steering wheel, drivers are increasingly diverting their attention away from their primary activity of driving and toward secondary tasks unrelated to driving. However, certain activities might also include combining two or even three different distractions (using a cell phone, for instance, while reading, messaging, or mulling about something like a text while chatting), which increases the chance of an accident. Added varieties of driving vibratory stimuli can cause mixed vocal and perceptual disorientation. Because it is tricky to see what a driver's intellect (as contrasted with palms or vision) is doing and how moving, intellectual distractions may be the most problematic sort of inspiration to examine.

Many investigations that used simulations were conducted out in the field and were carried out in order to investigate and comprehend accidents from a psychological point of view and to come up with ways to decrease their consequences. As illustrated [6], utilizing data acquired either by phenomenology on the road study, a detailed assessment of the implications of driver indifference on accident rates was conducted. Therefore, results demonstrate that risk is significantly higher when a driver is fatigued or engaged in a challenging task while operating a vehicle, and risk is significantly lower when the driver looks away from the road to scan the surrounding area [8]. SVM and CNN are preferably and frequently used methods in the field of DDDT, based on the research papers we reviewed [13] (Fig. 1).

These models are used as a result of comparative analysis. Our paper is novel in two ways - firstly, a comparative analysis between SVM and CNN models on the basis of the loss function and accuracy function. Secondly, researchers can further study in the field of diver drowsiness technique using SVM and CNN models for better enhancements.

This paper is also sub-classified into seven main sections in which Sect. 2 represents the Related Work which includes. Driver dozy discernment method and techniques. Section 3 represents the Data and Evaluation Metrics which comprises Data preprocessing, Dataset Description and Evaluation metrics which includes two types of functions i.e. Accuracy and Loss Function. Section 4 represents the Model Description which includes classification models for comparative analysis between CNN and SVM on the basis of certain performance metrics. Section 5 represents the Implementation and

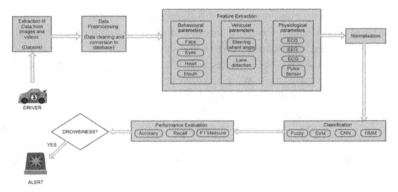

Fig. 1. Block Diagram of Driver Drowsiness Detection

Results which includes Computing environment specifications and a description of the result table graph. Section 6 represents the conclusion and the below content represents the References respectively.

2   Related Work

Driver dozy discernment method and techniques- In this, car owners who will not take regular breaks, on the report of specialists, are very much susceptible to falling unconscious while behind wheels. Thus, according to the evidence, fatigue operators who are required to take a break, produce more car crashes than alcohol-poisoning drivers. Concentration Aid, which seems to have programmable intensity, may advise drivers about their present level of drowsiness in addition to the quantity of time they have been traveling during their prior break. When a threat is perceived, the Priority Assistant will also alert the Positif guidance system to nearby service areas. Focus Assistance can assess indifference and exhaustion all throughout a diverse range of speeds and this is proposed by Deng et al. [5]. Albasrawi et al. [1] proposed the implementation of the drowsiness protection system (Fig. 2).

The above dissertation examines the progress of ai - enhanced or rather collaborative automobiles that really can warn or repudiate the client throughout awful circumstances but rather distribute essential data through crisis situations towards the owner's motor, the officials, maybe both. Tiredness resulting from insomnia is a significant component of the escalating number of fatalities along today's modern roadways. Researchers demonstrate a real-world hazard experiment in just this research which inhibits the speed of the car so when the driver is exhausted. An aim of this type of strategy is to establish a system that can detect traffic incident indicators and restrict vehicle speed in order to avoid accidents [2]. We describe a method for identifying driving drowsy that incorporates detectors including eye movement sensors to estimate a driver's level of exhaustion in the paper. If indeed the passenger is determined to be napping, the beep starts sounding and the car combustion is switched off, as proposed by Kumar et al. [10]. Utilizing Sensors to recognize Drowsiness Detection: Studies have employed the essential methods for evaluating drowsiness detection: (1) vehicular approaches, (2) cognitive approaches,

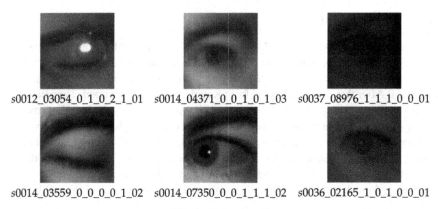

s0012_03054_0_1_0_2_1_01   s0014_04371_0_0_1_0_1_03   s0037_08976_1_1_1_0_0_01

s0014_03559_0_0_0_0_1_02   s0014_07350_0_0_1_1_1_02   s0036_02165_1_0_1_0_0_01

Fig. 2. Dataset used

and (3) physiological strategies. A comprehensive review of such measures might shed some light just on existing technologies, any obstacles they are undergoing, and indeed the enhancements that need to be addressed in order to formulate a sustainable operation [3]. Our research explored all various controller parameters and considers their benefits and drawbacks. It has also been established that drowsiness has been experimentally modulated in numerous ways. It is asserted that applying the integrated fatigue detection and recognition, particularly merging anti-physiological indicators with other parameters, can accurately estimate a performer's level of tiredness. Numerous car accidents might be prevented using these detection techniques. Vesselenyi et al. [15] proposed, actively monitoring Drowsiness Detection Screening and Collision Avoidance: This study demonstrates how and where to establish an interaction for monitoring driving weariness using dip al within continuous eye monitoring. Nanometer sleepers, or brief periods of awake lasting up to three seconds, have emerged as a significant predictor of overall depletion symptoms. As a consequence, the vehicle's lethargy can sometimes be diagnosed by persistently scrutinizing his or her eyes with a camera, and an immediate warning can indeed be conveyed. The main purpose of this project is to develop systems that should optimize automobile pedestrian safety whilst also implementing advanced image recognition and microprocessor innovation. This gadget recognizes sluggish steering, generates an alarm as well as decreases the position of the vehicle. Fouzia et al. [7] proposed that the methodology of diagnosing fatigue is supplemented through continuous range surveillance by both the Infrared sensor. The sensor module recognizes the obstruction as well as cautions safety for passengers, while also minimizing the speed and position.

3   Data and Evaluation Metrics

This section provides a brief description about data preprocessing, data description and evaluation metrics i.e., accuracy and loss function. We have considered these functions for performing comparative analysis between SVM and CNN.

3.1 Data Preprocessing

We made use of the eye images in the MRL Eye data. A sizable collection of images of the naked eye makes up the MRL Data. This dataset contains low-high-resolution infrared light images that were all captured using multiple tools and in diverse illumination conditions. After the dataset is loaded from Kaggle, the file is unzipped. Then the data is divided into two classes, that is open eyes and closed eyes classes. This is done by using libraries such as 'os' and 'cv2'. Then the training data is made by putting Open eyes and Closed eyes images in an array for training data. Then the length of the dataset for training is checked. The dataset is randomly shuffled to reduce redundancy in the data using the 'random' library. The images are reshaped with the 'NumPy' library. Next, the modeling is done through the 'MobileNet' model of 'TensorFlow' and 'Keras' libraries.

3.2 Dataset Description

We used the MRL Eye dataset which contains eye images. The detection of eyes and their features, such as gaze estimate and eye-blinking frequency, is a crucial job in computer vision. [17] A sizable collection of images of the human eye makes up the MRL Eye Dataset. This dataset comprises low-resolution and high-resolution infrared images that were all captured using various tools and lighting conditions. For this dataset, testing various features or learnable classifiers is feasible. The photos are divided into a wide range of groups to make it simpler to compare different approaches, which also makes them appropriate for creating and evaluating classifiers. The dataset is available publicly. In the dataset, the pictures are labeled, and the properties are annotated in the dataset labels. The labels are as follows: Subject ID: There are 37 people, 33 men, and 4 women ie. The subjects have their own IDs. 2. Image ID: the dataset consists of an ID for every image. 3. Gender: It is encoded as 0 for men and 1 for women. 4. Glasses: It contains information on whether the subject has glasses or not, 0 for no glasses, and 1 for glasses. 5. Eye-state: The eye state of the subject is represented by the eye-state encoding which is 0 for closed eyes and 1 for opened eyes. 6. Based on the types of reflections: 0 for no reflections and 1 for small reflections and 2 for bigger ones. 7. Lighting conditions: If the lighting of the image is good, the encoding for the same is 1, if the lighting conditions are bad, the encoding for that is 0. 8. Sensor ID: The different sensors through which the videos are captured, ie, 1 for RealSense, 2 for IDS and 3 for Aptina sensor. Furthermore, images from three separate sensors are included in the collection (Intel RealSense, IDS Imaging sensor, and Aptina). The gaze sensor is built on the histogram of oriented gradients (HOG) and the classifier is used to yield eye pictures. The suggested dataset includes pictures that can be utilized to build gaze detection. This dataset can be found on Kaggle. We took an API extension from Kaggle so that we could use it efficiently on Google Colab. We had to first upload and then unzip the file which had closed-eye and open-eye datasets.

3.3 Evaluation Metrics

The evaluation of a model can only be done if there are metrics that can be used to compare them. This is where a confusion matrix comes to help. A confusion matrix has

4 different cells for actual and predicted values for a classifier which are typically True Positive, True Negative, False Positive, and False negative. True Positive is all the values that were accurately predicted as true and were actually true, similarly, False negatives are all the values that were predicted as false by the classifier but were actually true values. Here True and False are merely classes for a binary classifier that can be used for calculating comparative metrics for classifiers.

Accuracy

The model's performance across all classes is often described by its accuracy metric. It is beneficial when each class is given equal priority. It is calculated by dividing the predictions by all the forecasts.

$$Formula\ for\ accuracy = \frac{(TP + TN)}{(TP + TN + FP + FN)}$$

where TP stands for true positive and FN stands for False negative and so on.

Loss function

The total of the errors in our model is represented by a value called loss. It assesses how well our model performed. If the errors are extensive, the model is not performing effectively because of a huge loss. Conversely, the lesser it is, the better the model performs. Using a loss or cost function, the loss is calculated. There are numerous cost functions to choose from. The best one to utilize depends on the problem because they each penalize mistakes differently. The most often employed metrics for classification and regression issues, respectively, are Cross-Entropy and Mean Squared Error. Formulae-

$$H_r(p) = -\frac{1}{N}\sum_{i=1}^{N} y_i.\log(p(y_i)) + (1 - y_i).\log(i - p(y_i))$$

yi and j stand for the true value, which is 1 if sample i is a member of class j and 0 otherwise. pi,j stands for the likelihood that sample i belongs to class j as predicted by the model.

The Mean Squared Error is calculated as:

$$MSE = \frac{1}{n}\Sigma(actual - forecast)^2$$

where: n is a sample size, actual is the actual data value and a forecast is the predicted data value.

4 Model Description

The classification models that have been used for the study are Convolutional Neural Network (CNN) and Support Vector Machine (SVM). These classifier models have been used for the study for finding the better classifier for driver drowsiness detection. CNN is a section of Deep Learning while SVM is a Machine learning model. We will see the comparative study of both models on the basis of certain performance metrics.

4.1   Convolutional Neural Network (CNN) Model

The Convolutional Neural Network (CNN), which is a subset of Deep Learning, has now been employed as the initial model in this instance [4]. CNNs are a subset of deep neural networks that are often employed in deep learning to analyze visual data. Deep convolutional networks are composed of many levels of neurons and Artificial neurons are numerical operations such that, in a manner similar to that of their native counterparts, calculate the weight value of a number of input and output an activation value. Whenever an input is supplied, the layer of a CNN generates a set of activation functions which are then transferred onto the subsequent layer. Edges that run horizontally or diagonally are typical primary elements that are extracted from the first layer. This result is sent on to the next layer, which finds more distinctive features like vertices and numerous edges. Objects, faces, and other complex elements may be recognized by the network as we delve further into it. Depending on the activation maps of the last convolution layer, the TensorFlow location layer gives a sequence of confidence scores that show how likely it is for the image to belong to a class. Each convolutional layer may be followed by a maximum pooling layer. Small rectangular pieces of the convolutional layer are subsampled by the pooling layer to provide an output out of each block. The maximum or mean or a learned combination of the neurons can be used as a starting point for this pooling, among other methods. [16] Convolutional layer: Assume that the convolutional layer is followed by an N × N square neuron layer. Our convolutional layer output will have a size of (Nm + 1) × (Nm + 1) if we employ an m × m filter. We must add the contributions (weighted by the filter components) from the preceding layer cells in order to calculate the pre-nonlinearity input to a particular unit xl in our layer:

$$x^l{}_{ij} = \sum_{a=0}^{m-1} \omega_{ab} \; y^{l-1}_{(i+a)(i+b)}. \tag{1}$$

The above represents a convolution. Max pooling layers: The simple max-pooling layers don't learn for themselves. For some k × k regions, they only generate a single number that is the highest in that region. For instance, since the max function condenses each k × k block to a single value, it would produce an N × k × N × k layer assuming the input layer is an N × N layer.

Transfer Learning: It is a term used in machine learning to describe how a model that has been pre-trained on some datasets before can work on another specific dataset using its previous knowledge about the other datasets. In transfer learning, a machine uses the information gained from previous work to improve prediction about a new task. For example, while training a classifier to determine whether an image contains a person, you may use the knowledge you learned to distinguish facial features. Reduced training time, enhanced neural network performance (in most cases), and the absence of a significant amount of data are three of transfer learning's most significant benefits. One of the famous models in 'TensorFlow' library of python for CNN transfer learning is MobileNet which has been used to create the model. MobileNet: MobileNets are a family of CNN models that can be used for classification, detection, etc. They are small, low-latency, and low-power models. Because of their small size, these are considered to be suitable deep-learning models for use on mobile devices. Although there is a slight accuracy trade-off, MobileNets are faster and more efficient than other models like VGG16 because they are compact, lower in size, and contain fewer parameters.

4.2 Support Vector Machine (SVM) Model

One of the most well-known and frequently used models is the SVM (Support Vector Machine) model. This approach can be applied to problems involving classification and regression. It is, however, overused in machine learning to make classifications. The foundation of classification is comprehending the features and grouping them into several groups [12]. The goal of the SVM algorithm is to create a decision boundary or best-fit line that can divide n-dimensional space into classes and enable us to recognize or locate non-trainable points in their ideal feature. In n-dimensional space, each item is represented by a point, and its coordinates are referred to as features. By constructing the hyperplane (which can be 2-Dimensional or 3-Dimensional) in such a way that all of the TensorFlows belongs to the same category of CNN are on one side of the hyperplane and various varieties are on the other, SVM is able to turn the TensorFlow [18]. Different varieties of numerous hyperplanes will be present and will be divided into two groups. SVM searches for the category separation that optimizes the distance between points in either category. A margin is a name for this separation. The supporting vectors are the points that fall precisely on the margin. This model needs the training set in order to find the hyperplane. The convex optimization issue that maximizes the margin is resolved by SVM.

$$Max(\theta, b) \, ||\theta||^{-2}$$

where the constraint says the points should be on the correct side i.e.,

$$\theta^T x + b \leq 1, \forall x \in C_1$$

$$\theta^T x + b \geq 1, \forall x \in C_2$$

SVM comes in two varieties:

Linear SVM: Linear SVM usually deals with simple data that can be separated by linear separation. Here data is divided into two classes based on their properties. This sometimes works as regression.

Non-linear SVM: NLSVM (Non-Linear Support Vector Machine) deals with high-level data that cannot be separated in a straight line. If data contains more than two classes. This works by mapping data non-linearly. Nonlinear mapping: In non-linear mapping sometimes, it is difficult to do mapping, so we use shortcuts called kernels, which can be linear or non-linear mapping, polynomials, and bases function called RBF. RBF is the most widely used function. Each of these has its own characteristics. It finds the difference between two inputs x and x's.

That is called a support vector. RBF kernel has a parameter then the RBF function will be

$$f(\text{x, x}) = \exp(a|x - x2|)$$

Hyperplane: In n-dimensional space, several lines or a decision boundary may be used to divide classes; however, the optimal decision boundary for classifying the data points must be identified. The hyperplane of SVM is a name for this optimal boundary. We always build a hyperplane with a maximum margin or the greatest possible separation between the data points i.e., we have to find w and b such that

$$\phi(\theta) = \frac{1}{2\theta\theta^T}$$

Should be minimized for all values of $(x_i, y_i) : y_i(\theta^T x_i + b)$ where $\theta^T x + b + 1$, are the support vectors and most probably we have made a hyperplane that should be maximum distance i.e., $\theta x + b + 1$ will most probably the hyper line parameter.

Support Vectors: Support vectors are the data points or vectors that are closest to the hyperplane and have the greatest influence on where the hyperplane is located. These vectors are called support vectors because they support the hyperplane.

5 Implementation and Results

This section describes the implementation and the result description. The implementation is done on a python IDE with two models for comparison which are Convolutional Neural Network and SVM i.e., Support Vector Machine. After the pre-processing steps that are mentioned earlier, the classifier model is built and trained with training data, and afterward, it is tested for its performance. The models are compared on the basis of several evaluation metrics to achieve the best classifier model for identifying drowsiness Detection.

5.1 Computing Environment Specifications

Hardware Specifications Operating System: Windows 11 Processor: 9th Generation intel core i5 processor Chipset: Intel Integrated SoC Graphics: NVIDIA GEFORCE GTX 3050 Laptop GPU Internal Storage: 256 GB Software specifications [9]. Python environment on laptop and IDE if work is done on the local environment, otherwise cloud platforms like google Colab. Python packages like NumPy, pandas, matplotlib, cv2, OS, TensorFlow, and Keras.

5.2 Result Table Graph and Their Description

The result discussion will be made on the basis of the performance of the two models that are CNN and SVM, made by certain evaluation metrics that are discussed in the section above which are Accuracy and Loss functions. For the first model which is transfer learning on MobileNet, the accuracy vs. val-accuracy is as follows (Figs. 3 and 4):

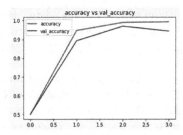

Fig. 3. Accuracy of CNN model

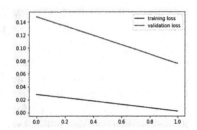

Fig. 4. Loss Graph of CNN model

For the training set, the accuracy reaches from 95% to 98% in just 5 epochs, whereas for the validation set accuracy, it is 93.15%. This shows us how transfer learning is more efficient than traditional learning and can achieve better accuracy in a shorter period of time and epochs due to its past knowledge of datasets. As for the loss function, for the training set, the loss reaches from 0.03 to 0.01 in just five epochs whereas the validation set loss reaches 0.10. This shows us that the loss component of the model is very low, so we can say that the performance of the model is great since the lesser the loss, the more accurate its prediction capability.

For the second model which is SVM, the accuracy vs validation accuracy graph is as follows:

As for the loss, it goes from 0.80 to 0.72 in 5 epochs whereas validation loss goes from 0.85 to 0.76. The loss is much higher than that of the CNN MobileNet classifier which is one of the latest technologies and works efficiently. So, we can see how the SVM classifier has greater loss overall which doesn't make it fit enough for predicting driver drowsiness efficiently and accurately (Figs. 5 and 6).

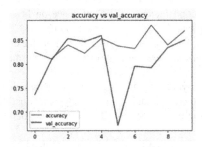

Fig. 5. Accuracy of SVM Model

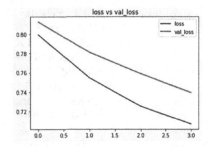

Fig. 6. Loss Graph of SVM Model

6   Conclusion

Drivers' lack of concentration or distraction causes major road accidents. To rectify this problem, several existing research papers have been accomplished, and on the basis of the study we evaluate effective methods, such as physiological and behavioural methods. This paper has outlined the project's contribution to the monitoring of drowsy driving in motor vehicles. We used SVM and CNN models in Machine Learning and the dataset used to perform these models is the MRL eyes dataset which contains a large number of images of drivers from different angles with opened and closed eyes. The comparative analysis of SVM and CNN in terms of accuracy function is 85.0% and 93.1%, respectively, and in loss function is 1 and 4, respectively. After comparative analysis using these models, we conclude that CNN provides better accuracy and loss function using the MRL dataset as compared to SVM. Hence, this paper indicates that CNN outperformed SVM in terms of performance metrics. Therefore, CNN works better in a real-world scenario which will help the researchers to analyse better outcomes in the field of driver drowsiness.

References

1. Albasrawi, R., Fadhil, F.F., Ghazal, M.T.: Driver drowsiness monitoring system based on facial landmark detection with convolutional neural network for prediction. Bull. Electr. Eng. Inform. **11**(5), 2637–2644 (2022)
2. Bajaj, J.S., Kumar, N., Kaushal, R.K.: Comparative study to detect driver drowsiness. In: 2021 International Conference on Advance Computing and Innovative Technologies in Engineering (ICACITE), pp. 678–683 (2021)
3. Caryn, F.H., Rahadianti, L.: Driver drowsiness detection based on drivers' physical behaviours: a systematic literature review. Comput. Eng. Appl. J. **10**(3), 161–175 (2021)
4. Chaabene, S., Bouaziz, B., Boudaya, A., Hökelmann, A., Ammar, A., Chaari, L.: Convolutional neural network for drowsiness detection using EEG signals. Sensors **21**(5), 1734 (2021)
5. Deng, W., Ruoxue, W.: Real-time driver-drowsiness detection system using facial features. IEEE Access **7**, 118727–118738 (2019)
6. Ferreira, P.M., et al.: AUTOMOTIVE: a case study on automatic multimodal drowsiness detection for smart vehicles (2021)
7. Roopalakshmi, R., Rathod, J.A., Shetty, A.S., Supriya, K.: Driver drowsiness detection system based on visual features. In: 2018 Second International Conference on Inventive Communication and Computational Technologies (ICICCT), pp. 1344–1347 (2018)
8. Horberry, T., Anderson, J., Regan, M.A., Triggs, T.J., Brown, J.: Driver distraction: the effects of concurrent in-vehicle tasks, road environment complexity and age on driving performance. Accid. Anal. Prev. **38**(1), 185–191 (2006)
9. Kiashari, S.E.H., Nahvi, A., Bakhoda, H., Homayounfard, A., Tashakori, M.: Evaluation of driver drowsiness using respiration analysis by thermal imaging on a driving simulator. Multimed. Tools Appl. **79**(25–26), 17793–17815 (2020). https://doi.org/10.1007/s11042-020-086 96-x
10. Kumar, A., Patra, R.: Driver drowsiness monitoring system using visual behaviour and machine learning. In: 2018 IEEE Symposium on Computer Applications Industrial Electronics (ISCAIE), pp. 339–344 (2018)
11. LaRocco, J., Le, M.D., Paeng, D.G.: A systemic review of available low-cost EEG headsets used for drowsiness detection. Front. Neuroinform. **14**, 42 (2020)
12. Moujahid, A., Dornaika, F., Arganda-Carreras, I., Reta, J.: Efficient and compact face descriptor for driver drowsiness detection. Expert Syst. Appl. **168**, 114334 (2021)
13. Niloy, A.R., Chowdhury, A.I., Sharmin, N., et al.: A brief review on different driver's drowsiness detection techniques. Int. J. Image Graph. Sign. Proc. **10**(3), 41 (2020)
14. Stancin, I., Cifrek, M., Jovic, A.: A review of EEG signal features and their application in driver drowsiness detection systems. Sensors **21**(11), 3786 (2021)

15. Vesselenyi, T., Moca, S., Rus, A., Mitran, T., Tǎtaru, B.: Driver drowsiness detection using ANN image processing. In: IOP Conference Series: Materials Science and Engineering, vol. 252(1), p. 012097 (2017)
16. Victoria, D.R.S., Mary, D.G.R.: Driver drowsiness monitoring using convolutional neural networks. In: 2021 International Conference on Computing, Communication, and Intelligent Systems (ICCCIS), pp. 1055–1059 (2021)
17. You, F., Gong, Y., Tu, H., Liang, J., Wang, H.A.: fatigue driving detection algorithm based on facial motion information entropy. J. Adv. Transp. **2020**, 1–17 (2020)
18. Zandi, A.S., Quddus, A., Prest, L., Comeau, F.J.: Non-intrusive detection of drowsy driving based on eye tracking data. Transp. Res. Rec. **2673**(6), 247–257 (2019)

An Efficient (MFFPA-2) Multiple Fuzzy Frequent Patterns Mining with Adjacency Matrix and Type-2 Member Function

Mahendra N. Patel[1]([⊠]) [iD], S. M. Shah[2] [iD], and Suresh B. Patel[3] [iD]

[1] Gujarat Technological University, Ahmedabad, India
mnpatel32@gmail.com
[2] Computer Engineering, GEC Rajkot, Rajkot, India
[3] Information Technology Department, GEC Gandhinagar, Gandhinagar, India

Abstract. Fuzzy technique is used to handle quantitative datasets and to generate meaningful representations of the dataset. The study proposes efficient mining of numerous fuzzy frequent itemsets from quantitative datasets using a type-2 membership function. The adjacency matrix and the Tid-fuzzy-list structure are built to mine (MFFI) several fuzzy frequent itemsets by accessing or reading the dataset only once. An effective pruning mechanism has been developed to condense the search space and minimize candidate itemsets, thus reducing execution time. Extensive experiments are conducted to verify efficiency regarding runtime, memory use, and join counts with different minsupport thresholds. Experimental results demonstrate that the designed approach MFFPA-2 achieved superior performance compared to cutting-edge techniques.

Keywords: Type-2 membership function · Data mining · Multiple fuzzy frequent itemsets mining · Adjacency matrix · Tid-fuzzy-list structure

1 Introduction

In the last decades, online shopping and mall shopping have been rustically increasing. For increasing the business, discovering valuable information from datasets is very important. Data mining techniques require finding knowledge from a large volume of the dataset. AR mining [1, 2], clustering [3], and classification [4, 5] are the three primary categories of KDD [1] methods. In frequent itemsets mining, mining of association rule is frequently employed. The Apriori algorithm is the first and fundamental data mining algorithm for association rule mining to mine common itemsets in a level-by-level (level-wise) methodology presented by Agrawal et al. [2]. Before finding FIs level by level, it generates candidate itemsets and prunes them. This method requires a time-consuming computation involving repeated database scanning and creating numerous candidate itemsets. Since Apriori requires multiple-time scanning of the database and generates more candidate sets, Han et al. [6] presented an FP-growth mining method to create frequent itemsets (FIs) without creating candidate itemsets and scanning the database only two times.

In real-world scenarios, quantitative databases are used in frequent itemsets mining for decision-making. It is challenging to manage the quantitative database. Mostly all authors used fuzzy set theory to manage quantitative databases. In fuzzy set theory, quantitative values of an item in the transaction are transformed into linguistic terms using a pre-defined membership function [7]. In [8], the authors used the max cardinality value in level by level approach to mining fuzzy frequent itemsets (FFIs). Using maximum cardinality generating FFIs cost is minimized, but not generate all possible FFIs. Authors present multiple fuzzy frequent itemsets (MFFIs) techniques for generating complete fuzzy frequent itemsets. Several tree-based methods were provided by Lin et al. for mining MFFIs [9–12]. Following fuzzy list structure and enumeration tree-based techniques for mining MFFIs [13, 14] presented by Lin et al. In [14], authors applied pruning approaches to minimize the searching space of the enumeration tree. There is still more calculation cost for discovering MFFIs due to more node join counts. In [15], authors presented adjacency matrix-based AMFFI mining techniques. The AMFFI technique performs a single database scan, resulting in fewer node join counts.

The methods described above generally employ type-1 membership function of fuzzy-set theory to mine MFFIs. Mendel et al. [16] Proposed Type-2 membership to involve uncertain factors to fuzzy itemsets mining for making a decision. Using the Type-2 membership function, Chen et al. [17] proposed a level-wise approach to mine FIs. This approach used only an item's max cardinality value (single linguistic term), resulting in FIs that are not completed. The mining of a list-based strategy for multiple fuzzy frequent itemsets was developed by Lin et al. [18] with a type-2 membership function. In comparison to the level-wise technique, it provides better performance. Next, Lin et al. [19], using a type-2 membership function, proposed a compressed list structure-based MFFPs (multiple fuzzy frequent patterns) mining technique. In this approach to increasing the performance, authors used two efficient pruning strategies; still, many unpromising node joins are identified. Thus generating more candidate itemsets.

This article presents an adjacency matrix-based mining method called MFFPA-2 (Multiple Fuzzy Frequent Patterns with Adjacency matrix and Type-2 Membership function). This approach first scans the database and creates a fuzzy list and adjacency matrix. By determining from the adjacency matrix directly whether two nodes should be joined if two fuzzy linguistic phrases co-occur more frequently than or equally as frequently as the minsupport threshold, fewer nodes will be joined, resulting in fewer candidate itemsets being produced by the MFFPA-2-miner algorithm, thereby reducing the amount of search space needed. Utilize the adjacency matrix to find L2 (2-frequent itemsets) directly. The rate of computing mining MFFIs can be significantly reduced as a result. An experimental result shows that our approach performs better than alternative methods.

2 Related Work

Delgado et al. presented an approach in [20] for finding association rules from quantitative and relational databases. Hong et al. [21] developed the technique using the Apriori Tid data structure for extracting FFIs to grow itemsets in quantitative datasets. Lin [9, 22, 23] and Hong [24] employed a structure similar to an FP tree to extract FFIs in quantitative datasets. To expedite mining, the author adopted a superset pruning mechanism in this method [9, 21–23]. The authors find MFFIs to obtain comprehensive details for every linguistic term in the fuzzy set. The MFFP-tree structure and MFFP-growth mining approaches were proposed by Hong et al. [10] to find MFFIs. To find MFFIs, Lin et al. [14] created an MFFI-miner algorithm and structure for Fuzzy-list. To cut down on the search area, run time, and run area, the authors of this method employed two distinct pruning strategies. Patel et al. [15] presented an adjacency matrix-based AMFFI method for discovering MFFIs. In this method, scan database only ones and superset are generated if possible, which is predicted from adjacency matrix, so shortening the running time and nodes join count value.

The solution, as mentioned above, solely counters type-1 fuzzy-set theory, which ignores uncertainty. Since the fuzzy-set theory of type 1 membership functions is crisp, it is insufficient for managing uncertainty models in practical applications. The fuzzy-set idea with type-2 membership function [16, 25], and [26] was then put out and improved to more effectively present the acquired information with uncertainty. To merge pattern mining and type-2 fuzzy sets, Chen et al. [17] applied the standard level-wise like-Apriori method for mining level-wise fuzzy type-2 frequent patterns. However, the procedure necessitates generating large numbers of candidates, which is ineffective for the mining task. Additionally, it retrieves only one linguistic term from the item using the maximal scalar cardinality approach, which may result in insufficient information for making decisions.

To store the information for the mining process, a list-based approach proposed by Lin et al., the strategy still needs to investigate a large number of candidates for determining the true FFIs due to inefficient pruning algorithm and loose upper bound value on the pattern, which are not frequent. After that, Lin et al. provided a list-based approach to keep the necessary information for discovering frequent items [18]. This approach still needs to investigate many candidates to get the actual FFIs because it requires more effective search space trimming strategies and a flexible upper bound measure on the patterns that could be more promising. The authors of [19] employed a complex fuzzy list (CFL)-structure to find MFFIs, and that was similar to the fuzzy list structure from [14].

3 Preliminary

A limited number of m unique items is called the Itemset-I $= i1, i2,...,im$. The quantitative dataset D has n transactions, where $D = T1, T2, T3,... Tn$. Every transaction contains the notation $Tq \in D$ and $Tq \in I$. Every transaction also includes a TID, which stands for a unique identifier. Each transaction Tq consists of an item and the value of the buy quantity; let's call it wiq. The term "k-itemset" refers to an itemset of length K $= i1, i2,...,ik$.

In the example below, Table 1 displays a sample quantitative dataset of seven transactions; let's call it D. Think about minimum support $\emptyset = 1$. The use of the Type-2 member function £ in the example is demonstrated in Fig. 1.

Mining fuzzy, frequently occurring itemsets typically involves the following three steps.

Table 1. Quantitative dataset

TID	Item with Quantity
1	A-4, B-3, C-2, D-2
2	B-3, C-2, E-3
3	A-5, B-3, C-4, E-4
4	A-2, C-1, D-3
5	A-4, B-2, C-5
6	B-3, C-3, D-2, E-2
7	C-3, E-2

Step 1: Determine the item's (Linguistic variable) fuzzy terms.

Consider the dataset D and item i (i ⊆ I), and the value of i is the collection of fuzzy terms. The built-in type-2 membership function produces it £ seen in Fig. 1. Fuzzy terms are represented as li1, li2,...,lih, where h is the membership degree. Figure 1 shows the 3-term membership function means here h = 3. It may differ as 4-term or 5-term as per requirements. Three linguistic concepts are employed in this example: High-H, Middle-M, and Low-L.

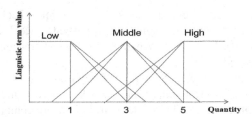

Fig. 1. Type-2 Membership Function

Term Viq for transaction Tq is the quantitative value of i (item) in the transaction. Term fiq is the linguistic term of item i. Fiq generated from item i quantity value Viq using membership function £. Fiq is a set of three linguistic terms $fiq1^{lower}$, $fiq1^{upper}$/li1 for membership value low, $fiq2^{lower}$, $fiq2^{upper}$ /li2 for membership value middle, and $fiq3^{lower}$, $fiq2^{upper}$ /li3 for membership value high as shown in following Eq. 1. Where $li1$ shows l-th linguistic (fuzzy) terms, $fiql^{lower}$, and $fiql^{upper}$, respectively, show lower and upper membership values of Viq for item i.

$$fiq(viq) = \frac{fiq1^{lower}, fiq1^{upper}}{li1} + \frac{fiq2^{lower}, fiq2^{upper}}{li2} \cdots + \frac{fiqh^{lower}, fiqh^{upper}}{lih} \quad (1)$$

Table 2 shows the resultant fuzzy dataset say D', after applying the membership function £ on given an example.

Table 2. Fuzzy Dataset

TID	Original Dataset	Fuzzy dataset
1	A-4, B-3, C-2, D-2	$\underline{0.5,0.62} + \underline{0.5,0.62}$, $\underline{0.0.25} + \underline{1.1} + \underline{0.0.25}$. $\underline{0.5,0.62} + \underline{0.5,0.62}$, $\underline{0.5,0.62} + \underline{0.5,0.62}$ AM AH BL BM BH CL CM DL DM
2	B-3, C-2, E-3	$\underline{0.0.25} + \underline{1.1} + \underline{0.0.25}$. $\underline{0.5,0.62} + \underline{0.5,0.62}$. $\underline{0.0.25} + \underline{1.1} + \underline{0.0.25}$ BL BM BH CL CM EL EM EH
3	A-5, B-3, C-4, E-4	$\underline{0.0.25} + \underline{1.1}$. $\underline{0.0.25} + \underline{1.1} + \underline{0.0.25}$. $\underline{0.5,0.62} + \underline{0.5,0.62}$. $\underline{0.5,0.62} + \underline{0.5,0.62}$ AM AH BL BM BH CM CH EM EH
4	A-2, C-1, D-3	$\underline{0.5,0.62} + \underline{0.5,0.62}$. $\underline{1.1} + \underline{0.0.25}$. $\underline{0.0.25} + \underline{1.1} + \underline{0.0.25}$ AL AM CL CM DL DM DH
5	A-4, B-2, C-5	$\underline{0.5,0.62} + \underline{0.5,0.62}$, $\underline{0.5,0.62} + \underline{0.5,0.62}$, $\underline{0.0.25} + \underline{1.1}$ AM AH BL BM CM CH
6	B-3, C-3, D-2	$\underline{0.0.25} + \underline{1.1} + \underline{0.0.25}$. $\underline{0.0.25} + \underline{1.1} + \underline{0.0.25} + \underline{0.5,0.62} + \underline{0.5,0.62}$ BL BM BH CL CM CH DL DM
7	C-3, E-2	$\underline{0.0.25} + \underline{1.1} + \underline{0.0.25}$. $\underline{0.5,0.62} + \underline{0.5,0.62}$ CL CM CH EL EM

Step 2: Find the support count of each fuzzy item.

Fuzzy itemset Lik's support count (scalar cardinality) is shown by the symbol sup (Lik). Find each fuzzy itemset's support in this stage. According to this definition,

$$\text{Sup(Lik)} = \sum_{q=0,\, Lik \subseteq Tq \,\wedge\, Tq \in D'}^{n} \text{fiqk} \tag{2}$$

In fuzzy dataset D', fuzzy item Lik 's fuzzy value is fiqk. Check each fuzzy item's sup (Lik); if the minimum support requirement is satisfied, place the item in FL1.

FL1 = FL1 ∪ (sup (Lik) > = Ø). Where FL1 is fuzzy 1-frequent itemsets and Ø is the minsupport threshold.

Step 3: Finding the Sup of each frequent fuzzy itemsets:

The following-level frequent itemsets are fuzzy k-itemsets with k = 2, produced by fuzzy 1-frequent itemsets (FL1). Fuzzy items from FL1 are combined using the join procedure to create a candidate set, such as FC2 (fuzzy 2-candidate itemsets). Consider the itemset X that was produced by merging the FL1 itemsets A and B. Sup(X) = Support of itemset: X, where $X \subseteq Tq$ and $Tq \in D'$, is determined by adding the lowest fuzzy values of fuzzy itemsets A and B from truncation Tq. According to this definition,

$$\text{Sup(X)} = X \in Li / \sum_{q=0,\, X \subseteq Tq \,\wedge\, Tq \in D'}^{n} \min(\text{faq, fbq}) \tag{3}$$

From FC2 itemsets, store in fuzzy 2-frequent itemsets (FL2) if they meet the minimal support Ø. Similarly, fuzzy k-frequent itemsets are discovered later.

A list-based structure was created in 2016 by Lin et al. [18] to mine the various type-2 fuzzy-set-based common fuzzy patterns. Next, in 2020 Lin et al.[19]Employed complex fuzzy list (CFL)-structure to locate MFFIs using a type-2 fuzzy set. However, many unpromising candidates still need to be considered as it is an inefficient trimming mechanism to narrow the search field. For efficiently discovering the MFFP, this study considers the type-2 membership functions of fuzzy-set theory. It is necessary to devise an effective pruning algorithm that can decrease the search space and enhance the performance of pattern mining. An effective adjacency matrix structure is employed to keep the complete information.

4 Proposed MFFPA-2

This section suggests a two-step process for creating many fuzzy frequent itemsets. Using the quantitative dataset D, generate an adjacency matrix and Tid-fuzzy list in phase 1. Using approach MFFPA-2, quickly extract numerous fuzzy frequent itemsets from the adjacency matrix and Tid-fuzzy list discussed in phase 2. The suggested technique effectively creates entire MFFIs from a single database scan.

4.1 Adjacency Matrix and Tid-Fuzzy List Construction

In this phase, first construct adjacency matrix M. Size of matrix M is $(I * h) \times (I * h)$. In the membership function, h is the number of linguistic terms, and I is the number of items in D. Total items in D' (fuzzy dataset) is a product of I in D, and h say m. So the total items in D' are m and $m = I * h$—this adjacency matrix is written below. The corresponding required adjacency matrix is shown in Fig. 2.

$$AdjMat(M) = m \times m \tag{4}$$

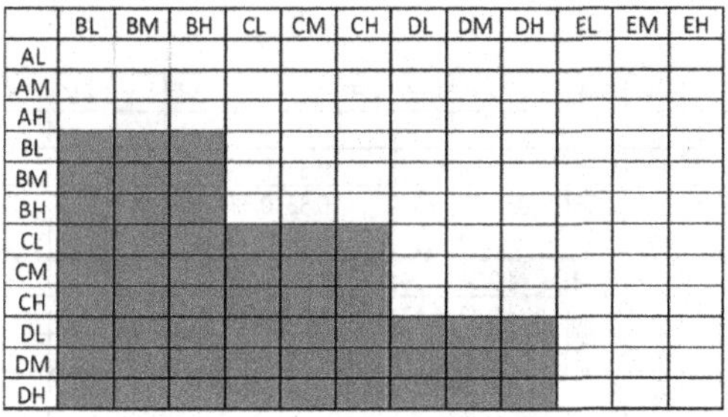

Fig. 2. Adjacency Matrix

Next, in this phase, scan transaction Tq from quantitative database D and apply a pre-defined type-2 membership function, which generates fuzzy linguistics terms, as shown in Table 2. Here get two values of each fuzzy linguistic term, the first value associated with a lower boundary and the second value associated with a higher boundary of membership function, as shown in Fig. 1. For example, the first transaction item A with quantity four generates two fuzzy linguistics terms, AM (A-middle) and AH (A-high), each linguistic term with fuzzy value 0.5 and 0.62 as lower and higher value respectively, as shown in Table 2. Thus it is a complex task to mine MFFIs from two values of each fuzzy linguistic term. So take the fuzzy interval value by taking an average of it, and use the centroid type-reduction approach [17] to reduce the complexity. Getting the interval value using the following formula

$$fiq1 = \frac{fiq1^{lower} + fiq1^{upper}}{2} \tag{5}$$

Table 3. Final first fuzzy transaction

TID	Final Fuzzy dataset
1	$\dfrac{0.56+0.56}{AM\ \ \ \ AH}, \dfrac{0.13+1}{BL\ \ \ \ BM} +\dfrac{0.13}{BH}, \dfrac{0.56+0.56}{CL\ \ \ \ CM}, \dfrac{0.56+0.56}{DL\ \ \ \ DM}$

So, get 0.56 internal fuzzy values of linguistic terms AM and AH according given formula. This final transformed first transaction from D is displayed in Table 3.

Add value to the adjacency matrix, and the following step is to build a pair of converted fuzzy itemsets for various fuzzy variables from transaction Tq. The value of the corresponding cell in the adjacency matrix can be changed by entering the corresponding fuzzy list and adding the least fuzzy value of each pair.

$$AM(Li, Lj) = AM(Li, Lj) + \min(fwiq, fwjq) \quad (6)$$

fwiq and fwjq are the fuzzy value of the fuzzy items Li and Lj, respectively. Create a Tid-fuzzy list for Li and Lj if it does not exist. A minimum of fwiq and fwjq was added with transaction ID q to the Tid-fuzzy list. Completing the first-row adjacency matrix and Tid-fuzzy list is shown in Fig. 3 and Fig. 4, respectively.

	BL	BM	BH	CL	CM	CH	DL	DM	DH	EL	EM	EH
AL												
AM	0.13	0.56	0.13	0.56	0.56		0.56	0.56				
AH	0.13	0.56	0.13	0.56	0.56		0.56	0.56				
BL				0.13	0.13		0.13	0.13				
BM				0.56	0.56		0.56	0.56				
BH				0.13	0.13		0.13	0.13				
CL							0.56	0.56				
CM							0.56	0.56				
CH												
DL												
DM												
DH												

Fig. 3. Adjacency Matrix after the first-row scan

AM-BL		AM-BM		AM-BH		AH-BL		AH-BM		AH-BH		AM-CL		AM-CM	
TID	FUZZY VALUE	TID	FUZZY VALUE	TID	FUZZY VALUE	TID	FUZZY VALUE	TID	FUZZY VALUE	TID	FUZZY VALUE	TID	FUZZY VALUE	TID	FUZZY VALUE
1	0.13	1	0.56	1	0.13	1	0.13	1	0.56	1	0.13	1	0.56	1	0.56

AH-CL		AH-CM		AM-DL		AM-DM		AH-DL		AH-DM		BL-CL		BL-CM	
TID	FUZZY VALUE	TID	FUZZY VALUE	TID	FUZZY VALUE	TID	FUZZY VALUE	TID	FUZZY VALUE	TID	FUZZY VALUE	TID	FUZZY VALUE	TID	FUZZY VALUE
1	0.56	1	0.56	1	0.56	1	0.56	1	0.56	1	0.56	1	0.13	1	0.13

BM-CL		BM-CM		BH-CL		BH-CM		BL-DL		BL-DM		BM-DL		BM-DM	
TID	FUZZY VALUE	TID	FUZZY VALUE	TID	FUZZY VALUE	TID	FUZZY VALUE	TID	FUZZY VALUE	TID	FUZZY VALUE	TID	FUZZY VALUE	TID	FUZZY VALUE
1	0.56	1	0.56	1	0.13	1	0.13	1	0.13	1	0.13	1	0.56	1	0.56

BH-DL		BH-DM		CL-DL		CL-DM		CM-DL		CM-DM	
TID	FUZZY VALUE	TID	FUZZY VALUE	TID	FUZZY VALUE	TID	FUZZY VALUE	TID	FUZZY VALUE	TID	FUZZY VALUE
1	0.13	1	0.13	1	0.56	1	0.56	1	0.56	1	0.56

Fig. 4. Tid-Fuzzy list after the 1st-row scan.

The steps to generate an adjacency matrix and a tid-fuzzy list are shown in Algorithm 1 below. The application of algorithm 1 to the example is shown in Table 1, and the final adjacency matrix and Tid-fuzzy list are shown in Figs. 5 and 6, respectively.

Algorithm 1: Build the AM- Adjacency matrix and F- Tid-fuzzy list

Input: D-Crisp database, M- Total Items
Output: AM and F: Tid-Fuzzy-list
1: Create AM for m rows and m columns, where m is total fuzzy linguistic terms
2: Set Transaction ID Tid=1
3: From database D, scan line Tq
4: While Tq is in D, proceed to step 9 again
5: Apply the member function type-2 to the quantitative values of each item in Tq to generate the fuzzy linguistic terms fl[] for all   of the items. In fl[] contains the internal fuzzy value of the linguistic term.
6: Define FV= minimum (FV of fl[i], FV of fl[j]), where FV stand for fuzzy value
*   Insert fuzzy values for all co-occurrences of fuzzy linguistic phrases in the adjacency matrix AM.*
7: Build Tid-Fuzzy List of "fl[i]- fl[j] " if not exist
8: Build an element using (TID, FV) and add it to the Tid-fuzzy list "fl[i]-fl[j]."
9: Add 1 to Tid and scan the next line from D to Tq.
10: End.

	BL	BM	BH	CL	CM	CH	DL	DM	DH	EL	EM	EH
AL				0.56	0.13		0.13	0.56	0.13			
AM	0.82	1.25	0.26	0.69	0.95	0.69	0.69	1.12	0.13		0.13	0.13
AH	0.82	2.12	0.26	0.56	1.25	1.12	0.56	0.56			0.56	0.56
BL				0.39	0.65	0.82	0.26	0.26		0.13	0.26	0.26
BM				1.25	2.81	1.25	1.12	1.12		0.13	1.56	0.69
BH				0.39	0.52	0.26	0.26	0.26		0.13	0.26	0.26
CL							0.82	1.69	0.13	0.26	0.69	0.13
CM							1.25	1.25	0.13	0.69	1.68	0.69
CH							0.13	0.13		0.13	0.69	0.56
DL												
DM												
DH												

Fig. 5. Adjacency Matrix after all row scan

4.2 From Adjacency Matrix Mining MFFIs Using MFFPA-2 Method

Using Tid-fuzzy lists, row-by-row extraction of MFFIs from the adjacency matrix (M) produced results above 4.1 points in this phase. Select the cell with a value greater than or equal to in the row of Adjacency matrix M. (minsupport). Declared identified cell row-column combination as fuzzy 2-frequent itemsets (FL$_2$) and fetched Tid-fuzzy list for identified cell. For the example in the first row in Fig. 5 with headed AL, there is no cell with value $> = \emptyset$; here, minsupport $\emptyset = 1$. Scan the second row in which two cells, namely AM-BM and AM-DM, find which satisfies minsupport threshold $\emptyset$ so

AH-CL

TID	FUZZY VALUE
1	0.56

AH-DL

TID	FUZZY VALUE
1	0.56

AH-DM

TID	FUZZY VALUE
1	0.56

BL-EL

TID	FUZZY VALUE
2	0.13

BM-EL

TID	FUZZY VALUE
2	0.13

BH-EL

TID	FUZZY VALUE
2	0.13

CL-EH

TID	FUZZY VALUE
2	0.13

AM-EM

TID	FUZZY VALUE
3	0.13

AM-EH

TID	FUZZY VALUE
3	0.13

AH-EM

TID	FUZZY VALUE
3	0.56

AH-EH

TID	FUZZY VALUE
3	0.56

CH-EH

TID	FUZZY VALUE
3	0.56

AL-CL

TID	FUZZY VALUE
4	0.56

AL-CM

TID	FUZZY VALUE
4	0.13

AL-DL

TID	FUZZY VALUE
4	0.13

AL-DM

TID	FUZZY VALUE
4	0.58

AL-DH

TID	FUZZY VALUE
4	0.13

AM-DH

TID	FUZZY VALUE
4	0.13

CL-DH

TID	FUZZY VALUE
4	0.13

CM-DH

TID	FUZZY VALUE
4	0.13

CH-DL

TID	FUZZY VALUE
6	0.13

CH-DM

TID	FUZZY VALUE
6	0.13

CH-EL

TID	FUZZY VALUE
7	0.13

AM-BH

TID	FUZZY VALUE
1	0.13
3	0.13

AH-BH

TID	FUZZY VALUE
1	0.13
3	0.13

AM-CL

TID	FUZZY VALUE
1	0.56
4	0.13

AM-DL

TID	FUZZY VALUE
1	0.56
4	0.13

AM-DM

TID	FUZZY VALUE
1	0.56
4	0.56

BL-DL

TID	FUZZY VALUE
1	0.13
6	0.13

BL-DM

TID	FUZZY VALUE
1	0.13
6	0.56

BM-DL

TID	FUZZY VALUE
1	0.56
6	0.56

BM-DM

TID	FUZZY VALUE
1	0.56
6	0.56

BH-DL

TID	FUZZY VALUE
1	0.13
6	0.13

BH-DM

TID	FUZZY VALUE
1	0.13
6	0.13

BL-EM

TID	FUZZY VALUE
2	0.13
3	0.13

BL-EH

TID	FUZZY VALUE
2	0.13
3	0.13

BM-EM

TID	FUZZY VALUE
2	1
3	0.13

BM-EH

TID	FUZZY VALUE
2	0.13
3	0.13

BH-EM

TID	FUZZY VALUE
2	0.13
3	0.13

BH-EH

TID	FUZZY VALUE
2	0.13
3	0.13

CL-EL

TID	FUZZY VALUE
2	0.13
7	0.13

CL-EM

TID	FUZZY VALUE
2	0.56
7	0.13

CM-EL

TID	FUZZY VALUE
2	0.13
7	0.56

CM-EH

TID	FUZZY VALUE
2	0.13
3	0.56

AM-CH

TID	FUZZY VALUE
3	0.13
5	0.56

AH-CH

TID	FUZZY VALUE
3	0.56
5	0.56

BH-CH

TID	FUZZY VALUE
3	0.13
6	0.13

CH-EM

TID	FUZZY VALUE
3	0.56
7	0.13

AM-BL

TID	FUZZY VALUE
1	0.13
3	0.13
5	0.56

AM-BM

TID	FUZZY VALUE
1	0.56
3	0.13
5	0.56

AH-BL

TID	FUZZY VALUE
1	0.13
3	0.13
5	0.56

AH-BM

TID	FUZZY VALUE
1	0.56
3	1
5	0.56

AH-CM

TID	FUZZY VALUE
1	0.56
3	0.56
5	0.13

BL-CL

TID	FUZZY VALUE
1	0.13
2	0.13
6	0.13

BM-CL

TID	FUZZY VALUE
1	0.56
2	0.56
6	0.13

BH-CL

TID	FUZZY VALUE
1	0.13
2	0.13
6	0.13

CL-DL

TID	FUZZY VALUE
1	0.56
4	0.13
6	0.13

CL-DM

TID	FUZZY VALUE
1	0.58
4	1
6	0.13

CM-DL

TID	FUZZY VALUE
1	0.56
4	0.13
6	0.13

CM-DM

TID	FUZZY VALUE
1	0.56
4	0.13
6	0.56

CM-EM

TID	FUZZY VALUE
2	0.56
3	0.56
7	0.56

BL-CH

TID	FUZZY VALUE
3	0.13
5	0.56
6	0.13

BM-CH

TID	FUZZY VALUE
3	0.56
5	0.56
6	0.13

AM-CM

TID	FUZZY VALUE
1	0.56
3	0.13
4	0.13
5	0.13

BH-CM

TID	FUZZY VALUE
1	0.13
2	0.13
3	0.13
6	0.13

BM-CM

TID	FUZZY VALUE
1	0.56
2	0.56
3	0.56
5	0.13
6	1

BL-CM

TID	FUZZY VALUE
1	0.13
2	0.13
3	0.13
5	0.13
6	0.13

Fig. 6. Tid-Fuzzy list after all row scan from dataset D

fetched Tid-fuzzy list of AM-BM and AM-DM. Next, recursively create fuzzy k-frequent itemsets, such as FL_k ($K > 2$), in a subsequent step by intersection-operating TIDs on FL_{k-1}. The binary search method can be used to find combined fast fuzzy lists. To create the Tid-fuzzy list (FL_k) for k-frequent itemsets ($k > 2$), existing FL_{k-1} Tid-fuzzy lists are combined. Elements in a newly created Tid-fuzzy list are those with a common Tid in an existing Tid-fuzzy list.

Only joining fuzzy itemsets that can create their superset, known directly from the adjacency matrix M, will reduce the search space and candidate set. As found, FL2 is from the second row AM-BM and AM-DM, so the subsequent possible superset is AM-BM-DM. If the BM-DM value from the BM row and DM column cell value satisfy the minsupport threshold, generate the AM-BM-DM Tid-fuzzy list by joining the AM-BM Tid-fuzzy list and AM-DM Tid-fuzzy list using the intersection operation on it. In the fifth row headed by BM, five cells satisfy Ø, so generated FL2 from this row is BM-CL, BM-CM, BM-CH, BM-DL, BM-DM. Next subsequent possible FL3 are BM-CL-DL, BM-CL-DM, BM-CM-DL, BM-CM-DM, BM-CH-DL and BM-CH-DM. Before joining BM-CL-DL, check the CL-DL value, which does not satisfy Ø, so ignore this set directly without generating its candidate itemsets or not join BM-CL-DL. This way drastically minimizes join operation or candidate itemsets, ultimately improving running time efficiency. For this MFFPA-2 method is shown in Algorithms 2 and 3.

Algorithm 2: MFFPA-2

Input: Minsupport, FLs: Tid-fuzzy list, M: Adjacency Matrix
1: Do for every row in M
FL ← null;   \\ Tid-fuzzy list Initialization
2: for each row's cell
      If cell value >= minsupport
      Get fuzzy-list of ([row.id]-[col.id]) from FLs into FL
      Call MFFPA-2-miner with FL and row-id

Algorithm 3: MFFPA-2-miner

Input: Minsupport, FL: Tid-fuzzy list, M: Adjacency Matrix, row-id
Output: MFFI,
1: do for every fuzzy-list P in FL
2:   MFFIs ← P ∪ MFFIs.
3:   TFL ← null; //temporary Tid-fuzzy list
4:   do for every fuzzy-list Q after P in FL
          If P.product = Q.product, then
                skip
          else If M[P.product][Q.product] >= minsupport then
       Join P, Q and insert into TFL;
5:   MFFPA-2-Miner(TFL);
6: Return MFFIs.

After reading all rows according to the algorithm-generated candidate Tid-Fuzzy list shown in Fig. 7. Many candidate sets are possible, but this approach generates only seven Tid-fuzzy sets out of these five fuzzy frequent itemsets of length 3. Next, it does not generate a candidate set for length 4, which know directly from the adjacency matrix. So using the MFPPA-2 method and Adjacency matrix generates fewer candidate sets than the state-of-art method.

AM-BM-DM		AH-BM-CM		AH-BM-CH		BM-CL-DM		BM-CM-DL		BM-CM-DM		BM-CM-EM	
TID	FUZZY VALUE	TID	FUZZY VALUE	TID	FUZZY VALUE	TID	FUZZY VALUE	TID	FUZZY VALUE	TID	FUZZY VALUE	TID	FUZZY VALUE
1	0.56	1	0.56	3	0.56	1	0.56	1	0.56	1	0.56	2	0.56
		3	0.56	5	0.56	6	0.13	6	0.56	6	0.56	3	0.56
		5	0.13										

Fig. 7. Tid-Fuzzy list after all row scan from Matrix M

5 Assessment of Experimental Findings

Here, we contrast the MFFPA-2 performance of the recommended method with that of the list-based techniques put forward by Lin et al. [18] and EFM [19]. We Implement the proposed MFFPA-2, EFM, and Lin's method in Java. The outcomes are examined using two real datasets, chess and mushroom [27], and one artificial T10I4D100k dataset [27].

The quantities of objects in the datasets provide at random intervals between 1 and 7. The outcome of the experiment runtime, join count, and memory usage were all examined.

5.1   Runtime Analysis

We implemented MFFPA-2, EFM [19], and Lin's [18] methods using the type-2 member function with 3-term fuzzy linguistic terms. To compare the execution time of the implemented methods, we have used different minimum support threshold values. Figures 8, 9 and 10 show the findings of the execution running time evaluation on the chess dataset, mushroom dataset, and T10I4D100k dataset.

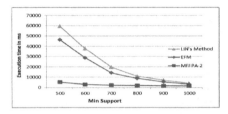

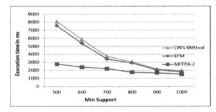

Fig. 8. Comparisons of execution times: Chess dataset

Fig. 9. Comparisons of execution times: Mushroom dataset

From the results, it is observed that the proposed MFFPA-2 approach works better than the alternative method. A lower minimum support criterion is also used to show how resilient the MFFPA-2 technique is.

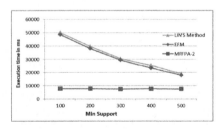

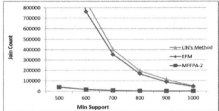

Fig. 10. Comparisons of execution times: T10I4D100k dataset

Fig. 11. Comparisons of Join Counts: Chess dataset

5.2   Number of Join Counts Analysis

The amount of joins made during the formation of MFFIs is considered while evaluating performance in this area. The results of the number of join count evaluations on the chess, mushroom, and T10I4D100k dataset are displayed in Figs. 11, 12 and 13.

The outcome demonstrates that the MFFPA-2 approach produces fewer join counts (candidate itemsets). It was noted that the MFFPA-2 method's join count performance is by far the most impressive. The proposed MFFPA-2 method produces fewer candidate itemsets than cutting-edge techniques.

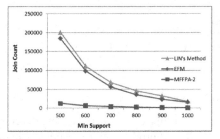

Fig. 12. Comparisons of Join Counts: Mushroom dataset   **Fig. 13.** Comparisons of Join Counts: T10I4D100k dataset

5.3 Memory Utilization Analysis

Here, effectiveness is measured by how extensively memory was used in the studies. The results of the memory usage on the chess dataset, mushroom dataset, and T10I4D100k dataset are displayed in Figs. 14, 15 and 16.

The outcome demonstrates that the MFFPA-2 method uses less memory than the compared approach on the chess and mushroom datasets. It was noted that the MFFPA-2 method uses more memory than the compared approach on the artificial T10I4D100k dataset. We may deduce from additional trials with other datasets that in a special case where a dataset has more than 1000 items, the suggested MFFPA-2 will need a larger memory.

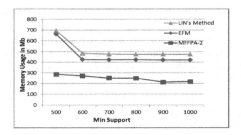
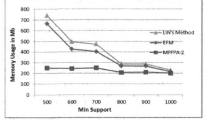

Fig. 14. Comparisons of Memory Usage: Chess dataset   **Fig. 15.** Comparisons of Memory Usage: Mushroom dataset

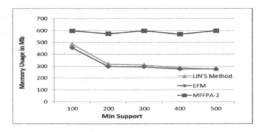

Fig. 16. Comparisons of Memory Usage: T10I4D100k dataset

6 Conclusion

This study introduces the MFFPA-2 technique and an adjacency matrix as a data structure for efficiently mining large, frequent fuzzy itemsets. It produces fewer candidate itemsets than cutting-edge techniques because it uses an efficient search methodology. The execution time performance is enhanced because it generates fewer candidate itemsets (join counts). The memory requirement is related to the number of items in the dataset, as an example is given in Sect. 5.3; for everything else, it consumes substantially less memory.

References

1. Agrawal, R., Imielinski, T., Swami, A.: Database mining: a performance perspective. IEEE Trans. Knowl. Data Eng. **5**(6), 914–925 (1993)
2. Agrawal, R., Srikant, R.: Fast algorithms for mining association rules (1994)
3. Berkhin, P.: A survey of clustering data mining techniques. In: Kogan, J., Nicholas, C., Teboulle, M. (eds.) Grouping Multidimensional Data. Springer, Berlin, Heidelberg (2006). https://doi.org/10.1007/3-540-28349-8_2
4. Antonelli, M., Ducange, P., Marcelloni, F., Segatori, A.: A novel associative classification model based on a fuzzy frequent pattern mining algorithm. Expert Syst. Appl. **42**(4), 2086–2097 (2015)
5. Hu, K., Lu, Y., Zhou, L., Shi, C.: Integrating classification and association rule mining: a concept lattice framework. In: Zhong, N., Skowron, A., Ohsuga, S. (eds.) RSFDGrC 1999. LNCS (LNAI), vol. 1711, pp. 443–447. Springer, Heidelberg (1999). https://doi.org/10.1007/978-3-540-48061-7_53
6. Han, J., Pei, J., Yin, Y., Mao, R.: Mining frequent patterns without candidate generation: a frequent-pattern tree approach. Data Min. Knowl. Discov. **8**(1), 53–87 (2004)
7. Friedman, J., Zadeh, L.A.: Fuzzy, "Similarity Relations and Fuzzy Orderings". North-Holland Publishing Company, Amsterdam (1968)
8. Hong, T.P., Kuo, C.S., Chi, S.C.: Mining association rules from quantitative data. Intell. Data Anal. **3**(5), 363–376 (1999)
9. Lin, C.W., Hong, T.P.: Mining fuzzy frequent itemsets based on UBFFP trees. J. Intell. Fuzzy Syst. **27**(1), 535–548 (2014)
10. Hong, T.-P., Lin, C.-W., Lin, T.-C.: The MFFP-tree fuzzy mining algorithm to discover complete linguistic frequent itemsets. Comput. Intell. **30**(1), 145–166 (2014)
11. Lin, J.C.-W., Hong, T.-P., Lin, T.-C.: A CMFFP-tree algorithm to mine complete multiple fuzzy frequent itemsets. Appl. Soft Comput. **28**, 431–439 (2015)

12. Lin, J.C.W., Hong, T.P., Lin, T.C., Pan, S.T.: An UBMFFP tree for mining multiple fuzzy frequent itemsets. Int. J. Uncertainty, Fuzziness Knowl. Based Syst. **23**(6), 861–879 (2015)
13. Lin, J.C.W., Li, T., Fournier-Viger, P., Hong, T.P.: A fast algorithm for mining fuzzy frequent itemsets. J. Intell. Fuzzy Syst. **29**(6), 2373–2379 (2015)
14. Lin, J.-W., Li, T., Fournier-Viger, P., Hong, T.-P., Wu, J.-T., Zhan, J.: Efficient mining of multiple fuzzy frequent itemsets. Int. J. Fuzzy Syst. **19**(4), 1032–1040 (2016). https://doi.org/10.1007/s40815-016-0246-1
15. Patel, M.N., Shah, S.M., Patel, S.B.: An adjacency matrix-based multiple fuzzy frequent itemsets mining (AMFFI) technique. Int. J. Intell. Syst. Appl. Eng. **10**(1), 69–74 (2022)
16. Mendel, J., John, R.: Type-2 fuzzy sets made easy. IEEE Trans. fuzzy Syst. **10**(2), 117–127 (2002)
17. Chen, C.-H., Hong, T.-P., Li, Y.: Fuzzy association rule mining with type-2 membership functions. In: Nguyen, N.T., Trawiński, B., Kosala, R. (eds.) ACIIDS 2015. LNCS (LNAI), vol. 9012, pp. 128–134. Springer, Cham (2015). https://doi.org/10.1007/978-3-319-15705-4_13
18. Lin, J.C.W., Lv, X., Fournier-Viger, P., Wu, T.Y., Hong, T.P.: Efficient mining of fuzzy frequent itemsets with type-2 membership functions. In: Nguyen, Ngoc Thanh, Trawiński, Bogdan, Fujita, Hamido, Hong, Tzung-Pei. (eds.) ACIIDS 2016. LNCS (LNAI), vol. 9622, pp. 191–200. Springer, Heidelberg (2016). https://doi.org/10.1007/978-3-662-49390-8_18
19. Lin, J.C. W., Wu, J.M.T., Djenouri, Y., Srivastava, G., Hong, T.P.: Mining multiple fuzzy frequent patterns with compressed list structures. In: IEEE International Conference on Fuzzy Systems, vol. 2020 (2020)
20. Delgado, M., Marín, N., Sánchez, D., Vila, M.A.: Fuzzy association rules: general model and applications. IEEE Trans. Fuzzy Syst. **11**(2), 214–225 (2003)
21. Hong, T.P., Kuo, C.S., Wang, S.L.: A fuzzy AprioriTid mining algorithm with reduced computational time. Appl. Soft Comput. J. **5**(1), 1–10 (2004)
22. Lin, C.W., Hong, T.P., Lu, W.H.: Linguistic data mining with fuzzy FP-trees. Expert Syst. Appl. **37**(6), 4560–4567 (2010)
23. Lin, C.W., Hong, T.P., Lu, W.H.: An efficient tree-based fuzzy data mining approach. Int. J. Fuzzy Syst. **12**(2), 150–157 (2010)
24. Hong, T.P., Lin, C.W., Wu, Y.L.: Incrementally fast updated frequent pattern trees. Expert Syst. Appl. **34**(4), 2424–2435 (2008)
25. Hagras, H.: Type-2 fuzzy logic controllers: a way forward for fuzzy systems in real world environments. In: Zurada, J.M., Yen, G.G., Wang, J. (eds.) WCCI 2008. LNCS, vol. 5050, pp. 181–200. Springer, Heidelberg (2008). https://doi.org/10.1007/978-3-540-68860-0_9
26. Castillo, O., Melin, P.: Introduction to type-2 fuzzy logic. Stud. Fuzziness Soft Comput. **223**, 1–4 (2008)
27. Frequent Itemset Mining Dataset Repository. http://fimi.ua.ac.be/data

Student Personality, Motivation and Sustainability of Technology Enhanced Learning: A SEM-Based Approach

Rohani Rohan[1] , Subhodeep Mukherjee[2] , Syamal Patra[3] , Suree Funilkul[1] ,
and Debajyoti Pal[1(✉)]

[1] School of Information Technology, King Mongkut's University of Technology Thonburi,
Bangkok, Thailand
`debajyoti.pal@mail.kmutt.ac.th`
[2] Department of Operations Management, GITAM (Deemed to be University), Visakhapatnam,
Andhra Pradesh, India
[3] Department of Computer Science and Engineering, Camellia Institute of Technology, Kolkata,
West Bengal, India

Abstract. Technology enhanced learning (TEL) has come into prominence and become more relevant after the onset of COVID-19 pandemic. However, it is not known that whether TEL will be socially sustainable, and what factors can affect its sustainability. Therefore, in this work we propose a new research model based on UTAUT2 and the Big 5 Personality Framework that considers several motivational factors together with the different personality traits of the students. Data is collected from two Asian countries and analyzed using a Covariance-based SEM method. Results suggest that motivational factors of performance expectancy, hedonic motivation, social influence, price value and habit significantly affect the social sustainability of TEL. Likewise, the personality traits of agreeableness and neuroticism are also relevant. Suitable theoretical and practical implications are discussed based on the results.

Keywords: Big 5 Personality Framework · Social Sustainability · Structural Equation Modelling · Technology Enhanced Learning · UTAUT2

1 Introduction

Technology Enhanced Learning (TEL) is defined as the learning process that is conducted, supported, and assessed by using different forms of educational technologies [1]. In TEL, the teaching-learning process maybe entirely online or in a hybrid mode that involves blending of the online learning format with the traditional in-person learning. Emerging technologies like Internet of Things (IoT), 5G communication capabilities, Metaverse, and application of artificial intelligence (AI) into various educational processes have provided impetus to the growth of TEL. Flexibility, ease of use, variety and higher student engagement are some of the advantages of TEL [2].

© The Author(s), under exclusive license to Springer Nature Switzerland AG 2023
M. Singh et al. (Eds.): ICACDS 2023, CCIS 1848, pp. 516–528, 2023.
https://doi.org/10.1007/978-3-031-37940-6_42

It is often argued that the onset of the COVID-19 pandemic has resulted in the adoption of TEL within a very short time-span. There is a wide body of literatures that investigates the student's experience of online learning during the pandemic [3, 4], together with the challenges faced [5]. Several factors have been discussed in this context ranging from student engagement [6], academic self-concept [7], satisfaction [4] to even the quality of the learning platforms [8]. However, the more important question is whether TEL will have a long-term sustainability in this post-pandemic world. To the best of our knowledge not much research has been done on this sustainability aspect of TEL that makes this topic worth investigating.

Social, economic, and environmental are the three fundamental types of sustainability. In this work we consider the aspect of social sustainability since it focuses directly on the two prominent stakeholders of TEL – the teachers and students. We define the social sustainability of TEL as the degree to which the social issues (e.g., social support, social development, cultural competency, or social responsibility) are addressed by TEL. Although the aspect of TEL in the pandemic and post-pandemic world is increasing rapidly, yet there are very limited number of studies that investigate the determinants affecting its social usage [9]. Since, social sustainability emerges from human-computer interaction (HCI), it becomes important to analyze the relevant factors that will guarantee such long-term social viability.

In terms of HCI research, individual personality traits have been found to significantly affect human behavior and perceptions, especially in case of emerging technologies [10]. Strangely, however, the current studies related to TEL do not consider how the personality traits of the students might affect its social sustainability. Human behavior is complex, and the way students react to TEL may depend on their respective personality traits. It is important to understand how different students react to TEL because it will be useful to design better applications, creative evaluation schemes and build an engaging teaching-learning environment. Thus, through this study we aim to answer the following research objective: *"What effect does different personality traits of the students have in predicting the social sustainability of TEL?"*.

For answering the above research question, we propose a research model based on two theoretical frameworks: Unified Theory of Acceptance and Use of Technology (UTAUT2) model, together with the Big 5 Human Personality Framework (FFM). UTAUT2 not only helps to explain why people adopt new technologies, but it is also able to predict the user's behavior. Likewise, the FFM model helps us to classify the student's personality into 5 distinct types: agreeableness, openness to experience, conscientiousness, extraversion, and neuroticism. By combining these two frameworks, we are able to successfully predict the social sustainability of TEL.

2   Related Work, Hypotheses and Research Model

2.1   Unified Theory of Acceptance and Use of Technology (UTAUT2)

UTAUT was first proposed by Venkatesh et. al. [11] by combining four primary factors: performance expectancy (PE), effort expectancy (EE), facilitating conditions (FC), and social influence (SI). This model has been validated across several application areas, although later it has been extended by the same authors to include three additional factors of hedonic motivation (HM), price value (PV), and habit (HT) [12]. We decided to use UTAUT2 as the core framework of our research model because of the following main reasons. First, it is a well- known research model that has been validated across several different types of technologies, and covering a wide range of demographics [13]. Second, UTAUT2 is a considerable improvement over its earlier version, since by adding three additional factors the predictive capability of the model increases from 56% to 74% for behavioral intention (BI), and 40% to 52% for actual usage (AU) [11, 12]. This makes it a suitable candidate to explore new factors like the social sustainability of TEL. Third, the factors of hedonic motivation and price value have been used extensively to examine adoption of various IoT related services [14], smart products [15], and even in online learning platforms [16]. All these reasons make UTAUT2 a suitable candidate for the present case.

PE is defined as *"the extent to which any technology will provide benefit to the users to satisfy their needs and activities"* [11]. This factor is highly correlated to perceived usefulness of the Technology Acceptance Model (TAM). A higher degree of usefulness is associated with positive adoption perceptions that has been reported by current research [17, 18], and such positive perceptions will ultimately help a technology to be socially sustainable.

EE is defined as *"the extent of easiness with which the users can use any technology in their daily life"* [11]. This factor is highly correlated with perceived ease of use of TAM. If the students intend to use TEL they must need a certain level of knowledge, understanding and skill in handling the different technologies of TEL. If they believe that using smart-wearables, AR/VR devices or even online learning platforms like MS Teams or Zoom are easy, then they will continue using and adopting TEL. Current research also supports that higher the EE value, greater will be the usage scenario of a particular technology [17, 18].

FC is defined as *"the degree to which the user believes that the existing infrastructure (organizational and technical) is sufficient to support the use of technology"* [11]. The entire TEL ecosystem contains different types of technological tools (hardware and software), e.g., laptops, desktops, AR/VR devices, smart-wearables, fast internet connectivity, online learning platforms, and much more. It becomes evident that if the students perceive that the available resources ae sufficient for using such a virtual environment, they will be more willing to adopt TEL [19].

The fourth factor is that of social influence (SI) which we define as *"the belief of an user that some important person/group of person close to him//her will approve and support their use of a particular technology"* [11]. Existing research has shown that in case of any new technology, trust formation is an important aspect that leads to continuance usage of that technology [20]. Social influence plays an extremely important role in such trust formation [21].

HM is defined as *"the extent to which the students perceive that the use of TEL will be a pleasure and provide them with fun and enjoyment"*. Several empirical studies in literature have reported that hedonic motivation is positively associated with technology usage [8, 15, 17]. If the students perceive TEL usage to be fun and full of excitement, the adoption is supposed to increase, which in turn should improve its social sustainability.

PV refers to the *"cognitive trade-off of the users with regards to the perceived benefits vs. the cost of using a technology"* [12]. In case of TEL, the cost may include the hardware and software costs, data charges, charges associated with network usage, and several others. The price value will be positive if the students feel that the benefits and value, they are getting from using TEL outweigh the cost incurred. A higher level of PV is expected to positively impact the social sustainability of the TEL.

The last factor of habit (HT) from the UTAUT2 framework is defined as *"the extent to which the users tend to perform behaviors automatically"* [12]. Most likely the users would repeat a particular behavior if they had a satisfactory outcome previously. Several studies in multiple contexts have shown that habit helps to understand the behavior of the users based on their past experiences that may produce positive or negative feelings towards the behavior [18, 22, 23]. Therefore, habit helps to enhance the desire of the students and motivate them to engage more with TEL.

Based on the above discussion of the UTAUT2 framework, we propose the following 7 hypothesis:

- H_1: Performance expectancy positively affects the social sustainability of TEL
- H_2: Effort expectancy positively affects the social sustainability of TEL
- H_3: Facilitating conditions positively affects the social sustainability of TEL
- H_4: Social influence positively affects the social sustainability of TEL
- H_5: Hedonic motivation positively affects the social sustainability of TEL
- H_6: Price value positively affects the social sustainability of TEL
- H_7: Habit positively affects the social sustainability of TEL

2.2 The Big 5 Personality Framework (FFM)

Commonly referred to as the Five Factor Model (FFM), this was developed using a psycholexical approach, and has been one of the most popular models depicting human personality traits [24]. 5 unique traits are proposed that make each individual unique, and different from each other. In the context of education, current research has examined the effects of individual differences towards course design [25], second language learning [26], and even in the context of online learning [27]. However, one major drawback of these current studies is they regard individual differences as a single factor due to which how the different personality traits of the students might affect their adoption perception becomes unclear. We decided to examine this under-researched aspect of individual differences from the human personality perspective by incorporating concepts from the FFM. FFM is chosen because not only it has proven to be a good candidate for predicting individual preferences, but it has also been applied specifically in the educational scenario that makes it more relevant for the present study.

The personality trait of openness (ON) to experience refers to the intellectual characteristics of the students. These types of students tend to look for new experiences, and they have the capability to adapt quickly to changes. Typically, they are the early adopters of any new technology. Research has shown that students who are open to experiences, are likely to adopt new forms of technology and adopt it in their lifestyle [28]. Therefore, it is reasonable to assume that students having this openness personality will embrace and interact with TEL positively that will have a positive impact on its social sustainability.

The personality trait of conscientiousness (CT) refers to a person being organized and responsible. Previous research indicates that this is an important trait that affects the student performance [29]. Likewise, individuals who are conscientious have positive performances in their job [28]. Overall, individuals having this personality type exhibit goal-oriented behavior and have a high level of thoughtfulness. Since they are disciplined and can plan ahead, it is expected that students having a high level of conscientiousness will interact more with TEL, explore its details, and develop a positive mindset that should help to improve the social sustainability of TEL.

The third trait of extraversion (EV) captures how sociable, active, assertive, and ambitious the individuals are [24]. Since extraverted individuals are highly sociable and can easily mix with the society, they can act as a positive catalyst to spread the advantages of using a specific technology like TEL by convincing their friends and close acquaintances. Research has shown that students having high levels of extraversion have a greater knowledge sharing intention [30]. Therefore, such types of individuals are more likely to spread the positive effects of TEL that will improve its social sustainability.

Agreeableness (AB) refers to personality traits like honesty, kindness, co-operative, helpful, and tolerant [24]. Individuals having this type of personality are caring by nature, and always assist others who are in need of help. Therefore, if the students having this type of personality have a positive experience with TEL, they will share it in their social network that will ensure social sustainability of TEL.

The final trait of neuroticism (NC) symbolizes negative personality like depression, moodiness and overall emotional instability [24]. These types of individuals worry a lot, experience a lot of stress, and get upset easily. Research has shown that these type of students often have to face discontinuity in their education [26, 31]. Thus, neurotic students might negatively affect the social sustainability of TEL.

Based on the above discussion of FFM, we present the following 5 hypothesis:

- H_8: Openness to experience positively affects the social sustainability of TEL
- H_9: Conscientiousness positively affects the social sustainability of TEL
- H_{10}: Extraversion positively affects the social sustainability of TEL
- H_{11}: Agreeableness positively affects the social sustainability of TEL
- H_{12}: Neuroticism negatively affects the social sustainability of TEL

Figure 1 presents our proposed research model.

3 Research Methodology and Data Collection

Data collection is done by conducting an online survey in Google Forms from two large public universities in India and Thailand that focus on STEM- science, technology, engineering, and mathematics education. The target population consists of those students who have previous experience in online learning as they are ideally suited to understand the concept of TEL. 264 participants responded in the survey from both the countries. After initial data screening, corrosponding to 23 participants we observed either incomplete or missing data that was removed, thereby yeilding a final sample size of 241 participants. The participants were contacted via university mailing lists, or from the personal contacts of the authors, thereby using a combination of convinience and random sampling. Majority of the participants are female (61.41%) having an age range between 18-32 years (Mean = 24.8 years, SD = 9.3 years). Further, 65.56% of the participants studied in the undergraduate level, and were dominated by the engineering discipline (51.45%).

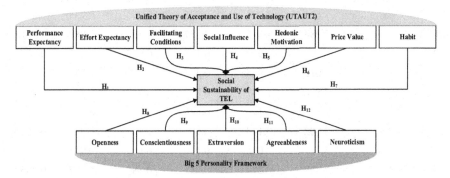

Fig. 1. Proposed Research Model based on UTAUT2 and Big 5 Personality Framework

We designed the survey into 3 distinct parts. The first part captured the demographic data of the participants, the second part contained items related to the UTAUT2 constructs, whereas the third part contained items related to the Big 5 personality traits. The items for UAUT2 are adopted from [12, 27] with the wordings being modified to suite the present context. Particularly, 4 items are used for measuring PE, 4 items for EE, 3 items for FC, 3 items for SI, 4 items for HM, 3 items for PV, and 3 items for HT. Likewise, the items corresponding to the BIG 5 traits are adopted from [24, 31] and include 10 items for openness, 9 items for conscientiousness, 8 items for extraversion, 9 items for agreeableness, and 8 items for neuroticism. The 3 items corresponding to social sustainability are adapted from [32]. All the items are measured on a 5-point Likert scale from (1 = strongly disagree to 5 = strongly agree).

Before analyzing the data, we investigated two types of biases: non-response and common method bias (CMB) that are typically associated with empirical survey-based research. The results from a one-way ANOVA indicate that there are no significant differences between the early and late responders, therefore non-response bias is not an issue. Additionally, a Harman's single factor test is conducted. The highest variance contributed by a single factor is 21.4% of the total variance, which is less than the recommended value of 50%, therefore CMB not being a problem presently.

Table 1. Reliability and normality measurements

Factor	Mean	Skewness	Kurtosis	α Value	CR	AVE
PE	3.794	1.013	0.146	0.864	0.940	0.797
EE	3.556	1.040	0.123	0.792	0.837	0.719
FC	3.621	1.024	0.235	0.806	0.933	0.823
SI	3.984	1.012	0.157	0.755	0.883	0.791
HM	3.166	1.929	0.128	0.845	0.955	0.877
PV	4.012	0.945	0.334	0.932	0.879	0.708
HT	3.828	1.042	0.267	0.862	0.931	0.773
ON	3.547	−0.116	-0.548	0.904	0.836	0.725
CT	3.702	0.144	-0.506	0.913	0.932	0.736
EV	3.181	0.035	-0.180	0.889	0.915	0.781
AB	3.829	0.229	-0.341	0.884	0.947	0.856
NC	3.447	-0.356	0.299	0.911	0.944	0.810
Social sustainability	3.254	0.098	0.245	0.864	0.909	0.715

Table 2. Correlation matrix and discriminant validity

Factor	1	2	3	4	5	6	7	8	9	10	11	12	13
PE	0.89												
EE	0.66	0.85											
FC	0.51	0.64	0.91										
SI	0.48	0.55	0.58	0.89									
HM	0.32	0.58	0.59	0.65	0.94								
PV	0.31	0.38	0.43	0.48	0.44	0.84							
HT	0.52	0.41	0.38	0.41	0.52	0.50	0.88						
ON	0.60	0.55	0.60	0.33	0.61	0.57	0.45	0.85					
CT	0.29	0.30	0.28	0.27	0.29	0.36	0.31	0.22	0.86				
EV	0.33	0.27	0.25	0.26	0.26	0.41	0.39	0.21	0.44	0.88			
AB	0.25	0.32	0.36	0.32	0.35	0.33	0.44	0.30	0.30	0.29	0.92		
NC	0.30	0.29	0.28	0.30	0.33	0.36	0.32	0.37	0.28	0.41	0.66	0.90	
SS	0.28	0.33	0.25	0.35	0.41	0.36	0.35	0.39	0.32	0.40	0.51	0.42	0.84

4 Results

We used a covariance-based Structural Equation Modelling (CB-SEM) approach for the purpose of data analysis. CB-SEM is best suited when the theory being tested is a well-established one, and when the assumptions of data normality are met. For the present case both UTAUT2 and the Big 5 model are well-established theoretical frameworks. With regards to the data normality requirements, we present the result in Table 1. All the skewness and kurtosis values lie in the range of $\pm$ 2, which makes the data suitable to be analyzed by CB-SEM technique.

The reliability of the survey items is assessed from the Cronbach's alpha (α) values (Table 1). The validity of the instrument is checked from two aspects: convergent and discriminant validity. For checking the convergent validity, we use two criteria – Composite Reliability (CR) and Average Variance Extracted (AVE). Both the CR and AVE values are reported in Table 1 and are found to be greater than their corresponding threshold values 0f 0.70 and 0.50 respectively. For checking the discriminant validity, we use the Fornell Larcker criterion that is presented in Table 2 in the form of correlation matrix. Discriminant validity is also satisfied since the square-root of AVE (diagonal elements in Table 2) of each of the latent constructs is greater that its correlation with any other construct in the model.

Table 3. Structure model and hypotheses testing

Hypothesis	βvalue	tvalue	pvalue	Status
Performance expectancy - > social sustainability (H_1)	0.22	5.19	< 0.01	✓
Effort expectancy - > social sustainability (H_2)	0.02	0.14	> 0.05	✗
Facilitating conditions -> social sustainability (H_3)	−0.04	−0.18	> 0.05	✗
Social influence -> social sustainability (H_4)	0.24	5.26	< 0.01	✓
Hedonic motivation -> social sustainability (H_5)	0.09	2.88	< 0.05	✓
Price value -> social sustainability (H_6)	0.31	7.33	< 0.001	✓
Habit -> social sustainability (H_7)	0.75	14.91	< 0.001	✓
Openness -> social sustainability (H_8)	0.09	2.86	< 0.05	✓
Conscientiousness -> social sustainability (H_9)	0.05	0.28	> 0.05	✗
Extraversion -> social sustainability (H_{10})	−0.07	−0.30	> 0.05	✗
Agreeableness -> social sustainability (H_{11})	−0.12	−3.33	< 0.05	✓
Neuroticism -> social sustainability (H_{12})	−0.15	−3.58	< 0.05	✓

After evaluating the measurement model, in the second phase we analyze the structural model, together with the process of hypothesis testing. The results of hypothesis testing are presented in Table 3. The results indicate that performance expectancy ($\beta = 0.22, p < 0.01$), social influence ($\beta = 0.24, p < 0.01$), hedonic motivation ($\beta = 0.09, p < 0.05$), price value ($\beta = 0.31, p < 0.001$), and habit ($\beta = 0.75, p <$

0.001) have significant positive effects on the social sustainability of TEL, thereby supporting hypothesis H_1, H_4, H_5, H_6, and H_7 respectively. However, the effects of effort expectancy ($\beta = 0.02, p > 0.05$) and facilitating conditions ($\beta = -0.04, p > 0.05$) on social sustainability of TEL is found to be non-significant, thereby hypothesis H_2 and H_3 are not supported. With respect to the personality traits, openness to experience ($\beta = 0.09, p < 0.05$) positively affects the social sustainability of TEL, whereas agreeableness ($\beta = -0.12, p < 0.05$) and neuroticism ($\beta = -0.15, p < 0.05$) have significant negative effects on the same, thereby supporting hypotheses H_8, H_{11} and H_{12} respectively. However, the effects of conscientiousness ($\beta = 0.05, p > 0.05$) and extraversion ($\beta = -0.07, p > 0.05$) on social sustainability of TEL are non-significant, thereby hypotheses H_9 and H_{10} being not supported. The results indicate that not all the personality traits are effective in ensuring the social sustainability of TEL.

5 Discussion

Although the concept of TEL is not new and involves the amalgamation of various types of technologies like AR, VR, smart-wearables, and educational robots, yet it has become more prominent and relevant with the onset of the COVID-19 pandemic that resulted in the closure of all face-to-face education. Now, in the post-pandemic period it becomes important to examine the student's perspective on the social sustainability of TEL, and to what extent its psychological and social effects will be. Accordingly, in this study we have examined the effect of the student's personality traits together with other motivational factors that will determine the social sustainability of TEL.

We based the motivational factors on the UTAUT2 framework, and observe that performance expectancy, social influence, hedonic motivation, price value and habit significantly affect the social sustainability of TEL. These findings agree with our current knowledge of TEL and online learning in particular [4, 6, 8, 28]. Performance expectancy refers to the ease of use, and current research has shown that people will accept a technology when it can be used easily. Social influence is another factor that affects the social sustainability of TEL. Past research has shown that user's tend to adopt a new technology when they are influenced by the society [11]. If the students get positive feedback about TEL from their peers, it will help to improve its social acceptance. Hedonic motivation is another aspect that results in feelings of pleasure and fun after using a technology [7]. Greater sense of perceived enjoyment will lead to higher satisfaction levels, and such cognitive satisfaction results in a greater system usage [15]. Therefore, if TEL provides the students with hedonic motivation, it will have a positive effect of its social sustainability. Price value is also found to affect TEL usage. Similar to current findings, our results also suggest that the students have a paradox in their mind, where they compare the price paid vs. the benefits obtained [3, 15]. If the benefits outweigh the price, it will lead to a greater use of TEL. Finally, we observe habit to be the strongest predictor of the social sustainability of TEL. This suggests that currently students mostly have a positive satisfaction interacting with online learning and TEL, which may in turn help to inculcate a new habit in them.

Apart from the motivational factors, several interesting findings are also obtained from the personality traits. We observe that agreeableness and neuroticism have a negative relationship with the social sustainability of TEL, whereas openness has a positive effect. Prior research has reported that students having greater neurotic levels are associated with greater negative behavioral intentions towards using educational technology, and reaping benefits from the online environment [26, 28, 29]. These types of students are introvert, stubborn, and have low self-confidence [28]. However, students who have agreeableness type of personality have a positive perception.

5.1 Theoretical Contributions

This study contributes to the theory advancement in multiple ways. First, this is one of the first studies to investigate the role of motivational factors and personality trait of the students towards the social sustainability of TEL. This is done by integrating factors from the UTAUT2 framework and the Big 5 personality framework. Although UTAUT2 has been used previously in the context of online education, still our results confirm the model's ability to predict a new dependent variable like social sustainability. In fact, our research model can explain 81.4% of the variance in social sustainability, whereas the original UTAUT2 framework could explain 74% of the variation in behavioral intention [12], indicating the suitability of the model for the present scenario. Third, although most of our proposed hypotheses are supported, yet, the effect of effort expectancy, facilitating conditions, conscientiousness, and extraversion on social sustainability is not supported. This shows that the applicability and validity of any theoretical framework cannot be pre-determined.

5.2 Practical Implications

The results confirm the opportunity provided by TEL to the educational community. Since the results show that performance expectancy, hedonic motivation, social influence, price value, habit, openness, agreeableness, and neuroticism affect social sustainability of TEL, therefore, the educational technology companies while designing the infrastructure or even the educational contents must keep these factors in mind. Quiet strangely, our results suggest a non-significant relationship between effort expectancy and social sustainability of TEL. However, keeping in mind the importance of effort expectancy in the overall adoption process of any technology, it is important for the companies to give proper training sessions to the students so that they become familiar and confident using these. The objective should be to improve the digital literacy of the students, especially those who are not "digital natives", by undertaking several government as well as academia-industry partnerships. The other non-significant relationship between facilitating conditions and social sustainability of TEL shows that in developing countries of Asia, the technological infrastructure that is needed for implementing TEL might be lacking.

6   Conclusion and Future Work

In this study we have investigated the joint effects of various motivational factors and personality traits of the students. However, we must acknowledge certain limitations of the present study. First, we conducted this study in an Asian context covering two countries. Future research should collect data from different countries across the globe for including wider demographics because the adoption of TEL might have cultural variations. Additionally, we employed a cross-sectional research design, however, future research may focus on longitudinal as well as qualitative data.

Acknowledgement. This work has been partially supported by the Thailand Science Research and Innovation (TSRI) Basic Research Fund under grant no FRB650048/0164.

References

1. Almpanis, T., Joseph-Richard, P.: Lecturing from home: exploring academics' experiences of remote teaching during a pandemic. Int. J. Educ. Res. Open. **3**, 100133 (2022). https://doi.org/10.1016/j.ijedro.2022.100133
2. Rohan, R., Pal, D., Funilkul, S.: Gamifying MOOC's a step in the right direction? In: Proceedings of the 11th International Conference on Advances in Information Technology, pp. 1–10. ACM, New York, NY, USA (2020). https://doi.org/10.1145/3406601.3406607
3. Nayak, B., Bhattacharyya, S.S., Goswami, S., Thakre, S.: Adoption of online education channel during the COVID-19 pandemic and associated economic lockdown: an empirical study from push–pull-mooring framework. J. Comput. Educ. **9**, 1–23 (2021). https://doi.org/10.1007/s40692-021-00193-w
4. Pal, D., Vanijja, V., Patra, S.: Online learning during COVID-19. In: Proceedings of the 11th International Conference on Advances in Information Technology, pp. 1–6. ACM, New York, NY, USA (2020). https://doi.org/10.1145/3406601.3406632
5. Mushtaha, E., Abu Dabous, S., Alsyouf, I., Ahmed, A., Raafat Abdraboh, N.: The challenges and opportunities of online learning and teaching at engineering and theoretical colleges during the pandemic. Ain Shams Eng. J. **13**, 101770 (2022). https://doi.org/10.1016/j.asej.2022.101770
6. Martin, F., Xie, K., Bolliger, D.U.: Engaging learners in the emergency transition to online learning during the COVID-19 pandemic. J. Res. Technol. Educ. **54**, S1–S13 (2022). https://doi.org/10.1080/15391523.2021.1991703
7. Rohan, R., Dutsinma, F.L.I., Pal, D., Funilkul, S.: Applying the stimulus organism response framework to explain student's academic self-concept in online learning during the COVID-19 Pandemic. In: Tiwari, S., Trivedi, M.C., Kolhe, M.L., Singh, B.K. (eds.) Advances in Data and Information Sciences. Lecture Notes in Networks and Systems, vol. 522, pp. 373–384 Springer, Singapore (2023).https://doi.org/10.1007/978-981-19-5292-0_35
8. Perera, R.H.A.T., Abeysekera, N.: Factors affecting learners' perception of e-learning during the COVID-19 pandemic. Asian Assoc. Open Univ. J. **17**, 84–100 (2022). https://doi.org/10.1108/AAOUJ-10-2021-0124
9. Visvizi, A., Daniela, L.: Technology-Enhanced learning and the pursuit of sustainability. Sustainability **11**, 4022 (2019). https://doi.org/10.3390/su11154022
10. Dutsinma, F.L.I., Pal, D., Roy, P., Thapliyal, H.: Personality is to a conversational agent what perfume is to a flower. IEEE Consum. Electron. Mag. 1–1 (2022). https://doi.org/10.1109/MCE.2022.3180183
11. Venkatesh, V., Morris, M.G., Davis, G.B., Davis, F.D.: User acceptance of information technology: toward a unified view. MIS Q. **27**, 425 (2003). https://doi.org/10.2307/30036540
12. Venkatesh, T.: Xu: consumer acceptance and use of information technology: extending the unified theory of acceptance and use of technology. MIS Q. **36**, 157 (2012). https://doi.org/10.2307/41410412

13. Tamilmani, K., Rana, N.P., Wamba, S.F., Dwivedi, R.: The extended unified theory of acceptance and use of technology (UTAUT2): a systematic literature review and theory evaluation. Int. J. Inf. Manage. **57**, 102269 (2021). https://doi.org/10.1016/j.ijinfomgt.2020.102269

14. Pal, D., Arpnikanondt, C., Razzaque, M.A., Funilkul, S.: To trust or not-trust: privacy issues with voice assistants. IT Prof. **22**, 46–53 (2020). https://doi.org/10.1109/MITP.2019.2958914

15. Pal, D., Arpnikanondt, C.: An integrated TAM/ISS model based PLS-SEM approach for evaluating the continuous usage of voice enabled IoT systems. Wirel. Pers. Commun. **119**(2), 1065–1092 (2021). https://doi.org/10.1007/s11277-021-08251-3

16. Al-Azawei, A., Alowayr, A.: Predicting the intention to use and hedonic motivation for mobile learning: a comparative study in two Middle Eastern countries. Technol. Soc. **62**, 101325 (2020). https://doi.org/10.1016/j.techsoc.2020.101325

17. Sewandono, R.E., Thoyib, A., Hadiwidjojo, D., Rofiq, A.: Performance expectancy of E-learning on higher institutions of education under uncertain conditions: Indonesia context. Educ. Inf. Technol. **28**, 4041–4068 (2022). https://doi.org/10.1007/s10639-022-11074-9

18. Al-Emran, M., AlQudah, A.A., Abbasi, G.A., Al-Sharafi, M.A., Iranmanesh, M.: Determinants of using AI-based chatbots for knowledge sharing: evidence from PLS-SEM and fuzzy sets (fsQCA). IEEE Trans. Eng. Manag. 1–15 (2023). https://doi.org/10.1109/TEM.2023.323 7789

19. Bouzguenda, I., Alalouch, C., Fava, N.: Towards smart sustainable cities: a review of the role digital citizen participation could play in advancing social sustainability. Sustain. Cities Soc. **50**, 101627 (2019). https://doi.org/10.1016/j.scs.2019.101627

20. Pal, D., Arpnikanondt, C., Razzaque, M.A.: Personal information disclosure via voice assistants: the personalization–privacy paradox. SN Comput. Sci. **1**(5), 1–17 (2020). https://doi. org/10.1007/s42979-020-00287-9

21. Fang, H., Li, X., Zhang, J.: Integrating social influence modeling and user modeling for trust prediction in signed networks. Artif. Intell. **302**, 103628 (2022). https://doi.org/10.1016/j.art int.2021.103628

22. Yoo, D.K., Cho, S.: Role of habit and value perceptions on m-learning outcomes. J. Comput. Inf. Syst. **60**, 530–540 (2020). https://doi.org/10.1080/08874417.2018.1550731

23. Nazir, S., Khadim, S., Ali Asadullah, M., Syed, N.: Exploring the influence of artificial intelligence technology on consumer repurchase intention: the mediation and moderation approach. Technol. Soc. **72**, 102190 (2023). https://doi.org/10.1016/j.techsoc.2022.102190

24. John, O.P., Srivastava, S.: The Big-Five trait taxonomy: history, measurement, and theoretical perspectives (1999)

25. Kim, K.-J., Liu, S., Bonk, C.J.: Online MBA students' perceptions of online learning: benefits, challenges, and suggestions. Internet High. Educ. **8**, 335–344 (2005). https://doi.org/10.1016/ j.iheduc.2005.09.005

26. Wallace, M.P.: Individual differences in second language listening: examining the role of knowledge, metacognitive awareness, memory, and attention. Lang. Learn. **72**, 5–44 (2022). https://doi.org/10.1111/lang.12424

27. Osei, H.V., Kwateng, K.O., Boateng, K.A.: Integration of personality trait, motivation and UTAUT 2 to understand e-learning adoption in the era of COVID-19 pandemic. Educ. Inf. Technol. **27**, 10705–10730 (2022). https://doi.org/10.1007/s10639-022-11047-y

28. Watjatrakul, B.: Online learning adoption: effects of neuroticism, openness to experience, and perceived values. Interact. Technol. Smart Educ. **13**, 229–243 (2016). https://doi.org/10. 1108/ITSE-06-2016-0017

29. Waldeyer, J., et al.: A moderated mediation analysis of conscientiousness, time management strategies, effort regulation strategies, and university students' performance. Learn. Individ. Differ. **100**, 102228 (2022). https://doi.org/10.1016/j.lindif.2022.102228

30. Lv, M., Sun, Y., Shi, B.: Impact of introversion-extraversion personality traits on knowledge-sharing intention in online health communities: a multi-group analysis. Sustainability. **15**, 417 (2022). https://doi.org/10.3390/su15010417
31. Tavitiyaman, P., Ren, L., Guan, J., Chung, K.-H.M.: How personality affects flow experience and performance in online classes: a cross-regional comparison among hospitality and tourism students. J. Hosp. Tour. Educ. 1–16 (2022). https://doi.org/10.1080/10963758.2022.2109479
32. Pelletier, L.G., Legault, L.R., Tuson, K.M.: The environmental satisfaction scale. Environ. Behav. **28**, 5–26 (1996). https://doi.org/10.1177/0013916596281001

Spear Phishing Using Machine Learning

Aditya Mahesh Hegde$^{(\boxtimes)}$, S.P. Bharath Kumar , R. Bhuvantej ,
R. Vyshak , and V. Sarasvathi

Department of CSE Bengaluru, PES University, Bengaluru, India
adityahegde0011@gmail.com , sarsvathiv@pes.edu

Abstract. In this digital age, personal data is more valuable than an actual human being. It corresponds to a person's daily internet activity. Open-source intelligence can be used to collect data in a variety of ways, which is later put to use for social engineering attacks. The purpose of this paper is to assess the level of data that is freely available online and its potential consequences for personal privacy. This paper discusses the use of machine learning algorithms and tools for automating email phishing attacks. The personal data of the victim is gathered from open-source websites in order to analyze their online activity. This information is then analyzed in an attempt to learn more about the victim's interests and a relevant email template is created based on this information. The machine learning algorithm is then provided with the constructed template, which predicts how successful a phishing attack would be if launched. This paper uses a machine learning algorithm that is composed of support vector machines (SVM) and logistic regression (LR). This hybrid algorithm, which is a combination of SVM and logistic regression, achieves a peak accuracy of 99.69% when compared to using only one type of classification method, such as SVM or LR. The purpose of this paper is to increase the effectiveness of phishing attacks by automating the data extraction process and to analyse the success rate of attack using machine learning before launching it. This paper will serve the interests of institutions/companies by providing a convenient way to conduct automatic phishing as part of Cyber-Security training to educate employees, giving them a practical experience of social engineering attack.

Keywords: Machine Learning · Support Vector Machine · Logistic Regression · Phishing · Social engineering attack · Open Source Data · Web Scrapping · Selenium · Python-Django · Gophish · Spear phishing

1 Introduction

Since the 1990s, data has become increasingly important in today's economy. With the growth of internet usage and technology, this information is now dispersed across billions of people worldwide. The process of data collection involves the use of computers to collect and store qualitative and quantitative information

© The Author(s), under exclusive license to Springer Nature Switzerland AG 2023
M. Singh et al. (Eds.): ICACDS 2023, CCIS 1848, pp. 529–542, 2023.
https://doi.org/10.1007/978-3-031-37940-6_43

in an electronic form. There are two ways to collect data online, one of which is using bots, often referred to as crawlers, which are popularly used in web crawling, also known as indexing, to index the content on the website. Search engines essentially crawl the web in order to render top articles in first page. It ultimately comes down to examining and indexing a page as a whole. A bot that crawls a website looks through every page, every link, and even the last line in search of any information. Web crawlers are mostly utilised by huge web aggregators, statistics organisations, and well-known search engines like Google, Bing, and Yahoo. While web scraping focuses on specific data set fragments, web crawling typically captures general information. Similar to web crawling, web scraping (also known as web data extraction) detects and locates the desired data from web pages. The main distinction is that with online web scraping, we are able to precisely identify the data set identifier, such as an HTML element structure for web pages that are being fixed and where data should be extracted from. We can see its importance clearly from this paper authored by S. Asiri et al. in [9]. Web scraping uses bots, commonly referred to as "scrapers," to automatically extract particular data sets from websites. Once the appropriate data has been gathered, it can be compared, verified, and analysed in accordance with the requirements and objectives of a certain organisation. So data collection plays a major role in executing a successful phishing attack.

Phishing is a type of social engineering attack wherein the attacker acts as a reputable source and sends a fraudulent message luring the victim in clicking a link designed to trick them to gain their personal sensitive data. So in order to prevent phishing many Artificial Intelligence techniques are discussed in the paper authored by M. F. Ansari in [15]. So it becomes really important to know the repercussions of sharing data online. These are the following Phishing methods that have been majorly used

- Bulk phishing: type of phishing attack wherein the messages are delivered in bulk to a group of users by email and are not personalized or targeted to a specific individual. Streaming services, banks, and other financial institutions, as well as email and cloud productivity providers, are frequent targets for impersonation.
- Spear phishing: It involves an attacker studying his subject using his/her browser activity and directly targeting a specific individual entity by essentially sending email and messages in order to obtain their sensitive personal information.
- Vishing: It is kind of phishing wherein the attacker uses the human psychology to trick the victim over a phone call. In this method the attacker manually contacts the victim and systematically asks for their sensitive information pretending to be a trust entity.

2   Literature Survey

Phishing has become a major issue for internet users all over the world, and it is something we must be concerned about. Phishing attacks are on the rise every day, so we need to provide users with education both theoretically and practically. Machine learning algorithms were used to detect phishing. While they have their own advantages and disadvantages, these techniques are essential in combating this threat.

The authors Alsariera, Adeyemo, Balogun and Alazzawi in [1] have discussed about detection of phishing websites using Artificial Intelligence. This study proposed four meta-learner models (AdaBoost-Extra Tree (ABET), Bagging – Extra tree (BET), Rotation Forest – Extra Tree (RoFBET) and LogitBoost-Extra Tree (LBET)) developed using the extra-tree base classifier. The AI-based meta-learners were used to learn from phishing website data and their performance was evaluated. The models achieved an accuracy of 97% or higher with a very low false-positive rate of less than 0.028%. The proposed models were found to be more effective than existing machine learning-based models when it comes to phishing attack detection. The paper suggests the use of meta learners when building detection models for phishing attacks. The authors Salloum, Gaber, Vadera and Shaalan in [8] uses natural language processing techniques to determine whether emails are phishing scams or not.

The authors T. Peng, I. Harris and Y. Sawa in [2] used a combination of Natural Language Processing and Machine Learning to detect phishing attacks. They analyzed the semantic content of the email that was sent to the victim. This process is used to ensure that each sentence in the email text is appropriate and then ranks it accordingly. These scores are then used to designate the email as spam or ham. The detection algorithm used in this paper was named as "SEAHound" which processes the document one sentence at a time and then evaluates to determine if there exists a social engineering attack. If the email contains a link, then an anti-spam analysis will be conducted to validate the URL. This includes using commercial tools like netcraft's anti-phishing toolbar. The "SEA Hound" algorithm used in this paper gave a total precision of 95% which offered comparatively better performance with other existing techniques of phishing detection.

The authors J. Lee, Y. Lee, D. Lee, H. Kwon and D. Shin in [3] discussed about profile phishing techniques that are focused on targeting high-profile individuals, whether they are in the government or a company. These criminals try to gain access to personal information by posing as someone who is trusted and has authority. In this paper a few types of phishing email attacks have been discussed with respect to the Kimsuky attack group. The hacking groups main technique or method used in collecting information was by using distributing document files such as vulnerable Hangul documents created by the Hangul Word Processing (HWP) application that disguises malicious code inside phishing e-mails in an attempt to gain the attention of victims with a social engineering attack technique residing within it. The authors Baig, F. Ahmed and A.M. Memon in [7] talked about spear-phishing campaigns, which discuss how malicious links can lead to phishing attacks.

The authors J. Feng, L. Zou, O. Ye and J. Han in [4] aims to detect and report phishing webpages and links that leads to malicious websites. In this paper, a model called 'Web2Vec' was created to identify online phishing websites/Webpages. This model utilizes representation learning and deep learning algorithms to convert website content, the URL, and DOM elements into text. This technology is then used to teach a model how to represent these websites. Deep learning model is hybridized with other algorithms and used to get high accuracy and precision score. This hybridization fuses convolution neural network (CNN), Bidirectional Long short term memory (Bi-LSTM) an artificial neural network based approach, attention mechanisms and other deep learning methods which ensures better performance and result.

The authors Kunju, E. Dainel, Anthony and S. Bhelwa in [5] have evaluated phishing attacks based on machine learning. They have done a survey about phishing and different methods to detect it. There are different kinds of phishing attack like spear, whaling and clone phishing. There are also different methods for conducting phishing attack like click jacking, cross-website scripting, drive-by-download etc. Some of the ML algorithms used to detect phishing attack are SVM, Random forest, adaboost, KNN, K means clustering and logistic regression. Each algorithm had its own advantages and disadvantages with respect to detect phishing attacks. According to literature survey carried out by authors M. Khonji, Y. Iraqi and A. Jones in [6] the k-means clustering algorithm was highly effective and gave reliable results among all the ML algorithms tested.

3   Proposed Methodology

This paper discusses various techniques for extracting valuable information about a victim's web activity. This data will be used to create phishing email templates that are sent to the victim. This type of attack can be classified as a spear-phishing attack, which is based on exploiting the open source data that's freely available online. In order to achieve a higher success rate, hybrid machine learning algorithms are often used which include the use of Support Vector Machine combined with the Logistic Regression. The goal of this research is to collect publicly available data on potential victims and analyze whether a successful phishing attack can be made. If the email is opened by a victim, it will appear similar to an email from a legitimate domain that embeds landing page asking for login credentials. The template will be used as a foundation for machine learning algorithms that can verify and develop phishing attacks.

The proposed methodology is based on three modules: data extraction using selenium, machine learning models (SVM and LR) and launching the phishing attack. Figure 1 illustrates the workflow of this application. The first module (data collection) involves collecting information about the victim's online activity. When a potential victim is chosen, their online data is collected as a data-set. This collection includes features like the website they were browsing and their username and passwords (if applicable), as well as any interests that may have been displayed on that site. The data is used to create an email template and landing page which are then checked for spam using the machine learning algorithms. If they pass the spam detection systems, the created emails are then sent

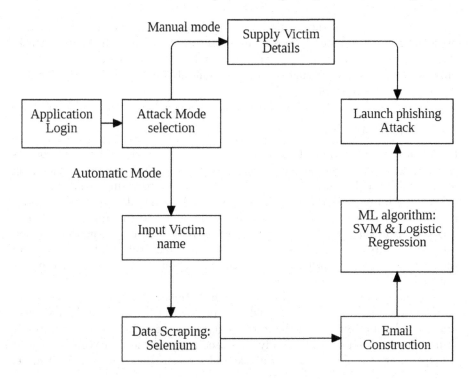

Fig. 1. Flow chart of Proposed Methodology

through the email service providers. These algorithms were chosen for this study: the Support Vector Machine and Logistic Regression. Once the ML algorithms granted clearance for the attack, the attacker can launch the email phishing. Once the victim clicks on a fake website link and enters their credentials, data theft is successfully carried out through spear-phishing.

3.1   User Interface

The User Interface is designed in a simplistic way that makes it easy for the attacker, who is the primary application user, to understand and participate in launching an automated phishing attack. They are given the role of an attacker, who carries out the attack himself and learns about social engineering techniques through trial and error. The primary goal of this application is to provide an educational tool for institutions and organizations to provide an automation tool that can train their employees and students on common social engineering attacks. This will help increase awareness about the risks posed by digital technologies, making it easier for people to protect themselves from unwanted contact or manipulation. Once the attacker has been verified by administrators, they are able to login and carry out a phishing attack in either an automated or manual fashion. The manual approach is the traditional way of launching an attack, which requires the attacker to complete all steps while the automatic mode of phishing includes every step necessary to pull off a sophisticated attack, starting with gathering open source data and moving on to constructing the phishing email and harvesting sensitive credentials from cloned web pages sent to victim via phishing emails.

3.2   Data Extraction

Data collection is essential to a successful phishing attack, and Fig. 2 provides an overview of the process. The various steps are outlined in Fig. 2, including data acquisition (the act of collecting information), processing (analyzing that information), and presentation (displaying it). Once the attacker has located a potential victim, they provide this individual's name to the application which accesses open-source data on their recent activities. This information is then used to activate the malicious code that scrapes confidential information from the internet. The data will include information about the victim's social media accounts and other personal profiles.

This data will be formatted to a python dictionary that contains the "category", "site-name" and "site-link" fields. According to this data, a list is compiled that provides an overview of the most frequently accessed category on websites visited by the victim. Additionally, the user is given the opportunity to visit a website that profiles victims of abuse and verify if their target is legitimate.

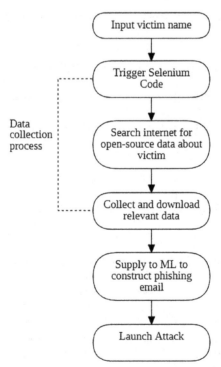

Fig. 2. Flow chart of Data Extraction Process

The user has the ability to choose which email category their phishing attack falls into, based on this data. The application will use this information to create appropriate email templates for sending to a hybridized phishing detection model to predict the corresponding class of email. A practical example of detection can be seen in the paper authored by M. A. Syafiq Rohmat Rose et al. in [14] wherein extensions can be effectively used for the prevention of phishing attacks.

The representation in Fig. 3 is a sample of the data-set used to train and test the machine learning models. The HAM class is meant to be used with valid and legal emails, if an email is classified as HAM, it results in a higher chance of success when attempting to phish. The SPAM class is assigned to emails that are likely sent with the intention of stealing someone's personal information, however, poorly constructed phishing templates will eventually be put in a folder labeled "SPAM" on the victim's computer. Based on the model accuracy score and email classification output, attacker decides to launch the attack or re-frame the phishing template.

Input e-mail content to Hybrid SVM-LR model	Output e-mail Class
Hi Coder, You have registered for November Circuits '22 currently happening on HackerEarth. November Circuits '22 is a rated contest that helps you to increase your HackerEarth rating and stand up among your peers. Please click on the below button to participate in this challenge.	HAM
Had your mobile 11 months or more? You are entitled to Update to the latest colour mobiles with camera for Free! Call The Mobile Update Co FREE on 08002986030	SPAM
SIX chances to win CASH! From 100 to 20,000 RUPEES. Text CSH11 and send to 87575. Cost 150p/day, 6day.	SPAM
Did you catch the bus? Are you frying an egg? Did you make a tea? Are you eating your mom's left-over dinner? Do you feel my Love?	HAM
Dear Participants, As many of you have your final exams scheduled this week, we have considered your request of extending the deadline for submission of Phase II entries.	HAM
Congrats! One-year special cinema pass for 2 is yours. call 09061209465 now! And enjoy Matrix3, StarWars3, etc all for FREE! Don't miss out!	SPAM

Fig. 3. Representation of Sample Data-set

3.3 ML Algorithms

When data is extracted from open-source websites, it is used to compile a list of different domains which the victim was interested online. Once the domain has been selected, an email phishing template is constructed using data scraped from websites. So machine learning algorithm is the best in place for this which can also be seen in the paper authored by K. Sushma et al. in [12]. So ML algorithms, SVM and logistic regression models are used to identify spam emails. If the ML model classifies email as spam, then the attack is more likely to fail and if it is determined that the model classifies email into a more desirable category (ham), then they are guaranteed to land an email in the victim's inbox. According to the literature survey, using two ML algorithms over one is more effective. In Fig. 4, the Machine Learning algorithm is shown in action when an attack is launched.

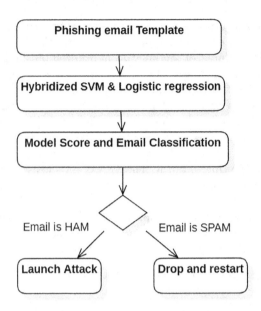

Fig. 4. Flow chart of ML Algorithm

Support Vector Machine. SVM is one of the most popular supervised learning algorithm which is a linear classifier. Support vector machines are primarily used for classification problems. In this paper, support vector machine is trained with good samples of data-set with around 5000 features. This model is trained with 73% of data-set as well as tested with 27% of entire data-set and provides a total accuracy of more than 94%. With every 100 emails, the model can detect and correctly classify 94 emails to spam or ham.

Algorithm 1. Support Vector Machine (SVM)

INPUT: X, Y loaded with labelled data from data-set, D.
$T \leftarrow$ Phishing email template
$S \leftarrow 0$, partially trained SVM
repeat
   for $\{Xi, Yi\}$ in *TrainingData* $\{Text, Label\}$ **do**
     Optimize Si
   end for
until no changes in S or training data exhausted.
Fit the input data T in trained model S.
if T falls above the support vector S **then**
   return HAM
else
   return SPAM
end if
OUTPUT: Email classified as HAM or SPAM

Logistic Regression. A logistic regression model is an effective machine learning algorithm for binary classification. It estimates the probability of an event occurring based on a given data-set of independent variables. This paper shows that by training and testing the model with enough data, it can achieve accuracy rates exceeding 96%. After predicting email content using this model, it's important to evaluate its goodness of fit. So out of every 100 emails, the model could detect and classify 96 emails as spam or ham accurately.

Algorithm 2. Logistic Regression (LR)

INPUT: X, Y loaded with labelled data from data-set, D.
$T \leftarrow$ Phishing email template
$L \leftarrow 0$, partially trained LR
repeat
    for $\{Xi, Yi\}$ in *TrainingData* $\{Text, Label\}$ **do**
        Features $\leftarrow$ feature.fit(Xi, Yi)
        Train *Li* with *Features*
    end for
until no changes in *L* or training data exhausted.
Fit the input data T in trained model L and classify email.
return *emailclass*
OUTPUT: Email classified as HAM or SPAM using trained LR model.

Hybridized SVM-LR. The hybridized version of Support Vector Machine and Logistic Regression is the most effective for email classification. The model was found to achieve a 98.4% accuracy rate, which surpassed the results achieved when each algorithm was used on its own. So the paper authored by R. K. Shah et al. in [13] is a good reference for utilizing unique methods.

Algorithm 3. Hybridised SVM-LR

INPUT: Models S (SVM) from *Algorithm−1*, Models L (LR) from *Algorithm−2*.
1. $C \leftarrow$ None, a classifier for input template, T
2. Supply input email to model S and L
3. Compute the accuracy of each model, say *Sa* for SVM accuracy and *La* for LR accuracy.
4. Final model accuracy, $A \leftarrow$ (*Sa+La*)/2.
5. Let the predicted class by each model be *Sc* and *Lc* respectively.
if at least one of *Sc* and *Lc* is HAM
    $C \leftarrow$ HAM
else
    $C \leftarrow$ SPAM
end if
return C, A
OUTPUT: Final model accuracy, A and resultant email class, C.

3.4   Phishing Attack

The study has implemented two different ways to carry out this phishing attack - manually or through an automated method using Gophish. When performing an attack manually, the user must specify each detail required to carry out the attack like - victim's name, email address, and a written link to their landing page. Automation can help reduce this process by providing these details automatically. The Gophish unit can reduce the amount of work required to conduct an attack and get statistics about the attack after it has been conducted. User groups are created with information on a victim's name (first name, last name), email address, and other contact details. Based on data collected, the email template can be customized to a victim's interests by selecting from an existing list of templates. For example, if Instagram is a popular interest for the victim, then that information will be taken into account when creating the email. The landing page can be chosen based on collected data. The user has the option to choose a Simple Mail Transfer Protocol (SMTP) settings for an attack or go with default setting to launch the attack.

3.5   Safety and Precautions

In order to stay safe and aware in the digital world, staying alert is key. This vigilance comes in handy while browsing the internet where attackers are always looking for a way to gather information they can use. In order to stay safe, we should avoid visiting any unfamiliar web-pages and be vigilant in examining the domain of URLs before trusting them. This is one line of defense against phishing attacks. If a data breach occurs, individuals are advised to change their username and password credentials in order to prevent the hacker from logging back into their account. Ensuring internet sanity while browsing and trusting entities is essential, especially when accessing information from untrusted sources. So new technologies as mentioned by G. -G. Geng et al. in [10] and T. Nathezhtha et al. in [11] should be used and links should be verified before using them.

4   Implementation and Result

The developed application stores user credentials in a SQLite database. The data is collected about the target victim for a brief period of time, and then it is deleted once the session ends. The machine learning models are blended together, using a support vector machine and logistic regression, to give the most accurate results. The application is coded in Python with Django as its framework; it also uses HTML and CSS for presentation of front-end. Python's SQLite is used as the database. The modules that are implemented have been extensively tested for expected results and adherence to coding standards, in order to ensure accuracy in application development. Proper object-oriented techniques are also followed throughout the implementation phase, guaranteeing correctness of the end product.

Table 1. Model Accuracy comparison table for different data-set size

Dataset Size	SVM	LR	SVM+LR
2500	95.87%	92.74%	98.0%
3000	96.95%	93.99%	98.83%
3500	97.49%	95.07%	97.49%
4000	98.30%	95.78%	99.69%
4500	98.21%	98.32%	98.07%
5000	97.76%	96.59%	99.28%

Table 1 summarizes the observed accuracy of each model in relation to data set size. It is found that with SVMs and LR models, accuracy peaks at 96% only. However, when these models are hybridized, an increase in accuracy is seen to reach 98.4%.

Figure 5 is a graphical representation of the data that was analyzed. The SVM curve begins from 96% for 2500 sample space and attains the maximum score of 98.3% and drops down to 97.76% for larger data-set. While SVM offers good accuracy, the LR model began with low score value of 92.74% for smaller data-set and maintained a relatively good increase with increase of sample space. From the graph, it is clear that the hybridized model achieved a peak accuracy of 99.69%. This score declares that phishing attacks conducted by the application

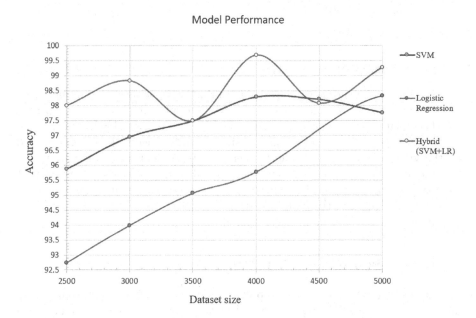

Fig. 5. Comparison of SVM and Logistic Regression

Fig. 6. Phishing status

via hybridised SVM-LR will have an approximate success rate of 99.69%, with very less chance of failure.

4.1 Research Output

The Fig. 6 diagram shows the sensitive data that was disclosed to the thief when he or she successfully logged into a victim's account using one of the phishing email templates sent to victim's email. The email template is loaded with duplicate login pages from social media websites, such as Facebook. When accessed by the victim to login, this collects their sign-in credentials along with timestamp and sends it to the database where the attacker can use it. This statistic will provide institutions with information on how many employees are being phished and give them the ability to study their awareness of social engineering techniques, so they can be better prepared.

5 Conclusion and Future Work

The objective of this paper was to provide information about open source data and its consequences. The research found that sharing open-source data increases the likelihood of being phished, as attackers can use this information to psychologically manipulate people. It is important for individuals to be aware of the data they are sharing online in order to avoid these types of attacks. The use of a hybridized version of SVM and Logistic Regression was found to produce the highest score, with 98.4% accuracy, compared to using only one classifier which achieved 96% accuracy. To ensure the accuracy of collected victim data, attackers had to check for any discrepancies in it themselves. This was done by verifying that each username matched up with the information stored on a separate website. The application is based on the practice of spear phishing and can be upgraded to a more effective group phishing campaign. By using the same data for each victim, flaws in collection could also be avoided. This software can be used to build attack vectors, detect intrusions and prevent them. Furthermore, this application can be extended to help with the creation of effective penetration tests.

References

1. Alsariera, Y.A., Adeyemo, V.E., Balogun, A.O., Alazzawi, A.K.: AI meta-learners and extra-trees algorithm for the detection of phishing websites. IEEE Access **8**, 142532–142542 (2020). https://doi.org/10.1109/ACCESS.2020.3013699
2. Peng, T., Harris, I., Sawa, Y.: Detecting phishing attacks using natural language processing and machine learning. In: 2018 IEEE 12th International Conference on Semantic Computing (ICSC), pp. 300–301 (2018). https://doi.org/10.1109/ICSC.2018.00056
3. Lee, J., Lee, Y., Lee, D., Kwon, H., Shin, D.: Classification of attack types and analysis of attack methods for profiling phishing mail attack groups. IEEE Access **9**, 80866–80872 (2021). https://doi.org/10.1109/ACCESS.2021.3084897
4. Feng, J., Zou, L., Ye, O., Han, J.: Web2Vec: phishing webpage detection method based on multidimensional features driven by deep learning. IEEE Access **8**, 221214–221224 (2020). https://doi.org/10.1109/ACCESS.2020.3043188
5. Kunju, M.V., Dainel, E., Anthony, H.C., Bhelwa, S.: Evaluation of phishing techniques based on machine learning. In: International Conference on Intelligent Computing and Control Systems (ICCS), pp. 963–968 (2019). https://doi.org/10.1109/ICCS45141.2019.9065639
6. Khonji, M., Iraqi, Y., Jones, A.: Phishing detection: a literature survey. IEEE Commun. Surv. Tutorials **15**(4), 2091–2121 (2013). https://doi.org/10.1109/SURV.2013.032213.00009
7. Baig, M.S., Ahmed, F., Memon, A.M.: Spear-phishing campaigns: link vulnerability leads to phishing attacks, spear-phishing electronic/UAV communication-scam targeted. In: 2021 4th International Conference on Computing & Information Sciences (ICCIS), pp. 1–6 (2021). https://doi.org/10.1109/ICCIS54243.2021.9676394
8. Salloum, S., Gaber, T., Vadera, S., Shaalan, K.: A systematic literature review on phishing email detection using natural language processing techniques. IEEE Access **10**, 65703–65727 (2022). https://doi.org/10.1109/ACCESS.2022.3183083
9. Asiri, S., Xiao, Y., Alzahrani, S., Li, S., Li, T.: A survey of intelligent detection designs of HTML URL phishing attacks. IEEE Access **11**, 6421–6443 (2023). https://doi.org/10.1109/ACCESS.2023.3237798
10. Geng, G.-G., Yan, Z.-W., Zeng, Y., Jin, X.-B.: RRPhish: anti-phishing via mining brand resources request. In: 2018 IEEE International Conference on Consumer Electronics (ICCE), Las Vegas, NV, USA, pp. 1–2 (2018). https://doi.org/10.1109/ICCE.2018.8326085

11. Nathezhtha, T., Sangeetha, D., Vaidehi, V.: WC-PAD: web crawling based phishing attack detection. In: 2019 International Carnahan Conference on Security Technology (ICCST), Chennai, India, pp. 1–6 (2019). https://doi.org/10.1109/CCST.2019.8888416

12. Sushma, K., Jayalakshmi, M., Guha, T.: Deep learning for phishing website detection. In: 2022 IEEE 2nd Mysore Sub Section International Conference (MysuruCon), Mysuru, India, pp. 1-6 (2022). https://doi.org/10.1109/MysuruCon55714.2022.9972621

13. Shah, R.K., Hasan, M.K., Islam, S., Khan, A., Ghazal, T.M., Khan, A.N.: Detect phishing website by fuzzy multi-criteria decision making. In: 2022 1st International Conference on AI in Cybersecurity (ICAIC), Victoria, TX, USA, pp. 1–8 (2022). https://doi.org/10.1109/ICAIC53980.2022.9897036

14. Rose, M.A.S.R, Basir, N., Heng, N.F.N.R., Zaizi, N.J.M., Saudi, M.M.: Phishing detection and prevention using chrome extension, 2022 10th International Symposium on Digital Forensics and Security (ISDFS), Istanbul, Turkey, pp. 1-6 (2022). https://doi.org/10.1109/ISDFS55398.2022.9800826

15. Ansari, M.F., Panigrahi, A., Jakka, G., Pati, A., Bhattacharya, K.: Prevention of phishing attacks using AI algorithm. In: 2022 2nd Odisha International Conference on Electrical Power Engineering, Communication and Computing Technology (ODICON), Bhubaneswar, India, pp. 1–5 (2022). https://doi.org/10.1109/ODICON54453.2022.10010185

An Interpretable Deep Learning Model for Skin Lesion Classification

Avinash Jha$^{(\boxtimes)}$ ⓘ and V. S. Ananthanarayana ⓘ

Department of Information Technology, National Institute of Technology Karnataka, Surathkal, Mangalore, India
{avinashjha.213it001,anvs}@nitk.edu.in

Abstract. Skin Cancer is a dangerous issue in society. Early diagnosis and therapy are two of the most crucial steps in preventing the onset of a disease. Dermatologists primarily use visual methods to identify the skin lesions which may cause skin cancer. With the development of technology, methods for classifying skin lesions, like deep learning and computer vision, are gaining popularity. A hybrid model is proposed using a pre-trained DenseNet architecture integrated with Convolutional Block Attention Module (CBAM) for feature refinement. The HAM10000 dataset, which includes 10015 dermoscopic pictures with seven distinct skin disease types, was used in our research. The proposed approach outperforms the original pre-trained DenseNet models, with an average accuracy of 93% . The LIME framework is used to assess predictions further and produce visual explanations, which can assist in decreasing the drawback of black-box models in aiding medical decision-making.

Keywords: Deep Learning · Skin Lesion Classification · CBAM · LIME

1 Introduction

A common irregularity that causes misery is skin lesions, some of which can have catastrophic repercussions for millions of individuals worldwide [1]. A skin lesion is the unrestrained proliferation of abnormal skin cells. It develops when unrepaired DNA damages skin cells, which causes those cells to change or inherit abnormalities. Environmental and hereditary causes can also cause skin illnesses. Cancerous lesions are referred to as malignant, and non-cancerous as benign. Basal Cell Carcinoma (Bcc), Actinic Keratosis (Akiec), and Melanoma (Mel) are the lesions that can cause cancer. At the same time, Dermatofibroma (Df) Vascular (Vasc), Melanocytic Nevi (Nv) and Benign Keratosis (Bkl) are non-cancerous lesions. Around 1% of all skin cancer cases are melanoma [2], yet it is the cause of a significant number of fatalities. Cancer death rates can decrease if found and treated in its early stages. Conventional skin cancer detection by visual inspection requires a skilled dermatologist. Dermatologists frequently employ dermoscopy as an imaging method. The most frequent usage is to identify pigmented skin lesions. It enlarges the skin lesion's surface, making the dermatologist inspecting its structure easier. Even a skilled dermatologist has trouble recognizing one skin lesion

M. Singh et al. (Eds.): ICACDS 2023, CCIS 1848, pp. 543–553, 2023.
https://doi.org/10.1007/978-3-031-37940-6_44

from another because of the unique patterns and shapes of the lesions. The identical dermatoscopic results picture sample might be misinterpreted and incorrectly categorized by medical experts as belonging to several types of skin cancer. To mimic the expertise of medical specialists and exceed them, an automated computational system must engage in significant amounts of visual exploration utilizing historical data, known as a data-driven technique.

2 Related Work

Computer-aided techniques have generated more study interest and given researchers possibilities to address the issues. Various straightforward computer-aided diagnostic (CAD) techniques, including edge recognition and line fitting, aid the diagnosis. Using attributes gathered by humans, skin lesion classification has been possible with advances in new cutting-edge technologies such as machine learning and pattern recognition. Initially, different grades of skin lesions were classified using machine learning techniques such as decision trees, Bayesian classifiers, and SVM. Recent advancements in deep learning have helped to classify skin cancer using medical imagery analysis. To categorize the HAM10000 dataset, [3] presented a comparative study of the most recent CNN models, Inception-ResNet-v2, Inceptionv3, DenseNet201, and ResNet50, where Inception-ResNet-v2 achieved the highest accuracy of 81.62%. [4] suggested two deep CNN-based lesion subsystems, which produced a merged, segmented picture for segmentation. Using the HAM10000 and ISIC2019 datasets, a 30-layered architecture was created to carry out the classification task. Initially, the features were retrieved and then downsampled using the suggested selection approach, achieving a classification accuracy of 88.39% on the HAM10000 dataset. [5] proposed a system for identifying skin lesions using an ensemble of improved CNN architectures on the HAM10000 dataset with a test accuracy of 83.6%. [6] introduced a selection of pre-trained CNN architectures for categorization on the ISIC2018 dataset. It utilized a three-level fusion technique. Models developed using different picture sizes were fused with the prediction vectors of SeResNext-50, EfficientNetB0, and Efficient Net B1 before being integrated. [7] proposed model used two different neural networks, such as Unet, to identify regions of interest and an updated EfficientNetB5 for the classification of skin lesions. [8] proposed model comprised two branches having pre-trained DenseNet201 and the parallel branch consisting of CNN and spatial attention. The features extracted from both branches were concatenated before prediction on the HAM10000 dataset, where they achieved the best accuracy of 82.57%. Despite substantial advancements, skin cancer classification remains difficult since there are few annotated data sets and slight inter-class variance. The scale of the skin lesion and contrast differences (color, form, etc.) further complicate the process.

3 Proposed Methodology

The efficacy of a lightweight deep learning architecture for classifying skin lesions is examined in this research. Our system is built on DenseNet201 and includes an attention mechanism integrated with the prior architecture for additional information feature refining.

3.1 Dataset

For the purpose of experimentally validating the proposed method, we will utilize the HAM10000 dataset [9], which has 10015 dermatoscopic pictures with a size of 450 × 600. DF, VASC, BCC, AKIEC, NV, BKL, and MEL lesions are the seven diagnostic categories that make up this condition. Figure 1 shows HAM10000 dataset images.

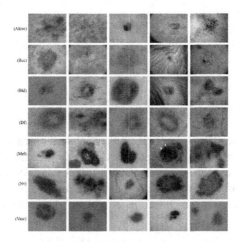

Fig. 1. HAM10000 Dataset Example [9]

3.2 Pre-processing

To increase the effectiveness of training and classification, the dataset must be pre-processed before being fed into the network. Because the dataset contains photos from various sources, the stages of normalization and rescaling are performed to standardize images in identical resolution and bit depth. In order to avoid learning bias, which has the unfavorable impact of generalization, duplicate images are also removed. After that, class imbalances are removed from the dataset's data by cleaning them; this is done by over-sampling the data to ensure sufficient amount of images in each class. Multiple combining methods, such as rotating, flipping, and shearing the pictures, apply various data augmentation approaches to the training set of images after splitting the original data. The training set consisted of 9187 skin lesion images, and the test set comprised 828 images having no duplicate images. Oversampled image details are given in Table 1.

3.3 DenseNet

A recently suggested CNN design is DenseNet, which stands for densely linked convolutional networks. With fewer parameters and less over-fitting for smaller datasets, [10] showed that the design is resilient to the vanishing gradient issue. It is made up of four substantial blocks. A dense block (DB) consists of interconnected layers with

Table 1. Shows the description of oversampled images of different categories.

Skin Lesion	Initial training images	Oversampled training images
akiec	304	6992
bcc	488	7858
bkl	1033	7931
df	109	5877
mel	1079	7903
nv	6042	8042
vasc	132	7096
Total	9187	51699

batch normalization, rectified linear unit, and convolution operation to modify the input non-linearly. The method allows the reuse of features by sending the preceding feature mappings to the subsequent layers. Every dense block is followed by the Transition Block (TB), which comprises batch normalization, average pooling, and convolution operation. Figure 2 shows the architecture of a 5-layer dense block of DenseNet.

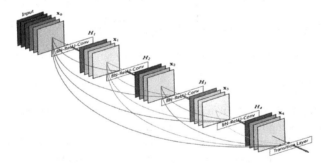

Fig. 2. A 5-layer dense block of DenseNet [10]

3.4   Convolutional Block Attention Module (CBAM)

Channel and Spatial-based attention mechanisms are combined in CBAM, an attention mechanism module. In terms of enhanced performance and more carefully managed feature production, CBAM [11] has been successful. There are two submodules in it:

3.4.1   Channel Attention Module

In order to complete channel attention, the channel attention module initially runs average and maximum pooling procedures on each layer's feature map. The gathered feature data is then sent into a shared neural network. The MLP, a single hidden layer multi-layer perceptron (MLP) in the shared neural network, provides feature information that is added and activated to produce the final channel attention feature. Figure 3 depicts the channel attention module's organizational structure. It is expressed as:

$$M_C(F) = \sigma(X_1(X_0(F_{avg}^C) + X_1(X_0(F_{max}^C))) \tag{1}$$

M_C corresponds to the feature map generated after channel attention. F corresponds to the input feature. σ is the sigmoid function, X_0 and X_1 are MLP weights that are shared to both the the the inputs. F_{avg}^C are the features after average pooling and F_{max}^C are the features after max pooling.

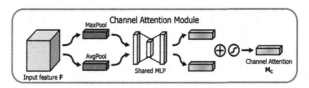

Fig. 3. Channel attention module structure [11]

3.4.2 Spatial Attention Module

The channel attention module's output is sent into this attention module. The maximum and average pooling procedures are applied to the feature maps of each layer in the orientation of the channel axis, creating two feature maps. The spatial attention information is first extracted using a convolution layer by concatenating the two feature maps, after which the spatial attention features are eventually retrieved using an activation function. Figure 4 depicts the spatial attention module's organizational structure. It is expressed as:

$$M_s(F) = \sigma(p^{7*7}([F_{avg}^s; F_{max}^s])) \tag{2}$$

M_s corresponds to the feature map generated after spatial attention. σ is the sigmoid function. p^{7*7} denotes convolution operation having kernel size 7*7. F_{avg}^s are the features after average pooling and F_{max}^s are the features after max pooling across the channels.

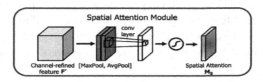

Fig. 4. Spatial attention module structure [11]

A dual attention mechanism CBAM model is created by combining the channel and spatial attention modules. The input and output dimensions are uniform, and CBAM is a compact model. Any convolutional network may incorporate the CBAM module to augment the weight information of the feature map and enhance the deep learning model's classification performance. Figure 5 shows CBAM action principle. It is expressed as:

$$F_a = M_C(F) * F$$
$$F_b = M_s(F_a) * F_a \tag{3}$$

Fa is the features after channel attention. F_b is the final refined feature. Here, '*' represents element-wise multiplication.

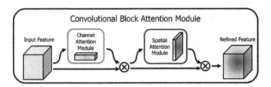

Fig. 5. Action principle of CBAM [11]

3.4.3 DenseNet+CBAM

After contrasting many credible pre-trained models, DenseNet201 was chosen from the DenseNet family. The suggested model comprises a CNN architecture as a backbone and an attention module. The input fundus pictures were initially processed using DenseNet201 as a feature extractor, and then CBAM was employed to refine the features further. CBAM is placed after the fourth dense block, where the feature map size is 7*7, before the classification head of DenseNet201 for further feature enhancement. We only used 28 out of 32 convolutional blocks in the fourth Dense Block (DB) of DenseNet201 after experimenting with different network depths to choose the best one. Feature maps obtained from the last dense blocks of the DenseNet201 network are given as input to the CBAM module. Feature map then passes through channel attention where average and maximum pooling operations are done. A channel attention map is generated after features are passed through a shared network and elementwise summation. The spatial attention module receives output from the previous attention module as input. The feature maps are subjected to maximum and average pooling in the direction of the channel axis. Spatial information is extracted after concatenating the feature maps obtained during the pooling operations. Feature vectors obtained after concatenation are passed through convolution operation to generate a feature map.

F_b, the final refined features are then converted to a one-dimensional array before being classified using Global Average Pooling (GAP) to average each feature map created by the attention module. Figure.6 shows the representation of the proposed network.

3.4.4 Evaluation Metrics

1. **Accuracy:** The average accuracy is calculated as a ratio between correct predictions and total samples.

$$Accuracy = TP + TN/TP + TN + FP + FN \tag{4}$$

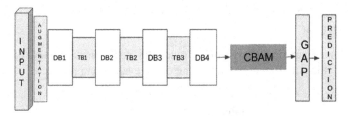

Fig. 6. Description of the Proposed Model

2. **Precision**: It provides information on the percentage of positive predictions that turn out to be true.

$$Precision = TP/TP + FP \tag{5}$$

3. **Recall**: It is the percentage of positively predicted outcomes among actual positive results.

$$Recall = TP/TP + FN \tag{6}$$

where True positives (TP) belong to the categories that the algorithm precisely identified; true negatives (TN) are the outcomes that were accurately predicted to be negative; false positives (FP) are the outcomes that were mistakenly identified as belonging to a positive class; and false negatives (FN) are the cases that were mistakenly identified as related to a negative class.

4   Results and Discussion

Our study's CNN models were pre-trained using ImageNet. For each experiment, an Adam optimizer with a 0.01 learning rate is employed to train the networks. Categorical cross-entropy loss function is used to optimize the neural network, a principal loss function in multiclass classification learning. Since there are seven skin cancer classes of the HAM10000 dataset [9] a softmax activation unit is constructed after an output layer with seven hidden units. Early stopping, an ideal control strategy for reducing overfitting while retaining model accuracy, is also used throughout the training phase. The training procedure would end after 48 epochs if the validation accuracy did not increase since the patient factor for early halting is set at 48. Initially, the proposed model was developed to run for 150 epochs, but the model training stopped after 65 epochs due to an early stopping checkpoint as performance did not improve. 85% of the dataset is the training set, while the remaining 15% is utilized to assess performance after eliminating duplicates from the test set with a batch size of 16. We had 828 testing images and 51699 training images after data augmentation. The training progress of all models generally achieves outstanding results exceeding 90%. The Keras framework was used for all of the experiments.

Here, we experimented with three DenseNet architecture models being the model's backbone. Models are namely:

- DenseNet121+CBAM

- DenseNet169+CBAM
- DenseNet201+CBAM

Table 2 summarizes the experiment outcomes, including accuracy, precision, and recall, to provide a thorough understanding of the evaluation process.

Table 2. Experimental Results

Model	Accuracy (%)	Precision (%)	Recall (%)
DenseNet121+CBAM	90.21	90.30	90.21
DenseNet169+CBAM	91.06	91.93	91.06
DenseNet201+CBAM	93	93	93

The weighted average accuracy of the test set for our suggested model, DenseNet201+CBAM, is most remarkable at 93%, followed by DenseNet169+CBAM at 91.06% and DenseNet121+CBAM at 90.21%. The proposed model outperforms [12] and [13], where they used baseline DenseNet architecture with data augmentation by at least 3% in terms of accuracy because of further feature enhancement done by the attention module. In the proposed method, DenseNet201+CBAM feature maps obtained from the last dense blocks of the DenseNet201 network are given as input to the CBAM module, where features are passed through channel attention and then to spatial attention in a serial manner. The goal of channel attention is to adjust each channel's weight adaptively. Channel attention determines what to pay attention to within an image by serving as an object selector. Spatial attention chooses where to focus on an image through an adaptive spatial area selection procedure. Attention mechanisms give DL algorithms greater flexibility to concentrate on crucial information. DenseNet201+CBAM outperformed [8], where they used DenseNet201 along with spatial attention, but the proposed model used both channel and spatial attention. Overall, the proposed network DenseNet201+CBAM outperforms competing techniques [8, 12, 13] in terms of practically all criteria mentioned in Table 3. Additionally, the confusion matrices are shown in Fig. 7, so a detailed analysis may be performed in each class.

Here, different skin lesions are depicted as follows [class0-akiec, class1-bcc, class2-bkl, class3-df, class4-mel, class5-nv, class6-vasc]. True positive samples for seven classes in HAM10000 dataset are [akiec:14, bcc:23, bkl:53, df:5, mel:21, nv:645, vasc:9]. It depicts that the DenseNet201+CBAM model has correctly identified the respective skin lesion classes with a good percentage out of the 828 testing samples.

4.1 Interpretability of Proposed Model

We created a heat map of our suggested model using the Grad-CAM [14] visualization mechanism and applied it to the target dataset. The relevant region of interest that emphasizes the aberrant part in skin lesion pictures is highlighted in the heat map of the image. These visual observations show that the suggested model is reliable and efficient for skin lesion classification. Moreover, the LIME (Local interpretable model-agnostic

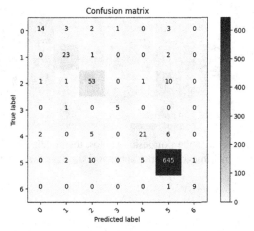

Fig. 7. Confusion Matrix for DenseNet201+CBAM

Table 3. Performance Comparison on HAM10000 dataset

Methodology	Accuracy	Precision	Recall
DenseNet121[12]	82.1	82.58	82.94
DenseNet201+Spatial Attention [8]	82.57		
DenseNet121+Data Augmentation [13]	90	77.71	72.57
DenseNet201+CBAM(Proposed)	**93**	**93**	**93**

explanations) framework presents helpful explanations to support rational decisions. A well-liked method for interpreting and explaining the judgments produced by ML algorithms is the LIME [15] method. It is a post-hoc technique used following model training. It helps to highlight the infected region within lesion images. With the assistance of visual classification, dermatologists may comprehend the workings of models and progressively develop confidence in them. Figure 8 shows the visual explanation of the model on skin lesions.

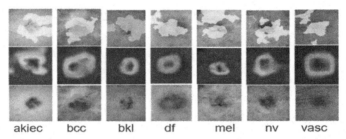

akiec   bcc   bkl   df   mel   nv   vasc

Fig. 8. The bottom row contains the skin lesion photos, the middle row lists the heat map and its coupling with the original image, and the upper row displays the LIME explanation on our predicted lesion labels.

5 Conclusion

A deep learning approach based on attention is proposed in this research, where the model first extracted features utilizing pre-trained DenseNet architectures with transfer learning. Then the features were refined using the CBAM attention mechanism, which was integrated with the DenseNet architectures. Despite the architecture's consistent performance for various imported models, the highest level of accuracy noted was 93% for the DenseNet201+CBAM model. The suggested model has also been compared to other cutting-edge techniques and demonstrated comparable performance. Further, we used the model interpretation method, where dermatologists can easily understand the model's functioning mechanism due to the visual explanations, which are beneficial for building their trust and fostering man-machine cooperation.

References

1. Tarver, T.: Cancer facts & figures 2012. American Cancer Society (ACS). J. Consum. Health Internet. **16**, 366–367 (2012). https://doi.org/10.1080/15398285.2012.701177
2. Gong, A., Yao, X., Lin, W.: Classification for dermoscopy images using convolutional neural networks based on the ensemble of individual advantage and group decision. IEEE Access. **8**, 155337–155351 (2020). https://doi.org/10.1109/ACCESS.2020.3019210
3. Singhal, A., Shukla, R., Kankar, P.K., Dubey, S., Singh, S., Pachori, R.B.: Comparing the capabilities of transfer learning models to detect skin lesion in humans. Proc. Inst. Mech. Eng. H. **234**, 1083–1093 (2020). https://doi.org/10.1177/0954411920939829
4. Khan, M.A., Muhammad, K., Sharif, M., Akram, T., Kadry, S.: Intelligent fusion-assisted skin lesion localization and classification for smart healthcare. Neural Comput Appl. (2021). https://doi.org/10.1007/s00521-021-06490-w
5. Thurnhofer-Hemsi, K., Lopez-Rubio, E., Dominguez, E., Elizondo, D.A.: Skin lesion classification by ensembles of deep convolutional networks and regularly spaced shifting. IEEE Access. **9**, 112193–112205 (2021). https://doi.org/10.1109/access.2021.3103410
6. Mahbod, A., Schaefer, G., Wang, C., Dorffner, G., Ecker, R., Ellinger, I.: Transfer learning using a multi-scale and multi-network ensemble for skin lesion classification. Comput. Meth. Programs Biomed. **193**, 105475 (2020). https://doi.org/10.1016/j.cmpb.2020.105475
7. Oniga, M., Micu, R.-F., Griparis, A.: Deep neural networks for classification of dermatological images with multiple skin lesions. In: 2022 14th International Conference on Communications (COMM), pp. 1–6 (2022). https://doi.org/10.1109/COMM54429.2022.9817233

8. Uddin, M.R., Mahmud, T.I.: Dense-par-AttNet: an attention based deep learning model for skin lesion classification by transfer learning approach. In: 2022 IEEE International Conference on Artificial Intelligence in Engineering and Technology (IICAIET), pp. 1–5 (2022). https://doi.org/10.1109/IICAIET55139.2022.9936758

9. Tschandl, P., Rosendahl, C., Kittler, H.: The HAM10000 dataset, a large collection of multi-source dermatoscopic images of common pigmented skin lesions. Sci. Data. **5**, 1–9 (2018). https://doi.org/10.1038/sdata.2018.161

10. Huang, G., Liu, Z., Weinberger, K.Q.: Densely connected convolutional networks. In: 2017 IEEE Conference on Computer Vision and Pattern Recognition (CVPR), pp. 2261–2269 (2016)

11. Woo, S., Park, J., Lee, J.-Y., Kweon, I.S.: CBAM: convolutional block attention module (2018). https://doi.org/10.48550/ARXIV.1807.06521

12. Neeshma, A., Nair, C.S.: Multiclass skin lesion classification using Densenet. In: 2022 Third International Conference on Intelligent Computing Instrumentation and Control Technologies (ICICICT), pp. 506–510 (2022). https://doi.org/10.1109/ICICICT54557.2022.9917913

13. Anand, V., Gupta, S., Nayak, S.R., Koundal, D., Prakash, D., Verma, K.D.: An automated deep learning models for classification of skin disease using Dermoscopy images: a comprehensive study. Multimed. Tools Appl. **81**, 37379–37401 (2022). https://doi.org/10.1007/s11042-021-11628-y

14. Selvaraju, R.R., Cogswell, M., Das, A., Vedantam, R., Parikh, D., Batra, D.: Grad-CAM: visual explanations from deep networks via gradient-based localization. In: 2017 IEEE International Conference on Computer Vision (ICCV), pp. 618–626 (2017). https://doi.org/10.1109/ICCV.2017.74

15. Ribeiro, M.T., Singh, S., Guestrin, C.: "Why should I trust you?": explaining the predictions of any classifier. In: Proceedings of the 22nd ACM SIGKDD International Conference on Knowledge Discovery and Data Mining, pp. 1135–1144. Association for Computing Machinery, New York (2016). https://doi.org/10.1145/2939672.2939778

Unsupervised Ambient Vibration-Based Feature Extraction for Structural Damage Detection

Ashuvendra Singh$^{(\boxtimes)}$ ⓘ and Smita Kaloni

National Institute of Technology, Srinagar, Uttarakhand, India
singh.ashuvendra14@gmail.com, smitakaloni@nituk.ac.in

Abstract. Structural damage detection (SDD) is one of the major components of the Structural health monitoring (SHM) which essentially indicates the sustainability of industries, buildings, and bridges. Manual field surveys have traditionally been used to evaluate gaps. However, this approach is typically unreliable, time-consuming, and dangerous to surveyors. To overcome these challenges, researchers have been developing automated structural damage methods which include data collection, processing, structural damage identification and diagnosis. Deep learning (DL), a subcategory of ML approaches based on multiple neural networks, has advanced rapidly in recent years. Unlike standard ML approaches, DL does not necessitate a preset feature-based stage and, by providing additional data, may train more broad and robust models. As a consequence of the success of ML approaches, several specific goals in mind in ML-based fracture detection have been undertaken. This study aims to provide a solution using artificial intelligence-based damage detection research in order to guide the SHM. The proposed model achieved 98% accuracy with better testing results on unseen datasets.

Keywords: Structural Damage Detection (SDD) · Deep Learning · Unsupervised Learning · Machine Learning

1 Introduction

Structural health monitoring (SHM) is a vast area to investigate the health condition of structural systems, (such as high-rise constructions, highways, bridges, dams, etc.) by observing the all-possible aspects like loading effect, environmental conditions, and dynamic response of the structures. In order to anticipate the serviceable life and requirement of maintenance for the structures, the initial step is to determine the existence, position, and intensity of faults in the structure. One basic feature of SHM is damage assessment in order to monitor abnormality and acquire real-time structural health status. Traditionally, instead of conducting visual inspections of the structures, which are seen as being ineffective, dangerous, and unreliable; SHM permits civil engineers to undertake observations for monitoring the structural states and diagnosing real threats. To resolve these problems, various techniques depending on vibration have been employed to identify internal damage in the structures. The variations in modal parameters between the intact and damaged states of structures can be determined by directly or indirectly interpreting the measured vibration characteristics acquired from structural vibration outputs

M. Singh et al. (Eds.): ICACDS 2023, CCIS 1848, pp. 554–565, 2023.
https://doi.org/10.1007/978-3-031-37940-6_45

[1]. The crucial recorded data for obtaining dynamic structural features are vibration signals (natural frequency, damping ratio, and mode shape etc.). Direct responses as time-domain vibration and structural vibration characteristics such as natural frequencies and mode shapes, can be used to detect structural damage. A large amount of vibration responses is recorded to access damage detection for complex structures such as high-rise buildings, bridges etc. But the feature extraction and obtaining the damage indices from these ambient vibration responses is a big challenge for damage detection. Many researchers have worked on this problem and with the passage of time new methods have also been developed [2–6]. Machine learning plays a vital role to solve this problem because of its thinking ability and fast calculation [2, 4–6, 10].

A branch of Artificial Intelligence (AI) called machine learning focuses on "Teaching" systems how to behave without having to explicitly design every instance. The core idea behind machine learning is creating algorithms that can educate themselves by learning from a wide variety of inputs [4, 10]. In many areas, like statistical feature extraction and object recognition, where it is challenging or impractical to create traditional algorithms to complete the required tasks, machine learning methods are used.

Sometimes, it can take more time and resources to train properly. Machine learning can analyze enormous amounts of data even more efficiently by integrating perceptive technologies and AI [4].

In other terms, computational statistics particularly emphasizes computer-assisted prediction, and machine learning is intimately associated on the basis of "signal" or "feedback" offered to the learning model. There are some past works done with the help of machine learning.

In the last decades, significant discoveries and advancements are carried out in the field of vibration-centric damage recognition, which is the focus of this re- search. The analysis focuses on the various damage indices and introspective methodologies employed, emphasizing their benefits and drawbacks. The goal of this review is to support scientists and practitioners in making optimal use of current damage detection algorithms and in the future development of more effective and useful approaches for civil engineering structures [5].

Deep learning method for identifying structural damage, especially the Variational Auto-encoder (VAE). When it employs this technique, the VAE may uncover important characteristics concealed inside the high-dimensional structural sample statistics by first reducing it to a low-dimensional feature map and then restoring the original data. The suggested method's capacity to precisely identify structural damage was tested on a bridge below a moving car. This methodology, in contrast to other approaches, does not call on preliminary in- formation or a flexural and shear model, thus making it a data-driven and design solution appropriate for useful uses in SHM [6]. Using information-based anomaly identification for both long-term and short- term monitoring, this research suggests a unique unsupervised machine learning approach for SHM. The approach relies on local concentration, unsupervised variable selection, local exclusion distance, and training set value to define an anomaly score. A probabilistic method based on moderately stochastic theory is presented to estimate a threshold limit for detecting damage, while a non-parametric approach is employed to calculate the local density and closest neighbours. A new unsupervised learning technique, a new anomaly score for SHM, an unsupervised closest neighbour selection, and a modern probabilistic threshold work. Two full-scale bridges were used to test the suggested technology, and the results revealed that it is effective [7]. This work provides a novel method for structural damage diagnosis using vibration data based on Dense Nets, a recent advancement in the field of computer vision. This method makes advantage of the convolutional neural network architecture's dense connections, which is appropriate for this study's usage of acceleration responses. During training, low-level and high-level features are both learnt and reused, facilitating information flow and maintaining all feature levels. With this approach, the number of parameters is decreased while the resilience and precision of damage identification are improved. The effectiveness of the pro- posed methodology is evaluated using numerical simulations and testing, and the results show its accuracy and resilience in terms of damage localization and quantification [8, 10].

Finding deterioration in a structure and assessing its health are the objectives of SHM. In recent years, researchers have concentrated on creating machine learning specifically deep learning based vibration-based damage detection systems. However, many of these deep learning techniques need training data from both the intact and damaged states of the structure, which can be difficult to gather for substantial civil structures. This study suggests a novel convolutional autoencoder-based unsupervised deep learning technique for structural damage identification (CAEs). The technique somehow doesn't event that occurred from damage states throughout training; instead, it uses raw vibration signals received from the structure's healthy condition to train the CAE network. The CAE network is selected to benefit [9]. This research aims goal of this research is to provide an overview of structural damage detection strategies based on machine learning (ML) and deep learning (DL). They studied 68 ML-based fracture detection methods and determined that pixel-level structural damage segmentation is the current trend. The researchers next put eight ML-based structural damage segmentation models to the test on 3D pavement images, showing that deeper backbone networks in FCN models and skip connections in U-Net models improved performance.

The main contribution of this paper is to develop an approach to extract the features from ambient vibration signals using machine learning to identify structural damage.

The paper is organized as Sect. 2 discusses data and methodological design, while Sect. 3 expands on testing results. Section 4 closes the proposed investigation with a discussion of the future scope.

2   Products and Suggestive Methodology

2.1   Dataset Description

Fig. 1. KW51 Bridge, Leuven, Belgium [11].

Physical Property of Monitored Real Bridge Structure [11]:

1. Type of Bridge: Bow string steel bridge as shown in Fig. 1
2. Length: 115 m
3. Width: 12.4 m
4. Usage: Railway bridge (Two lane, Up and Down)
5. Top speed: 160 km/h
6. Identity: KW51
7. Location: Leuven, Belgium
8. Coordinates: latitude 50.9004 N, longitude 4.7066 E

Sensor Arrangement on KW51 bridge: To record the acceleration data on this bridge, 12 uniaxial accelerometers were installed in different locations of the bridge as shown in Fig. 2 [11].

2.2   Proposed Methodology

An encoder is an artificial neural network (ANN) which frequently employed to construct efficient reconstructions of input data by nonlinearly integrating their unique properties. Back-propagation and gradient descendant-based optimisers are commonly applied to train autoencoders, which are generally unsupervised and hence do not require damaged examples during training. Most autoencoders' primary goal is to compress data and lower its dimensionality while keeping its fundamental properties (feature extraction). Data that has been shortened can be evaluated more efficiently and faster, with less computing load.

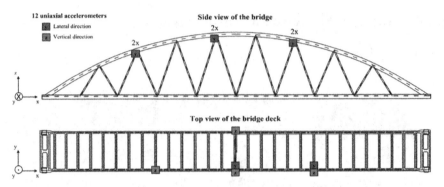

Fig. 2. Side and top view of KW51 Bridge, Leuven, Belgium [11].

An autoencoder network's design generally consists of two things encoder as well decoder. The encoder converts the input data and outputs crammed representations of it, decoder reconstructs the input data from either the compressed pictures. An autoencoder's production is meant to be identical to its input. The number of neurons present in the encoder's final layer can be altered to adjust the compressed data size of the compressed data.

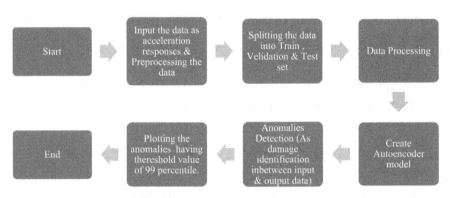

Fig. 3. Steps involved to detect the anomalies using Autoencoder

Table 1 illustrates the structure of the Autoencoder architecture layers. Autoencoders are used in various applications, such as image compression, feature learning, and anomaly detection. By using an autoencoder, researchers can build a more efficient neural network capable of processing large amounts of data quickly and accurately.

The steps for the proposed work is as follows and also shown in Fig. 3:

1. Data Preparation: The first step involves gathering vibration data from the structure under various conditions such as temperature, load, and humidity. The data is then cleaned, normalized, and prepared for the next steps.

2. Model Training: A machine learning model, such as an autoencoder, is trained using preprocessed data to understand the relationship between the vibration data and structural damage.

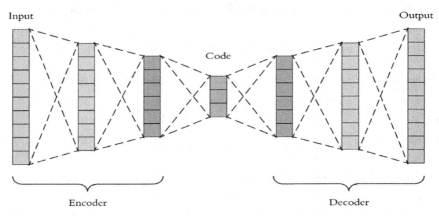

Fig. 4. Autoencoder framework

3. Autoencoder: The autoencoder is used to compress the vibration data into a low-dimensional representation that can be used for damage detection. The framework of Autoencoder is shown in Fig. 4.
4. KL Divergence: The difference between the damaged and undamaged states of the structure is measured using the Kullback-Leibler (KL) divergence, a measure of the difference between two probability distributions.
5. Loss Function: A loss function is defined based on the KL divergence to evaluate the performance of the autoencoder in reconstructing the input data.
6. Damage Indicator: There are two goals: restore the original data and make sure that the output distribution from latent space follows the distribution of input data. Thus, to measure the performance of the model, the metric used is KL Divergence and the loss function chosen is Mean Absolute Error (MAE). After getting the prediction from the model, measure the distance between the input data points and output data points using distance formula for N-dimension (N = 6) shown in Eq. 1.

$$\sqrt{(a2 - a1)^2 + (b2 - b1)^2 + (c2 - c1)^2} \tag{1}$$

Considering that out of entire data points anomalies should make up only 1 percentile of data, therefore distance values greater than the 99th percentile value of the distance column are considered anomalies. Hence, the distance measured between two datasets is considered a Damage Indicator for this process.

7. Anomaly Detection: The autoencoder is then used to detect structural damage by computing the KL divergence between the low-dimensional representation of the

vibration data and the normal, undamaged state of the structure. If the KL divergence exceeds a certain threshold, it indicates damage.

This is a summary of a proposed approach for detecting structural damage using vibration data and machine learning. Further refinement and improvement may be made by exploring alternative algorithms, loss functions, and threshold values.

Table 1. Structure of the Autoencoder architecture layers

Operation	Layer	Input	Output	Activation
Encoder	Input Layer	6	64	ReLU
Encoder	Layer 1	64	32	ReLU
Encoder	Layer 2	32	16	ReLU
Encoder	Layer 3	16	8	ReLU
Embedding	Layer 4	8	16	ReLU
Decoder	Layer 5	16	32	ReLU
Decoder	Layer 6	32	64	ReLU
Decoder	Layer 7	64	6	ReLU

3 Experimental Results

3.1 Visualization of Input Acceleration Responses

Figures 5, 6, 7 and 8 show the acceleration responses of April 30, 2019 and May 03, 2019 corresponding to train passage twice a day (before & after noon). The below graph shows the plot between the observed acceleration and the recording time on each sensor. These figures show that all sensors measured the responses in the multiplicity range and they do not follow the same ranges. Figure 9 shows the plots of all observed acceleration of 30 days as input. It can be observed that the acceleration value measured by the all six sensor has different ranges.

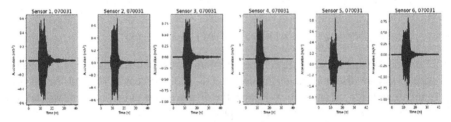

Fig. 5. Representation of Accelerations of the train crossing the bridge on April 30, 2019, 07:00:31 UTC

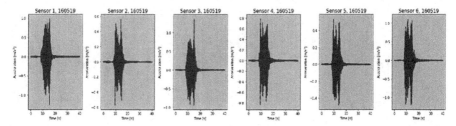

Fig. 6. Representation of Accelerations of the train crossing the bridge on April 30, 2019, 16:05:19 UTC.

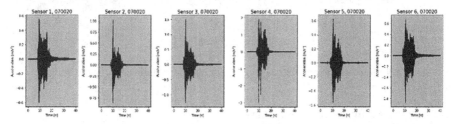

Fig. 7. Representation of Accelerations of the train crossing the bridge on May 03, 2019, 07:00:20 UTC

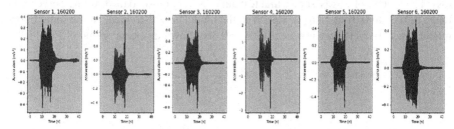

Fig. 8. Representation of Accelerations of the train crossing the bridge on May 03, 2019, 16:02:00 UTC.

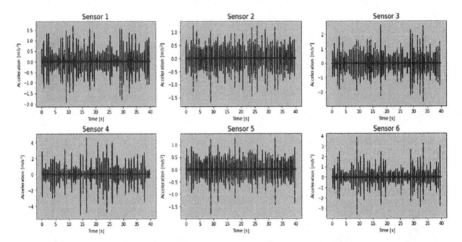

Fig. 9. Representation of 30 days Accelerations of the train crossing the bridge.

3.2 Model Accuracy and Loss Curve

The learning curve of the model represents the accuracy and loss of the developed model. In Fig. 10, the first graph shows the model metric and second graph shows the model loss. The first graph easily shows that test curve is very close to the training curve in the range of 0.00 to 0.01 at 12 to 17 Epoch so the model is well-trained. At the same Epoch model loss (MAE) is very low i.e., 0.2% to 0.3% shown in second graph of Figs. 10.

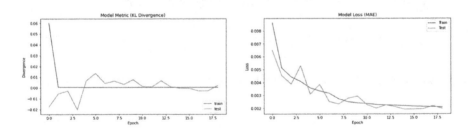

Fig. 10. Model matric & loss curve in terms of KL divergence

3.3 Plots of Anomaly/Outliers Detected

Figures 11, 12, 13, and 14 show the acceleration values recorded by the sensor for one incident, however, since there are 6 sensors installed in total so 6 graphs for the one-time incidents are shown in one incident. In a day, values are recorded 2 times by the sensors when a train passes over the bridge. The blue line plots the observed acceleration values and the Red "+" are the anomaly detected by the model. By observing the graphs at different time incidents for a given sensor, we can infer that the model is showing a large number of anomalies when the acceleration values rise or fall significantly from the usual values. Hence, it can be concluded that the autoencoder model is able to identify incidents that can be considered anomalies. Figure 15 shows all the anomalies corresponding to a train passage of 30 days.

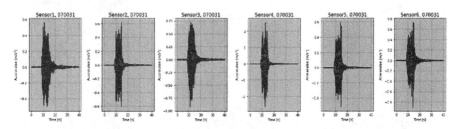

Fig. 11. Anomaly on April 30, 2019, 07:00:31 UTC

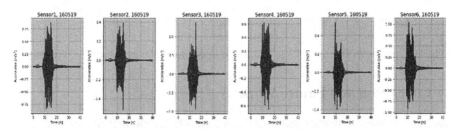

Fig. 12. Anomaly on April 30, 2019, 16:05:19 UTC.

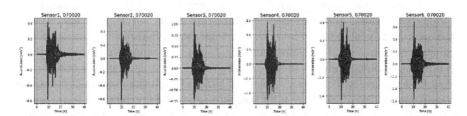

Fig. 13. Anomaly on May 03, 2019, 07:00:20 UTC.

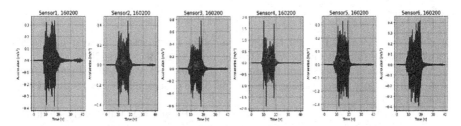

Fig. 14. Anomaly on May 03, 2019, 16:02:00 UTC.

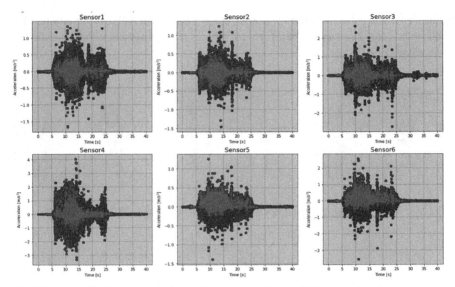

Fig. 15. Anomalies/Damages on the bridge corresponding to 30 days acceleration responses.

4   Conclusion and Future Work

Machine Learning (ML) has proven to be a game-changer in reducing computational time and detecting damage points in datasets. The proposed method takes advantage of unsupervised feature extraction, which is the process of extracting damage-sensitive features without using damage labels. This is achieved through the use of unsupervised learning, which doesn't require the labelling of complex data, but instead inputs measurement data directly into a neural network for training and automatically extracts effective features from the original accelerometer data. The method has proven to be effective in identifying structural damages in real complex structures as anomalies or outliers with 98 % model accuracy. For future studies, the plan is to validate this model through an experimental setup to generalize it for other structures. Additionally, the team plans to explore different approaches for calculating reconstruction error and finding a way to set a threshold to detect anomalies in structures.

References

1. Amezquita-Sanchez, J.P., Adeli, H.: Signal processing techniques for vibration- based health monitoring of smart structures. Arch. Comput. Meth. Eng. **23**(1), 1–15 (2016)
2. Malekjafarian, A., Golpayegani, F., Moloney, C., Clarke, S.: A machine learning approach to bridge-damage detection using responses measured on a passing vehicle. Sensors **19**(18), 4035 (2019)
3. Ghiasi, A., Moghaddam, M.K., Ng, C.T., Sheikh, A.H., Shi, J.Q.: Damage classification of in-service steel railway bridges using a novel vibration-based convolutional neural network. Eng. Struct. **264**, 114474 (2022)
4. Zhang, Y., Yuen, K.V.: Review of artificial intelligence-based bridge damage detection. Adv. Mech. Eng. **14**(9), 16878132221122770 (2022)

5. Santos, A., Figueiredo, E., Silva, M.F.M., Sales, C.S., Costa, J.C.W.A.: Machine learning algorithms for damage detection: kernel-based approaches. J. Sound Vib. **363**, 584–599 (2016)
6. Ghiasi, R., Torkzadeh, P., Noori, M.: A machine-learning approach for structural damage detection using least square support vector machine based on a new combinational kernel function. Struct. Health Monit. **15**(3), 302–316 (2016)
7. Müller, A.C., Guido, S.: Introduction to Machine Learning with Python. O'Reilly Media Inc., Sebastopol (2021)
8. Géron, A.: Hands-on Machine Learning with Scikit-Learn, Keras, and Tensor-Flow, 2nd edn. O'Reilly Media Inc., Sebastopol (2021)
9. Azimi, M., Eslamlou, A.D., Pekcan, G.: Data-driven structural health monitoring and damage detection through deep learning: state-of-the-art review. Sensors **20**(10), 2778 (2020)
10. Awadallah, O., Sadhu, A.: Automated multiclass structural damage detection and quantification using augmented reality. J. Infrastruct. Intell. Resilience **2**(1), 100024 (2023)
11. Maes, K., Lombaert, G.: Monitoring railway bridge KW51 before, during, and after retrofitting. J. Bridg. Eng. **26**(3), 04721001 (2021)

A New Grey Correlational Compromise Ranking Approach for Portfolio Selection for Investment in ESG Stocks

Sanjib Biswas[1](✉) ⓘ, Sayan Gupta[2], Arun Upadhyay[3], Gautam Bandyopadhyay[4], and Rahul Shaw[1]

[1] Calcutta Business School, South 24 Parganas, Bishnupur, West Bengal 743503, India
sanjibb@acm.org
[2] Dr. B. C. Roy Engineering College Academy of Professional Courses, Fuljhore, Paschim Barddhaman, West Bengal Durgapur 713206, India
[3] Department of Business Management, NSHM Business School, Durgapur Arrah, Shibtala Via Muchipara, West Bengal Durgapur 713212, India
[4] Department of Management Studies, National Institute of Technology, Mahatma Gandhi Road, A-Zone, West Bengal 713209 Durgapur, India

Abstract. In recent years environmental, social, and governance (ESG) aspects have gained a notable emphasis. The present paper primarily aims to select a portfolio of the stocks listed to ESG index at BSE, India based on their market performance. To this end the ongoing research proposes a new extension of compromise ranking of alternatives from distance to ideal solution (CRADIS) model by using grey correlational approach to foster a flexible decision-making under uncertainty. Through comparison with other multi-criteria decision making (MCDM) models it is revealed that the grey correlational CRADIS (GC-CRADIS) provides a reasonably reliable result. The sensitivity analysis (with flexible changes in the given conditions) shows the stability in the result. A possible portfolio of ESG stocks is resulted using GC-CRADIS. Further, the current paper attempts to delve into the association of the stock performance with ESG score. However, no significant association of stock performance with others is noticed. The findings of the paper help in formulating a portfolio of ESG stocks for investment.

Keywords: Stock selection · Market performance · ESG · Grey correlation · Compromise ranking of alternatives from distance to ideal solution (CRADIS) · Logarithmic Percentage Change-driven Objective Weighting (LOPCOW)

1 Introduction

Making investment decisions involves multiple factors, complicated endeavor. Before making an investment, investors take into account a number of factors, including their financial ambitions, the state of their finances at the time, their investing goals, and the financial instruments they choose in relation to their expected results [1]. The main purpose of financial investing is to create money in order to obtain financial stability,

© The Author(s), under exclusive license to Springer Nature Switzerland AG 2023
M. Singh et al. (Eds.): ICACDS 2023, CCIS 1848, pp. 566–580, 2023.
https://doi.org/10.1007/978-3-031-37940-6_46

independence, and the desired financial goals [2, 3]. The fundamental goal is to create a portfolio of financial instruments to maximize total return at a manageable risk [4]. Over the past few decades, the equities stock market has been a desirable investment option among all financial products [5].

Over the last decade, there has been a significant push towards sustainable investing to meet ESG criteria in businesses. The existing literature demonstrates the positive impact of ESG scores on stock performance, the integration of ESG components into investment portfolios, and the utilization of ESG as a risk management tool [6]. Researchers, such as [7] and [8], have highlighted a positive correlation between ESG and stock performance. Moreover, recent studies suggest that investments with an ESG focus exhibit low negative risks and minimal price volatility during turbulent periods. Empirical evidence gathered by researchers like [9] indicates that ESG participation can reduce organizations' exposure to negative risk factors and potential hazards. Equities with high ESG ratings exhibit better returns, reduced return volatility, and higher trading volumes than other stocks [10]. Furthermore, according to the research of Broadstock et al. [11], portfolios with high ESG ratings typically perform better than those with low ESG ratings and can reduce financial risk during financial crises.

However, although there has been an increasing focus on ESG investment, the review of extant literature reveals that there is a scantiness of research conducted to formulate portfolio of ESG stocks for Decision-making regarding investments. Most of the past work concentrated on investigating the influence of ESG practices on firm performance using causal models and establishing framework to measure ESG performance. In this regard, the current work intends to carry out a multi-aspect evaluation of the performance of ESG stocks for formulating a portfolio. Since assessment of stock performance is an intricate task subject to the effect of a number of market indicators, a MCDM based analysis is used. Therefore, the present problem is to select the ESG stocks (to form a portfolio) from a sample of size m based on their composite market performance score subject to n number of indicators. To this end the primary objective of the ongoing work is to propose a novel hybrid framework of LOPCOW (for determining criteria weights) and extended CRADIS method with grey correlational analysis aka GCA (for performance-based ranking of the stocks). The application of GCA enables to deal with impreciseness of information and provide flexibility to the decision makers while working with objective information.

The "LOPCOW" method has been invented by Ecer and Pamucar [12] to thrive on the benefits like parity in the criteria weights, reliable and reasonably efficient calculation grounded on fundamental statistical concepts, ability to work with negative values in the decision matrix and a larger number of alternatives and criteria. There has been an increasing number of multi-dimensional applications of LOPCOW evident in the extant literature (for example, [13-18]). To advance the growing strand of volume of MCDM models, Puška et al. [19] proposed a new hybridization of ARAS (Additive Ratio Assessment), MARCOS (Measurement of Alternatives and Ranking according to Compromise Solution), and TOPSIS models to develop the framework of CRADIS. The CRADIS method supports the decision-makers with the advantages like consideration for extreme ideal solutions and relative utility values, reasonable accuracy and stability of the results and freeness from the rank reversal phenomenon. The newly introduced CRADIS have been increasingly applied several complex real-life problems (for example, [20-23]).

In this regard, the current work fills the gap in the literature with two-fold contributions. First, the current work is apparently a rare MCDM based portfolio selection using ESG stocks. Second, the present paper provides a new extension of CRADIS method using grey correlational analysis that fosters a flexible decision-making under uncertainty. The remaining part of this paper is exhibited in the following manner. Section 2 describes the research methodology step by step. In Sect. 3 major findings are recorded. A brief discussion on the result is included in Sect. 4. Section 5 concludes the paper while mentioning some of the future scopes.

2   Materials and Method

The present paper aims to select a portfolio of the stocks listed to ESG index at BSE, India. The selection is done based on a comparative analysis of the stock performance at the marketplace. In this section a brief description of the step by step research methodology is presented.

2.1   Sample

The current work selects leading stocks under ESG index of BSE based on their market capitalization and ESG performance. Accordingly, the primary sample consists of 55 stocks for possible formation of the portfolio.

2.2   Criteria for Comparing Stock Performance

Over the years there has been a plethora of research work (for instance, Gupta et al. [3], Gupta et al. [5], Biswas et al. [15], Tan et al. [24], Asadi and Mohammadi [25], Chen et al. [26], Arif and Sohail [27], Bernal et al. [28], Nazneena et al. [29], Bermejo et al. [30], Nguyen et al. [31], Suroso et al. [32], Narang et al. [33]) conducted on evaluating stock performance. The researchers have emphasized on risk and return based analysis, earning prospects, and intrinsic value, fundamental and technical performances of the organizations for stock selection. The current work is focused on evaluation of stock performance based on the market indicators as given in Table 1..

2.3   Data

The data was gathered from both the BSE website and the 'CMIE-Prowess IQ (version 1.96)' database, with the decision matrix being presented in Appendix A. In this work the financial year 2021–22 has been taken into account.

Table 1. Criteria for comparing stock performance

S/L	'Criteria'	Effect
C1	Annual Return	(+)
C2	EPS	(+)
C3	P/B	(+)
C4	Yield	(+)
C5	Beta	(-)

2.4 LOPCOW Method

Let, $X = [x_{ij}]_{m \times n}$ represents the *decision-matrix* where, m is the number of stocks under comparison and n is the number of indicating criteria for comparing the alternatives. The procedural steps of LOPCOW method as described by [12] are as follows.

Step 1. Normalization of the decision-matrix

Let, the normalized decision matrix is $R = [r_{ij}]_{m \times n}$.

The elements of the normalized decision matrix is calculated (using linear max-min normalization) as

$$r_{ij} = \frac{x_{ij} - x_{j(\min)}}{x_{j(\max)} - x_{j(\min)}} \quad \text{(For the desired effect: maximization)} \tag{1}$$

$$r_{ij} = \frac{x_{j(\max)} - x_{ij}}{x_{j(\max)} - x_{j(\min)}} \quad \text{(For the desired effect: minimization)} \tag{2}$$

Step 2. Calculation of the Percentage Value (PV) scores for the alternatives with respect to the influence of the criteria

At this step, LOPCOW brings parity in the weight calculation by using statistical fundamental concepts and the PV scores are obtained as

$$P_j = \left| \ln \left(\frac{\sqrt{\frac{\sum_{i=1}^{m} r_{ij}^2}{m}}}{\sigma} \right) .100 \right| \tag{3}$$

σ: The standard deviation for the performance values of the alternatives for a given criterion.

Step 3. Compute the criteria weights

Based on the proportional contributions, the weight for the jth criterion is calculated as

$$w_j = \frac{P_i}{\sum_{j=1}^{n} P_j} \tag{4}$$

where, $\sum_{j=1}^{n} w_j = 1$

2.5   GC-CRADIS Method

The concept of grey correlational (GC) analysis stems from the requirements for estimating the strength of relationship among the data sequences subject to variations in the given conditions (e.g., low volume) and dealing with impreciseness. GC finds its genesis in grey relational analysis (GRA) grounded on the fundamental concepts of grey theory [34]. There has been a number of applications using GC in conjunction with MCDM models for solving real-life problems. For example, Das and Chakraborty [35] developed a hybrid GC-EDAS model for application into production engineering. Following the concept of [35], the researchers [36] very recently have proposed a picture fuzzy GC-EDAS model for mobile wallet service provider selection problem. The current work advances the stream of GC based MCDM models by modifying the CRADIS framework [19] with the use of GC based analysis. The procedural steps are given below.

Step 1. Normalization of the decision matrix

The normalization of the elements of the decision matrix is done by using the expression (1) and (2).

Step 2. Formulate weighted normalized decision matrix

Let, $V = [v_{ij}]_{m \times n}$ represents the weighted normalized decision matrix where the elements are found by the following expression.

$$v_{ij} = r_{ij} w_j \tag{5}$$

$$(j = 1, 2, ...n; \ i = 1, 2, ...m)$$

Step 3. Derive the ideal and anti-ideal solution points for the alternatives
The ideal and anti-ideal solutions are determined as

$$t_j^+ = \max(v_{ij}) \tag{6}$$

$$t_j^- = \min(v_{ij}) \tag{7}$$

Step 4. Derive the deviations from the ideal and anti-ideal solutions
The deviations are calculated as follows

$$d_{ij}^+ = t_j^+ - v_{ij} \tag{8}$$

$$d_{ij}^- = v_{ij} - t_j^- \tag{9}$$

Step 5. Find out the grey correlational deviation (GCD) values

Following the demonstration of [35] the GCD values with respect to ideal and anti-ideal solutions are obtained as

$$g_{ij}^+ = \frac{d_{ij(\min)}^+ + \zeta d_{ij(\max)}^+}{\left| d_{ij(\max)}^+ - d_{ij}^+ \right| + \zeta d_{ij(\max)}^+} \tag{10}$$

$$g_{ij}^- = \frac{d_{ij(\min)}^- + \zeta d_{ij(\max)}^-}{\left| d_{ij(\max)}^- - d_{ij}^- \right| + \zeta d_{ij(\max)}^-} \tag{11}$$

Here, ζ is the identification coefficient such that $0 \le \zeta \le 1$. However, in this paper we keep $\zeta = 0.5$ following the previous work of [35] and [36].

Step 6. Determine the total grey deviations from the ideal and anti-ideal solutions by the alternatives

The corresponding deviations are found as

$$GS_i^+ = \sum_{j=1}^n g_{ij}^+ \tag{12}$$

$$GS_i^- = \sum_{j=1}^n g_{ij}^- \tag{13}$$

Step 7. Derive the grey utility function values for the alternatives with respect to their deviations from the optimal alternative

The optimal alternative has least deviation from the ideal value (S_0^+) and/or greatest distance from the anti-ideal solution (S_0^-).

Accordingly, the grey utility function values are derived as

$$GK_i^+ = \frac{GS_0^+}{GS_i^+} \tag{14}$$

$$GK_i^- = \frac{GS_i^-}{GS_0^-} \tag{15}$$

Step 2.7. Calculate the final grey appraisal scores for the alternatives

The grey final appraisal score for the alternatives are calculated as

$$GQ_i = \frac{GK_i^+ + GK_i^-}{2} \tag{16}$$

Decision rule: Higher is the grey appraisal score, better is the alternative over others.

3   Results

This section presents the major findings of the data analysis. Initially, the criteria weights are computed using the decision matrix presented in Appendix A. After obtaining the normalized decision matrix (using Eq. (1) and (2)), the next steps (see expressions (3) and (4)) are followed to calculate the criteria weights. Table 2. provides the summary of calculations of criteria weights.

Table 2.   Criteria weights

	C1	C2	C3	C4	C7
Mean Square	0.3814	0.0382	0.0353	0.0248	0.4408
SD	0.1683	0.1602	0.1641	0.1412	0.2301
PV	129.9753	19.8849	13.5009	11.0253	105.9819
Wj	0.4636	0.0709	0.0482	0.0393	0.3780

After obtaining the criteria weights, the stocks are ranked based on their relative performance using the decision matrix given in Appendix A. Table 3. shows the final calculations and ranking obtained by using the expressions (6) to (14) at $\zeta = 0.5$.

Reliability and stability of the solution are essential considerations for MCDM based analysis [37]. In this regard, the extant literature (for instance, [5, 38]) followed comparison of results obtained by various MCDM models. To ascertain the reliability of the result obtained by GC-CRADIS the current work carries out a comparison with classical CRADIS, GC-EDAS, PIV and CoCoSo methods. After that the Spearman's rank correlation test (SRCT) is performed. Table 4. supports that GC-CRADIS provides considerably reliable result. To examine the stability aspect, a sensitivity analysis (SA) is performed following the scheme suggested in the literature (for example, [39, 40]). Under this scheme A, 18 experimental cases are designed wherein for the first nine cases, the weight of C1 is reduced by 10 percent at each step while the total reduced weight value is proportionately adjusted with the other criteria. For next nine cases, the weights of both C1 and C5 are reduced by 10 percent and the total reduced weight value is proportionately adjusted with the other criteria. In addition, in the ongoing work an additional scheme is followed. Under scheme B, the value of ζ is varied from 0.2 to 0.9 including the original value of 0.5. Figure 1 and 2 provide the outcomes of SA which reflect that GC-CRADIS provides a stable result.

Table 3. Final ranking of the stocks

S/L	STOCK	INDUSTRY TYPE	GSI+	GSI−	GKI +	GKI−	GQI	RANK
A1	A C C LTD.	CEMENT	3.5366	2.2040	0.7375	0.6689	0.7032	27
A2	A U SMALL FINANCE BANK LTD.	BANKING/NBFC	3.7320	2.4805	0.6989	0.7528	0.7258	17
A3	AMBUJA CEMENTS LTD.	CEMENT	3.7660	2.0700	0.6926	0.6282	0.6604	45
A4	AVENUE SUPERMARTS LTD.	MULTI BRAND RETAIL	3.5312	2.2362	0.7386	0.6786	0.7086	23
A5	AXIS BANK LTD.	BANKING/NBFC	3.8376	2.1031	0.6797	0.6383	0.6590	46
A6	BAJAJ FINANCE LTD.	BANKING/NBFC	3.6306	2.0423	0.7184	0.6198	0.6691	43
A7	BAJAJ FINSERV LTD.	BANKING/NBFC	3.5316	2.1094	0.7386	0.6402	0.6894	34
A8	BHARAT PETROLEUM CORPN. LTD.	OIL AND GAS - OMC	3.7170	2.1048	0.7017	0.6388	0.6703	41
A9	BHARTI AIRTEL LTD.	TELECOM	3.6742	2.3228	0.7099	0.7049	0.7074	24
A10	BRITANNIA INDUSTRIES LTD.	FMCG	2.9543	2.7155	0.8829	0.8241	0.8535	3
A11	CIPLA LTD.	PHARMACEUTICALS	3.6824	2.4092	0.7083	0.7312	0.7197	18
A12	COLGATE-PALMOLIVE (INDIA) LTD.	FMCG	3.0621	2.7370	0.8518	0.8306	0.8412	4
A13	DR. REDDY'S LABORATORIES LTD.	PHARMACEUTICALS	3.4567	2.3920	0.7546	0.7259	0.7402	15
A14	EICHER MOTORS LTD.	AUTO OEM	3.6259	2.0546	0.7194	0.6235	0.6715	40
A15	G A I L (INDIA) LTD.	OIL AND GAS - GAS	3.5457	2.2531	0.7356	0.6838	0.7097	21
A16	GODREJ CONSUMER PRODUCTS LTD.	FMCG	3.5263	2.5154	0.7397	0.7634	0.7515	10
A17	GRASIM INDUSTRIES LTD.	DIVERSIFIED	3.7836	2.0420	0.6894	0.6197	0.6545	47
A18	H D F C BANK LTD.	BANKING/NBFC	3.5885	2.1864	0.7269	0.6635	0.6952	30
A19	H D F C LIFE INSURANCE CO. LTD.	INSURANCE	3.7095	2.1799	0.7031	0.6616	0.6824	38
A20	HAVELLS INDIA LTD.	CONSUMER ELECTRICALS	3.6656	2.1071	0.7116	0.6395	0.6755	39
A21	HERO MOTOCORP LTD.	AUTO OEM	3.2892	2.3074	0.7930	0.7003	0.7466	13
A22	HINDALCO INDUSTRIES LTD.	METAL	4.4624	1.7676	0.5845	0.5364	0.5605	55
A23	HINDUSTAN UNILEVER LTD.	FMCG	3.3081	2.6969	0.7885	0.8185	0.8035	8

(*continued*)

Table 3. (*continued*)

S/L	STOCK	INDUSTRY TYPE	GSI+	GSI−	GKI +	GKI−	GQI	RANK
A24	HOUSING DEVELOPMENT FINANCE CORPN. LTD.	BANKING/NBFC	3.5603	2.1543	0.7326	0.6538	0.6932	33
A25	I C I C I BANK LTD.	BANKING/NBFC	3.8137	2.0180	0.6839	0.6124	0.6482	48
A26	INDIAN OIL CORPN. LTD.	OIL AND GAS - OMC	3.4492	2.3431	0.7562	0.7111	0.7336	16
A27	IRCTC	INTERNET	3.7940	2.0037	0.6875	0.6081	0.6478	49
A28	INDUSIND BANK LTD.	BANKING/NBFC	4.1256	1.9929	0.6322	0.6048	0.6185	53
A29	INFOSYS LTD.	IT	3.3969	2.4093	0.7678	0.7312	0.7495	12
A30	J S W STEEL LTD.	METAL	3.6379	2.2178	0.7170	0.6731	0.6950	31
A31	JUBILANT FOODWORKS LTD.	FOOD RETAIL	3.5540	2.2820	0.7339	0.6925	0.7132	20
A32	MAHINDRA & MAHINDRA LTD.	AUTO OEM	3.6833	2.1706	0.7081	0.6587	0.6834	37
A33	MARICO LTD.	FMCG	3.3060	2.8451	0.7890	0.8634	0.8262	6
A34	MARUTI SUZUKI INDIA LTD.	AUTO OEM	3.4896	2.1072	0.7474	0.6395	0.6935	32
A35	N T P C LTD.	POWER THERMAL	3.5227	2.3004	0.7404	0.6981	0.7193	19
A36	NESTLE INDIA LTD.	FMCG	2.6083	3.2369	1.0000	0.9823	0.9912	2
A37	P I INDUSTRIES LTD.	CHEMICAL	3.6225	2.1526	0.7200	0.6533	0.6866	35
A38	PAGE INDUSTRIES LTD.	CONSUMER PRODUCTS	2.6221	3.2951	0.9948	1.0000	0.9974	1
A39	POWER GRID CORPN. OF INDIA LTD.	POWER T&D	3.3759	2.5374	0.7726	0.7701	0.7713	9
A40	RELIANCE INDUSTRIES LTD.	DIVERSIFIED	3.7191	2.1026	0.7013	0.6381	0.6697	42
A41	S B I CARDS & PAYMENT SERVICES LTD.	FINANCIAL SERVICES	3.7469	2.2220	0.6961	0.6743	0.6852	36
A42	SHREE CEMENT LTD.	CEMENT	3.2906	2.3045	0.7927	0.6994	0.7460	14
A43	SIEMENS LTD.	INDUSTRIAL AND CAPITAL GOODS	3.5527	2.2567	0.7342	0.6849	0.7095	22
A44	STATE BANK OF INDIA.	BANKING/NBFC	3.8719	1.9792	0.6736	0.6006	0.6371	51
A45	SUN PHARMACEUTICAL INDS. LTD.	PHARMACEUTICALS	3.6010	2.2500	0.7243	0.6828	0.7036	26

(*continued*)

Table 3. (*continued*)

S/L	STOCK	INDUSTRY TYPE	GSI+	GSI−	GKI +	GKI−	GQI	RANK
A46	TATA CONSULTANCY SERVICES LTD.	IT	3.0689	2.5199	0.8499	0.7647	0.8073	7
A47	TATA MOTORS LTD.	AUTO OEM	4.2244	1.8802	0.6174	0.5706	0.5940	54
A48	TATA STEEL LTD.	METAL	3.6642	2.0602	0.7118	0.6252	0.6685	44
A49	TECH MAHINDRA LTD.	IT	3.3410	2.3765	0.7807	0.7212	0.7510	11
A50	TITAN COMPANY LTD.	CONSUMER PRODUCTS	3.5021	2.1658	0.7448	0.6573	0.7010	28
A51	ULTRATECH CEMENT LTD.	CEMENT	3.4530	2.1708	0.7554	0.6588	0.7071	25
A52	VEDANTA LTD.	METAL	3.1652	2.7619	0.8241	0.8382	0.8311	5
A53	VOLTAS LTD.	CONSUMER DURABLE	3.8547	2.0151	0.6766	0.6115	0.6441	50
A54	WIPRO LTD.	IT	3.6693	2.2557	0.7108	0.6846	0.6977	29
A55	ZEE ENTERTAINMENT ENTERPRISES LTD.	MEDIA	3.9353	1.9279	0.6628	0.5851	0.6239	52

Table 4. Result of SCRT

	Coefficient	GC_EDAS	CRADIS	PIV	CoCoSo
GC_CRADIS	Spearman's rho	.997**	.929**	.999**	.998**
	Sig. (2-tailed)	0.000	0.000	0.000	0.000

'** Correlation is significant at the 0.01 level (2-tailed).'

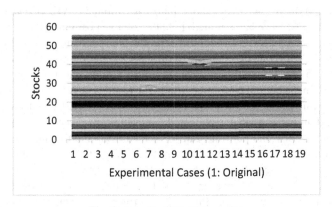

Fig. 1. Result of SA (Scheme A)

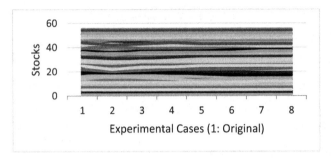

Fig. 2. Result of SA (Scheme B)

4   Conclusion

The present paper is an attempt to discern the market performance of a list of 55 ESG stocks listed to BSE India. The study has been conducted based on the stock performance during FY 2021–22. To compare the performance of the stocks, five criteria (i.e., market performance indicator variables) such as Annual Return, EPS, P/B, Yield and Beta have been considered in tune with the discussions made in the extant literature. To deal with the effect of multiple criteria an integrated MCDM framework using LOPCOW and CRADIS has been utilized wherein a new extension of the CRADIS method with GCA has been proposed as a measure of analysis of imprecise data. It is seen that mostly FMCG and consumer product based stocks perform well while banks and auto belong to the bottom performance list. It is observed from the result that annual return and systematic risk (beta) obtain the higher weights as calculated by using LOPCOW method. The finding is in tune with the modern portfolio theory as well as behavioral finance theory.

From the ranking outcome it is observed that A38 (Page Industries limited), A36 (Nestle India limited), A10 (Britannia Industries limited), A12 (Colgate-Palmolive (India) Ltd.) and A52 (Vedanta limited) hold the top five positions. Hence, top performers are predominantly FMCG and consumer products as per their stock performance. On the other hand, A22 (Hindalco Industries Ltd.), A47 (Tata Motors Ltd.), A28 (Indusind Bank Ltd.), A55 (Zee Entertainment Enterprises Ltd.), A44 (State Bank of India) secured their positions in the bottom performer group. The result is in sync with the past research exploring the impact of COVID-19 on the sectors of economy. The outcome is a reflection of the extant literature that explained the impact of the recent pandemic on various sectors. Hence, a possible portfolio can be formed using the findings of this paper. A further dig down analysis has revealed that the stock performance is not significantly associated with ESG score. It is also observed that GC-CRADIS provides a considerably stable and reliable solution. However, the model is limited to computational complexity and non-compensation of criteria. Proceeding further, an association among the stock performance and ESG score is investigated. It is noticed that there is no significant association between ESG score and stock performance.

5 Future Scope

The ongoing work has a number of future scopes. First, it may be an interesting analysis to compare the performance of ESG stocks over a period of several consecutive years. Secondly, a future work may contrast the fundamental and market performance. Third, there is a possibility to explore the impact of financial stability and long-term growth prospect on the stock performance. Fourthly, it provides an opportunity to compare ESG stocks and cryptocurrencies from investors' point of view. Sixth, the impact of COVID-19 on ESG stock performance may also be planned. Seventh, the GC-CRADIS framework may further be extended using fuzzy and rough numbers. Eighth, a possible future work may use Maximum Error Normalized Algorithms [41] to compare with our model. Nevertheless, the current work is a distinct attempt that has both technical and policy implications for common investors, strategists, and technical analysts.

Acknowledgments. 'The authors express their sincere thanks to all anonymous referees for their valuable comments to improve the quality of the paper'.

Appendix A: Decision Matrix

Stock	C1	C2	C3	C4	C5	ESG Score	Stock	C1	C2	C3	C4	C5	ESG Score
A1	0.140	43.220	3.290	2.430	0.970	59	A29	−0.550	53.480	9.600	2.110	0.610	76
A2	3.640	20.250	3.950	0.080	1.360	67	A30	1.980	28.650	3.050	2.280	1.390	57
A3	−1.210	9.030	4.550	1.210	1.040	61	A31	0.710	7.940	15.010	0.230	0.890	51
A4	0.670	36.930	14.460	0.000	0.990	50	A32	1.590	44.710	3.990	0.880	1.380	66
A5	1.390	48.000	2.300	0.110	1.530	71	A33	2.790	9.990	17.190	1.250	0.500	71
A6	0.070	135.800	7.970	0.330	1.540	67	A34	−0.530	197.630	4.640	0.710	1.250	60
A7	−0.440	5.470	42.960	0.030	1.360	64	A35	−0.040	15.180	1.240	4.160	0.780	58
A8	−0.240	−9.940	1.810	3.150	1.110	58	A36	0.470	249.240	103.200	1.070	0.580	60
A9	0.800	−0.270	5.480	0.390	0.790	63	A37	−1.040	65.900	7.500	0.180	0.920	58
A10	1.770	65.710	56.720	1.290	0.670	66	A38	3.000	657.310	36.760	0.740	0.880	58
A11	−1.310	32.120	3.640	0.470	0.530	68	A39	0.180	20.750	1.960	5.780	0.540	59
A12	0.740	37.770	35.120	2.600	0.460	59	A40	−1.460	64.090	3.430	0.320	0.960	61
A13	−1.040	143.780	3.710	0.690	0.610	66	A41	−1.890	10.050	8.130	0.330	0.710	60
A14	−1.010	78.440	7.590	0.660	1.180	58	A42	−0.250	380.340	5.000	0.370	1.090	61
A15	1.500	15.680	1.100	4.100	1.140	63	A43	1.200	42.990	9.360	0.260	1.030	57
A16	2.600	13.410	11.180	0.000	0.830	66	A44	−0.160	32.560	1.950	1.200	1.540	68
A17	−0.170	42.130	2.220	0.600	1.290	63	A45	−0.600	12.480	9.570	0.960	0.780	54
A18	−0.440	70.430	3.480	0.940	0.920	72	A46	0.220	104.110	18.950	3.560	0.610	72
A19	−0.680	6.700	8.190	0.280	0.870	60	A47	−0.490	−1.170	6.870	0.000	1.890	62
A20	0.290	16.870	11.770	0.370	1.220	64	A48	−0.070	19.990	1.180	4.130	1.340	62
A21	1.110	133.280	3.370	3.430	1.020	64	A49	0.210	51.160	4.410	4.560	0.730	72
A22	−5.040	22.410	2.020	0.800	1.560	58	A50	0.500	35.230	20.450	0.310	1.180	57

(continued)

(continued)

Stock	C1	C2	C3	C4	C5	ESG Score	Stock	C1	C2	C3	C4	C5	ESG Score
A23	1.660	41.300	13.050	1.360	0.500	68	A51	−0.550	203.880	4.110	0.530	1.070	60
A24	0.120	81.070	4.010	1.110	1.090	72	A52	1.180	47.580	2.050	24.640	1.470	52
A25	0.030	39.300	3.390	0.580	1.430	69	A53	−1.840	14.550	3.960	0.700	1.090	64
A26	1.490	6.230	0.930	6.090	0.970	60	A54	−1.440	22.390	3.780	1.490	0.700	74
A27	−1.870	11.210	10.350	0.550	1.150	56	A55	−1.960	12.99	2.16	1.34	1.34	50
A28	0.930	86.560	1.790	0.710	2.010	66							

References

1. Asad, A.L., Clair, M.: Racialized legal status as a social determinant of health. Soc. Sci. Med. **199**, 19–28 (2018)
2. Goyal, P., Gupta, P., Yadav, V.: Antecedents to heuristics: decoding the role of herding and prospect theory for Indian millennial investors. Rev. Behav. Finance. **15**, 79–102 (2021). https://doi.org/10.1108/RBF-04-2021-0073
3. Gupta, S., Bandyopadhyay, G., Biswas, S., Upadhyay, A.: A hybrid machine learning and dynamic nonlinear framework for determination of optimum portfolio structure. In: Saini, H.S., Sayal, R., Govardhan, A., Buyya, R. (eds.) Innovations in Computer Science and Engineering. LNNS, vol. 74, pp. 437–448. Springer, Singapore (2019). https://doi.org/10.1007/978-981-13-7082-3_50
4. Ren, F., Lu, Y.N., Li, S.P., Jiang, X.F., Zhong, L.X., Qiu, T.: Dynamic portfolio strategy using clustering approach. PLoS ONE **12**, e0169299 (2017). https://doi.org/10.1371/journal.pone.0169299
5. Gupta, S., Bandyopadhyay, G., Bhattacharjee, M., Biswas, S.: Portfolio selection using DEA-COPRAS at risk–return interface based on NSE (India). Int. J. Innov. Tech. Expl. Eng. (IJITEE) **8**(10), 4078–4086 (2019). https://doi.org/10.35940/ijitee.J8858.0881019
6. de la Fuente, G., Ortiz, M., Velasco, P.: The value of a firm's engagement in ESG practices: are we looking at the right side? Long. Range. Plann. **55**(4), 102143 (2022)
7. Yu, H.: Does sustainable competitive advantage make a difference in stock performance during the COVID-19 pandemic? Finance Res. Lett. **48**, 102893 (2022)
8. Shanaev, S., Ghimire, B.: When ESG meets AAA: the effect of ESG rating changes on stock returns. Finance Res. Lett. **46**, 102302 (2022)
9. Pedersen, L.H., Fitzgibbons, S., Pomorski, L.: Responsible investing: the ESG-efficient frontier. J. Financ. Econ. **142**(2), 572–597 (2021)
10. Albuquerque, R., Koskinen, Y., Yang, S., Zhang, C.: Resiliency of environmental and social stocks: an analysis of the exogenous COVID-19 market crash. Rev. Corp. Financ. Stud. **9**(3), 593–621 (2020)
11. Broadstock, D.C., Chan, K., Cheng, L.T., Wang, X.: The role of ESG performance during times of financial crisis: evidence from COVID-19 in China. Finance Res. Lett. **38**, 101716 (2021)
12. Ecer, F., Pamucar, D.: A novel LOPCOW-DOBI multi-criteria sustainability performance assessment methodology: an application in developing country banking sector. Omega **112**, 102690 (2022)

13. Bektas, S.: Evaluating the performance of the Turkish insurance sector for the period 2002–2021 with MEREC, LOPCOW, COCOSO, EDAS CKKV Methods. J BRSA Bank Financ. Markets **16**(2), 247–283 (2022)
14. Biswas, S., Bandyopadhyay, G., Mukhopadhyaya, J.N.: A multi-criteria framework for comparing dividend pay capabilities: evidence from Indian FMCG and consumer durable sector. Decis. Mak. Appl. Manag. Eng. **5**(2), 140–175 (2022)
15. Biswas, S., Bandyopadhyay, G., Pamucar, D., Joshi, N.: A multi-criteria based stock selection framework in emerging market. Oper. Res. Eng. Sci. Theory Appl. **5**(3), 153–193 (2022)
16. Biswas, S., Bandyopadhyay, G., Pamucar, D., Sanyal, A.: A decision making framework for comparing sales and operational performance of firms in emerging market. Int. J. Knowl.-Based Intell. Eng. Syst. **26**(3), 229–248 (2022)
17. Biswas, S., Bandyopadhyay, G., Mukhopadhyaya, J.N.: A multi-criteria based analytic framework for exploring the impact of COVID-19 on firm performance in emerging market. Decis. Anal. J **5**, 100143 (2022)
18. Biswas, S., Chatterjee, S., Majumder, S.: A Spherical fuzzy framework for sales personnel selection. J. Comput. Cogn. Eng. (2022). https://doi.org/10.47852/bonviewJCCE2202357
19. Puška, A., Stević, Ž, Pamučar, D.: Evaluation and selection of healthcare waste incinerators using extended sustainability criteria and multi-criteria analysis methods. Environ. Dev. Sustain. **24**, 11195–11225 (2022). https://doi.org/10.1007/s10668-021-01902-2
20. Puška, A., Božanić, D., Nedeljković, M., Janošević, M.: Green supplier selection in an uncertain environment in agriculture using a hybrid MCDM model: Z-numbers–Fuzzy LMAW–Fuzzy CRADIS model. Axioms **11**(9), 427 (2022)
21. Puška, A., Nedeljković, M., Prodanović, R., Vladisavljević, R., Suzić, R.: Market assessment of pear varieties in Serbia using fuzzy CRADIS and CRITIC methods. Agriculture **12**(2), 139 (2022)
22. Puška, A., Nedeljković, M., Stojanović, I., Božanić, D.: Application of Fuzzy TRUST CRADIS Method for Selection of Sustainable Suppliers in Agribusiness. Sustainability **15**(3), 2578 (2023)
23. Starčević, V., Petrović, V., Mirović, I., Tanasić, L. Ž. Stević, Ž., Đurović Todorović, J.: A novel Integrated PCA-DEA-IMF SWARA-CRADIS model for evaluating the impact of FDI on the sustainability of the economic system. Sustainability **14**(20), 13587 (2022)
24. Tan, Z., Yan, Z., Zhu, G.: Stock selection with random forest: an exploitation of excess return in the Chinese stock market. Heliyon. **5**(8), e02310 (2019). https://doi.org/10.5267/j.dsl.2021.4.001
25. Asadi, M., Mohammadi, S.: Selection of fuzzy multi-purpose portfolios based on the cross-sectional return model of data envelopment analysis in Tehran stock exchange. Financ. Eng.Portfolio Manage. **10**(41), 338-365 (2020). https://dorl.net/dor/20.1001.1.22519165.1398.10.41.15.7
26. Chen, B., Zhong, J., Chen, Y.: A hybrid approach for portfolio selection with higher-order moments: empirical evidence from Shanghai stock exchange. Expert Syst. Appl. **145**, 113104 (2020). https://doi.org/10.1016/j.eswa.2019.113104
27. Arif, U., Sohail, M.T.: asset pricing with higher co-moments and CVaR: evidence from Pakistan stock exchange. Int. J. Econ. Financ. Issues. **10**(5), 243 (2020). https://doi.org/10.3479/ijefi.10351
28. Bernal, M., Anselmo Alvarez, P., Muñoz, M., Leon-Castro, E., Gastelum-Chavira, D.A.: A multicriteria hierarchical approach for portfolio selection in a stock exchange. J. Intell. Fuzzy Syst. **40**(2), 1945–1955 (2021). https://doi.org/10.3233/JIFS-189198
29. Nazneena, S.: A study on portfolio management WRT textile industry. Turkish J. Comput. Math. Educ. (TURCOMAT) **12**(9), 2552–2556 (2021). https://doi.org/10.17762/turcomat.v12i9.3741

30. Bermejo, R., Figuerola-Ferretti, I., Hevia, T., Santos, A.: Factor investing: a stock selection methodology for the European equity market. Heliyon. **7**(10), e08168 (2021)
31. Nguyen, P.H., Tsai, J.F., Kumar, V.A.G., Hu, Y.C.: Stock investment of agriculture companies in the Vietnam stock exchange market: an AHP integrated with GRA-TOPSIS-MOORA approaches. J. Asian Financ. Econ. Bus. **7**(7), 113–121 (2021)
32. Suroso, A., Tandra, H., Syaukat, Y., Najib, M.: The issue in Indonesian palm oil stock decision making: sustainable and risk criteria. Decis. Sci. Lett. **10**(3), 241–246 (2021). https://doi.org/10.5267/j.dsl.2021.4.001
33. Narang, M., Joshi, M.C., Bisht, K., Pal, A.: Stock portfolio selection using a new decision-making approach based on the integration of fuzzy CoCoSo with Heronian mean operator. Decis. Mak. Appl. Manag. Eng. **5**(1), 90–112 (2022). https://doi.org/10.31181/dmame0310 022022n
34. Xia, X., Govindan, K., Zhu, Q.: Analyzing internal barriers for automotive parts remanufacturers in China using grey-DEMATEL approach. J. Clean. Prod. **87**, 811–825 (2015)
35. Das, P., Chakraborty, S.: Application of grey correlation-based EDAS method for parametric optimization of non-traditional machining processes. Sci. Iran **29**(2), 864–882 (2022). https://doi.org/10.24200/SCI.2020.53943.3499
36. Biswas, S., Pamucar, D.: A modified EDAS model for comparison of mobile wallet service providers in India. Financ. Innov. **9**(1), 1–31 (2023)
37. Deveci, M., Gokasar, I., Pamucar, D., Biswas, S., Simic, V.: An integrated proximity indexed value and q-rung orthopair fuzzy decision-making model for prioritization of green campus transportation. In: Garg, H. (eds.) q-Rung Orthopair Fuzzy Sets: Theory and Applications, pp. 303–332. Springer, Singapore (2022). https://doi.org/10.1007/978-981-19-1449-2_12
38. Biswas, S., Majumder, S., Pamucar, D., Dawn, S.K.: An extended LBWA framework in picture fuzzy environment using actual score measures application in social enterprise systems. Int. J. Enterp. Inf. Syst. (IJEIS) **17**(4), 37–68 (2021)
39. Pamucar, D., Torkayesh, A. E., Biswas, S.: Supplier selection in healthcare supply chain management during the COVID-19 pandemic: a novel fuzzy rough decision-making approach. Ann. Oper. Res. 1-43 (2022). https://doi.org/10.1007/s10479-022-04529-2
40. Pamucar, D., Žižović, M., Biswas, S., Božanić, D.: A new logarithm methodology of additive weights (LMAW) for multi-criteria decision-making: application in logistics. Facta Univ. Ser. Mech. Eng. **19**(3), 361–380 (2021)
41. Zıa Ur Rahman, M., Vardhan, B.V., Jenith, L., Rakesh Reddy, V., Surekha, S., Srinivasareddy, P.: Adaptive exon prediction using maximum error normalized algorithms. In: Mathur, G., Bundele, M., Lalwani, M., Paprzycki, M. (eds.) Proceedings of 2nd International Conference on Artificial Intelligence: Advances and Applications: ICAIAA 2021, pp. 511–523. Springer, Singapore (2022). https://doi.org/10.1007/978-981-16-6332-1_44

Decision Tree Based Test Case Generation Using Story Board and Natural Language Processing

Nishant Gupta[1]([✉]), Vibhash Yadav[2], and Mayank Singh[3]

[1] Department of CSE, Sharda University, Greater Noida, India
nishantlira@gmail.com
[2] Department of IT, REC, Banda, India
[3] Department of CSE, JSSATE, Noida, India
dr.mayank.singh@ieee.org

Abstract. It is true that generating test cases manually can be time-consuming and may result in incomplete coverage of user requirements. Our approach of using a story board, natural language processing, and decision tree to generate test cases is a novel solution that can save time and improve the quality of software. By processing user acceptance criteria with natural language processing and using decision trees to generate test cases, our approach can help ensure that all user requirements are covered. The activity flow diagram and encoded graph that are generated can provide a visual representation of the test cases, which can make it easier to identify any gaps in coverage. We have evaluated our technique using a case study and compared it with other state-of-the-art methods. The significant reduction in overall effort proves that our technique is effective. Overall, our approach of generating test cases using a story board, natural language processing, and decision tree is a promising development that has the potential to improve the quality of software and reduce the effort required to generate test cases.

Keywords: Regression testing · Test case generation · Natural language processing · Decision tree · Story board

1 Introduction

Traditional methods in the software industry typically require manual utilization of source code for test case generation. However, the development of agile technology and frequent changes in software development have led to a shift towards gathering requirements in human understandable language, also known as Natural Language. To simplify the process, various modeling diagrams are used in the industry to help the software development team understand it. Early testing using user requirements and specifications can reduce project delays and be cost-effective. Among the different graphical representations in UML, the Activity diagram is the simplest and most useful. It illustrates the workflow of the scenario and aids in comprehending the sequential actions [1]. Utilizing NLP in a given scenario can help generate the elements required for creating an activity diagram. Implementing regression test cases requires an efficient testing team, as

M. Singh et al. (Eds.): ICACDS 2023, CCIS 1848, pp. 581–591, 2023.
https://doi.org/10.1007/978-3-031-37940-6_47

it demands a considerable amount of effort and time. The purpose of generating these test cases is to minimize execution time and cover as many faults as possible. In industries where test cases are manually generated based on customer requirements, it is essential to automate functionalities. Nevertheless, as the testing team lacks direct involvement with users, it may pose challenges in comprehending the customers' requirements [2].

Agile development has replaced the waterfall model, which was the most popular model followed in the industry. The waterfall model relied on the sequential process of gathering requirements, followed by design, implementation, and testing [3]. Due to its dynamic behavior and interactive environment, most industries are now adopting agile development for software development. In the waterfall model, where the client has very low interaction during the later stages of software development, there is a lack of trust in the product by the customer.

Creating a story line in agile development is essential to help the customer understand their requirements and functionalities better. The exchanges between the customer and analyst should be documented as a Software Requirements Specification (SRS) and shared with the testing team. However, automating the process of developing a story board and generating test cases using scenarios can be challenging. The preferred method for generating a story board is natural language. Natural language processing is a widely recognized technique employed to program computers in order to comprehend and replicate natural language. In an Agile environment, the focus is on creating test cases using the story board. Generating an activity diagram can help understand the NLP output extraction, and decision trees can be utilized to generate test cases. Decision trees are helpful in making decisions based on probability models and analyzing their consequences.

2 Background

In regression testing, if modifications have been demanded in continuation during the development of software, then more effort and time will be required to test the system. The test case written manually for regression testing will also require more effort and time, which may also affect the quality of the software. In [4], the test cases have been generated using Natural language processing. In [5], user stories have been created using Natural language processing from SRS. The user stories have been generated and transformed in UML use case using NLP [6]. The NLP has been used which transforms the given user story file into UML case diagram. In [7], it has been shown that the user stories have been created with constraints and then, with the proposed technique, test cases have been generated. The selenium tool has been used to execute the test cases.

Chen et al. [8] has introduced a model that accepts requirements based on consumer experience and transforms these requirements into operation diagrams. Sanz and Mishra [9] proposed the generation of test case based on the requirements and UML operation diagram. They also prioritized the test cases produced. Linzhang et al. [10] has suggested a technique that incorporates black box and gray box testing for the generation of test cases. In the process, scenarios were created based on the activity diagram.

A fully automated tool capable of utilizing Natural Language Processing (NLP) techniques to convert test cases from provided requirements is currently lacking [11].

In [12] authors have generated test cases but requires a test data for complete execution of the test cases. In [13], researchers have generated the test cases where there is certain restriction on the dictionary and changing every time the new project will run. Another work to convert the user stories to use cases has been done by [14]. The framework has been introduced which uses the NLP techniques to transform the user stories into use cases.

Different agile techniques are used in the industry, such as JIRA, Versionone and Rally, which help to interact with various activities and incorporate agile development methods. For the implementation of regression testing, there are a range of methods that have a management suite like Versionone. The most commonly used JIRA tool is also combined with other tools such as selenium used for test automation. In the context of agile technology, a user story is created with the expectation that it will be considered complete once all acceptance tests have been successfully cleared.. When completed, it can be analyzed to further produce test suites and deliver the project into small pieces.

3 Methodology

Manual generation of test cases is a time-consuming task and requires a considerable amount of effort. It is important to cover as many user requirements as possible in order to ensure the quality of the software. Our objective is to generate test cases by utilizing the user's created story board. The story board will undergo NLP processing to extract essential details from the user's requirements. The test cases will be derived from the user story, acceptance criteria, and the generated scenario. These scenarios will be employed to generate an activity diagram, which will further guide the test case generation process through the utilization of decision trees.

To initiate the process, the initial action involves creating the story board in a predefined format, utilizing the user-provided parameter values. Additionally, the user can specify acceptance criteria and requirements that will contribute to the formulation of the test scenario. Subsequently, a dictionary will be established to store and organize all the defined acceptance criteria. This entails incorporating relevant keywords from the acceptance criteria and requirements into the dictionary. The generation of test case scenarios relies on the user's requirements and the corresponding test scenario, which is obtained from the database in an automated manner. NLP techniques will be employed to parse the available test scenario, facilitating the generation of the necessary test cases. This pre-processing step is essential to utilize the dictionary and decision tree approach for test case generation.

Figure 1 illustrates the process of utilizing the NLP technique and decision tree for generating test cases based on user-defined acceptance criteria, requirements, and various parameters for test scenario generation.

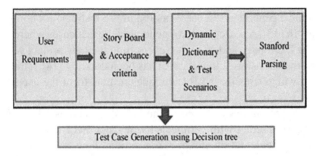

Fig. 1. Process of Test case generation

4 Proposed Technique

Our proposed method for generating test cases utilizes NLP and decision tree techniques. NLP is employed to parse the user story and save test scenario keywords in a dynamic dictionary. The next subsection explains the implementation and process of our proposed test case generation method.

4.1 Story Board and Acceptance Criteria

The story board is a format that defines the system's features and customer requirements resulting from communication between the business analyst and the customer. It presents the system's features and facilitates understanding the customer's needs. The acceptance criteria are defined as the outcome of user stories [15].

4.2 Dynamic Dictionary

A story board is generated when a user provides acceptance criteria and requirements based on predefined parameters. The dictionary is created by extracting common keywords from both the story board and acceptance criteria and is dynamically updated. This facilitates the generation of multiple test cases based on the customer's chosen requirements and parameters. The dictionary can be expanded by adding more keywords, allowing for additional functionality to be included.

4.3 Generation of Test Scenarios

To provide the system requirements, the flow of conversation is recorded in test scenarios. This gives a comprehensive understanding of the system functioning and the desired output. The generated test scenarios are saved in a database. In case of any repetition of the test scenario, the corresponding data can be retrieved from the database. The relevant test scenario is selected and utilized for further processing in generating the test cases.

4.4 Stanford Parsing

Numerous studies have employed NLP to gain a deeper comprehension of text data. In our proposed approach, we leverage NLP to analyze user stories and extract keywords for our dynamic dictionary. These extracted keywords are then employed to generate test cases using the decision tree method. As depicted in Fig. 2, the preprocessing of user stories is achieved through Stanford parsing. The output from parsing is utilized to create a dictionary, which is then fed to the decision tree to produce test cases.

4.5 Decision Tree Approach for Test Case Generation

The parsing output will be utilized to construct a dictionary, which will, in turn, assist in creating a decision tree to identify the test cases. A decision tree is a graphical structure where each internal node represents a specific attribute test (e.g., determining if a coin lands on heads or tails), each branch represents the outcome of the test, and each leaf node represents a classification decision made after evaluating all attributes. The paths from the root to the leaf nodes follow the classification rules..

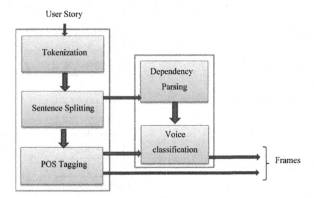

Fig. 2. Preprocessing through NLP

The decision tree can aid in guiding the testing team to specific test cases. Figure 3 illustrates an activity flow diagram created using the keywords in the dictionary. The probability of each acceptance criteria generated by the decision tree will determine the likelihood of each test case. As each test case is tested or resolved, it will be removed and the remaining test cases will be prioritized based on their likelihood of occurrence. Afterwards, the table variables need to be transformed into integer variables, enabling their utilization in an algorithm for constructing the decision tree.

5 Implementation

We have included all the essential files needed for pre-processing and decision classification. The three main libraries that have been utilized are: "Pandas", "Label Encoding", and "Decision Tree Classifier".

5.1   Generation of Test Cases

A function has been developed to receive the encoded value of a test and predict the first test case. The function returns the details of the test case with its index value, considering the parameter "Index = 1". For instance, if the parameter "Password" is selected, the function will provide the index value of the first test from the dataset. To avoid repetition of the same test, the selected test is removed from the dataset.

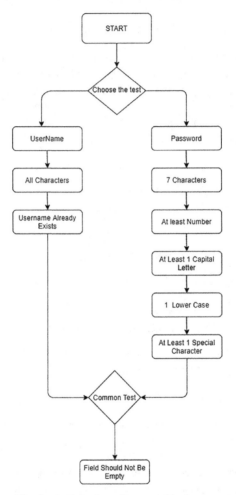

Fig. 3.   Activity flow diagram of login page

Using the decision tree, we extract multiple paths and corresponding test cases for the selected path. The generated test cases are based on the encoded graph through decision tree and user stories, as shown in Fig. 4. For the given acceptance criteria and requirements, six paths have been identified. The first path is associated with user name validity and registration. The remaining paths are focused on password creation and validation.

6 Technique Evaluation

The test case generation technique has been assessed through the evaluation of test coverage criteria and effort calculation, using a case study of cash withdrawal from an ATM. To implement our technique, we utilized the previously designed cash withdrawal from ATM case study [16]. The activity flow diagram was generated by our tool, which includes activities, decision nodes, and transitions.The corresponding Encoder graph was also generated from the activity diagram produced by our tool. To evaluate the test coverage criteria, we considered the following coverage criteria [17]. The following criteria of test coverage is taken into consideration for evaluating the coverage of test cases.

- Activity coverage
- Transition coverage
- Path coverage
- Branch coverage

In addition to evaluating the coverage, the tool's second evaluation was conducted by estimating the effort. Effort is determined by estimating the average time required to generate a test case [18], as given below.

Path: 0-->1-->2-->3-->4
user open welcome page-->Click at register user-->Enter username by selecting all characters-->click at register button-->Error message user is already exist please choose another username-->end

Path: 0-->1-->2-->3-->4-->5-->6
user open welcome page-->Click at register user-->enter username by selecting all characters-->enter password using 6 small case characters only-->click at register button-->error message that password should be 7 characters long along with the password creation rules on UI-->end

Path: 0-->1-->2-->3-->4-->5-->6
user open welcome page-->Click at register user-->enter username by selecting all characters-->enter password using 7 small case characters only-->click at register button-->error message that password does not meet the expected criteria-->end

Path: 0-->1-->2-->3-->4-->5-->6-->7
user open welcome page-->Click at register user-->enter username by selecting all characters-->enter password using 7 small case characters with single number-->click at register button-->error message that password does not meet the expected criteria-->end

Path: 0-->1-->2-->3-->4-->5-->6-->7
user open welcome page-->Click at register user-->enter username by selecting all characters-->enter password using 6 small case characters with single number-->click at register button-->error message that password does not meet the expected criteria-->end

Path: 0-->1-->2-->3-->4-->5-->6-->7-->8-->9-->10
user open welcome page-->Click at register user-->enter username by selecting all characters-->enter password combination of 2 numbers, 1 capital letter,3 small letters and one special character-->click at register button-->user should register successfully-->end

Fig. 4. Generated paths and corresponding test cases

Effort = Time required to generate test cases / total number of generated test cases.

We compared the effort of our technique using decision tree with the technique used in [4], which utilized depth first search for generating test cases for ATM cash withdrawal.

6.1 Results

Our technique is being evaluated using the test coverage criteria, which is shown in Fig. 5 for the ATM cash withdrawal activity diagram. The coverage criteria we have considered are activity coverage, path coverage, transition coverage, and branch coverage. Table 1 displays the test coverage criteria obtained using our technique, which we have compared with other elements presented in [16].

Figure 6 illustrates the test cases generated from our tool for the given case study. Our proposed technique, with its constraints, achieves a 100% test coverage. All coverage criteria have been covered, indicating that the proposed technique provides the best possible test coverage and generates sufficient test cases.

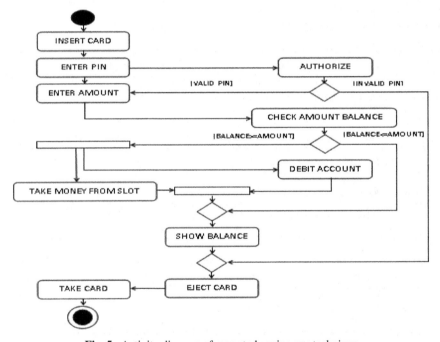

Fig. 5. Activity diagram of case study using our technique

The evaluation of our proposed tool includes the assessment of effort required to generate the test cases. This process involves several steps such as storyboarding, feeding acceptance criteria, dynamic dictionary creation, and test scenario generation. The time taken to complete the entire process was approximately 3 min. Using our tool, we were able to generate 8 test cases.

The effort approximately has been calculated as = 23 s/ test case (approx.)

A comparison was conducted between the effort estimation of our proposed technique and the effort presented in [4] for the ATM cash withdrawal case study. Our technique utilizes the decision tree within our tool, while the depth-first search algorithm was used for comparison. The results of this comparison are presented in Table 2.

Table 1. Test coverage criteria

Element	Number of Element present	Number of Elements Covered	Coverage (%)
Activity	10	10	100
Basis Path	3	3	100
Transitions	14	14	100
Branches	2	2	100
Conditions	4	4	100

Verify that card will be successfully authorize-->Enter the ATM card-->Enter the Valid Pin-->ATM card should be Authorize successfully and No error message should appear on UI
Verify that card will not be successfully authorize-->Enter the ATM card-->Enter the Invalid Pin-->ATM card should not be Authorize successfully and error message state that please enter the valid PIN should appear on UI
Verify Application message for ejecting the ATM card-->Enter the ATM card-->Enter the Invalid Pin-->ATM card should not be Authorize successfully and ATM card should be eject from ATM machine with a message state please collect your card.
Verify the Balance check Functionality-->Enter the ATM card-->Enter the Valid Pin-->Click Balance Enquiry button-->Available Account Balance should appear Successfully on Screen
Verify the Money Withdrawal functionality-->Enter the ATM card-->Enter the Valid Pin-->Click Withdrawal Button-->Enter the Valid Amount in INR-->Click Enter button to Withdrawal money-->Required Amount should be comes out from the ATM slot with a message states that please coint your amount before leaving.
Verify that ATM card will be eject successfully-->Enter the ATM card-->Enter the Valid Pin-->Click Withdrawal Button-->Enter the Valid Amount in INR-->Click Enter button to Withdrawal money-->Money should be taken out from ATM slot-->Card will be successfully eject after successful transaction with the message as Please collect your card.
Verify tha Money Withdrawal functionality in case user enter the invalid Amount-->Enter the ATM card-->Enter the Valid Pin-->Click Withdrawal Button-->Enter the Invalid Amount in INR-->Click Enter button to Withdrawal money-->Verify that an error message should be appear on Screen and verify that No money should be taken out from ATM slot.
Verify that remaining account balance will appear on screen after successful transaction-->Enter the ATM card-->Enter the Valid Pin-->Click Withdrawal Button-->Enter the Valid Amount in INR-->Click Enter button to Withdrawal money-->Verify that now available balance of bank account should appear on ATM screen.

Fig. 6. Test cases

Table 2 demonstrates that the time required for test case generation is decreased when using decision tree with NLP in comparison to Depth First Search. Therefore, this approach also reduces the overall effort needed for the process.

Table 2. Comparison of effort

Technique	Effort
Technique using Depth first search [4]	34 s/test case
Our proposed tool using decision tree	23 s/test case

7 Threats to Validity

Our proposed tool is based on the acceptance criteria defined by the customer. The decision tree approach has been implemented on a case study and test cases have been generated. The activity flow diagram generated from tool is also dependent on acceptance criteria. The story board and the acceptance criteria should be as per pre-defined format so that tool can understand the initial inputs. The activity flow diagram has been generated with limited features which can show activities and transition without any loops. However, when implemented, it provides the result which are prominent.

8 Conclusion

Our proposal is a novel approach for generating test cases through the integration of NLP and decision tree methods, which involves the creation of a storyboard to help identify test scenarios. The decision tree method is employed to determine test paths through the encoded graph. This tool has the capability to generate test cases using the storyboard and scenarios as input.

Our proposed method has reduced the effort required compared to other state-of-the-art methods. Moreover, we have demonstrated that decision tree is a more efficient method for implementing NLP in test case generation as compared to depth-first search. Our technique has the potential to be further evaluated with other state-of-the-art methods used for test case generation.

References

1. Fan, X., Shu, J., Liu, L., Liang, Q.: Test case generation from UML sub-activity and activity diagram. In: 2009 Second International Symposium on Electronic Commerce and Security, vol. 2, pp. 244–248. IEEE (2009)
2. Broek, R.V., Bonsangue, M.M., Chaudron, M., van Merode, H.: Integrating testing into agile software development processes. In: Proceedings of the 2nd International Conference on Model-Driven Engineering and Software Development (MODELSWARD), pp. 561–574. SCITEPRESS (2014)
3. Ambler, S.W.: The Object Primer: Agile Model-Driven Development with UML 2.0. Cambridge University Press, Cambridge (2006)
4. Rane, P.P.: Automatic generation of test cases for agile using natural language processing (2017)
5. Pereira, A.C.: Using NLP to generate user stories from software specification in natural language (2018)

6. Elallaouia, M., Nafilb, K., Touahni, R.: Automatic transformation of user stories into UML use case diagrams using NLP techniques. Procedia Comput. Sci. **130**, 42–49 (2018). https://doi.org/10.1016/j.procs.2018.04.010.7

7. Masud, M., Iqbal, M., Khan, M.U., Azam, F.: Automated user story driven approach for WebBased functional testing. World Acad. Sci. Eng. Technol. Int. J. Comput. Inf. Eng. **11**(1), 91–98 (2017). https://doi.org/10.5281/zenodo.1339932

8. Chen, Y., Probert, R.L., Sims, D.P.: Specification based regression test selection with risk analysis. In: Proceedings of the Conference of the Centre for Advanced Studies on Collaborative research (CASCON 2002), pp. 1–14. IBM Press (2002)

9. Sanz, L.F., Misra, S.: Practical application of UML activity diagrams for the generation of test cases. Proc. Rom. Acad. Ser. A: Math. Phys. Tech. Sci. Inf. Sci. **13**(3), 251–260 (2012)

10. Linzhang, W., Jiesong, Y., Xiaofeng, Y., Xuandong, L., Guoliang, Z.: Generating test cases from UML activity diagram based on Gray-Box Method. In: Proceedings of the 11th Asia-Pacific Software Engineering Conference, pp. 284–291. IEEE (2004). https://doi.org/10.1109/APSEC.2004.55

11. Sinha, A., Sutton Jr, S.M., Paradkar, A.: Text2Test: automated inspection of natural language use cases. In: 2010 Third International Conference on Software Testing, Verification and Validation, pp. 155–164. IEEE (2010). https://doi.org/10.1109/ICST.2010.199

12. Escalona, M.J., et al.: An overview on test generation from functional requirements. J. Syst. Softw. **84**(8), 1379–1393 (2011). https://doi.org/10.1016/j.jss.2011.03.051

13. Carvalho, G., et al.: Test case generation from natural language requirements based on SCR specifications. In: Proceedings of the 28th Annual ACM Symposium on Applied Computing, pp. 1217–1222. ACM (2013). https://doi.org/10.1145/2480362.2480591

14. Azzazi, A.: A framework using NLP to automatically convert user-stories into use cases in software projects. IJCSNS Int. J. Comput. Sci. Netw. Secur. **17**(5), 71–76 (2017)

15. Solis, C., Wang, X.: A study of the characteristics of behavior driven development. In: 2011 37th EUROMICRO Conference on Software Engineering and Advanced Applications, pp. 383–387. IEEE (2011)

16. Oluwagbemi, O., Hishammuddin, A.: Automatic generation of test cases from activity diagrams for UML based testing (UBT). Jurnal Teknologi **77**(13), 1–12 (2015)

17. Mingsong, C., Xiaokang, Q., Xuandong, L.: Automatic test case generation for UML activity diagrams. In: Proceedings of the 2006 International Workshop on Automation of Software Test, pp. 2–8. ACM (2006)

18. Elghondakly, R., Moussa, S., Badr, N.: Waterfall and agile requirements-based model for automated test cases generation. In: 2015 IEEE 7th International Conference on Intelligent Computing and Information Systems, pp. 607–612. IEEE (2015)

Author Index

M. Singh et al. (Eds.): ICACDS 2023, CCIS 1848, pp. 593–595, 2023.
https://doi.org/10.1007/978-3-031-37940-6

Printed in the United States
by Baker & Taylor Publisher Services